BIOLOGY

A HUMAN APPROACH

FIRST EDITION

BIOLOGY

First Edition
BSCS Human Version

STUDENT EDITION

BSCS

Biological Sciences Curriculum Study
Pikes Peak Research Park
5415 Mark Dabling Boulevard
Colorado Springs, CO 80918-3842

A HUMAN APPROACH

KENDALL/HUNT PUBLISHING
Dubuque, Iowa

BSCS DEVELOPMENT TEAM

Rodger W. Bybee, *Principal Investigator*
Michael J. Dougherty, *Project Director, 1995–1996*
Janet Carlson Powell, *Project Director, 1993–1995*
Gordon E. Uno, *Project Director, 1992*

Randall K. Backe, *Staff Associate, 1993–1996*
Wilbur Bergquist, *Staff Associate, 1992–1993*
William J. Cairney, *Staff Associate, 1993–1994*
Michael J. Dougherty, *Staff Associate, 1993–1995*
B. Ellen Friedman, *Staff Associate, 1992–1995*
Philip Goulding, *Staff Associate, 1992–1995*
David A. Hanych, *Staff Associate, 1996*
Laura J. Laughran, *Staff Associate, 1993–1995*
Lynda B. Micikas, *Staff Associate, 1994–1995*
Jean P. Milani, *Senior Staff Associate, 1992–1996*
Josina Romero-O'Connell, *Staff Associate, 1992–1993*
Jenny Sigsted, *Staff Associate, 1992–1993*
Pamela Van Scotter, *Staff Associate, 1992–1996*

CONTRIBUTORS

Robert A. Bouchard, College of Wooster, Wooster, OH
Edward Drexler, Pius XI High School, Milwaukee, WI
Kim Finer, Kent State University, Canton, OH
Ann Haley-Oliphant, Miami University, Oxford, OH
Laura J. Laughran, New Directions, Tucson, AZ

CONSULTANTS

Susan Speece, External Evaluator, Fresno City College, Fresno, CA
Constance Bouchard, College of Wooster, Wooster, OH
Ted Dunning, New Mexico State University, Las Cruces, NM
Irene Pepperberg, University of Arizona, Tucson, AZ
Bert Kempers, Media Design Associates, Inc., Boulder, CO
Will Allgood, Media Design Associates, Inc., Boulder, CO
Mark Viner, Media Design Associates, Inc., Boulder, CO
Larry N. Norton, Quantum Technology, Inc., Evergreen, CO
Chester Penk, Quantum Technology, Inc., Evergreen, CO
Ward's Natural Science Establishment, Inc., Rochester, NY

BSCS PRODUCTION STAFF

Jan Chatlain Girard, *Art Coordinator*
Sariya Jarasviroj, *Research Assistant*
Teresa Powell, *Senior Executive Assistant*
Joy L. Rasmussen, *Research Assistant*
Judy L. Rasmussen, *Senior Executive Assistant*
Barbara C. Resch, *Production Manager*
Donna Vammen, *Executive Assistant*
Linda Ward, *Executive Assistant*
Lee Willis, *Production Coordinator*
Yvonne Wise, *Assistant to the Principal Investigator*

ARTISTS AND EDITORS

Susan Bartel
Wendy Demandante
Jan Chatlain Girard
Angela Greenwalt
Mark Handy
Becky Hill
Marjorie C. Leggitt
Audrey Penk
Brent Sauerhagen
Mary Snyder
Paige Louis Thomas
Linn Trochim

BSCS ADMINISTRATIVE STAFF

Timothy Goldsmith, *Chair, Board of Directors*
Joseph D. McInerney, *Director*
Larry Satkowiak, *Chief Financial Officer*

Editorial, design, and production services provided by
Learning Design Associates, Columbus, Ohio.

Copyright ©1997 by BSCS
Library of Congress Catalog Card Number: 96-75717
ISBN 0-7872-0368-8

This material is based on work supported by the National Science Foundation under Grant No. ESI 9252974. Any opinions, findings, conclusions, or recommendations expressed in this publication are those of the authors and do not necessarily reflect the views of the granting agency.

Printed in the United States of America
10 9 8 7 6 5 4 3 2

ACKNOWLEDGMENTS

ADVISORY BOARD

Judy Capra
Jefferson County Public Schools, Golden, CO

Mack Clark
Academy School District 20,
Colorado Springs, CO

Diane Ebert-May
Northern Arizona University, Flagstaff, AZ

Philip R. Elliott
The Colorado College, Colorado Springs, CO

April Gardner (*Executive Committee*)
University of Northern Colorado,
Greeley, CO

Michele Girard
Peyton High School, Peyton, CO

Eville Gorham
University of Minnesota, Minneapolis, MN

Joseph Graves
Arizona State University—West, Phoenix, AZ

Ann Haley-Oliphant (*Executive Committee*)
Miami University, Oxford, OH

Paul DeHart Hurd, Prof. Emeritus
Stanford University, Stanford, CA

Mary Kiely
Stanford University, Stanford, CA

Douglas Kissler
Douglas County High School,
Castle Rock, CO

Carole Kubota
University of Washington, Seattle, WA

Douglas Lundberg
Air Academy High School,
United States Air Force Academy, CO

Michael E. Martinez (*Executive Committee*)
University of California, Irvine, CA

Donald E. Mason
Mitchell High School, Colorado Springs, CO

Laurence McCullough
Baylor College of Medicine, Houston, TX

Martin K. Nickels
Illinois State University, Normal, IL

Floyd Nordland, Prof. Emeritus
Purdue University, West Lafayette, IN

S. Scott Obenshain
University of New Mexico,
Albuquerque, NM

William O'Rourke
Harrison School District,
Colorado Springs, CO

Ann Pollet
Pueblo County High School, Pueblo, CO

Jerry Resnick
Clara Barton High School, Brooklyn, NY

Parker A. Small, Jr.
University of Florida, Gainesville, FL

Gordon E. Uno
University of Oklahoma, Norman, OK

Betty M. Vetter
Commission on Professionals in Science and
Technology, Washington, DC

Bruce Wallace (*Executive Committee*)
Virginia Polytechnic Institute and State
University, Blacksburg, VA

Harry Zimbrick
School District 11, Colorado Springs, CO

REVIEWERS

Douglas Allchin
University of Texas, El Paso, TX

Tom Anderson
University of Illinois, Champaign, IL

James Botsford
New Mexico State University, Las Cruces, NM

Robert A. Bouchard
College of Wooster, Wooster, OH

Jack Carter, Prof. Emeritus,
The Colorado College, Colorado Springs, CO

Frank Cassel, Prof. Emeritus,
North Dakota St. University, Fargo, ND

Angelo Collins
Vanderbilt University, Nashville, TN

Robert Cook-Degan
Institute of Medicine, Washington, DC

David Corbin
Monsanto Company, Chesterfield, MO

Jorge Crisci
Museo de La Plata, Argentina

Mary Ann Cutter
University of Colorado, Colorado Springs, CO

Hans Dethlefs
The Neighborhood Health Center—South,
Omaha, NE

Edward Drexler
Pius XI High School, Milwaukee, WI

James Ebersole
The Colorado College, Colorado Springs, CO

Diane Ebert-May
Northern Arizona University, Flagstaff, AZ

Philip R. Elliott
The Colorado College, Colorado Springs, CO

Michael Fatone
United States Air Force Academy, CO

Kim Finer
Kent State University, Canton, OH

Steven Fleck
United States Olympic Center, Colorado
Springs, CO

Geoff Gamble
Washington State University, Pullman, WA

Barbara Grosz
Pine Crest School, Fort Lauderdale, FL

Topper Hagerman
Steadman-Hawkins Sports Medicine
Foundation, Vail, CO

Jerry Harder
NOAA Aeronomy Laboratory, Boulder, CO

Jeff Hays
United States Air Force Academy, CO

Werner Heim, Prof. Emeritus
The Colorado College, Colorado Springs, CO

Barry Hewlett
Washington State University, Pullman, WA

Michael Hoffman
The Colorado College, Colorado Springs, CO

Michael Keelan
Medical College of Wisconsin, Milwaukee, WI

Rich Kulmacz
University of Texas Health Science Center,
Houston, TX

Linda Lundgren
Bear Creek High School, Lakewood, CO

Thomas Manney
Kansas State University, Manhattan, KS

Cheryl Mason
San Diego State University, San Diego, CA

Jeffry Mitton
University of Colorado, Boulder, CO

Adrian Morrison
University of Pennsylvania, Philadelphia, PA

Jamie Nekoba
Waiákea High School, Hilo, HI

Gene O'Brien
Hartford Union High School, Hartford, WI

John Opitz
Montana State University, Helena, MT

Carl Pierce
Harrington Cancer Center, Amarillo, TX

Tracy Posnanski
University of Wisconsin, Milwaukee, WI

Ken Rainis
Ward's Natural Science Establishment, Inc.,
Rochester, NY

Barbara Saigo
Saiwood Biology Resources, Montgomery, AL

Orwyn Sampson, Brigadier General Retired
United States Air Force Academy, CO

James Short
Packer Collegiate Institute, Brooklyn, NY

James Siedow
Duke University, Durham, NC

Fran Slowiczek
San Diego City Schools, San Diego, CA

Susan Speece
Fresno City College, Fresno, CA

Sam Stoler
National Institutes of Health, Washington, DC

Richard Storey
The Colorado College, Colorado Springs, CO

Gordon E. Uno
University of Oklahoma, Norman, OK

Jeff Velten
New Mexico State University, Las Cruces, NM

Mariana Wolfner
Cornell University, Ithaca, NY

FIELD-TEST SITE CENTERS

Arizona

Northern Arizona University, Flagstaff, AZ
    Diane Ebert-May
    Julie McCormick
    Brownie Sternberg

Colorado

BSCS, Colorado Springs, CO
    Laura J. Laughran
    Randall K. Backe

University of Northern Colorado, Greeley, CO
    April Gardner
    Alan Lennon
    Brenda Zink

Florida

University of South Florida, Tampa, FL
    Barbara Spector
    Leslie Brackin
    Craig Holm

Kansas

Kansas State University, Manhattan, KS
    Gail Shroyer
    Carol Arjona

Ohio

Miami University, Oxford, OH
    Jane Butler Kahle
    Rick Fairman

Washington

University of Washington, Seattle, WA
    Carole Kubota
    Claire McDaniel Orner

FIELD-TEST SCHOOLS

Colombia, South America:

Haydée Bejardno de Cadena, Marcela Melendez, Monica Sarmiento, Colegio Los Nogales, Bogotá, Colombia

Arizona:

Marcia Fisher, Arcadia High School, Scottsdale, Arizona; Doug Davis, Dub Manis, Dee Schwartz, Chinle High School, Chinle, Arizona; Geri Fisher, Jo Quintenz, Desert View High School, Tucson, Arizona; Kathy Thayer, Ray High School, Ray, Arizona; Clyde Christensen, Scott Greenhalgh, Ray Pool, Mary Southall, Elizabeth Stone, Tempe High School, Tempe, Arizona; Willie Long Reed, Tuba City High School, Tuba City, Arizona; Jack Johnson, Williams High School, Williams, Arizona; Carlos Estrada, Karen Steele, Window Rock High School, Fort Defiance, Arizona

Colorado:

Don Born, Peggy Wickliff, Air Academy High School, United States Air Force Academy, Colorado; Linda Lynch, Douglas County High School, Castle Rock, Colorado; Doug Hewins, Liberty High School, Colorado Springs, Colorado; Barbara Andrews, Mitchell High School, Colorado Springs, Colorado; Rata Clarke, Ray Coddington, Jean Orton, Jim Snare, Palmer High School, Colorado Springs, Colorado; Rod Baker, Michele Girard, Peyton High School, Peyton, Colorado; Kathy Dorman, Malcom Hovde, Ponderosa High School, Parker, Colorado; Ann Pollet, Deborah Walters, Pueblo County High School, Pueblo, Colorado; Glen Smith, Sabin Junior High School, Colorado Springs, Colorado; Jeff Cogburn, B.J. Stone, Valley High School, Gilcrest, Colorado; Bill Bragg, Wasson High School, Colorado Springs, Colorado; Larry Jakel, Doug Steward, Weld Central High School, Keenesburg, Colorado; Jay Matheson, West Center for Intergenerational Learning, Colorado Springs, Colorado; Christy Beauprez, Glenn Peterson, Windsor High School, Windsor, Colorado

Florida:

James Happel, Constance Hopkins, Manatee High School, Bradenton, Florida; Scott MacGregor, Joe Martin, Palmetto High School, Palmetto, Florida; Barbara Grosz, Pine Crest School, Fort Lauderdale, Florida

Hawaii:

Jamie Nekoba, Waiákea High School, Hilo, Hawaii; Jennifer Busto, Maryknoll Schools, Honolulu, Hawaii

Illinois:

Shelly Peretz, Thornridge High School, Dolton, Illinois

Kansas:

J.D. Hand, Chuck Mowry, Gina Whaley, Junction City Senior High School, Junction City, Kansas

Minnesota:

Clyde Cummins, St. Paul Academy Summit School, St. Paul, Minnesota

Missouri:

David Jungmeyer, California R-1 High School, California, Missouri

New Jersey:

Judith Jones, Saint John Vianney High School, Holmdel, New Jersey; Margaret Sheldon, West Morris Central High School, Chester, New Jersey; Karen Martin, West Morris Mendham High School, Mendham, New Jersey

Ohio:

Barbara Blackwell, Susan Keiffer-Barone, Aiken High School, Cincinnati, Ohio; Scott Popoff, Sycamore High School, Cincinnati, Ohio

Texas:

Peter Mariner, Francis Mikan, Dean Mohlman, Tom Stege, St. Stephen's Episcopal School, Austin, Texas

Washington:

Kathleen Heidenrich, Vicky Lamoreaux, River Ridge High School, Lacey, Washington; Larry Bencivengo, Mary Margaret Welch, Mercer Island High School, Mercer Island, Washington; Mary Ketchum, Jeannie Wenndorf, Lindberg High School, Renton, Washington; Gro Buer, Carol Nussbaum, B.E.S.T. Alternative School, Kirkland, Washington; Connie Kelly, Diane Lashinsky, Patrick Taylor, Shorecrest High School, Seattle, Washington

Wisconsin:

Gene O'Brien, Hartford Union High School, Hartford, Wisconsin

Acknowledgments continue on page E271.

TABLE OF CONTENTS

Engage   Being a Scientist 2
　　　　　　　Thinking As a Scientist Thinks 4
　　　　　　　Communicating the Results of Scientific Thinking. 9
　　　　　　　You and the Science of Biology 10

U N I T   O N E

Evolution: Patterns and Products of Change in Living Systems 14

Chapter 1   The Human Animal. 16
　　　　　　　How Different Are We? 18
　　　　　　　Primates Exploring Primates 20
　　　　　　　Portraying Humankind 26
　　　　　　　A Long Childhood. 28
　　　　　　　What Does It Mean to Be Human? 31

Chapter 2   Evolution: Change across Time 32
　　　　　　　Lucy 35
　　　　　　　Modeling the Earth's History. 36
　　　　　　　Evidence for Change across Time 38
　　　　　　　Explaining Evolution 45
　　　　　　　Modeling Natural Selection 47
　　　　　　　A Cold Hard Look at Culture 49
　　　　　　　Evolution in Action 52

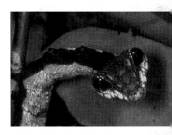

Chapter 3   Products of Evolution: Unity and Diversity 56
　　　　　　　Strange Encounters. 58
　　　　　　　Describing Life 59
　　　　　　　A Look at Diversity 61
　　　　　　　Adaptation, Diversity, and Evolution 63
　　　　　　　Using Unity to Organize Diversity 66
　　　　　　　Explaining the Zebra's Stripes 70
　　　　　　　First Encounter with Organism X 74

U N I T   T W O

Homeostasis: Maintaining Dynamic Equilibrium in Living Systems 76

Chapter 4   The Internal Environment of Organisms 78
　　　　　　　Can You Stand the Heat? 80
　　　　　　　Cells in Action 80
　　　　　　　A Cell Model 86
　　　　　　　Regulating the Internal Environment 88
　　　　　　　Can You Stand the Heat—Again? 90

Chapter 5   Maintaining Balance in Organisms 92
The Body Responds 94
What's Your Temperature Now? 95
Stepping Up the Pace 99
On a Scale of 0 to 14 102
How Do They Stay So Cool? 107
Homeostasis in Organism X 108

Chapter 6   Human Homeostasis: Health and Disease 110
Pushing the Limits 112
Hospital Triage 112
Self Defense! 119
Immunity Simulation 123
What's the Risk? 128
Health Care Proposal 131

U N I T   T H R E E

Energy, Matter, and Organization: Relationships in Living Systems 138

Chapter 7   Performance and Fitness 140
Thinking about Fitness 142
What Determines Fitness? 144
What Is in the Food You Eat? 146
You Are What You Eat 151
Structures and Functions 153
Marathon 155

Chapter 8   The Cellular Basis of Activity 164
Releasing Energy 166
Energy in Matter 166
Keep on Running! 171
Using Light Energy to Build Matter 176
Building Living Systems 180
Tracing Matter and Energy 182

Chapter 9   The Cycling of Matter and the Flow of
Energy in Communities 184
A Matter of Trash 186
Exploring the Cycling of Matter in Communities 187
Spinning the Web of Life 190
Generating Some Heat 193
What Have I Learned about Matter and Energy in Communities? . . . 196

Explain   Conducting Your Own Inquiry 200
Science All Around You 202
Being an Experimental Scientist 203

UNIT FOUR

Continuity: Reproduction and Inheritance in Living Systems. 210

Chapter 10   **Reproduction in Humans and Other Organisms**. 212

A Zillion Ways to Make More. 214
Making Sense of Reproductive Strategies 214
Making Sense of Human Reproduction. 218
Observing Reproductive Behavior in Nonhuman Animals 221
Cultural Influences on Human Mating Behavior 223
A Reproductive Strategy for Organism X. 225

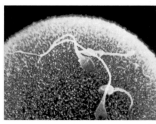

Chapter 11   **Continuity of Information through Inheritance** 228

Gifts from Your Parents. 230
Game of Chance. 230
Patterns of Inheritance 237
Understanding Inherited Patterns 241
Can You Sort It Out?. 243
The Genetic Basis of Human Variation 247
Continuity and Change 251

Chapter 12   **Gene Action** 254

The Stuff of Life. 256
Transferring Information 260
Modeling DNA. 261
Gene Expression. 266
A Closer Look at Protein Synthesis. 272
Genetic Technology 274
Prime Time 278

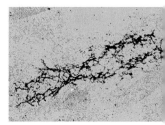

UNIT FIVE

Development: Growth and Differentiation in Living Systems. 282

Chapter 13   **Processes and Patterns of Development**. 284

One Hundred Years of Questions 286
A Start in Development 289
Processes that Generate Complexity 292
Development Gone Awry 297
Evaluating Where We Stand. 299

Chapter 14   **The Human Life Span** 304

A Century of Photographs. 306
Growing Up—What Does That Mean?. 307
A View of Life 311
Life-Span Development: Examining the Contexts 313
Cultural Diversity in the Human Life Span 315

Ecology: Interaction and Interdependence in Living Systems 318

Chapter 15   **Interdependence among Organisms in the Biosphere** 320

Interactions in the World around Us 322
Reindeer on St. Paul Island 323
On the Double 328
A Jar Full of Interactions 333
Changing Ecosystems 336
Organism X and Global Interdependence 340

Chapter 16   **Decision Making in a Complex World** 342
Calling the Question 344
The Sun and Life (or Death?) 345
Where Do We Go from Here? 354
The Ozone Layer: A Disappearing Act? 361
The Limits of Abundance 368

Evaluate   **Thinking Like a Biologist** 382
Recognizing Biology in Medicine 384
Chapter Challenges 392
Building a Portfolio of Scientific Literacy 403

Appendix A   **Laboratory Safety** 406

Appendix B   **Techniques** 414
Technique 1   Journals 414
Technique 2   Graphing 416
Technique 3   Measurement 422
Technique 4   The Compound Microscope 424

ESSAYS

U N I T   O N E

The Chimp Scientist. E2
Mapping the Brain E2
Do You Have a Grip on That? E5
On Being Human. E6
Brains and a Lot of Nerve E9
The Importance of Being Children E13
Fossils: Traces of Life Gone By E15
Technologies That Strengthen Fossil Evidence E17
Modern Life: Evidence for Evolutionary Change. E20
Primates Show Change across Time E22
Darwin Proposes Descent with Modification E23
Evolution by Natural Selection E25
Just a Theory?. E29
Describing Life: An Impossible Challenge? E31
Five Kingdoms E36
From Cell to Seed. E42
Evolution Produces Adaptations. E44
Organizing Diversity E47

U N I T   T W O

Compartments E50
Membranes E52
Molecular Movement E53
Making Exchanges throughout the Body E56
Disposing of Wastes E58
Homeostasis E60
Careful Coordination. E62
Regulation and Homeostasis E66
The Breath of Life E68
Behavior and Homeostasis. E70
Beyond the Limits E73
Coping with Disruptions: The Role of Medicine in Homeostasis E75
Avoiding Disruptions: The Immune System. E78
Self and Nonself E81
Immune System Memory. E82
Avoiding Disruptions: Behavior, Choices, and Risk. E84
Individual Behavior Can Affect Larger Groups E86
Ethical Analysis E87

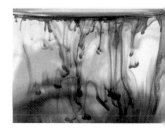

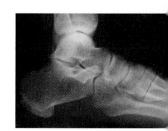

ESSAYS

UNIT THREE

Human Performance: A Function of Fitness E88
Food: Our Body's Source of Energy and Structural Materials E90
What Happens to the Food You Eat? E96
Anorexia Nervosa: Dying to be Thin E98
The Structural Basis of Physical Mobility E99
The Ant That Terrorized Milwaukee E103
Energy's Role in Making Structures Functional E104
Factors Influencing Performance E105
Matter and Energy Are Related E108
Energy is Converted and Conserved E110

Historical Connections between Matter and Energy E115
Controlling the Release of Energy and Matter E116
Cellular Respiration: Converting Food Energy into Cell Energy E117
Regulation and Energy Production E120
Whose Discovery Is This? E122
Getting Energy and Matter into Biological Systems E123
Metabolism Includes Synthesis and Breakdown E127
Garbage among Us—From Then until Now! E128
Matter in Nature Is Going Around in Cycles . . . What Next? E132
Worms, Insects, Bacteria, and Fungi—Who Needs Them? E135
Let's Ask Drs. Ricardo and Rita E136
Losing Heat E141

UNIT FOUR

Continuity through Reproduction E142
Making More People E144
Hormones and Sexual Reproduction E146
Infertility E149
Sexual Activity and Health Hazards E149
Mating Behaviors of Nonhuman Animals E151
Cultures and Mating Patterns E153
Phenotype and Genotype E155
Case Studies of Two Tragic Genetic Disorders E158
Meiosis: The Mechanism Behind Patterns of Inheritance E160
The Role of Variation in Evolution E164
Mutations Can Be Used as Tools for Biological Research E165

Genetic Complexity E168
A Model for How Multiple Genes Might Influence the Complex Trait
    of Skin Color E170
Incomplete Dominance E170
Genetic Information Is Stored in Molecular Form E172
DNA May Be the Stuff of Genes! E173
DNA Structure and Replication E173
Replication Errors and Mutation E178

ESSAYS

What's New? Why the Fuss about Watson and Crick? E179
The Expression of Genetic Information E180
Translating the Message in mRNA E182
News Flash! White-coated Sleuths Decipher Genetic Code. E184
Cellular Components in Protein Synthesis E184
News Flash! Extraordinary New Technique Changes Biology Forever E186
Manipulating Genetic Material E187
Ethical and Public Policy Issues Related to Genetic Engineering E192

UNIT FIVE

The Long and Short of Development E194
The Cell Cycle and Growth Control E195
Coordinating Growth E198
Differentiation and the Expression of Genetic Information. E199
Cloning People? E203
Development and Birth Defects. E203
Cancer: Unregulated Growth E206
Patterns of Development E210
Human Development 101 . . . . E214
Growing Up through Life's Phases E216
Physical Growth Influences Mental Growth E220
Physical Growth Influences Social and Emotional Growth. E223
All Phases of Life Require Self-Maintenance E224
Culture: The Great Shaper of Life. E226

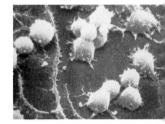

UNIT SIX

Interdependence Involves Limiting Factors and Carrying Capacity E232
Growing, Growing, Grown. E234
Inquiring Minds E237
Endless Interactions E238
The Ecology of Kirtland's Warbler E241
Yellowstone Burning: Is Fire Natural to Yellowstone National Park? E242
Forests of the Past E244
From Puddle to Forest. E247
Environmental Ethics and You E252
Ultraviolet Light and the Ozone Layer. E252
The Farmers Are Saved!. E255
Breakthrough Insecticide Benefits the Boys Overseas and Farmers Alike. E256
Conservationists in Hysterics over DDT E257
DDT Banned!. E257
Systems Analysis E258

Index E260

Dear Learners:

The staff at BSCS developed *BSCS Biology: A Human Approach* for students first. Not all of the books you use are developed this way, but we feel that focusing on the student is the only way to provide you with the best biology learning experience possible. As you glance through the book, notice that this is not a passive, encyclopedic approach to biology; you will not sit day after day in lectures. Instead, this book helps you learn biology through active involvement.

BSCS cares so much about the quality of its books because we care about students and teachers. We are a nonprofit organization where scientists and educators are dedicated to improving science education. This book is the result of BSCS's study of the research in biology and learning and our interpretation of what we think is a better way to help students learn biology. When we defined the word "better," we decided that a better high school biology program would mean the following:

- more emphasis on the big concepts of biology and less emphasis on the vocabulary words of biology,
- more opportunities to conduct investigations that you design, and
- more connections between biological concepts and your life.

In the program overview on the next several pages, you can read about how we put these differences into practice in *BSCS Biology: A Human Approach*. The overview describes the key features of the program that we think make it noticeably different from and better than other texts. To make sure our ideas worked in classrooms, we spent two and one-half school years testing our ideas in classrooms with approximately 80 teachers and 5000 students. In addition, many biologists and educators reviewed the materials to make sure they were accurate and current at the time we went to press. We list these contributors and reviewers at the front of the book.

Hundreds of people work together to create one new BSCS program. This process never ends as long as the program is in print. If you read something that you think is confusing or inaccurate, or if you have suggestions for improving an activity, please write to us. We will consider your comments when we revise the book in later editions. Our mailing address is

BSCS Biology: A Human Approach Revision Team
BSCS
5415 Mark Dabling Boulevard
Colorado Springs, CO 80918-3842

We hope you enjoy learning biology in this new way. We enjoyed putting this program together for you because learners are the most important people in schools.

Sincerely,

The Project Staff

The letter from the project staff states that *BSCS Biology: A Human Approach* is a better way to learn biology. Six key features of the program help explain why we think this program is better. We describe each feature below.

I. UNIFYING THEMES

We organized the program into three sections and six core units. The three sections are the Engage, Explain, and Evaluate Sections. They come at the beginning, middle, and end of the program and are described under the *Instructional Model* heading. We organized the six core units around six major biological concepts. These concepts are recurring themes that unify all of biology. Although you will see these themes in every unit, we focus on one theme in each unit.

Evolution: Patterns and Products of Change in Living Systems

What does it mean to be human? This is the central question of Unit 1. You will assess the unique qualities of humans and the diversity of life while trying to place humans in the scheme of living systems. Then you will consider characteristics that are common to all living systems as well as those that are unique to humans, and you will grapple with the question of whether life is definable. Unity, diversity, genetic variation, and evolution, including cultural evolution, are the major conceptual themes in Unit 1.

Homeostasis: Maintaining Dynamic Equilibrium in Living Systems

Unit 2 explores the controlled internal environment that all organisms require to function well. You will use familiar examples to develop an understanding of the concepts of response, regulation, and feedback. Then you will examine the division between internal and external conditions and the processes by which internal conditions are maintained in spite of changes in external conditions. In the final chapter of the unit, you will expand these concepts by analyzing the way health and disease affect both the individual human and society as a group.

Energy, Matter, and Organization: Relationships in Living Systems

Unit 3 begins by letting you explain the requirements of physical performance and consider the effects of fitness, drugs, and alcohol on performance. You will develop an understanding of the relationship between structure and function. You also will explore the interplay between energy and matter through such metabolic processes as photosynthesis and cellular respiration as well as through interactions in a community. Finally, you will consider the role of producers, consumers, and decomposers in the flow of energy and cycling of matter in a community.

Continuity: Reproduction and Inheritance in Living Systems

Unit 4 focuses on reproduction, patterns of inheritance, and the role of genes and DNA in inheritance. The discussion of human sexual reproduction includes reproductive systems and cycles, reproductive behavior, and ethical issues, such as contraception and sexually transmitted diseases. You will consider the role inheritance plays in continuity and variation and how genes are a source of coded information. In conclusion, you will study the dynamics of gene expression and replication at a molecular level, which provides a basis for understanding genetic engineering.

Development: Growth and Differentiation in Living Systems

As Unit 5 begins you will consider development as a process that involves differentiation and growth and that requires regulation. You will explore patterns of development, which appear in stages such as reproductive maturity, aging, and death. Development is affected by evolutionary history and provides opportunities for evolutionary change, and it depends on communication. Finally, you will consider human life stages, looking at biologically programmed events as well as the cultural environment in which they occur.

Ecology: Interaction and Interdependence in Living Systems

Unit 6 centers on issues in the area of ecology, including dilemmas about the interactions among populations, resources, and environments. You will examine the concepts involved in population dynamics, which sets the stage for studying the interactions between humans and their environment. Next, you will focus on how human actions can modify the environment, especially through the use of technology. The final emphasis is on the ethical issues raised by human actions and technology.

II. SUBTHEMES

Two subthemes, or background ideas, are woven through the entire program. They help to establish connections between biology and your life, and they improve your reasoning ability. The Science As Inquiry subtheme refers both to the discovery process by which information is obtained and evaluated and to the changing body of knowledge that characterizes scientific understanding. This theme systematically exposes you to the processes of science, including making observations, making inferences, assembling evidence, developing hypotheses, designing experiments, collecting data, analyzing and presenting results, and communicating and evaluating conclusions.

The Science and Humanity subtheme makes your study of biology more relevant and approachable by incorporating the critical elements of human culture; the history of science; the place of ethics, ethical analysis, and decision making in today's controversial science world; and the importance of human technology as a way of adapting. We define technology as the use of knowledge to achieve a practical solution to a perceived problem and recognize that the ultimate effects of the technological process or product on society and the biosphere may extend beyond the intended effects.

III. INSTRUCTIONAL MODEL

We organized the instruction of major concepts in this book around a model of learning that recognizes how individuals build or construct new ideas. We call this type of instructional model the "5Es" because the program is organized around five phases of learning that we best can describe using words that begin with E: Engage, Explore, Explain, Elaborate, and Evaluate. You will get an overview of the program by completing the Engage Section. Units 1 to 3 let you explore the big ideas of scientific inquiry, and then the Explain Section sets you up to conduct your own scientific inquiry. Units 4 to 6 are designed to help you elaborate your understanding of the processes of science, and the Evaluate Section provides several opportunities for you to evaluate your progress in learning biology.

IV. COOPERATIVE LEARNING

Cooperative learning is an educational strategy that helps you increase your responsibility for your own learning. Cooperative learning also models the processes that scientists use when collaborating and helps you develop the working relationship skills necessary for today's workforce.

V. ASSESSMENT

Assessment opportunities, which allow you to evaluate your progress, are embedded throughout the program, and the assessments themselves are learning experiences. The following assessment strategies are included in the program:

- assessments of your performance, such as experiments;
- written tests that have a variety of short-answer and essay questions;
- assessments of cooperative learning skills;
- debates;
- presentations, both by teams and by individuals;
- written assignments, both by teams and by individuals;
- journal assignments that include short-term and long-term work;
- projects, both ongoing and one-time;
- an ongoing activity about a newly discovered organism;
- opportunities for self-assessment and peer assessment; and
- discussions, both by teams and the whole class.

VI. EDUCATIONAL TECHNOLOGY

Educational technology is integral to the program and is used as a tool to enhance learning and understanding. The program includes three major electronic technologies:

- Videodiscs, which include interactive video activities for the chapters.

- The LEAP-System™, which is a computer interfacing system for microcomputer-based laboratories.

- Computer simulations, which allow you to explore complex biological interactions.

> "*The whole of science is nothing more than a refinement of everyday thinking.*"
>
> Einstein, *Out of My Later Years*, 1950

Engage
BEING A SCIENTIST

When you are faced with a problem or a puzzle, how do you figure things out? You might answer, "I think about them." But do you know how you think? Use the photograph on the left to answer the question, What is on the bottom face of each die? As you answer this question, focus on the way you think as you come up with an answer. Thinking is something all humans do, but it is a skill we each do in a slightly different manner. Use the following questions to help you think about thinking.

- How did observation help you answer the question?
- What pictures did you see in your mind that helped you solve the problem?
- How did remembering patterns help you answer the question?
- How did you use reasoning to solve the puzzle?

The biology program you are just beginning focuses on science as one of the ways humans understand and explain their world. The activities throughout the program encourage you to think as a scientist thinks. This short section is titled Engage because it is designed to engage you in the type of experiences you will participate in throughout this course. The goals of this section are

- to let you experience the nature of science by asking questions, gathering evidence, and proposing explanations and
- to provide an overview of the program.

ACTIVITIES

Engage
Explore   Thinking as a Scientist Thinks

Explain   Communicating the Results of Scientific Thinking

Elaborate
Evaluate   You and the Science of Biology

THINKING AS A SCIENTIST THINKS

**Engage
Explore**

The opening photograph and text encouraged you to think about your own style of thinking, and you will continue to do so during this program. You also will develop the skills of thinking like a scientist. To think as a scientist thinks you need to ask questions, gather information, and propose explanations.

You will practice these skills in this activity about the Tri-Lakes region. The Tri-Lakes situation is not real, but it is very similar to the situations faced in many communities. Residents around Lake Erie have successfully revitalized one of the Great Lakes that many thought was dead due to industrial pollution. The citizens of Sweden, Canada, Germany, and the northern United States have been studying the acidification of lakes in their countries. In some places professional and citizen scientists have been able to reverse what some thought was an irreversible course of events leading to dead lakes. The types of questions raised by the Tri-Lakes Association could be asked in any community about many science or technology issues.

You will use the Tri-Lakes problem to become acquainted with the basic methods of science; then you will build your understandings and abilities throughout the year until you can design and conduct your own independent investigations.

Materials (per team of 2)

data packet
dropping pipets
watch glasses
10-mL graduated cylinders
petri dish
beaker labeled *Culture Water*
beaker labeled *Used Culture*
nonmercury thermometers
hand lenses or stereomicroscopes
Detain™
forceps
pH strips
vinegar in dropping bottles
hot water
microorganisms such as *Daphnia* or *Gammarus*

Process and Procedures
Part A Asking Questions

1. Listen to or read to yourself the letter from the Tri-Lakes Association on page 5.

2. Create a table with four columns and these headings: *Questions, Information from the Data Packet, Evidence from Experimentation, Proposed Explanations.* See Figure En.1 for an example.

    The first step in thinking scientifically is to identify the questions you are trying to answer. You can start this process by thinking about the information you have. A table

TRI-LAKES ASSOCIATION

Dear Biology Students at Tri-Lakes High:

The members of the Tri-Lakes Association are very concerned about a perplexing problem we have in the Tri-Lakes region. In general the fishing is good, but people are not catching as many bass as they did years ago. As a result, our reputation as the bass-fishing capital of the world is suffering, and the reservations at the local resorts are down by 25 percent, a disturbing economic indicator for our area.

I know that the biology classes at Tri-Lakes have kept records on the water quality of the lakes for many years. The members of the association have noticed that the change in the number of fish being caught seems to have occurred along with, or as a result of, a number of other changes around the area. We hope that your proximity to the lake and your scientific abilities will help us determine what is happening, or at least what questions we need to pursue.

The members of the association have made the following observations, which may help you:

• Microorganisms such as *Daphnia* and *Gammarus* are less common in the lake.
• The lake is greener for more of the year than it used to be.
• The perch seem smaller and less colorful than in previous years. They also congregate in the hollows of discarded cinder blocks and hover in one place for a long time.

In addition to these observations, I have sent a number of data packets for your convenience. These packets represent information that the association members have pulled together in an attempt to understand what might be happening to our region. If some data look familiar, it is because we pulled it from your annual report to the association. We trust that this combination of local, national, and historical data will provide enough clues to identify our problems so that we can begin working on solutions.

Our next association meeting is in two weeks. We hope you will have additional information you can share with us at that time.

Because many of you and your families are involved in the fishing and resort industries, I am sure you understand the seriousness of this situation. I eagerly await your response.

Sincerely,

Chris Tackle

Chris Tackle
President, Tri-Lakes Association

---

is a useful tool for organizing data or information related to the question you are trying to answer. This step of organizing the information will help you think about the question in a productive manner.

Set up this table in your journal so you can use it throughout this activity.

Figure En.1 Sample table
When you draw your table in your journal, leave plenty of room to record information.

Questions	Information from the Data Packet	Evidence from Experimentation	Proposed Explanations

3. Based on what you learned from the letter on page 5, work with a partner to fill in the first column of your table, *Questions*.

You can use the questions in the letter or new questions that came to your mind as you thought about the Tri-Lakes issue. Record each question as precisely as possible in a new box on the table in your journal.

Part B Gathering Information

The process of answering questions scientifically requires the scientist to gather as much information as possible. This information may come from several sources; it may be a written record of the work of other scientists, or it may be the information, or **data**, gathered from experiments conducted by the scientist. In this part of the activity you will try both methods of gathering information.

1. Pick up a data packet from your teacher.

Each data packet contains information sheets.

2. Review the titles listed in Figure En.2 and choose the information sheets most likely to be helpful to you.

This list identifies the material contained in a data packet. You may use this and any other resource you have handy to gather information that may help you and your partner answer the questions listed in the first column of your table.

Figure En.2 Information sheets Study the information sheets of interest to you to help solve the mystery at Tri-Lakes. You also can use other available resources.

Information Sheet #	Title of Information Sheets
1	Tri-Lakes Advertisement
2	*Tri-Lakes Tribune* Article, 1989
3	*Tri-Lakes Tribune* Article, 1993
4	Location of Tri-Lakes Industries and Resorts
5	Zone Map of the Average Temperatures in the Tri-Lakes
6	Largemouth Bass
7	Yellow Perch
8	Graph of Total Number of Fish Caught Annually, 1975-1995
9	Number of Largemouth Bass and Yellow Perch Caught Annually, 1975-1995
10	Average Size of Perch and Bass Caught, with Comparison to Legal Limit, 1975-1995
11	Table of Dissolved Oxygen and pH for Tri-Lakes, 1975-1995
12	Algae and Cyanobacteria
13	*Daphnia*
14	*Gammarus*
15	Pesticides
16	Acid Precipitation

You do not have to read everything in the data packet. Compare the titles of the information sheets with the questions you raised. Which information sheets are likely to be helpful?

3. After you have chosen the information sheets that you and your partner will study, divide the work in half. Study your half of the packet, and tell your partner what you have learned. As you exchange information and you find things that might help answer a question you have raised, fill in the corresponding square in the column *Information from the Data Packet* on your chart.

   At the end of this activity you will rate yourself and your partner on how well you taught each other about your information sheets.

   Sometimes reading the available information does not answer all of a scientist's questions. For the Tri-Lakes situation, it is important to understand the microorganisms that Chris Tackle referenced. The next steps in this activity will lead you through a process to help you learn more about these microorganisms.

4. Identify a question about the microorganisms that you hope to answer by experimentation. Record this question in your table if it is different from the ones you already recorded.

5. Write in your journal a procedure to use to gather data about your question. After you have written the procedure, ask your teacher to approve it.

   The protocols in Figures En.3 and En.4 provide two examples of methods for studying the microorganisms Daphnia *and* Gammarus.

6. Conduct your experiment. Record your observations and results in the squares of the column *Evidence from Experimentation* that correspond to the questions you raised about *Gammarus* or *Daphnia*.

Protocol 1: Microorganisms and pH

a. Place about 10 mL of culture water in a watch glass. Use the dropping pipet to carefully transfer 2 or 3 microorganisms from the culture to the watch glass.

b. Observe the microorganisms with a hand lens or stereomicroscope long enough to determine their behavior.

c. Record this behavior in your journal but not in your table.

d. Measure the pH of the water in the watch glass, and record this value in your journal but not in your table.

e. Use the vinegar and pH strips to *gradually* change the pH of the water by 0.5 pH units. Stir the water after each drop of vinegar that you add and wait 30 seconds before adding the next drop.

   CAUTION: Vinegar is a *mild irritant*. Avoid eye contact. If contact occurs, flush affected area with water for 15 minutes; call the teacher.

f. Observe the microorganisms for 1 minute, and record your observations in your journal.

g. Transfer the microorganisms to the container labeled *Used Culture* when you are finished.

Figure En.3 Protocol 1: Microorganisms and pH

Figure En.4 Protocol 2: Microorganisms and Temperature

Protocol 2: Microorganisms and Temperature

a.   Place about 10 mL of culture water in a watch glass. Use the dropping pipet to carefully transfer 2 or 3 microorganisms from the culture to the watch glass.

b.   Observe the microorganisms with a hand lens or stereomicroscope long enough to determine their behavior.

c.   Record this behavior in your journal but not in your table.

d.   Measure the temperature of the water in the watch glass.

e.   Gradually add hot water to the watch glass until the temperature is 2°C different from the starting temperature.

   You may need to remove some water.

f.   Observe the microorganisms for 1 minute and record your observations in your journal.

g.   Repeat Steps d-f until you have changed the temperature of the water 10°C from its starting temperature.

h.   Transfer the microorganisms to the container labeled *Used Culture* when you are finished.

Helpful Background

•   pH (A measure of how acidic a solution is. The pH scale goes from 0 to 14; the lower the number, the more acidic the solution is.)

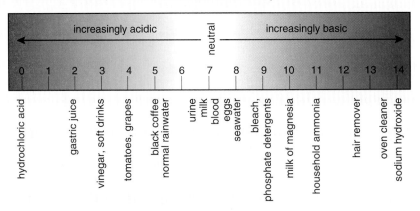

•   Detain™ (A substance that slows down the movements of microorganisms, enabling you to observe them.)

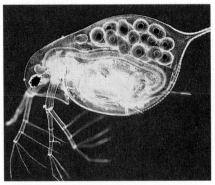

Daphnia Can you find the heart, gut, and brood pouch on your *Daphnia*? *Daphnia* are about 1 mm long.

Gammarus *Gammarus* are about 4 mm long. What features are easier to see because the photograph is enlarged?

7. If time permits, visit other teams and ask what they have learned from the data packet and their experiments.

Analysis

1. Use the information you gathered during this activity to develop an explanation for at least three of the questions on your chart. Record your explanations in the corresponding squares in the column labeled *Proposed Explanations.*

2. Rate your team on how well you taught each other about the packet of information. Use a scale of 1 = not very well to 5 = very well. Record your rating and two sentences of justification in your journal.

COMMUNICATING THE RESULTS OF SCIENTIFIC THINKING

Explain

As you explored the Tri-Lakes issue you used several methods of science: asking questions, gathering and assessing information by looking at a data packet and by conducting a laboratory experiment, and proposing explanations. Now you will prepare to communicate your results to someone else. You will complete your analysis of your work, draw conclusions based on your work so far, and prepare a letter to the Tri-Lakes Association.

Process and Procedures

1. Work with your partner to develop a response to the letter from the Tri-Lakes Association. Let the president of the association know

   • what you have determined so far,
   • how you learned what you know, and
   • what else you would like to know.

2. Exchange letters with another team as your teacher directs. Analyze this letter from the perspective of the association president. Begin your analysis by asking, Does their letter provide the answers that I want and need?

3. Provide the other team with at least three specific comments about their letter.

   Your feedback should be a mix of positive statements that indicate strengths of the letter and other statements that identify weak areas that need to be strengthened.

4. Revise your letter to reflect the feedback that you received.

Analysis

1. You and your partner should be prepared to read your letter aloud in class.

2. The Tri-Lakes situation that you just studied allowed you to participate in most of the methods of science. Think about your experiences and listen to the letters of your classmates. Then reflect on the following statements:

- Science is a way of knowing.
- Science can find all the answers.
- Science is hard and only smart people do science.
- Anyone can use the processes of science.

Choose two statements that are best illustrated by the work you have just completed in class. Write in your journal two short paragraphs that tell why each of the statements you chose summarizes the nature of your work in biology class.

YOU AND THE SCIENCE OF BIOLOGY

Imagine yourself doing one or more of the following:

- understanding the choices a doctor offers,
- deciphering nutritional information on a food package label,
- voting on an issue involving science and technology,
- serving on a jury that has to listen to an expert describe DNA evidence, or
- deciding whether or not to support the construction of a new dam.

These are examples of actions that occur in the United States. Will you be one of the people acting with information because you have learned to think scientifically? Or will you be one of the people who acts and hopes for the best, despite a lack of information and understanding? By participating in this biology program you are taking a big step toward joining the first group of people.

In this course we use six main ideas to organize your study of biology. We also integrate these ideas with opportunities to think about and use the methods of science. As a result, you have the opportunity to learn how to think scientifically while you learn biology.

Process and Procedures

1. Read *A Human Approach to Biology* on page 11, and identify at least one reason why each of the following is relevant:

- the study of biology now,
- the study of biology in the future, and
- this program.

2. Think about the question, Why should the residents of the Tri-Lakes region care about the issue that the association raised? Record your thoughts about this question in your journal.

A HUMAN APPROACH TO BIOLOGY

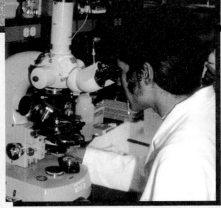

The scientist in the photograph above is the visual image many people have of scientists. While this is one example of what doing science looks like, there are many other examples. Think of the activities listed in the introduction to the activity *You and the Science of Biology*; all of those activities can be conducted using the methods of science.

Biology is a way of explaining by *scientific processes* what living organisms do and how they do it. There are other ways to look at life. You could describe living things in romantic poetry, by artistic paintings, or in a written story. These activities have great value and add an important dimension to human life. These ways of describing life, however, would not be a scientific approach. In a scientific approach, the explanations are based on asking questions that can be answered by gathering information and evidence and then analyzing what is known in a logical manner (see the summary of the processes of science in Figure En.5). In this program you will be practicing "Science As Inquiry" when you learn and use the processes of science.

Biology is a challenging scientific study because organisms are very complex and show a lot of variety. Fortunately, they also share some very important characteristics that can be summarized as large ideas. In this program these large and universal ideas are the topics of each of the six units.

This program is called *BSCS Biology: A Human Approach* because we have focused your activities around the interactions between humans and biology. You will experience this human approach in several ways. You will complete experiments that involve yourself. You will read examples of the unifying principles that point out the connection between human beings and evolution, homeostasis, matter, energy, organization, continuity, development, and interdependence. And you will understand "Science and Humanity" when you experience connections to biology content that point out the following:

- Asking questions

- Gathering information

- Proposing explanations

Figure En.5 The processes of science The processes of science help you learn more about biology.

- Humans use technology to help solve problems. Consequently, we are able to improve our ability to survive in the near future.
- Technology varies from culture to culture and within a historical period.
- Science takes place within a cultural and historical context and within a set of constraints often dictated by the culture and time period.
- Humans can use ethical analysis to make decisions and solve problems. This applies to both scientific investigations and questions because humans conduct these endeavors.

As you study the biology in this course, you will learn about living systems. If you can understand how you learn, you'll enjoy the process of learning more and be more successful at it. Scientists who study how people learn are called cognitive psychologists. These scientists have proposed an explanation for learning that suggests that in order to learn, one first must be **engaged** in an idea; then the learner must **explore** the idea; next, the learner develops an **explanation** of the idea. Finally, the learner **elaborates** his or her understanding and is able to **evaluate** what he or she has learned.

As you look through this book, you will notice the use of five words that start with the letter "E." This program uses these "E words" to organize the instructional flow in each chapter. This instructional flow is designed to help you construct an understanding of each big idea in biology.

We have developed flow charts to help you see the opportunities for learning in each chapter. Figure En.6 is the flow chart for this section of the program. The flow charts for all other chapters are included in your teacher's book for the program so they can be duplicated and distributed in class.

As you look at Figure En.6, you will notice that the flow chart shows you when you will be in each stage of the instructional flow, what concepts you should be learning, how science as inquiry is used, when you are studying the interaction between science and humanity, and what essays and assessment opportunities exist for your learning. The essays for most of the program are grouped together at the end of this book, while the assessment opportunities are embedded within the program. These physical placements reflect two aspects of our approach to science and learning. First, because the content information changes frequently, we designed a program that acknowledges this. The set of essays that we have included gives you a place to begin your research, but it is intended to be *one* resource. A second aspect of the approach is reflected in the embedded assessment. We think you should have continual opportunities to evaluate your understanding and growth, so you will notice frequent chances to assess yourself.

BSCS

Evolution: Patterns and Products of Change in Living Systems

Homeostasis: Maintaining Dynamic Equilibrium in Living Systems

Energy, Matter, and Organization: Relationships in Living Systems

Continuity: Reproduction and Inheritance in Living Systems

Development: Growth and Differentiation in Living Systems

Ecology: Interaction and Interdependence in Living Systems

Unifying Principles of Biology These six big ideas are one way to organize the discipline of biology. These ideas are also the titles of the six units in *BSCS Biology: A Human Approach.*

Figure En.6 Flow Chart for Engage Section This chart is a summary of the learning opportunities in this section of the book.

Flow Chart for Engage Section

Chapter Flow	Related Essays	Assessment	Concepts	Science As Inquiry	Science and Humanity
Engage/ Explore Thinking As a Scientist Thinks	Tri-Lakes letter	analysis questions, journal entries	science is a process	involves asking questions, gathering information, proposing explanations	
Explain Communicating the Results of Scientific Thinking		letter writing	results need to be communicated	analyzes available information, organizing conclusions for presentation	applies analysis to a community problem
Elaborate/ Evaluate You and the Science of Biology	*A Human Approach to Biology*	analysis questions	biology has both content and processes	is integral to this program	adds relevance to this program

Analysis

Use your experiences from the activities you have just completed and your general life experiences to answer the following questions in your journal. After each answer leave a few blank lines so that you have room to revise your response following a class discussion of the questions.

1. What role does science play in your life?

2. What role do you think science will play in your future?

3. How can science help you make decisions about yourself, your lifestyle, your community, and your planet?

4. How do the decisions that we make today influence future generations, generations that could include your children and grandchildren?

5. What issues exist in your community that are similar to the Tri-Lakes issue? (A community may be a subdivision, town, state, nation, or the world.)

EVOLUTION:
PATTERNS AND PRODUCTS OF CHANGE IN LIVING SYSTEMS

1   **The Human Animal**

2   **Evolution: Change across Time**

3   **Products of Evolution: Unity and Diversity**

How did the great diversity of life on earth come to be, and what do the organisms living on earth have in common with each other? As we reflect on ourselves as living organisms, what does it mean to be human, and how did humans come to be?

These questions do not have simple answers. Scientists and philosophers have been asking these questions for hundreds of years. In particular, biologists have been studying the diversity of living systems and looking for patterns. They have found patterns in the information, sequences of events, and chemical and physical structures of living things. Scientists learn more about these patterns by asking questions, gathering information, and proposing explanations.

In this unit you will explore change in living systems in much the same way. The following goals will help you learn the big ideas in this unit. By the end of Unit 1, you will understand

- that humans have characteristics that distinguish them from other organisms,
- that humans share many characteristics with other living systems,
- that there is a diversity of living systems on earth,
- how evolution provides the scientific explanation for this diversity,
- the methods of science, such as developing critical-thinking skills, making observations, asking questions, collecting data, and recording and analyzing data, and
- how to use evidence and inference appropriately.

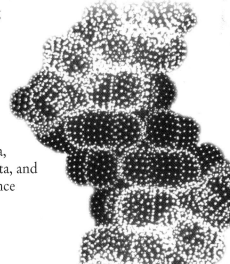

" . . . and this is the weaving of human living: of whose fabric each individual is a part . . . "

James Agee and Walker Evans, *Let Us Now Praise Famous Men*, 1941

1 THE HUMAN ANIMAL

Consider for a moment the vast differences among humans, as depicted by the people shopping in the photograph to the left. We differ in stature, in the shape of our eyes, in the texture of our hair, and in the color of our skin. We differ in our ways of living, the languages that we speak, the foods that we grow and eat, and our concept of the universe. Yet, we share the common bond of humanness, and we see some of ourselves reflected in each of a thousand other faces.

As you begin the first chapter of this program, you will find yourself looking at humans in a way that may be new to you: the way scientists do. In order to do this, you will build a repertoire of skills and processes that exemplify science as a way of knowing. You will have an opportunity to examine some physical characteristics that humans share with other primates, and you will examine the ways in which humans differ. The chapter also explores humans as social and cultural animals that use language and have an immense capacity for learning. Your experiences in these activities will help you begin to develop a scientific understanding of the human animal and what it means to be human.

ACTIVITIES

Engage   How Different Are We?

Explore   Primates Exploring Primates

Explain   Portraying Humankind

Elaborate A Long Childhood

Evaluate   What Does It Mean to Be Human?

Engage

HOW DIFFERENT ARE WE?

In the Engage Section you began to explore science as a way of learning about the natural world. You probably most enjoyed the experimental work that you did. Remember, however, that before you began your experiments, you thought about the Tri-Lakes problem, and you collected and organized the observations that many people had made about the area. These processes—observing and thinking—always precede experimental work. Both are techniques that enhance all the methods of science that you have used in the program so far. Careful observations help us collect information about the world around us, and thinking about these observations helps us understand the information that we have gathered. It also tends to point to other questions.

To begin your study of humans, try making some preliminary observations and doing some thinking about them. Just by looking around you, it is clear that humans are in some obvious ways different from other organisms. Just what is it that makes humans different?

Process and Procedures

1. With your classmates, conduct a brainstorming session to consider the question, What is it that makes humans different from other animals?

2. Take turns sharing your responses and recording them on the large piece of paper posted in the front of the room.

3. Listen to or read the narrative that describes Jane Goodall's observations of a family of chimpanzees at the Gombe Stream Chimpanzee Reserve in Tanzania, Africa. (See *Chimps at Gombe*, page 19.)

   Jane Goodall is a British scientist who began studying chimpanzees in the wild in the 1960s.

Figure 1.1 Jane Goodall As Goodall studied chimpanzees in the wild, she recorded their every move, interaction, gesture, and grunt.

4. With your classmates discuss what thoughts crossed your mind as you listened to this account. Consider the following questions:

CHIMPS AT GOMBE

Gombe Stream Chimpanzee Reserve, Tanzania, Africa

1980: Nope and Pom, two wild chimpanzees, are feeding quietly. Suddenly a chimp screams in the distance. Pom stares toward the sound, turns to Nope, and grimaces. Then the calls break out again. Now at least two chimps are screaming, one an infant. Above the screams, the observer can hear male barks and whoops of attack. Instantly, Pom leaps to her feet and charges toward the sounds of battle. By the time the observer arrives, it is quiet again and Pom and her mother, Passion, are grooming each other intently. Pom's little brother, Pax, is close beside them. Both Passion and Pax have fresh, bleeding wounds, but the family is safe and together.

1981: On a blustery morning, Pom's new baby clings tightly to her hair. As the wind starts to die down, he begins to play, venturing out on the limb that he and his mother are sharing. Just then a violent gust sweeps through the tree and his little body falls spread-eagle through the air. The observer hears a thud and then nothing but silence. Pom looks down at her infant son. Slowly she climbs out of the tree and gathers the tiny form into her arms. For the next two hours, she grooms and nurses him. He leans against her body with his eyes closed. Finally, she carries his battered body away. Three days later, he is dead.

1982: Pom is 17 and her brother Pax is still a youngster when their mother, Passion, dies. Passion has been ill for weeks, and now she trembles with every movement. One morning, she is dead. She must have fallen in the night; her body hangs in a tangle of vines. Pom and Pax sit staring at their mother's body. Little Pax repeatedly approaches her and tries to nurse from her cold breasts. Then he starts to scream and to pull at her dangling hand. So frantic are his efforts that finally he succeeds in pulling her loose. As Passion sprawls lifeless on the wet ground, her children inspect her body many times. Pax cries softly. At last, just before darkness, Pom and Pax move off together.

1983: Having lost both her baby and her mother, a weak and listless Pom finally leaves the community and is not seen at Gombe again. Pax, on the other hand, attaches himself to an older brother, Prof, who provides the care that a mother chimpanzee would. One day, Pax sneezes loudly. Before the astonished observer can get out the camera, Prof hurries over to Pax and stares at his runny nose. Then, picking up a handful of leaves, Prof carefully wipes the mess away.

a. Were you surprised by anything that Dr. Goodall observed? If so, what things?

b. Did you find yourself reflecting more on the similarities or on the differences between humans and chimps?

c. If you had been Goodall observing this group of chimpanzees, what questions would you now have about these animals?

Use the essay The Chimp Scientist *(page E2) as a resource for this discussion.*

Analysis

Analyze the list that you created with your classmates of characteristics that make humans different from other animals. Record your responses to the following tasks in your journal:

1. For each characteristic you have listed, identify a nonhuman animal that also displays that characteristic to some degree.

2. Do you think there is any one characteristic that makes humans human? Explain your answer.

Further Challenges

Ask your teacher for a list of resources if you are interested in learning more about scientists who have done or are doing work similar to Jane Goodall's research.

PRIMATES EXPLORING PRIMATES

In *How Different Are We?* you read about and thought about observations that someone else has made. You have just considered how humans are like other animals and how they differ from other animals. Now you will have a chance to explore these ideas in more depth and to make some observations of your own. You will observe humans and other related animals as they move about and use tools. You also will compare human brains to those of other organisms. What will your observations tell you? What questions might come to mind?

Materials (per team of 2)

assortment of objects to grip
masking tape
padlock and key
sheep brain

Process and Procedures
Part A Get a Grip!

1. Spend 10 minutes observing how humans move from place to place. Make a record of these observations in your journal.

    Take the time to notice such things as the different types of strides that are possible and the differences in the way humans use their arms, legs, and feet. You can record observations in a variety of ways: words, drawings, snapshots, or videotape.

2. Now work with a partner to observe how humans use their hands. Use the objects that your teacher provides to explore the different ways that humans hold and use objects. Record these observations in detail in your journal.

    CAUTION: Do not swing, throw, drop, or manipulate any of the objects in a way that might harm you or your classmates.

3. Watch the videodisc segment *Observing Primates.* Record your observations of how each primate moves about and uses its arms and legs and hands and feet. Make the same type of observations for each primate as you made for humans. As you observe humans in this segment on the videodisc, add any new observations to those that you made in Steps 1–2.

Primates are the group of animals that include humans, apes, and monkeys along with a few less well known animals. Record as much detail as you can and be sure to label your journal entries clearly so you can tell which observations go with each primate. Watch the segments several times if necessary.

4. With your partner have a brainstorming session to create a list of all the questions that come to mind about how primates move about and use their hands. Record this list in your journal.

   Your list should have at least five questions on it.

5. Science begins with observing and asking questions about the world around us, and then it moves to a stage in which we begin to answer these questions by using a combination of further observation and experimentation. Scientists often have many questions that they would like to answer, but usually they focus their efforts on those that are testable.

   Use the criteria in Figure 1.2 to determine which of the questions on your list are testable. Mark these questions with a "T." Continue to work with your partner.

Testable Questions

A question is likely to be testable if it

a. uses question words like *whether, when, where, what, how many, how much,* and *how often,* rather than question words like *why* or *how.*

   For example, the question How many fingers does a gorilla have on its right hand? *suggests an easy way to answer the question. But the question* Why does a gorilla have four fingers on its right hand? *does not.*

b. identifies the specific issue to be tested and the specific items that will be involved.

   The question How much do gorillas eat? *is not easily tested. A more easily tested version of this question is* How many pounds of bananas do adult male gorillas eat?

c. describes the conditions under which the test should be conducted.

   An even better version of the gorilla and banana question would be How many pounds of bananas do adult male gorillas gather and eat during one week in the wild? *This question specifies the conditions that we are interested in; we are not asking how many pounds of bananas gorillas would eat in captivity or during their lifetimes.*

d. describes the criteria that will be used to judge the outcome of the test.

   Does a half-eaten banana count as having been eaten? How about a banana that is three-quarters eaten? Can you phrase this gorilla and banana question so that it is even more easily tested than the others listed above?

e. can be tested using available resources and procedures.

   In the end, questions about gorillas and bananas, however testable by some people in some parts of the world, are not testable questions for us in the classroom. All researchers (students and teachers alike) are limited by the resources that they have available.

Figure 1.2 **Testable Questions**

6. To gain a clearer sense of the nature of a testable question, read these two questions:

   a. What is the importance of an opposable thumb for the ways in which humans use their hands?

   b. Without using your thumbs, can you use a key to open a lock?

   With your partner discuss which question is easier to test. Record your answer and an explanation for your answer in your journal.

7. Now that you have a testable question, you and your partner are ready to carry out an experiment.

   Use the following protocol in Figure 1.3 to test the question.

Figure 1.3 Protocol for Testing the Question

Protocol for Testing the Question

a. Obtain a lock, a key, and masking tape as your teacher directs.

b. Decide who will carry out the task first and who will keep track of the time and record observations.

c. Be certain that the lock is in the locked position, and place it and the key on the table in front of you and your partner.

d. When the person responsible for keeping track of the time says, "begin," the person performing the task will pick up the key and attempt to unlock the lock.

   This task lasts 30 seconds. The timer/recorder should keep track of the time and record whether or not the person was successful.

e. Switch places and repeat Steps c and d.

f. Have one person at a time use the masking tape to tape his or her thumbs to one side of each hand as shown in Figure 1.4.

   You and your partner may need to help each other tape one or both hands.

g. Repeat Steps c–e.

h. Contribute your data to a class table, and record the class data in your journal.

---

8. With your partner complete the following in your journal.

   a. Use the class results to formulate an answer and an explanation to the question you tested.

   b. In what ways do you think the thumb is important to the way humans use their hands?

   c. Look back in your journal to your observations of the way other primates use their hands.

   - In what ways do they use their thumbs?

   - How are these ways different from what you just discovered about how humans use their thumbs?

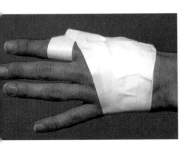

Figure 1.4 Ask your partner to help you tape your thumbs to the sides of your hands as shown. You should not be able to use your thumb at all, but you should be able to move and use each of your other fingers.

Part B All Brains on Board

In Part A of this activity you observed primates using observational science techniques similar to the ones that Jane Goodall used. In this part of the activity, you will observe the brains of various organisms and compare them to each other. You will use both drawings and the videodisc segment *Comparing Brains* for your observations and comparisons. *Mapping the Brain*, page E2 in the essay section, presents general information about the brain that will help you make your observations. Continue to work with your partner throughout this part of the activity; you will need to share sheep brains with other teams.

1. Examine the sheep brain that your teacher provides.

2. Use the drawing of the sheep brain (Figure 1.5) to study the two color-coded regions of the brain. Locate these regions on the sheep brain that you are examining.

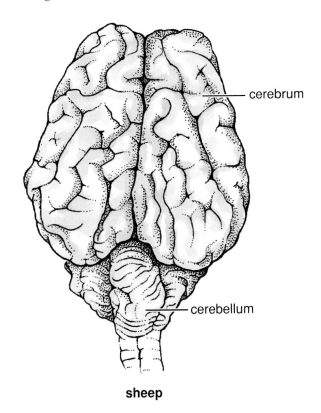

cerebrum

cerebellum

sheep

Figure 1.5 **Sheep brain (life-size)** Can you find each of these regions on the sheep brain that you are examining?

3. As you continue to study the sheep brain, read the essay *Mapping the Brain* (page E2) to learn a little bit about these regions of the brain.

4. Compare the color-coded regions of the sheep brain to the color-coded brains of six additional organisms. Record your ideas in your journal.

    Use the drawings in Figure 1.6 on pages 24 and 25 and the images of brains in the videodisc segment Comparing Brains *to make these comparisons.*

    a. Compare the relative size, shape, and position of each region. Record your observations in your journal.

        Create a table similar to the one in Figure 1.7 on page 25 in which to record your observations so that you can make comparisons more easily.

    b. Consider the sheep brain again, and decide where you would place it among the collection of other brains. That is, would you place it alongside the fish or alongside the human?

        Use the information you collected in Step 4a to help you. Make sure you explain your placement.

Figure 1.6 **Comparing brains** These drawings show the brains of a variety of animals from the top view. A side view of the human brain is included to show the cerebellum. All of the brains are life-size except the side view of the human brain, which is 90 percent of life-size. In addition to the cerebellum (in yellow) and the cerebrum (in blue), the optic lobes, which are associated with the sense of sight, and the olfactory bulbs, which are associated with the sense of smell, are labeled on some drawings.

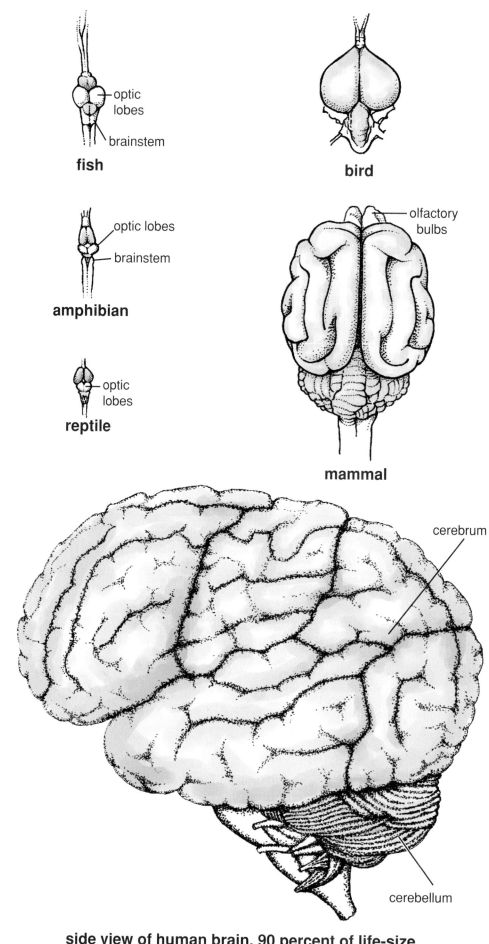

fish
— optic lobes
— brainstem

bird

amphibian
— optic lobes
— brainstem

reptile
— optic lobes

mammal
— olfactory bulbs

cerebrum

cerebellum

side view of human brain, 90 percent of life-size

left hemisphere   right hemisphere

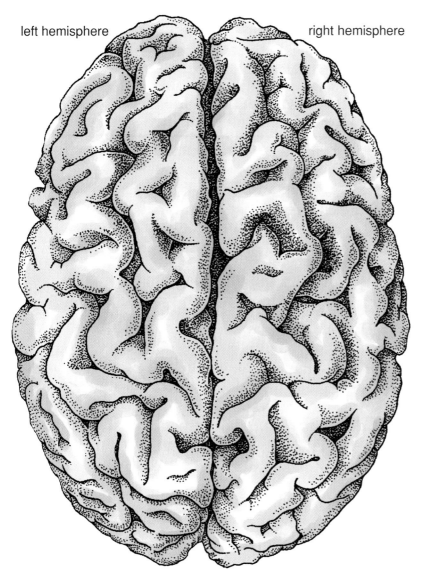

top view of human brain, life-size

	Cerebrum	Cerebellum
Sheep		
Fish		
Amphibian		
Reptile		
Bird		
Nonhuman mammal		
Human		

Figure 1.7 Sample table A table such as this one will help you organize the information so that you can make comparisons more easily.

c. Consider the cerebrum of each animal.

- Do you observe anything about the human cerebrum that makes it distinctive from the others? If so, what?

- What did you learn about the function of the cerebrum as you read the essay *Mapping the Brain* that helps you understand why the cerebrum makes humans distinctive?

d. Following these observations and reflections, what additional two or three questions come to mind?

e. Which of these questions are testable as phrased?

Use the criteria in Figure 1.2, Part A, to help you determine which questions are testable.

Analysis

Discuss the following with your partner, and then record your own answers in your journal.

1. How might the differences that you notice in the behaviors of the animals you studied be related to the differences that you noticed in their brains? Explain your answer.

    Use what you know about the behavior of fish, amphibians, reptiles, birds, sheep, other nonhuman mammals, and humans, along with what you have learned from the observations that you made during this activity, to answer this question.

2. In a sentence or two, describe why observation is important in scientific studies. Use your experience in Parts A and B to support your response.

Further Challenges

1. If you want to learn more about the way humans use their hands, read the essay *Do You Have a Grip on That?* (page E5). When you have finished reading, review the videodisc segment *Observing Primates* and see which grips you observe other primates using.

2. Visit a zoo and make detailed observations of the way various primates or other animals move.

3. Develop an experiment designed to answer one of your testable questions from this activity.

PORTRAYING HUMANKIND

Explain

By now, you have some idea of the range and subtlety of characteristics that make humans human. You have made observations of humans, other primates, and a collection of brains, and you have reflected on the similarities and differences that emerged. These observations and reflections led you to ask questions, some of which were testable. In this activity you will begin a special project that will help you to pull together some of your ideas and to construct your own understanding of what it means to be human.

Materials (per person or team of 2)

assorted materials that will help you create your project:
> poster board
> markers
> glue
> old photographs
> musical instruments
> video camera
> tape recorder
> magazines
> cardboard boxes

Process and Procedures

1. Think about designing a project to illustrate your understanding of what it means to be human. When you have a general idea in mind, share your idea with your partner.

   Choose a medium that you are comfortable with to create your project. Some possibilities include a poster, a diorama, a poem, a story, a report, a musical piece, a play, a TV show, or a video. As you think about and plan your project, use the criteria in Figure 1.8 to guide your work.

Criteria for Projects

Conceptual Understanding	Your project should demonstrate that you understand the idea that even though humans somehow are different from other animals, they share many characteristics.	Worth up to 1/4 of the total project points
Information Base	Your project should (1) identify characteristics that distinguish humans from other animals and (2) identify ways that humans are similar. The characteristics should include general physical characteristics, (e.g., a description of hand use and locomotion, a description of the brain) and behavioral characteristics (e.g., use of language, ability to learn, social and cultural behaviors).	Worth up to 1/4 of the total project points
Creativity	Your project should be engaging for others to observe, participate in, or listen to.	Worth up to 1/4 of the total project points
Connections	Your project should refer to ideas that you have explored in other activities of the program so far.	Worth up to 1/8 of the total project points
Presentation	You should present your project in a confident manner, be able to answer questions, and adhere to space or time limits.	Worth up to 1/8 of the total project points

Figure 1.8 Criteria for Projects
Use these criteria as you develop your project. Your teacher also will use these criteria to evaluate your understanding of the concepts in this chapter.

2. To meet the criteria listed in the rows labeled *Conceptual Understanding* and *Information Base*, be sure that you understand and can explain the following ideas to your partner:

- how humans are structured to be bipedal,
- how the human hand is similar to and different from the hands of other primates,
- how different parts of the cerebrum are associated with various behaviors,
- how different parts of the human brain are similar to and different from the brains of other primates, and
- how nerves transmit information.

The essays On Being Human *and* Brains and A Lot of Nerve *on pages E6–E13 and the videodisc segment* More About the Brain *will help broaden your understanding of these ideas. As you watch the videodisc segments, practice your observation skills. You may want to record notes in your journal.*

3. Write a short description of your project in your journal and list four or five concepts from the essays and the previous activities that you plan to incorporate into your project.

Analysis

Complete this *Analysis* according to your teacher's directions.

1. Meet with your partner and exchange written descriptions of your projects.

   a. Read your partner's description.

   b. If something about your partner's project is unclear to you, ask questions of him or her.

   c. Review the criteria for the project that are listed in Figure 1.8, and tell your partner how well you think his or her proposed project meets them.

   d. Offer your partner suggestions that you think would contribute to his or her project.

2. According to the feedback that you received from your partner, revise or add to your description.

   You will present your project to the class as the Evaluate Activity for this chapter.

A LONG CHILDHOOD

Elaborate

Have you noticed that puppies may be separated from their mother within two months of birth and a newborn foal will be up and walking around within minutes of its birth? A human child, however, usually will stay with its mother for years and does not begin to walk until it is close to a year old. Have you ever wondered why this is so?

The collection of physical characteristics that sets humans apart from other organisms results in some behavioral differences as well. You have

explored various aspects of the human brain and have compared it with the brains of other organisms. What does this really mean? How do the features of the human brain contribute to behaviors that make us different? Could these differences have anything to do with our long childhoods? In this activity you will explore some possible answers to these questions and no doubt will come up with more intriguing questions of your own.

Materials

resources on the development of other animals after birth

Process and Procedures

1. Participate in a class discussion and answer the following questions as they apply to humans.

   a. What is the length of pregnancy?

   b. At what age do the young stop nursing or are able to get food on their own?

   c. At what age are the young completely and independently mobile?

   d. At what age do individuals become sexually mature?

   Record the answers to these questions in your journal. You also will collect this information for other animals, so create a table in which to organize and record all of the information. You may want to arrange the questions across the top and list the animals in the left margin as in Figure 1.9.

	Length of Pregnancy	Get Food	Completely Mobile	Sexually Mature
Humans				

Figure 1.9 **Sample table of information on animals** A table similar to this one will help you organize your information in a way that makes it easy to use.

2. Select another animal and answer the questions in Step 1 for that animal.

   Use the resources that your teacher provides or resources from the library.

3. Contribute your answers to a class data table.

   Record the class data in your own data table.

4. With your classmates discuss what patterns emerge as you compare the data for humans with those of the other animals.

    Use the data to identify the animals that have long "childhoods."

5. With a partner discuss the question, What are some of the things that happen as a result of a long childhood in humans? Record your ideas in your journal.

    Use the essay The Importance of Being Children *(page E13) as a resource.*

Analysis

Think about how you will fold ideas that you have developed in this activity into the project that you began thinking about in the activity *Portraying Humankind.*

1. In order to help you do this, discuss the following questions with a classmate:

    a. How does a long childhood in humans help humans develop complex culture?

    b. How does our language ability help us develop complex culture?

    c. How does our ability to learn help us develop complex culture?

    d. Give three examples of different ways that humans learn. How do you learn best?

    e. Give two examples of other animals that exhibit a certain capacity for learning and for language and culture.

    f. How different and how alike are humans and other animals? Justify your answer with new information that you have learned in this activity.

2. Decide how you will fold some of these expanded ideas into your project and then add this information to your written description for the project.

3. Have your teacher approve your completed plan for the project.

4. Work on your project as your teacher directs.

Further Challenges

1. Together with other interested classmates, explore the existence of culture in other animals.

2. Do further library research on the learning abilities of other primates or dolphins.

3. Learn more about your own learning style. Your teacher may have some resources to get you started.

WHAT DOES IT MEAN TO BE HUMAN?

In this activity you will demonstrate what you have learned in Chapter 1 and how this new knowledge helped you answer the question, What does it mean to be human? from a scientific perspective.

Process and Procedures

1. Present the project that you began working on in *Portraying Humankind* to your classmates according to your teacher's directions.

2. Use the following criteria (Figure 1.10) to identify what you think are the three strongest projects in your class. List these in your journal and justify your choices.

   These are the same criteria that are listed in the activity Portraying Humankind, *page 27.*

Criteria for Projects

Conceptual Understanding	Your project should demonstrate that you understand the idea that even though humans somehow are different from other animals, they share many characteristics.	Worth up to 1/4 of the total project points
Information Base	Your project should (1) identify characteristics that distinguish humans from other animals and (2) identify ways that humans are similar. The characteristics should include general physical characteristics, (e.g., a description of hand use and locomotion, a description of the brain) and behavioral characteristics (e.g., use of language, ability to learn, social and cultural behaviors).	Worth up to 1/4 of the total project points
Creativity	Your project should be engaging for others to observe, participate in, or listen to.	Worth up to 1/4 of the total project points
Connections	Your project should refer to ideas that you have explored in other activities of the program so far.	Worth up to 1/8 of the total project points
Presentation	You should present your project in a confident manner, be able to answer questions, and adhere to space or time limits.	Worth up to 1/8 of the total project points

Figure 1.10 **Criteria for Projects**

Analysis

Use the same criteria to rate your own project and presentation. Explain your rating in your journal.

> *. . . just as he was about to quit searching for the day, he noticed part of a fossilized arm bone just lying on the ground partway up the slope of a gully.*

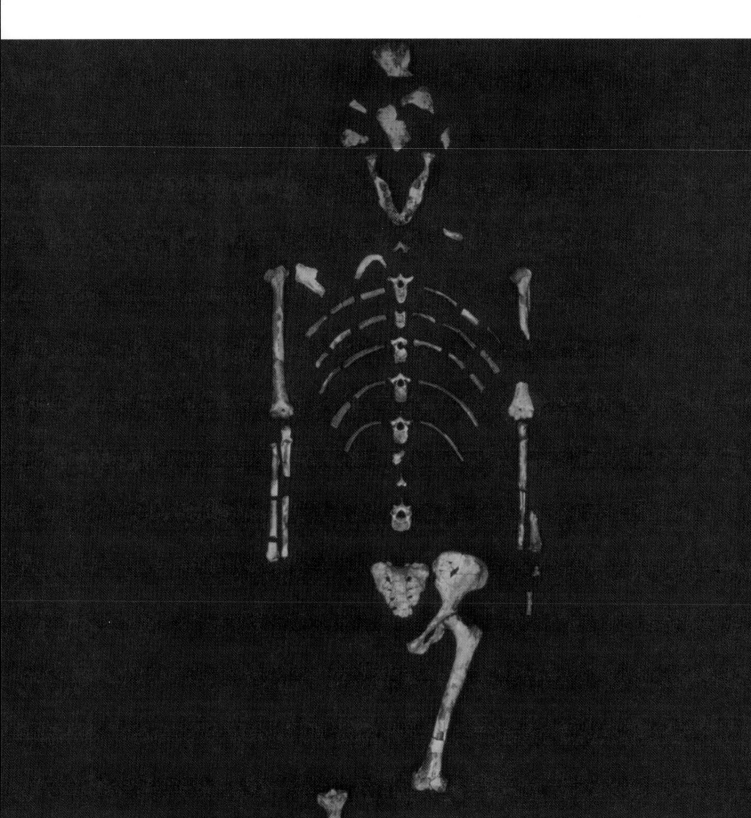

2 EVOLUTION: CHANGE ACROSS TIME

Fossil skeletons like the one shown at left provide a tremendous opportunity for scientists to study the physical characteristics of organisms that lived long ago. Such studies reveal an interesting fact: as you look further and further into the past, the skeletons of ancient organisms look less and less like organisms that are alive today. This is because living organisms change across time. In this chapter you will begin to understand how living organisms change and why change is important to your understanding of the relationships between biological systems.

We begin the chapter with a look at Lucy, the fossil skeleton of an ancient humanlike organism, which is shown in the opening photograph. We ask that you imagine how she may have looked and behaved at the time she lived. Next you will construct a time line of earth's history, and in doing so gain an appreciation for the vast time spans required for geological and evolutionary change. After modeling deep time, you will explore some of the biological, geological, and anthropological evidence for evolution by becoming a specialist in one of these scientific disciplines. Then in a team effort, you will present all of the available evidence to support an explanation of the changes that you observed.

As a newspaper reporter in the next activity, you will uncover the story behind the theory of evolution and its founding father, Charles Darwin. You also will model a mechanism for evolutionary change by simulating the interactions of predators and their prey. After this, you will elaborate your understanding of evolution by studying the archaeological remains of a well-preserved, 5000-year-old man. Your studies will lead you to think about cultural evolution—the rapid, nonbiological changes at which humans excel. Finally, you will evaluate your understanding of evolution by analyzing an important modern example of evolution in action.

ACTIVITIES

Engage Lucy

Explore Modeling the Earth's History

Explore
Explain Evidence for Change across Time

Explain Explaining Evolution

Elaborate Modeling Natural Selection

Elaborate A Cold Hard Look at Culture

Evaluate Evolution in Action

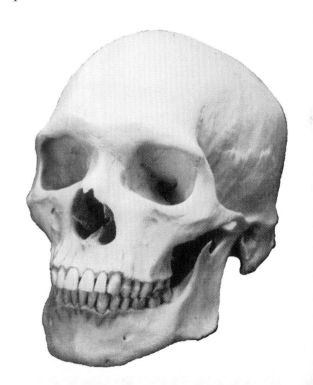

DIGGING UP THE PAST

No one knows how Lucy died. She apparently died quietly. If she had been killed by a lion or a leopard, her bones probably would have been splintered and crushed. She had not been scavenged by hyenas, or her skeletal parts would have been scattered over a wide area. Instead, she died by the edge of an ancient lake and was covered by mud and sand, where she remained buried for almost 3.5 million years.

Lucy lived in what is today the Afar desert, a remote region of northeastern Ethiopia in East Africa (see map in Figure 2.1). The large lake that once existed there has long since dried up, and the area is now hot and desertlike. Even though she was fully grown and probably in her twenties when she died, Lucy stood only three and one-half feet tall. Her head was a bit larger than a softball, and her brain could not have been much larger than that of a modern chimpanzee. The shapes of her knee joint and pelvis bones indicate that she walked upright on two legs. For more than 3 million years, Lucy remained buried in the ancient lake bed.

In 1973 Donald Johanson, pictured above, a young anthropologist from the United States, arrived in Ethiopia to look for fossils. Working with two Frenchmen, Maurice Taieb and Yves Coppens, Johanson found a primitive primate knee joint that had washed out of a slope during a rain. It seldom rains in the Afar desert, but when it does, the rain is torrential, cutting gullies into the gravel and bare rock. On very rare occasions, these rains uncover ancient fossils, and as luck would have it, such a rain occurred not long before Johanson's find. This fossil knee joint was intriguing because it was about 3 million years old, and its structure indicated that this individual had walked erect. Because of this find, Johanson decided to return to the same ancient lake bed the following year to continue the search. If any more humanlike fossils were embedded in the ground, they might be of a similar age. Johanson and his colleague, Tom Gray, had been searching all fall with no success when they parked their vehicle on the slope of a gully on the morning of November 30, 1974.

The temperature had reached 43°C (110°F), and they were about to return to camp when Johanson noticed part of a fossilized arm bone lying on the ground partway up the slope of a gully. As he searched further, he found pieces of a skull, thighbone, and pelvis, along with other skeletal parts but no evidence of any tools. Remarkably, the skeletal parts all seemed to be from one individual, and a very humanlike individual at that. The two scientists barely could contain their excitement. They named the skeleton "Lucy," after a song that was popular at the time. This skeleton (shown in the opening photo) has become one of the most famous fossils found of an early humanlike animal.

What did this new find mean in terms of human origins? Did gradual changes in animals such as Lucy eventually lead to modern humans? Some of the answers had to wait until Johanson returned to his lab at the Cleveland Museum of Natural History. There, he and his colleagues spent many months in painstaking detective work, comparing Lucy with other fossils of more recent human ancestors as well as with modern human and ape skeletons. Johanson's conclusions sharpened the debate within the scientific community about exactly when the human line split from the ape line, but most scientists agree that at a very early time, there was a single primate ancestor to both modern humans and modern apes. The controversy about when the split occurred still is not resolved.

Will more fossil remains of the links between humans and our most ancient ancestors be discovered, and if so, will they help to resolve the dispute about this timing? Perhaps a future anthropologist will discover more clues to clarify further the details of our origins.

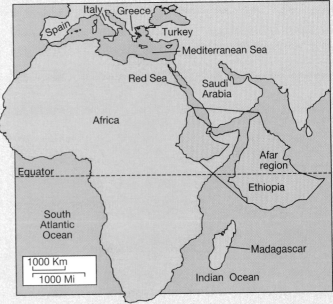

Figure 2.1 **Map of the Afar region in Ethiopia**

LUCY

Since the mid-1800s, scientists have been piecing together the puzzle of human evolution. Some of the most important pieces of this complex puzzle are the fossilized skeletal remains of individuals who lived millions of years ago. Dated at about 3.5 million years old, Lucy is one of the oldest and most complete hominid fossils (hominids are erect-walking primates that include modern humans, earlier human species, and early humanlike species).

In this activity you will think about how Lucy may have looked and behaved while she was alive, and you will begin to appreciate how much humans have changed across time.

Process and Procedures

1. Describe in your journal how Lucy may have looked and some of her physical behaviors.

   Use the photograph of the fossil skeleton on the opening pages of this chapter and your notes from Chapter 1 to help you develop your description. In your description, pay particular attention to Lucy's hands, feet, posture, and way of moving. Try to describe how she may have communicated with family members and others living in her group.

2. Consider the following question and record your answer in your journal.

   How might Lucy bridge the gap between modern humans and early nonhuman primates?

Analysis

Use the information from your description to answer the following questions as part of a class discussion.

1. Comparing hominids from Lucy's lifetime to your own, do you think there have been more changes in physical characteristics of the body (such as hands, feet, head, posture) or more changes in how hominids lived (types of shelter, ways of getting around, ways of gathering food)?

2. Use the paragraph below to help you answer these questions:

   a. Which aspects of your description were based on evidence?

   b. Which aspects of your description were inferences related to evidence?

   c. Which aspects of your descriptions were guesses?

When scientists find skeletal remains, they work carefully to gather as much information as possible from their findings. Often the skeletal remains are incomplete, but such remains are **evidence** from which the scientists can draw conclusions about the individual. Conclusions that follow logically from some form of direct evidence are known as **inferences**. (Conclusions that do not follow logically from evidence are just guesses; guessing is not an acceptable way to draw scientific conclusions.) For example, scientists have made inferences based on skeletal evidence about how tall an individual was and whether or not the individual was bipedal. Fossil evidence and inferences based on this evidence may support a current theory about human origins or may point to new ideas.

MODELING THE EARTH'S HISTORY

Explore

The Lucy fossil is extremely old, more than 3 million years old. According to geologists, however, the earth was formed 4.6 *billion* years ago, a time span that is difficult to comprehend because time for humans generally means tens and hundreds of years. One way to comprehend the immensity of the earth's history is through a time line. In this activity you will develop a time line of the earth's history to help you better understand when certain human events and major geological and biological events occurred. This time line should give you an appreciation of **deep time** and the changes that have occurred since the formation of the earth.

Materials (per team of 4)

10 clothespins or paper clips (optional)
event cards that your teacher provides

Process and Procedures

1. Discuss the following question with your teammates and answer it in your journal.

    How long ago do you think each of the following events occurred and in what sequence?

    *   first dinosaurs
    *   formation of Rocky Mountains
    *   first hominids
    *   first life (bacteria)
    *   first modern humans
    *   first oxygen in atmosphere
    *   first land plants

2. Compare your team's estimated times and sequence of these events with those of other teams and discuss the questions below with your teammates.

    a. Were your estimates logical conclusions based on evidence or were they guesses?

    b. Why did your team's estimates differ from those of other teams?

3. Examine the table *Major Events in the Earth's History* that your teacher provides and answer the following question in your journal.

    Which times or sequences of occurrence surprised you?

    Scientists use several tools to think about deep time, and one of these tools is the use of evidence and inference. This table is based on inferences from evidence that scientists have gathered about the history of the earth and of living organisms on the earth. It also is based on theories that geologists and paleontologists have developed about the time spans and patterns of change in the earth's history. For these reasons, the dates in this table are more accurate than guesses.

4. Study the marked clothesline that your teacher has prepared and discuss with your teammates how it might be used to represent the events listed in the copymaster *Major Events in the Earth's History.*

   *A second tool used by scientists to think about deep time is a **model**—that is, a simple system or situation that mimics a more complex system or situation.*

   You can make a more useful comparison of the events in the table by constructing a time line. A time line is a type of model that shows when events occurred in relationship to each other. Because the length of a time line corresponds to time, a time line offers a simple, visual way to picture how much time separated certain events in the past. Figure 2.2 is a small version of the time line that you'll be working with in class.

   The distance between the red marks on the clothesline represents 1 billion years. The distance between the black marks represents 100 million years. Your teacher has placed cards on the time line at 5 billion years ago and at the present.

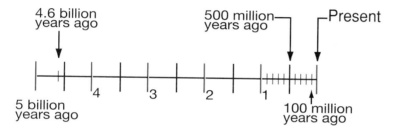

Figure 2.2 **Time line of earth's history**

5. Study the event cards that your teacher provides.

   Fold each card in half, crosswise, to form a tent.

6. Decide where on the time line each event card should go and place your cards in the appropriate locations.

   Use the information in the table of events to help you decide. It may help to write on each card the number of years ago the event occurred.

   You may want to secure your event cards with clothespins or paper clips.

Analysis

Use the time line that you just created to answer the following questions with your teammates and then participate in a class discussion.

1. Describe five or six patterns that you see in the time line. For example, describe a relationship between geological and biological events or a relationship between plant and animal events.

2. Is it likely that these patterns occurred separately or do you think they might be related? Explain.

3. What did the time line (a model) help you understand about the earth's history?

EVIDENCE FOR CHANGE ACROSS TIME

Explore
Explain

What has changed in your lifetime? Throughout history, people have been intrigued by the changes that take place around them. For thousands of years, people have been noticing, recording, and trying to explain the various changes they observe in nature. These changes include both routine changes, such as the pattern of day and night, the phases of the moon, and the changing seasons, as well as devastating changes, such as erupting volcanoes, earthquakes, and floods.

Changes may be rapid (a 30-second earthquake changes the environment), or they may be slow (a snowshoe rabbit's coat turns from brown in the summer months to white in the winter months). The biological changes of evolution generally are very slow. The story of Lucy provides us with an idea of some biological changes that have taken place during the past 3.5 million years. In order to appreciate these changes, we first must understand the process of biological change across time.

In this activity you will assume the role of a specialist and work with other specialists to study the evidence that scientists have accumulated about biological change. As you and your teammates study your data, you will begin to see how evidence from several branches of science can be combined to support an explanation. Your team of specialists will develop a presentation that is based on this collection of evidence.

Materials (per team of 4)

assorted materials for presentations
   poster board
   butcher paper
   felt-tipped markers
   plastic ruler

Process and Procedures

1. With your team decide who will be the embryologist (one who studies embryos), the evolutionary biologist (a specialist who studies change across time in living systems), the paleogeologist (a fossil and geology specialist), and the physical anthropologist (a specialist who focuses on how humans have changed across time).

   As you work with your team, make certain that everyone contributes ideas.

2. Meet with the members of other teams who have assumed the same role that you have and together read and study the information assigned to your role. Then go to the procedural section that follows Step 5.

   You will be responsible for sharing the information with your original teammates. When you complete your specialist work, return to your team and begin Step 3.

38

3. Share your information with the rest of your team and discuss how each set of information represents evidence that supports the concept of change across time. Also discuss how the collection of evidence is stronger than any of the separate pieces.

4. Create a data table or diagram, like the one shown in Figure 2.3, to help you keep track of this collection of information.

	Embryologist	Evolutionary Biologist	Paleogeologist	Physical Anthropologist
Evidence				

Figure 2.3 **Evidence for change across time** Draw a table like this in your journal to help you keep track of each specialist's evidence for change across time.

5. Agree on a format to present your findings and work on this presentation together. Your teacher will assess your presentation based on the following criteria.

Your presentation should include

- both verbal and visual material,
- ideas from each specialist group that are presented clearly,
- justification for the ideas you present (in other words, show the evidence supporting each idea), and
- contributions from and the participation of each team member.

You might agree to write a news release, to hold a press conference with visual aids, to develop a poster, or to write an article for a magazine or journal.

Procedures for the Specialists

Paleogeologists

Paleontologists are scientists who search for clues to biological change by studying fossils, much as detectives reconstruct a crime from the evidence left behind. **Geologists** study the origin, history, and structure of the earth. **Paleogeologists** are scientists who combine the study of paleontology and geology. In your role as a paleogeologist, you will examine both types of evidence because understanding biological change requires an understanding of geological change.

a. Complete the following tasks in your journal:

1. With the other paleogeologists in your group, identify 10 important concepts or ideas that support the use of fossil evidence in the study of change across time.

2. Explain how fossils are formed and how their positions in the geologic strata indicate their relative age.

You might use the essays Fossils: Traces of Life Gone By *(page E15) and* Technologies That Strengthen Fossil Evidence *(page E17) as resources.*

b. Study the drawings of past geological phases of the earth shown in Figure 2.4. Try to match all of the phases to the descriptions that follow. When you have completed this step, you should have a chronological sequence of how continental drift affected the earth in the distant past.

Figure 2.4 Changes in the continents during the earth's history These drawings of geological phases are not in chronological order. Can you determine the chronological order of the changes shown?

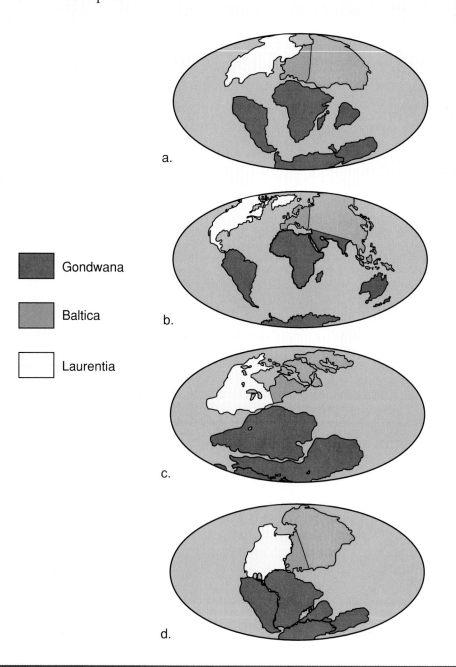

Gondwana

Baltica

Laurentia

1. Laurentia and Baltica had collided to form one continent. Sea levels were low. In some regions there were periods of drought. Devonian Earth: 408–360 million years ago.

2. Pangea, a single huge continent composed of all the earth's major land masses, had been formed by the collision of four major continents. Sea levels were low and there were many deserts. Appalachian mountains were formed. Triassic Earth: 245–208 million years ago.

3. North America and Europe began splitting apart and South America and Africa separated. The sea level was high. Formation of the Rocky Mountains began. Cretaceous Earth: 144–66 million years ago.

4. During each of the four ice ages, the sea level was very low. Warm periods followed periods of glaciation. The continents had drifted almost to their positions of today. Pleistocene Earth: 1.5 million years ago.

c. Explain how geological and paleontological evidence contributes to our understanding of how organisms have changed across time.

Evolutionary Biologists

Evolutionary biologists are scientists who study the physical evolution of living organisms.

a. Complete the following tasks in your journal:

1. With the other evolutionary biologists in your group, identify seven important concepts or ideas that support biological change across time.

2. Are you familiar with any examples of homologous structures other than those discussed in the essay? If so, what are they?

3. Explain how homologous structures support the idea that apparently diverse organisms may be related to one another.

4. Are you familiar with any examples of vestigial organs other than those discussed in the essay? If so, what are they?

5. Explain how vestigial structures support the idea that apparently diverse organisms may be related to one another.

6. Explain how comparisons of primate DNA support the idea that primates are related.

7. How does this information support the inference that change has occurred across time?

You might use the essay Modern Life: Evidence for Evolutionary Change *on page E20 as a resource.*

b. Describe a situation in which scientists were able to observe evolutionary change as it occurred in modern times.

Embryologists

Embryologists are scientists who study the pre-birth or pre-hatching developmental phases of living organisms.

a. Read the following paragraph in your group:

Embryology is the branch of science that focuses on the early development of organisms before they are born or hatched. During this pre-birth or pre-hatching stage, the developing organisms are called embryos. When scientists compare the developing embryos of organisms as diverse as fish, amphibians, reptiles, birds, and mammals, they find that the embryos of these vertebrate animals (animals that have backbones) resemble each other.

Make certain that everyone understands these concepts before continuing to Step b.

b. Study the individual drawings of embryos that your teacher provides. Try to arrange all of the embryonic stages in a developmental order for each animal. When you have finished, your arrangement should show three stages of embryonic development for a fish, a frog, a chicken, a calf, and a human.

These drawings depict various embryonic stages of five different vertebrates. These stages are relative; they do not represent the same point in time, but rather the same relative amount of development.

c. Compare your arrangement with the illustration that your teacher provides.

d. Discuss the following questions in your group and record your answers in your journal:

1. In general, which organisms have embryonic stages that are the most similar and the least similar? Explain your answer.

2. What do you think these similarities and differences tell scientists about how these organisms have changed across time and how they are related?

e. Study the videodisc segment *Embryology* and compare the images to the ones you have assembled by answering the following questions:

1. Do the images on the videodisc help you see other similarities and differences that are not apparent in the drawings? If so, record this information in your journal.

2. How does the information in Step a, the drawings, and the videodisc images help you to see relationships between diverse organisms?

Physical Anthropologists

Physical anthropologists are scientists who study the biological evolution of humans.

a. Complete the following task in your journal:

With the other physical anthropologists in your group, identify five important concepts or ideas that support the biological change of primates across time.

You might use the essay Primates Show Change across Time *on page E22 as a resource.*

b. When the videodisc player is available, view the sequence *Hominid Skulls*. As you compare the skulls, focus on the features listed below, make measurements where appropriate, and record the information in your journal.

- size of lower jaw
- prominence of brow ridges
- slope of face
- width of face
- size of forehead
- size of brain case
- size of molars

Use the sketches of skulls in Figure 2.5 to help you determine which measurements to make. Create a data table like that shown in Figure 2.6 in which to record your results. Review the sequence of skulls or individual frames as many times as necessary to make your comparisons.

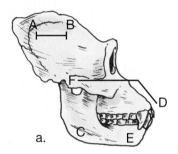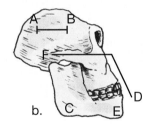

a.   b.

Figure 2.5 Compare the features on these gorilla (a.) and early hominid (b.) skulls.

Figure 2.6 Record of fossil measurements

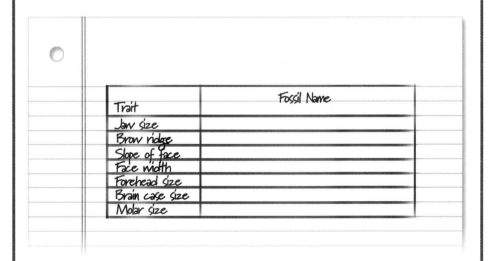

Trait	Fossil Name
Jaw size	
Brow ridge	
Slope of face	
Face width	
Forehead size	
Brain case size	
Molar size	

c. Discuss the following in your group and record your ideas in your journal:

1. What evidence did you find in the essay and in your observations of the hominid skulls that indicates change between early and modern hominids?

2. What evidence did you find in the essay and in your observations of the hominid skulls that indicates relatedness between early and modern hominids?

3. Summarize the pattern of changes that you observe in the hominid skulls from *Australopithecus afarensis* to *Homo sapiens*.

4. Use the evidence that you have examined to draw an inference about the ancestors of modern humans and modern primates.

Further Challenges

Continental drift and the ice ages had a tremendous effect on the distribution of living organisms throughout the earth's history. The movement of continents and the rising and falling of sea levels created and destroyed many land bridges—that is, connections between continents. Think about these concepts as you use the fossil data in Figure 2.7 to complete the following tasks:

1. Propose a place of origin and a migration route for marsupial mammals.

   Mammals are animals characterized, in part, by hair and sweat glands and the nursing of young with milk secreted by mammary glands. Marsupials are mammals that produce embryos that spend only a short time in the mother's uterus. At birth, the immature infants must crawl into the mother's pouch, where they continue to mature while nursing from the mother's teats.

2. Consider what the fossil record indicates about ancient distributions of marsupial mammals and propose an explanation for the present marsupial mammal distribution.

Figure 2.7 Distribution of marsupials

Continent	Oldest Marsupial Fossil (mya=million years ago)	Present Marsupial Distribution (number of families)
Australia	25-30 mya	10
Europe	52-58 mya	1
North America	63-135 mya	1
South America	63-75 mya	2
Asia	none	0
Africa	none	0
Antarctica	52-58 mya	0

EXPLAINING EVOLUTION

Explain

Land masses separate and come together again. The sea level rises and falls with each ice age. Species of organisms live in one environment, sometimes migrate to other environments, change gradually into related but physically distinct forms, and usually become extinct. You have explored much about the changes that occur across deep time. Now you will have the opportunity to explain *how* biological changes take place.

This activity provides an opportunity for you to develop an explanation of evolution. By assuming the role of a newspaper reporter, you will be able to uncover the story behind the theory of evolution. For example, you may have heard of Charles Darwin, the nineteenth century naturalist who is famous for writing *On the Origin of Species*[1], but do you really know who he was? What did he do, how did he do it, and why did it matter? The story you write will help answer some of the questions that you may have had about evolution, how it takes place, and why biologists are so interested in it.

Materials (per team of 2)

large piece of butcher paper

Process and Procedures

You may have heard that newspaper reporters are taught to include the most important information in the first few sentences of their news stories—the *who, what, when, where, why,* and *how* of the breaking story. Imagine that it is the year 2009, and you and your partner are rookie newspaper reporters. You just have been given your first assignment: put together a short article about the theory of evolution in recognition of the 150th anniversary of the publication of Darwin's book. Imagine as well that your teacher is your editor, pleased with your enthusiasm for journalism, but not at all sure that you have the scientific and investigative skills required to nose out the story, much less the reporting skills necessary to identify the highlights of the story, right up front.

1. Write the six question words listed above along the left side of your large piece of paper.

2. Develop these question words into full questions that are more specific. These are the questions that your team needs to answer to satisfy your skeptical editor.

   *For instance, you might develop your **who** question as follows: Who first proposed the theory of evolution?*

---

[1]The full title of Darwin's book is *On the Origin of Species by Means of Selection*, first published in 1859.

3. Decide how you will divide responsibility for finding the answers to each of these questions.

   Questions about who, when, where, and what are likely to be easier to answer than questions about how and why. You may wish to divide the easy and more difficult questions between you and your partner.

4. Work individually or with your partner (as determined by the team's decisions in Step 3) to gather and summarize the information required to answer your assigned question(s). Record notes in your journal.

   One source of information about Darwin's work can be found in the essays Darwin Proposes Descent with Modification *on page E23 and* Evolution by Natural Selection *on page E25.*

5. Work with your partner to develop short but interesting and informative answers to each of the questions and write those answers on the large paper.

   Share with your partner the information that you gathered for each of your questions. Discuss and agree upon the team answers before writing them on the paper.

6. Post your answers as your teacher directs.

7. How did your summary of the highlights of the story compare with those of your classmates? Be prepared to share your response to this question in a short class discussion.

Analysis

Your editor is impressed with your work, but has started to doubt whether the story is important enough to print. "After all," the editor argues, "we refer to it as the 'theory' of evolution. If it's just a theory, it means that we are not very sure about it. Why should we pay attention to it? What makes Darwin's theory any more convincing than a theory that you or I might suggest?" You, of course, are horrified at the editor's lack of understanding about the nature of scientific theories.

1. Work with your partner to prepare a rebuttal to the editor's criticism. Your response should include the following:

   a. a definition of a scientific theory that demonstrates the error in the editor's argument and

   b. an explanation of the importance of the theory of evolution to modern biology.

   Information in the essay Just a Theory? *on page E29 will help you develop your response.*

2. Choose one member of your team to offer the rebuttal orally to your editor (your teacher). Although one person should present the team's response, both of you should be prepared to answer further questions about the issue if your editor is not convinced by your initial comments.

Further Challenges

To create a real journalism article, take the summary of the highlights of your story and craft it into a smooth, informative article that answers for the reader all of the important questions about evolution.

MODELING NATURAL SELECTION

Darwin developed his theory of evolution, a theory that explains how organisms change across time, by analyzing his observations, including those from his voyage on the *HMS Beagle*, during a period of 20 years. In his theory, he proposed natural selection as the mechanism of evolution. You can model natural selection in the classroom in one or two days by limiting an investigation to just one of the many types of selective pressures, in this case the pressure imposed by predation. Predation is important to evolution because it places a limit on one of the key requirements of natural selection—the ability of organisms to survive long enough to reproduce.

In this activity you will model the effects of predation on a prey population. The prey consists of paper dots that represent a variation in the color of individuals in a species. You and your classmates will be the predators. By analyzing the selective effects of predation in this model, you will gain a better understanding of how natural selection can change the average characteristics of a population.

Materials (per team of 4)

3 petri dish halves
36-×-44-in piece of Fabric A or Fabric B
3 sheets of graph paper
6 colored pencils with colors similar to the dot colors
zip-type plastic sandwich bag containing 120 paper dots, 20 each of 6 colors
   (labeled *Starting Population*)
6 zip-type plastic sandwich bags, each containing 100 paper dots of a single
   color
watch or clock with a second hand
3 forceps (optional)

Process and Procedures

1. Decide which team member will be the game warden and which team members will be the predators.

   Three members of your team will play the role of predators of paper dots. As predators, each will hunt paper dots (the prey) in their habitat (the piece of fabric). The fourth member will be the game warden, who will keep track of the hunting.

2. Examine the paper dots in the bag labeled *Starting Population* and record the number of individuals (dots) of each color.

   The colored dots represent individuals of a particular species. The individuals of this species can be one of six colors.

The game warden should record the starting population in his or her journal and label this First Generation Starting Population.

3. Spread out the piece of fabric on a desk or table top.

    Half of the teams will have pieces of Fabric A, and half of the teams will have pieces of Fabric B.

4. Set up the model as follows:

    Predators: Obtain a petri dish half. Face away from the habitat.

    Game Warden: Spread the dots from the bag labeled *Starting Population* throughout the habitat.

    Spread the dots as uniformly as possible so that no dots are sticking together or covering other dots.

5. Begin to manipulate the model as follows:

    Game Warden: Direct the predators to face the habitat and begin picking up dots (prey); say "Stop" after 20 seconds.

    Predators: Pick up as many paper dots (prey) as possible until the game warden says "Stop."

    Use only your eyes to locate your prey; do not feel the fabric. Use only one hand (or one forceps). Pick up one dot at a time and put each dot in your petri dish half before taking another dot.

6. Finish Round 1 of predation in your model as follows:

    Predators: Collect the remaining paper dots from the fabric and sort them by color.

    Game Warden: Record the number of each color of the remaining paper dots.

    Label this First Generation Surviving Population.

7. Prepare for Round 2 of predation in your model as follows:

    Predators: Simulate reproduction among the paper dots by adding three paper dots for each remaining dot of the same color.

    The three paper dots of each color represent offspring. Obtain these offspring from the bags containing single colors of dots.

    Game Warden: Record the number of each color of paper dot in your journal as *Second Generation Starting Population.*

8. Repeat Steps 4–6 for Round 2 of predation.

9. Calculate the number of each color of paper dot, if each surviving paper dot were to produce three offspring. Record this information as *Third Generation Starting Population.*

    Each team member should record this information in his or her journal.

10. As a team use colored pencils and graph paper to prepare bar graphs that show the number of paper dots of each color in each of the three starting populations.

   Use colored pencils that correspond to the colors of paper dots. You should have three bar graphs when you are finished with this step. If you need information about how to make bar graphs, refer to Appendix B, Techniques.

11. Study the bar graphs of each generation. With your teammates consider the following questions and record your team's responses in your journal:

   a. Which, if any, colors of paper dots survived better than others in the second and third generation starting populations of paper dots?

   b. What might be the reason that predators did not select these colors as much as they did other colors?

   c. What effect did capturing a particular color dot have on the numbers of that color in the following generations?

12. Now that your manipulations are complete, clean up by sorting the colored dots into their respective plastic bags as you found them, and then return the bags to the materials station.

Analysis

At the location that your teacher indicates, post your team's bar graph for the third starting population beside the fabric that you used. Compare the bar graphs of teams who used Fabric A with the bar graphs of teams who used Fabric B. Complete the following tasks as a team. Record your team's responses in your journal.

1. How well do the class data support your team's conclusions in Step 11?

2. Imagine a real-life predator/prey relationship and write a paragraph that describes how one or more characteristics of the predator population or the prey population might change as a result of natural selection.

   Base your proposition on what you have learned in this model of natural selection by predation.

3. Write one paragraph that describes how the process of natural selection adds to the collection of evidence and inference that your team examined in the activity *Evidence for Change across Time*. Write a second paragraph that summarizes your understanding of biological evolution.

A COLD HARD LOOK AT CULTURE

Seahorses that look like marine plants, bacteria that live in hot springs, and giant pandas that feed almost exclusively on bamboo. Through biological evolution, all of these organisms have changed and passed their

unique characteristics to offspring in generation after generation. Many of these changes are adaptations that enable these organisms to survive in very specific habitats. Humans, however, are not restricted to a single habitat. We are able to survive in extreme heat, extreme cold, underwater, and even in space. What enables humans to exploit so many environments when most other organisms are limited to a narrow range of viable habitats?

The answer is culture. We have an exceptional ability to develop behaviors and technologies that permit us to cope with new situations in nonbiological ways. For instance, we can minimize the effects of an extremely hot and unpleasant environment by using the technology of air conditioning to modify that environment. This ability is called cultural adaptation or **cultural evolution**, and it is one of the distinguishing characteristics of humans. Cultural evolution differs significantly from biological evolution because acquired cultural changes can be transmitted from one generation to the next. In fact, the cultural transmission of information and values from person to person and generation to generation through communication and learning is the primary reason that the experience of being human has changed so dramatically in the last 10,000 years.

In this activity you and your teammates will be a team of archaeologists. An archaeological society has asked your team to help them study the remains of a human who died more than 5000 years ago. They also wanted help studying the items that were found with him. Because he was buried in ice and snow high in the Alps, his body and his belongings were unusually well preserved. Your assignment is to determine whatever you can about his life and his culture as well as to decide whether you think he was physically similar to or distinctly different from humans living today.

Process and Procedures

1. Read the following paragraphs by yourself.

A man died high in the Alps, above tree line. His body was frozen into a glacier and remained there for more than 5000 years. Then, in 1991, some hikers who had veered slightly off course discovered the well-preserved corpse. They assumed that it was a modern hiker who had been killed the previous winter, and they notified authorities. Because the authorities did not realize that they were dealing with something very old, they removed the body from the ice somewhat carelessly, in a way that resulted in the loss of valuable evidence.

Imagine the surprise of the authorities and the hikers when they found out that the body was really that of a human from an ancient time. He and his possessions are among the most valuable archeological evidence ever found.

Usually the types of evidence that are preserved from humans who lived 5000 years ago are hard substances, such as teeth or bones and the stone or metal parts of tools. Pottery, too, may survive, but soft tissues such as skin, hair, plant materials, and leather generally are destroyed by decay.

Just as scientists must infer ideas about the skin and outward appearance of a dinosaur from the fossil remains of the dinosaur, so must scientists infer details about the appearance or culture of early humans from whatever is preserved. Material remains, such as pieces of pottery, tools, and textiles, are the evidence that scientists use to infer the details of culture, and these items are called **artifacts**.

In this activity you will have the opportunity to use inference in a setting with many artifacts that usually do not preserve well. You also will have the opportunity to consider some particular aspects of biological evolution.

2. Examine the table in Figure 2.8, which lists many of the artifacts that authorities and scientists found with the Iceman and gives a general description of his body.

   This will help you get a general idea of the types of clothing and tools the Iceman had.

Artifact	Comments
Grass cape	Carefully stitched together; isolated repairs made with grass thread
Arrows and leather quiver (carrying case for arrows)	Oldest leather quiver ever found; some arrows with flint arrowheads; other arrows unfinished
Very long bow	Unstrung
Pouch with worked bits of flint (type of stone that flakes easily into sharp pieces)	
Clothing and boots	Made of leather; grass stuffed inside boots
Copper-bladed axe with wooden handle	One of the oldest axes of its type and first with handle intact; attached by leather ties and glue; copper blade
Mushrooms strung on a leather strip	A type of fungus with medicinal (antibiotic) properties
Flint dagger and grass sheath	Worked flint blade on wooden handle; first sheath of this age recovered
Bone needle	
Grass rope	
Stone disk threaded with a leather strap	
Bits of a primitive wheat and wheat pollen	This type of wheat grew only at low altitudes.

Figure 2.8 **Artifacts recovered with the Iceman**

General description of body: General age: 5300 years; mostly intact; apparently died of freezing; 5 feet, 2 inches tall; 20 to 30 years old at time of death; a series of markings on skin—several sets of blue parallel lines on lower back, stripes on right ankle, and a cross behind left knee; medium length hair (cut); physical features similar to humans today.

3. Do the artifacts listed in Figure 2.8 represent evidence or inference? Explain your answer in your journal.

4. Watch the video segment *A Glimpse of the Iceman.*

5. With your team of four discuss what each artifact is and how the Iceman might have used it. Record your ideas in a table in your journal. As part of your analysis, answer the following questions:

   a. What might the bow and the finished and unfinished arrows and arrowheads indicate about his way of life?

   b. What might the copper-bladed axe with the wooden handle indicate about his culture?

   c. What can you infer from the bits of wheat and wheat pollen present?

   d. What might the different markings on his body indicate?

Refer back to the descriptive paragraphs in Step 1 and to Figure 2.8 if you wish.

6. Did you use evidence or make inferences to answer the questions in Step 5? Explain.

7. If you could have three more pieces of evidence or three more bits of information to help you complete your work, what would you want them to be? Explain why you chose the pieces you did.

Analysis

Answer the following questions in your journal and be prepared to share your ideas in a class discussion.

1. What particular physical features of the Iceman would you compare with modern humans if you were looking for evidence of biological evolution? Explain.

2. What artifacts from the Iceman would you compare with artifacts from modern humans as evidence of cultural evolution?

3. Do you think there have been greater changes in humans physically or culturally in the last 5000 years? Explain your answer.

EVOLUTION IN ACTION

> The attendants wheel the teenage girl into the operating room as her mother waits anxiously in the sitting room at the end of the hall. The girl's appendix is so severely inflamed that her doctor worries that it might rupture before the operation can be performed. In spite of the danger, the operation goes smoothly, and the surgeon removes the girl's inflamed appendix without mishap. After surgery, nurses take the patient to the recovery room. In about 30 minutes, she regains consciousness and asks for her mother.
>
> All seems to be going well in the first 24 hours after surgery. However, on the following day, the girl begins to run a fever, which quickly rises. Her doctor realizes that she has contracted an internal infection from the surgery.
>
> The girl in this story has a bacterial infection. A strain of *Staphylococcus* bacteria contaminated the open wound during surgery, and it continued to multiply inside her body. Will she survive this infection? You will use your knowledge of evolution and your scientific thinking skills to propose an explanation for what happens next. As you complete this activity, you will evaluate what you have learned about the way living organisms change across time.

Process and Procedures

1. Read the following three descriptions of *possible outcomes* for the opening scenario. Each description takes place in a different time period in the history of Western medicine.

   ### Scenario 1

   The year is 1925: The girl becomes delirious from fever; in a few days, she dies.

   ### Scenario 2

   The year is 1945: The girl receives an injection of the antibiotic penicillin, followed by repeated doses. Within 24 hours her fever is reduced, and in a week she is released from the hospital, well on her way to recovery.

   ### Scenario 3

   The year is 1965: The girl receives an injection of the antibiotic penicillin, followed by repeated doses. Despite this treatment, her fever continues, and she becomes delirious. In a few days, she dies.

2. For each outcome, write one paragraph that explains *why* that outcome is possible at that time in history.

   Base your explanation on your experiences in this chapter. The information in Antibiotics *and in Figure 2.9 on page 55 in this activity may help you decide why the outcomes are different.*

Analysis

1. The example of a bacterial infection in this activity can serve as a model of evolution. In a short essay, explain how this example illustrates evolution in action in modern times. A good description of this model of evolution

   - describes the evolutionary change that occurs in this model,
   - identifies the factor in the bacteria's environment that exerts a pressure for natural selection,
   - explains the role that variation in individual characteristics plays in the evolution of resistant populations,
   - explains how new generations of offspring play a role in the evolution of resistant bacterial populations, and
   - uses specific examples and evidence to support your response.

2. Describe in two or three sentences the interaction between culture and medicine.

Further Challenges

1. Explain how the difference in generation time for humans (about 20 years) or bacteria (about 20 minutes) makes a difference in their rates of evolutionary change.

2. You might want to read the *Scientific American* article called "The Evolution of Virulence" (April 1993) and briefly summarize how human behavior can influence the evolution of certain disease-causing organisms (pathogens).

Antibiotics

An antibiotic is a medicine that is toxic to certain bacteria. Antibiotics are used to fight bacterial infections. You may have heard of antibiotics such as penicillin, amoxicillin, tetracycline, or erythromycin. A few antibiotics were discovered by researchers in the 1920s and some, such as penicillin, were in limited use by the late 1930s. Mass production of penicillin in the 1940s made this powerful therapy against bacterial infections a widespread tool in medicine. Before that time, doctors had few options to fight a bacterial infection once it started, and patients often died if they suffered from serious infections such as bacterial pneumonia.

Today doctors have a wide range of antibiotics from which to choose, and they use them to treat a variety of illnesses, such as sore throats (which are sometimes caused by the *Streptococcus* bacterium), bacterial pneumonia (a serious lung infection), and even acne. This range of antibiotic choices is relatively recent in the history of medicine and human illnesses. In fact, the dependence of Western medicine on technology is a relatively recent cultural development that has dramatically changed the way we think about health care and physicians. Many patients now demand an antibiotic prescription when they are sick, even if an antibiotic will not cure their problem.

An antibiotic is not effective against every type of bacterium. A particular antibiotic can kill only a limited number of bacterial species. In addition, the genetic material (DNA) of some bacteria can change in a way that allows these bacteria to resist the killing effects of an antibiotic. These bacteria are said to be resistant to that antibiotic. Such genetic changes do not occur very often, but because bacteria with these changes can survive in the presence of the antibiotic, they are more likely to reproduce than nonresistant bacteria. If the resistant bacteria pass the genetic changes along to future generations, an entire population of resistant bacteria can arise. When this happens, the antibiotic becomes ineffective against that bacterial population.

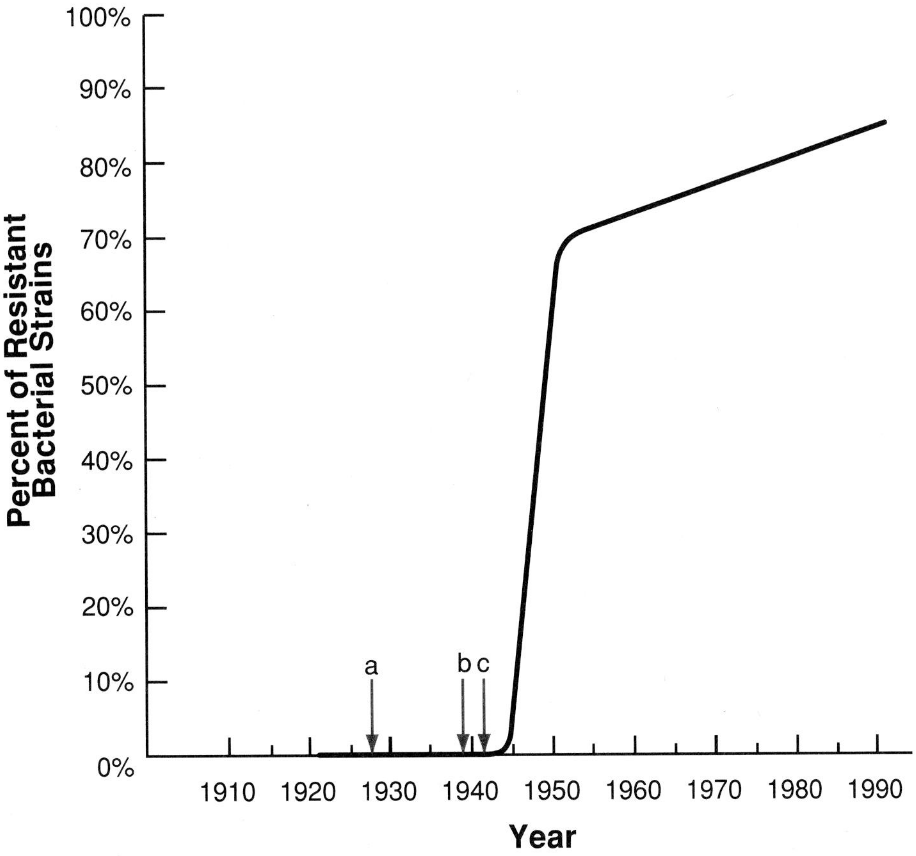

Figure 2.9 Increase in resistance of hospital strains of *Staphylococcus aureus*

a.   1928: In the laboratory, Sir Alexander Fleming observed that the mold *Penicillium* kills the bacterium *Staphylococcus*.

b.   1939: Ernest Chain, Sir Howard Florey, and researchers at Oxford University isolated the antibiotic penicillin from the mold *Penicillium*.

c.   1941: Large-scale production of penicillin begins.

3 PRODUCTS OF EVOLUTION: UNITY AND DIVERSITY

How many different types of life can you find in the photo to the left? How many different types do you think you would find if you actually went snorkeling among the soft coral, feather starfish, and fairy basslets of this coral reef?

One of the most remarkable aspects of life is its enormous diversity. Think of the differences, for example, between the tiny microorganisms that fill this ocean water and the huge blue whales that feed on them. Think as well of the differences between the towering trees that may grow along streets and yards in your neighborhood and the insects and worms that live in the soil that anchors the roots of the trees.

Your challenge in this chapter is to look at life's diversity as a scientist might look at it. To do this, you will consider not only the enormous differences that exist among organisms, but also the ways in which all living systems are alike. You will examine how the underlying unity of life helps scientists understand and explain life's diversity, and you will explore the ways in which scientists organize the knowledge that they have accumulated about living systems. And, as you complete these activities, you should find, as most scientists find, that a close study of the variety of life on earth has not lessened your appreciation for life's diversity and complexity, but has increased it.

ACTIVITIES

Engage	Strange Encounters
Explore/Explain	Describing Life
Explain	A Look at Diversity
Explain/Elaborate	Adaptation, Diversity, and Evolution
Elaborate	Using Unity to Organize Diversity
Elaborate	Explaining the Zebra's Stripes
Evaluate	First Encounter with Organism X

STRANGE ENCOUNTERS

In Chapter 2 you examined evolution, a process of change across time. You have looked at how long biological evolution has been going on and how evolution takes place, but have you ever thought about its remarkable results? Lions on the Serengeti, bacteria that live in near-boiling water, delicate roses in your grandmother's back yard . . .

In this chapter you will have the opportunity to observe and to think about all of the products of evolution. How is it that the life around us (and even in us) occurs in so many different forms, and how do scientists identify and keep track of all of life's variations?

As a starting point for your study of the diversity that resulted from evolution, let us consider how it is that we can easily recognize life, despite its sometimes unusual manifestations.

a.

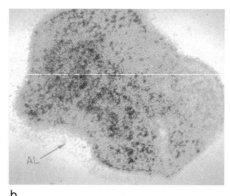

b.

c.

Figure 3.1 a. This pair of lions (*Panthera leo*) watches prey in the distance. b. This thermophile (*Sulfolobus brierlexi*) is part of a group of bacteria that thrive in hot water (60°–90°C). c. These roses (*Rosa* ssp.) can live in moderate to warm climates.

Materials (per person)

question card

Process and Procedures

1. Observe the videodisc segment *Unusual Creatures*. As you watch, keep in mind the question written on the card that your teacher has given you.

   You may wish to take some notes in your journal to help you answer the question on your card.

2. As a class, discuss your responses to the questions on the question cards.

3. Read the story *What Is It?* on page 59. Imagine that you are Katrina, suddenly confronted with a strange object, one that you have never seen before. Write in your journal a short two- or three-paragraph ending to her story that describes how she finally determines whether or not the object really *is* alive. Resolve the issue for her (that is, describe how she finally discovers the object's identity), be creative (for example, you may want to give the story a surprise ending), but be sure that your story addresses the following questions:

   • What characteristics make her think that it is alive?
   • What characteristics make her think that it is not alive?
   • How does she finally make her decision?

4. Share your ending as your teacher directs.

WHAT *IS* IT?

Katrina peered at the . . . thing . . . in front of her. Definitely unusual. Peculiar, even. She has walked this path a thousand times in the three months she has lived here and never had seen anything like this before.

She walked around it, once, slowly. It had a regular shape, sort of like a hexagon, but its surface didn't look shiny and synthetic like plastic or metal or glass. Instead, it looked almost biological: sort of scaly and wrinkled and, well,

something like a snake's skin. The object was colored strangely, too: sort of purplish-greenish, with amber highlights. It looked out of place in the weeds and low shrubs along the path, yet somehow it didn't look manufactured either.

What was it? An egg? A cocoon? A mushroom? Some type of dormant reptile? Or was it alive at all? It really didn't *look* alive, but neither did it look dead. She almost expected it to . . . but then again she also wouldn't have been surprised if it . . .

DESCRIBING LIFE

It is one thing for us to have such a good, intuitive sense of what is alive and what is not alive that we usually are able to distinguish one from the other *here on earth.* It is quite another thing, however, to describe life so clearly that we would be able to identify it in a very different setting. This is the problem that faced scientists in 1976 as they designed experiments to search for life on Mars. What characteristics of life are so general, so basic, that if we observed them, we would be ready to say that life really was present?

This activity will help you think about this question and gain an understanding of the fundamental characteristics that all organisms on earth share. Considered together, they provide the best general description of life that we have been able to construct.

Materials (per team of 4)

large sheet of paper
felt-tipped marker

Process and Procedures

1. Work with your team to develop a comprehensive list of characteristics that you think are common to all living systems on earth. Record your list on a large sheet of paper.

   Refer to the stories from the previous activity for ideas to get you started.

2. With your team identify the one characteristic from your list that you think would be easiest to look for if you were to mount a search for life on another planet. Mark this characteristic with a check mark and be prepared to explain why your team selected it.

The list that you have created is really a description of living systems as you understand them now. Notice, however, that your description also can be used as a check list to try to determine whether some unknown object is or is not alive.

3. Follow your teacher's instructions for displaying your team's list so that the other members of the class can see it.

4. In 1976 scientists actually did search for evidence of life on Mars. Work with your team to learn about the Mars experiments and to answer the questions below. Record your answers in your journal.

   The essay Describing Life: An Impossible Challenge? *on page E31 includes information about the Mars experiments that will help you answer these questions.*

   a. How do scientists describe life (that is, what list of characteristics do scientists associate with life)?

   b. How does this list compare with your list? to the lists developed by other teams?

   c. What specific characteristics of life did the scientists involved in the Mars project look for?

   d. Why do you think that they chose these characteristics and not others?

   e. How did their choice of characteristics to search for compare with the choice that you and your teammates made in Step 2?

Analysis

Complete the following tasks in your journal:

1. Evaluate each of the following statements as either *true* or *false.* Support each evaluation with a short explanation and an example drawn from this or the previous activity.

   a. We usually can recognize life on earth because our experience of life has taught us that living things possess certain characteristics that are not generally found in things that are not alive.

   b. The way in which scientists tried to recognize life on Mars was very similar to the way in which we recognize life, but it was more systematic.

   c. Our preconceived notions about an issue may affect the questions that we ask and the ways in which we are prepared to interpret the data that we collect.

2. Examine the list of unifying principles displayed in Figure 3.2. What justification can you offer for organizing a biology course around these principles?

Evolution: Patterns and Products of Change in Living Systems

Homeostasis: Maintaining Dynamic Equilibrium in Living Systems

Energy, Matter, and Organization: Relationships in Living Systems

Continuity: Reproduction and Inheritance in Living Systems

Development: Growth and Differentiation in Living Systems

Ecology: Interaction and Interdependence in Living Systems

Figure 3.2 Unifying principles of biology

A LOOK AT DIVERSITY

As you saw in the previous two activities, to say that an object is alive is to say something quite significant about it. Think of all the characteristics that living systems and species share in common: evolution, mechanisms for assuring internal homeostasis, complex organization, genetic continuity, growth and development, and interaction and interdependence with other organisms and with the nonliving world. If we know that an object is alive, then we know that it displays each of these features.

Recognizing the ways in which all organisms are alike—recognizing the underlying unity of life—gives us a common starting point from which we can ask and answer questions about life's diversity. Organisms on earth often are very different from one another, in appearance, structure, behavior, and the environments that they occupy. But they all exhibit certain fundamental characteristics, and they all face similar challenges. To understand diversity, then, we need to understand these characteristics and challenges, and we need to consider how different evolutionary pressures have resulted in organisms that have responded to these challenges in widely different ways.

Materials (per team of 6)

large sheet of paper
colored felt-tipped markers
6 unifying principle cards

Process and Procedures

1. Take turns reading the unifying principle that is listed on the card that your teacher has given to each of you and discuss it with the other members of your team. How would you explain this characteristic of life to someone else? What examples would you use to illustrate its importance to humans? to other organisms? Record your ideas in your journal.

   Reviewing the essay Describing Life: An Impossible Challenge? *on page E31 may help you think about your principle.*

2. Meet with the other members of the class who received cards listing the same principle and discuss your ideas briefly. Add any new ideas that this discussion raises to the notes in your journal.

   Make sure that you all share the same understanding of the principle that you have been assigned. Refer any questions that your group cannot answer to your teacher.

3. Create a table in your journal that lists the following types of organisms down the left side and that provides plenty of space for you to take notes about how different types of organisms display the principle that you will monitor. Title your table with the unifying principle that you have been assigned to track.

   • prokaryotes
   • protists
   • fungi
   • plants
   • animals

4. Watch the videodisc segment *A Diversity of Organisms*. As you watch the video, complete the following tasks:

   a. In the note section of your table, begin describing how different types of organisms display the principle that you are tracking.

      Avoid taking notes on all of the specific examples and details that the segment provides. Instead, concentrate on understanding in general how each major type of organism displays the principle you are tracking. For example, if you are tracking the principle of interaction and interdependence, you might record the fact that plants supply the oxygen to the air that both plants and animals require for life. You likely would not record the name of the specific places in the world in which each type of plant can be found.

      Do not be concerned if you are unable to complete your table from the information presented in the videodisc segment. Some principles are easier to observe in such images than others. Step 5 offers another opportunity to complete your table.

   b. List one organism from each group that you find particularly interesting or surprising with respect to the principle that you are tracking, and be prepared to explain why you chose each organism.

5. Discuss briefly your notes on each major group of organisms with your classmates who tracked the same unifying principle that you followed. Use this discussion to help you expand or modify the information in your notes so that you will be able to represent your principle effectively when you rejoin your original team.

   The essay Five Kingdoms *on page E36 contains information that may help you complete your table.*

6. Join the members of your original team and share the information that each of you has gathered. Then construct a summary diagram, table, or drawing that illustrates how each of the five major groups of organisms expresses or has expressed each of the six unifying principles of life.

7. Add to your diagram a few specific organisms in each kingdom and identify one interesting fact about how each displays one of the six unifying principles of life.

Analysis

Use the information presented in your summary diagram to answer the following questions. Record your answers in your journal.

1. Which of the six unifying principles seems to have been most useful to biologists in grouping organisms into different large categories, or kingdoms? Explain your answer.

   Think about the major difference that distinguishes prokaryotes from all other organisms on earth, how plants differ from both fungi and animals in the manner in which they obtain the energy required for life, and how organisms in each of the kingdoms reproduce.

2. How do these large categories by which biologists organize their thinking about life illustrate both its unity and diversity? Explain your answer.

Further Challenges

Read these observations about a virus, called a bacteriophage (see Figure 3.3), that infects bacteria.

A bacteriophage

- contains genetic material,
- reproduces only when inside another organism,
- has an outer case made of protein,
- injects genetic material into a bacterial cell, and
- uses the energy and the structure of the bacterial cell to make parts that assemble into copies of itself (and often kills the bacterial cell).

You may have noticed that neither the videodisc segment nor the essay contained any reference to viruses. Yet we often think about viruses in relation to life. For example, have you ever heard someone complain about being *attacked* by a virus?

Use your knowledge of the unifying principles of life to construct a well-reasoned argument to answer the question, Is the bacteriophage alive? If viruses were considered alive, into which (if any) of the five kingdoms would scientists likely categorize them? Support your answer with specific references to the unifying principles and to the information about the bacteriophage listed above.

ADAPTATION, DIVERSITY, AND EVOLUTION

Think back to the organisms you saw in the previous activity, *A Look at Diversity*. Each one shares the general characteristics of the kingdom in which it is categorized, yet each is uniquely its own creature, with peculiarities of structure or function or behavior that make it distinctly different from all of the others.

In Chapter 2 you learned about evolution, the process of change and selection that can introduce and perpetuate biological differences in a

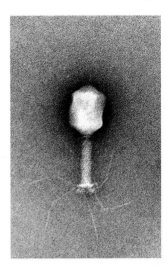

Figure 3.3 The bacteriophage (photographed at ×240,000) is a virus that attacks bacteria.

Explain
Elaborate

population of organisms. Can that process explain all of the structural, functional, and behavioral diversity that we find on earth?

To examine the relationship between evolution and diversity more closely, we need to consider the more fundamental relationship between a species and the environment in which it lives. Recall from Chapter 2 that inherited characteristics that help an organism survive and reproduce within a particular environment are called adaptations. What happens to organisms that are well-adapted to one set of environmental conditions when the environment changes? Is it possible that new characteristics might appear in a species that would allow organisms to live in new environments?

To discover some answers to these questions, consider life in two quite different settings: first, across a series of oceanic life zones and, second, across a series of eras in the earth's history.

Materials (per team of 4)

slides and cover slips
compound microscope
hand lens
forceps
dissecting needles with corks on tips
scalpel
large sheet of paper
4 felt-tipped markers
plant specimens

Process and Procedures
Part A Diversity and Adaptation in a Marine Environment

1. Observe the videodisc segment *Marine Life.* Pay particular attention to the conditions that are present in each of the different environments described. Indicate these conditions in a few words at the appropriate places on the copymaster *Oceanic Life Zones.*

2. Watch the videodisc segment again. As you watch, work with the other members of your team to write on your copymaster the names of several organisms that live in each life zone. Describe in a few words one adaptation for each species that appears to enhance its ability to survive in the particular environment in which it usually is found.

   Adaptations may be structural, functional, or behavioral, or may involve a combination of these. Examples of adaptations that you might use include lungs, gills, internal skeleton, external skeleton, structural protections against predation, behavioral protections against predation, and feeding habits.

   The information on the copymaster Marine Organisms *will help you complete this task.*

3. Work with the other members of your team to answer the following questions. Record your responses in your journal.

   a. In general, do most of these marine organisms appear to be well-suited to survival in their environments? Support your answer with specific examples.

b. Are all inherited characteristics adaptations? Explain your answer and support it with specific examples.

c. Name a characteristic that is an adaptation in one environment but would not be an adaptation in another environment. What explanation can you offer for the presence of characteristics that may not be particularly adaptive for a modern organism in its current environment?

Part B Diversity and Adaptation across Time

1. Work with the members of your team to examine the plant specimens that your teacher has provided.

   Manipulate, dissect, or otherwise examine the specimens to gather as much information as possible.

   CAUTION: Scalpel blades and needles are sharp; handle with care. Replace cork on needle tip after use.

2. Answer the following questions. Record your answers in your journal.

   a. Identify at least one characteristic that all of these organisms share.

   b. Are these organisms as diverse as those in the videodisc segment *Marine Life* or are they more similar to each other? Explain your answer.

3. Construct a list in your journal of the major adaptations that occurred during the evolution of modern plants.

   Refer to the essay From Cell to Seed *on page E42 for a short description of the evolution of modern plants. Construct your list of adaptations as you read.*

4. Use the list of adaptations to help you identify at least one characteristic of each organism that biologists might use to distinguish that organism from all of the others. Record these characteristics in your journal.

5. Do some of the characteristics that you identified in Step 4 match the evolutionary adaptations that you identified in Step 3? Explain how these matches might allow you to organize your specimens into an order based on probable time of evolutionary appearance.

6. As a team, create a branching pattern on your large sheet of paper that organizes these organisms as you think biologists would classify them according to their evolutionary divergence, or separation from an ancestral form.

   Be sure that your diagram would be understandable to someone not in your team. For example, you may wish to label your diagram with the names of the organisms as you have placed them. Include a short statement for each organism that identifies the major distinguishing characteristic that caused you to place it at that point in the diagram.

7. Post your diagram in your classroom as your teacher directs, and be prepared to share your ideas in a class discussion.

Analysis

Work with the members of your team to answer the following questions about the relationships among adaptation, diversity, and evolution. Record your answers in your journal.

1. What are adaptations? Illustrate your answer using examples drawn from the organisms that you studied in Part A and Part B.

2. How is the diversity of organisms on earth related to the diversity of environments in which organisms live? Illustrate your answer using examples drawn from the organisms that you studied in Part A and Part B.

3. How is the appearance of new adaptations related to changes in existing environments or to the ability of organisms to colonize new environments? Illustrate your answer using examples drawn from your work in Part B.

4. Propose a general explanation for the appearance of biological diversity on earth.

   Refer to the essay Evolution Produces Adaptations *on page E44 for background information that may help you answer this question.*

USING UNITY TO ORGANIZE DIVERSITY

Elaborate

The characteristics that different living systems share by virtue of their common ancestry provide biologists with a powerful set of guidelines for organizing the millions of different types of life into large categories of similar organisms. This organizing process, called **classification**, plays an important role in biologists' attempts to understand life. First, classification helps them subdivide the enormous numbers of different types of living (and extinct) organisms into groups of more or less similar types that they can study productively. And second, the act of creating classification categories helps them think about the evolutionary relationships that exist among different types of organisms. In fact, we might say that classification categories represent hypotheses that biologists develop about how different forms of life are related. As scientists change their thinking about these relationships, the ways in which they organize these categories also may change.

This activity offers you a chance to practice biological classification and, in the process, to develop further your growing understanding of biological diversity, its origins, and the ways in which scientists study and make sense of it.

Materials (per team of 4)

set of nonliving objects
set of organism cards

Process and Procedures
Part A Classification As a Tool

1. Work with your team to examine the set of nonliving objects. What, if anything, do all of these objects share in common? What characteristics do some objects possess that others do not? Record your observations in your journal.

2. Sort the objects in your set into different categories.

   Notice that the objects have different characteristics. Choose one or more of those characteristics as the basis for creating your categories.

3. Compare your classification scheme with that of one other team. Are the categories the same? What categories did the other team create?

4. Answer the following questions in your journal:

   a. Your team and the other team probably used a different basis for the classification schemes. Nevertheless, each team used *some* basis; that is, each team used some general **criteria** to determine what groups they would include in their scheme and what objects would go into each group. Is it possible to classify without establishing such criteria? Support your answer with a different example of classification drawn from your life experience.

   b. Consider the categories of objects that you created in your classification scheme. Could the objects *in any of these categories* have been further separated (classified) into smaller groups within the large category? Support your answer with specific references to the objects and to the classification scheme that you developed.

Part B Biological Classification

1. Examine the data provided on your organism cards. Work with the other members of your team to create a simple hierarchical scheme for these organisms.

   A classification scheme in which objects are sorted first into large categories and then into smaller groups within the large categories is said to be hierarchical.

   Ask yourselves what broad category all of these organisms fit into. Then ask yourselves what criteria you could use to sort them into smaller categories. Continue creating smaller categories until each category contains only one type of organism.

2. Figure 3.4 illustrates the hierarchical classification scheme that most biologists use. Compare this classification scheme to the one that you have developed for your organisms. Which levels on the scheme shown in Figure 3.4 are represented in the scheme that you developed? Explain your answer.

3. Complete the classification diagram shown on the copymaster *Biological Classification*.

   Fill in the names of the organisms that you classified, and add the names of other appropriate organisms to the remaining lines.

Figure 3.4 Biologists use this hierarchical scheme to classify organisms.

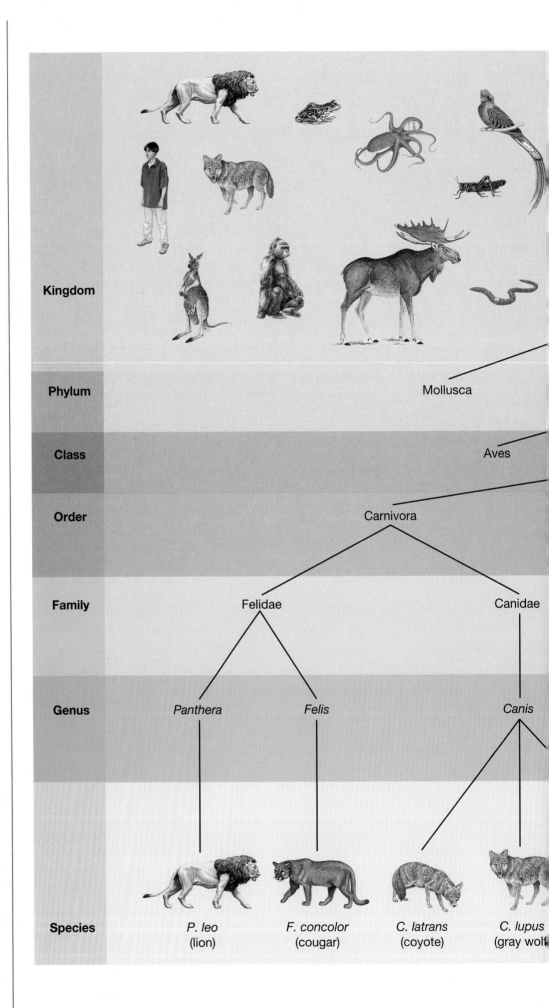

Kingdom

Phylum   Mollusca

Class   Aves

Order   Carnivora

Family   Felidae   Canidae

Genus   *Panthera*   *Felis*   *Canis*

Species   *P. leo* (lion)   *F. concolor* (cougar)   *C. latrans* (coyote)   *C. lupus* (gray wolf)

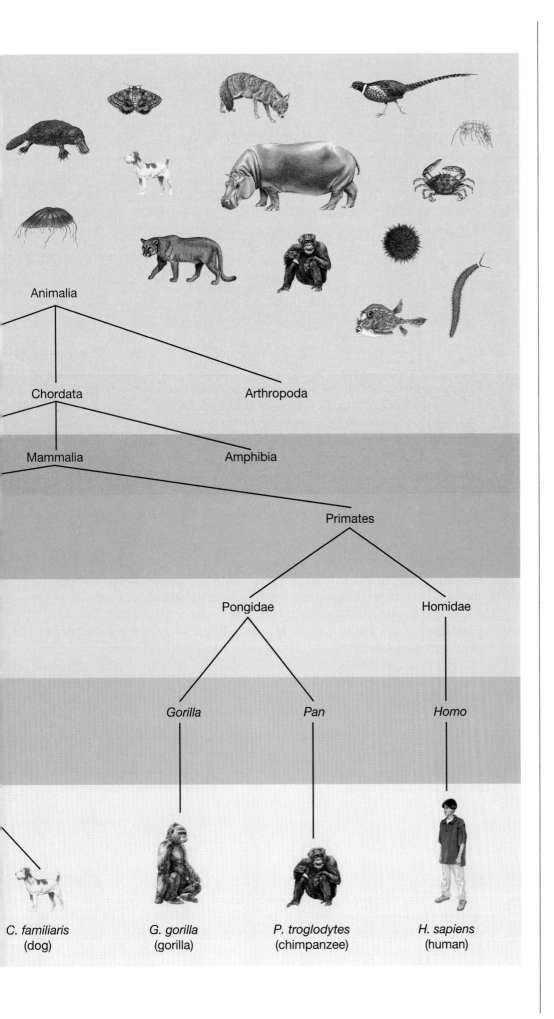

Animalia

Chordata   Arthropoda

Mammalia   Amphibia

Primates

Pongidae   Homidae

Gorilla   *Pan*   *Homo*

C. familiaris
(dog)

G. gorilla
(gorilla)

P. troglodytes
(chimpanzee)

H. sapiens
(human)

4. Answer the following questions in your journal:

   a. Which group in a hierarchical classification scheme contains

   • the most organisms?
   • the most different types of organisms (the greatest diversity)?
   • the fewest organisms?
   • the fewest different types of organisms (the least diversity)?

   b. What does this classification scheme suggest about evolutionary relationships that exist among these organisms? Explain your answer.

   Information in the essay Organizing Diversity *on page E47 may help you with this step.*

Analysis

Work with your team to complete the following tasks:

1. Imagine that you are a member of a large publishing company that is writing a new textbook for high school biology classes. Your team has been assigned to write the chapter on the nature and importance of biological classification. Before you can begin to write, of course, you need to decide what the students using the book should learn about your topic.

   Identify the *three most important ideas* about biological classification that you think a scientifically literate citizen should understand. Summarize these ideas in three concise statements and record them in your journal.

2. Now imagine that the editor-in-chief has decided that the size of the book must be reduced and that there will be no chapter on biological classification. Instead, everything you want the students to learn about the topic has to fit on one page, preferably in the form of a diagram or a drawing (although this can include words or short statements). Luckily, you already have distilled your understanding about the topic into three clear statements (Step 1).

   Create a one-page display that communicates and illustrates these ideas in an attractive and interesting manner.

3. Follow your teacher's instructions for posting your display so that members of other teams can view it. If you were a textbook editor, which display(s) would you choose to use in the book and why? Record your answer in your journal.

EXPLAINING THE ZEBRA'S STRIPES

Elaborate

In the last activity you examined the process of biological classification as a scientific tool that organizes and expresses the structural and evolutionary relationships that exist among all living things. This idea bears closer study: if classification is not an end in itself, but is a means that biologists use to express their understanding of biological diversity, then we might expect classification schemes to contain some areas of controversy and question, certainly at the points at which our knowledge still is incomplete. We also might expect these schemes to change across time, as scientists discover new information that increases or modifies their understanding of the significance of particular relationships and adaptations.

An important characteristic of scientific knowledge is this openness to change and modification. Scientific knowledge is not static because scientists continuously discover new information and test and reevaluate existing understandings. Usually, changes in scientific knowledge are not so great that we must discard all of our previous explanations in favor of new ideas. Nevertheless, as we gain more detailed information about the natural world, our explanations also grow and change to reflect that new information.

This characteristic of scientific knowledge is illustrated in this activity about the difficulties that scientists face as they sort through what you might think would be easily answered questions. Can you tell a zebra by its stripes? Or is one zebra just like the next?

Figure 3.5 What similarities and differences do you observe among these four zebras?

Chapman's zebra

Grant's zebra

Grevy's zebra

mountain zebra

Process and Procedures
Part A Looking Closely at the Concept of Species

1. Work with the other members of your team to compare the physical characteristics of the four zebras shown in the videodisc segment *Zebras* and in Figure 3.5.

    These zebras represent four populations of zebras that live in Africa; each population has a different common name. Record your observations about these animals in your journal in a table similar to the one in Figure 3.6.

2. Examine the map in Figure 3.7 and compare the ranges of each of these four populations of zebras. Add this information to the observations that you recorded in Step 1.

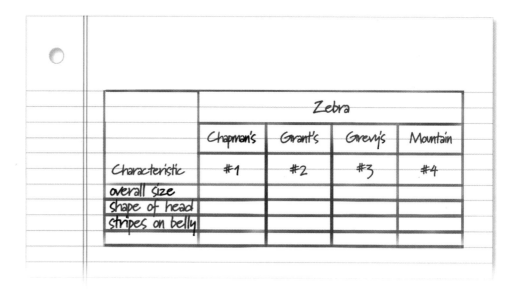

Figure 3.6 Sample table
Record your observations in a table similar to this one.

Figure 3.7 How do the ranges of the four zebras compare?

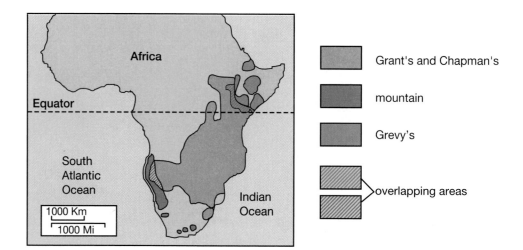

3. Discuss with your teammates how you might categorize the four zebras into different species. Record your answer in your journal.

   The fundamental question here is whether each of these populations of zebras represents a separate species or whether some of these zebras are members of the same species. Develop and support your answer using the information that you collected in Steps 1 and 2.

4. Examine the additional information available on the copymaster *African Zebras* to make further comparisons among the four populations of zebras. How might you categorize these zebras? Support your answer using all of the information you have available to you. Record your answer in your journal.

5. Contribute your ideas to a class discussion about the zebras and about the difficulties involved in assigning species distinctions.

Part B Explaining the Adaptive Significance of Structural Characteristics

One of the most important relationships that you have encountered in this chapter is the relationship between the particular adaptations that a species displays and its environment. In the activity *Adaptation, Diversity, and Evolution*, for example, you looked at some of the different adaptations that various types of marine organisms possess, and you matched these adaptations to the challenges that these organisms face in their particular environments. In the same activity you also looked at the history of the evolution of plants and saw how the appearance of new adaptations was correlated with the movement to land.

How far can we extend the statement that biological diversity results from random genetic change combined with natural selection? Is all variation among different species the result of selection for adaptive characteristics? Or are there limitations to the explanations that we can offer about some characteristics?

1. Hold a brainstorming session with the other members of your team about the significance of a zebra's stripes. How might these stripes be adaptations? List your ideas in your journal.

2. Read the two short descriptions on the copymaster *Ideas about the Zebra's Stripes* that offer explanations for the appearance and perpetuation of the zebra's stripes.

3. Work with the other members of your team to analyze the data that relate to each explanation.

   As you read and discuss each explanation, identify and mark or record each bit of evidence, each inference, and all of the assumptions that you recognize. Then decide whether each piece of data supports, does not support, or contradicts the explanation offered. Take notes in your journal about the results of your discussion. View the videodisc segment again, if you think it will help you.

4. Discuss the strengths and weaknesses of each of the two explanations.

5. Work with the other members of your team to determine which explanation best accounts for the data presented. Record your answer in your journal.

   Weigh the data that you have collected by asking yourself such questions as, What pieces of information are most relevant to each explanation, What are the most serious flaws of each explanation, and Might there be more than one explanation to account for the stripes?

6. Participate in a discussion and evaluation of the explanations.

Analysis

Answer the following questions in your journal:

1. Have you ever heard people say "I don't pay much attention to reports about new scientific findings. After all, scientists say one thing about x, y, or z this year, but they said something else last year, and something different from that the year before. It is clear that they don't know what they're doing"?

   Critique those statements in the light of what you have learned in this activity about the nature of science. In your answer, refer to your experiences in both Part A and Part B of this activity. Record your critique in your journal and be prepared to share it as your teacher directs.

2. In the essay *Five Kingdoms,* you read about the history of the classification scheme that biologists use to group all living systems into five large kingdoms.

   a. How does that history illustrate the statement that classification is not an end in itself, but is a means that biologists use to express their understanding about biological diversity?

   b. How does it illustrate the statement that science is characterized by its openness to change and modification?

FIRST ENCOUNTER WITH ORGANISM X

Evaluate

Throughout the activities and essays in this chapter, you have focused on how evolution has produced the tremendous diversity of living systems that exist (and have existed) on the earth. You also have seen that, in addition to explaining the diversity of life, evolution accounts for its unity. Evolution has produced organisms that are very different from one another, adapted to life in different habitats, and even adapted differently to the same habitat. Evolution *from a common ancestor,* however, also has resulted in organisms that show important similarities to each other.

This activity, the first of six, invites you to use your imagination as you evaluate your understanding of evolution—the process that explains both the unity and the diversity of life—in an unusual way. You will revisit Organism X in each unit of this program and add to your description of the organism as your understanding of the unifying principles of biology increases.

Materials (per person)

folder with 5 sheets of plain paper
felt-tipped marker
habitat card

Process and Procedures

It was a thrill to hear that the research funds came through; now you can embark on your long-awaited trip into the wilderness to observe and record data about several endangered species. The conservation organization that is sponsoring your trip has assigned you to a team that will be spending several weeks in one location. The organization also has given you a list of organisms that you should observe, but you are a born naturalist and you are determined to keep your eyes open for whatever you find of interest.

1. Study the habitat card that describes where you are going. Mount the card on a page in your folder as a permanent record of the environment in which you will work for the next few weeks.

2. After weeks of careful observation and mountains of detailed notes, you have located and studied several of the organisms on your list, although others have eluded you. Then the unexpected happens: you find an

organism that scientists have never seen before. You've done it! You have discovered a previously unknown species.

Describe your organism by drawing one or more diagrams of it (including labels of its distinctive structures) and by writing a paragraph that tells about its characteristics.

You will have to make up an organism. Use your imagination. The organism that you discovered may resemble others found in this habitat or another habitat or it may be quite different from any real species.

You will know that you have generated a good description if it meets the following criteria:

- *It describes how your organism resembles all known forms of life.*
- *It describes how your organism is adapted to the habitat in which you found it.*
- *It identifies the known species to which your new organism is most closely related, and it describes the evidence on which you base your answer.*

3. One of the most exciting privileges granted to the individual who discovers a new form of life is the privilege of naming it. Complete your initial description of this new species by giving it both a common name and a scientific name. (Remember that the scientific name must indicate the organism's relatedness to other organisms.) Add an appropriate caption(s) to the diagram(s) that you have drawn and record the organism's names in your folder.

Refer to the essay Organizing Diversity *on page E47 for information about how scientists name new species.*

Figure 3.8 U.S. researcher Karla MacEwan is collecting plant samples in Costa Rica.

Analysis

Which of the two statements below is most consistent with scientists' understanding of the process of evolution, and why? What is wrong with the other statement? Record your response in your folder.

Statement 1: The habitat that I worked in is very wet and salty. The organism that I discovered evolved a rubberlike skin to keep from shriveling up.

Statement 2: The habitat that I worked in is very wet and salty. The organism that I discovered has a rubberlike skin that may have evolved because it offered protection from the harmful effects of all that salt.

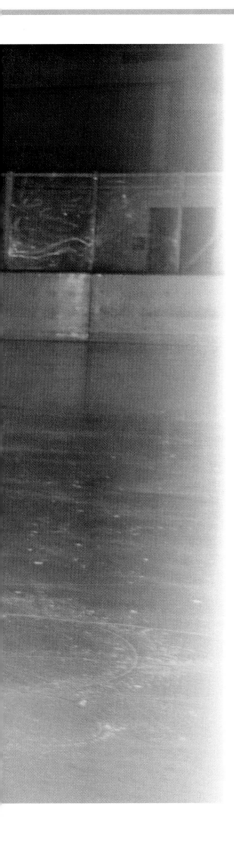

HOMEOSTASIS:
MAINTAINING DYNAMIC EQUILIBRIUM IN LIVING SYSTEMS

4   **The Internal Environment of Organisms**

5   **Maintaining Balance in Organisms**

6   **Human Homeostasis: Health and Disease**

Balance. Under normal conditions it is an intricate, dynamic state that your body attends to without much conscious effort. Think of a skater gliding across the ice: not only is he balancing on an edge of a blade, but his body is maintaining a balance in other less obvious ways. Although he is surrounded by ice and frigid air, his body temperature stays fairly constant. And as he exerts energy in spins and jumps, his heart rate and breathing rate accommodate these changes.

In this unit you will examine some of the processes involved in maintaining balance in the human body. You will consider the characteristics of the human life form that allow for stable internal conditions, and you will apply what you have learned by studying how the human body reacts when this balance is disrupted significantly. By the end of Unit 2, you should understand that

- all organisms have an internal and external environment and are affected by interactions between these environments,
- the interactions of systems that adjust the internal environment result in a dynamic balance called homeostasis,
- stressors may overwhelm the ability of organisms to maintain a balance in their internal environment, and
- individual and collective behavior may influence an individual's ability to maintain homeostasis.

You also will continue to
- collect, analyze, and graph data,
- make and test predictions,
- construct and use models, and
- perform ethical analyses.

> **'No man is an island,'** said John Donne.
> *I feel we are all islands—in a common sea.*

Anne Morrow Lindbergh, *A Gift from the Sea*, 1955

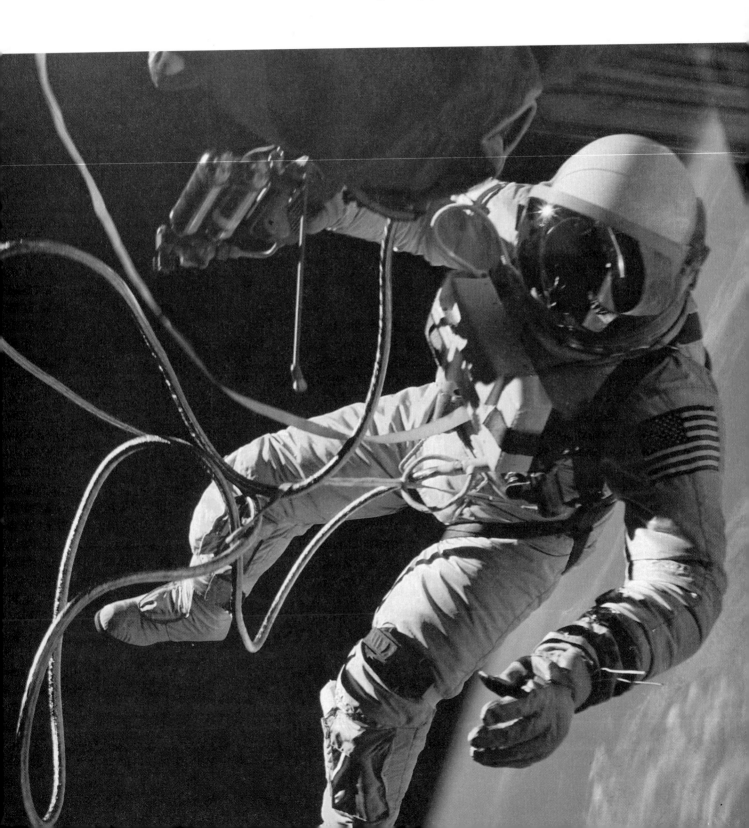

4 THE INTERNAL ENVIRONMENT OF ORGANISMS

What NASA spent millions of dollars to develop, your body does for free!

Consider for a moment the importance of a space suit for astronauts who walk on the surface of the moon or venture into space to retrieve a satellite. In such circumstances, a space suit serves as an effective boundary between the external environment of space and the internal environment within the space suit. Without such an effective boundary, the men and women in space would not be able to maintain the internal conditions necessary to sustain life.

Although astronauts in space must consciously attend to the differences between their external and internal environments, the rest of us do not have to pay much attention to these differences. As long as we eat and dress appropriately, intricate systems inside our bodies constantly monitor and attend to the differences between the external and internal environments. These processes occur in most organisms in a variety of ways. In this chapter you will explore different systems in the human body, from the level of the cell to the body as a whole, to learn ways the human body maintains an internal balance.

ACTIVITIES

Engage   Can You Stand the Heat?

Explore   Cells in Action

Explain   A Cell Model

Explain
Elaborate Regulating the Internal Environment

Evaluate   Can You Stand the Heat—Again?

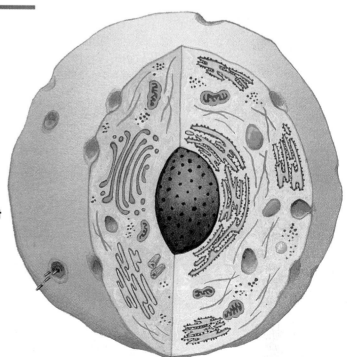

Engage

CAN YOU STAND THE HEAT?

On a scorching summer day, getting out of the hot sun and into the shade seems like the natural thing to do. On a blustery winter day, seeking shelter in a warm house seems an obvious way to restore physical comfort. Even the family pets exhibit similar behaviors to minimize stressful external conditions. Behavioral responses such as these help to relieve stresses placed on the body by the external environment. Are there other ways your body works to maintain its internal environment in the face of external stresses? As you complete this activity, you will begin to answer this question. In the activities that follow, you will develop a more complete understanding of how the body maintains its internal environment.

Process and Procedures

1. Read to yourself the story *A Pause That Refreshes?* (page 81).

2. With a partner, develop an explanation for Ralph's condition.

   Consider the evidence you have and what inferences you can make about how Ralph's body responded to external stresses.

Analysis

As a class, develop an answer to the following question:

How did Ralph's conscious behaviors affect the stresses placed on his body?

CELLS IN ACTION

Explore

"A box without hinges, key, or lid, yet golden treasure inside is hid."[1]

How can you determine the identity of the box in this riddle? These clues make you think about the contents of boxes or containers that hold all sorts of treasures. Your body is another type of container that holds valuable information. Thinking about the human body as a container with contents that are different from the outside environment can help you understand what happened to Ralph in the story *A Pause That Refreshes?* Because the body is a large complicated container, it is difficult to study in detail. You can begin to study it by examining cells, which are smaller containers of living systems, and by comparing the contents of the cells to what is outside them. Based on your examinations, you will begin to understand how living systems maintain an internal environment that is different from the external environment.

---

[1] J.R.R. Tolkien, *The Hobbit,* 1966, Houghton Mifflin Co., Boston.

A PAUSE THAT REFRESHES?

How much heat can you take? How might your body respond to the stress of a hot summer day, especially if you are involved in demanding physical activity? Perhaps you've experienced something like this and would not make the same mistake that Ralph made . . .

Ralph had just graduated from college in Minnesota and was visiting his parents in central Texas. He was planning to surprise his parents, who were going to be gone all day, by clearing a large area of brush to prepare for a barn they wanted to construct.

As his parents drove off, Ralph loaded the pickup truck with the gear he needed for the job. His parents' dog, Ranger, wanted to come along so Ralph let her hop into the truck as well. This June day was already hot, and the temperature climbed past 35°C (95°F) by late morning.

Ralph began the job enthusiastically. He set to work clearing brush, dislodging sharp-spined cacti from the ground, and raking debris. Even though he worked up a substantial sweat, the small brush and cacti were no match for his muscles and his tools. He thought that he could clear the entire area with one day of hard work, finishing before his parents returned.

After two hours had passed, Ralph acknowledged that the job might be bigger than he originally had thought. This was about as hot an environment as he ever had worked in. He was grateful for the ice-cold juice that he sipped from the plastic bottle he had brought. He drank often, although he kept reminding himself that he had to conserve what was there and make it last for the day. Ranger obviously was hot too. She tried to find what little shade there was and lay on the ground panting. Ralph was sorry he hadn't thought to bring water along for the dog. He vowed to himself that he and Ranger would get plenty to drink once he had cleared the brush and could go back to the house.

By mid-afternoon, Ralph was out of juice and began to notice the extreme dryness of his mouth. He wanted to stay and finish the work but decided that he should drive back to the house and get something more to drink. As Ralph opened the kitchen door, Ranger eagerly ran to her dog dish and rapidly lapped up what water was there. Ralph opened the refrigerator to look for a refreshing beverage. The first thing he spotted was a six-pack of his dad's beer. Thinking his dad wouldn't mind, he helped himself to a can of beer, then another. He still felt thirsty so he also drank a glass of water. He sat for awhile in the air-conditioned house to cool down, refilled Ranger's dish with water, made a bathroom stop, then headed out to the truck. "That beer sure went through me fast," Ralph mused as he drove back to the work site. Because he had less than an hour's work left, he didn't take anything along to drink.

Ralph had been working again for only a few minutes when he experienced some dizziness and a faint touch of nausea. Nothing much, he decided. Besides, he would be quitting shortly. But soon he noticed a pounding in his head and some changes in his vision. Instead of seeing in sharp color, Ralph began to feel as though his world was slowly becoming black and white. His muscles ached, and he suddenly felt very tired. His dizziness increased so that he had a difficult time driving the pickup back to the house.

Ralph's parents returned at the same time that he pulled into the yard. Later, when Ralph explained what had happened, they were pleased with the work he had accomplished but were not surprised at his condition . . .

Materials

Part A (per team of 4)

4 pairs of safety goggles
4 lab aprons
3 500-mL beakers
balance
plastic wrap
coffee filters

slotted spoon
paper towels
300 mL of corn syrup solution
300 mL of distilled water
3 shell-less eggs in a beaker of vinegar

Part B (per team of 2)

dropping pipet
microscope slide and cover slip
compound microscope
forceps
scalpel
dissecting needle with a cork on the tip
paper towels
5% salt solution in dropping bottle
onion wedge

Process and Procedures
Part A An Eggs-periment

How big is a cell? Most cells are so small you need a microscope to see them, but there are exceptions. A chicken egg is actually a single cell, although it is an unusually large one. It is protected by a hard shell that surrounds several soft membranes that you can see when you peel a hard-boiled egg. Placing a chicken egg in an acetic acid (vinegar) solution for three days causes the calcium in the hard shell to dissolve. As a result, what remains is a fragile chicken egg surrounded by a soft membrane. This membrane allows certain substances to pass in and out of the egg.

1. Use the information in the introduction to generate questions about what might happen if you place the shell-less chicken egg in different solutions.

    You have distilled water and corn syrup available as solutions.

2. Choose one of the questions you generated, and design a **controlled experiment** to gather information to answer your question.

    A controlled experiment is one in which you control all variables except one. See the background information on controlled experiments for more information.

    Your experiment should include a control and a record of the results. You may want to review the available materials and safety guidelines before you plan your experiment.

    Do not leave your eggs in any one solution for less than 30 minutes or for more than one day.

3. Have your teacher approve your design. Then conduct your experiment.

    Remember to record your experimental design and results in your journal. Your records could include how long the eggs were in the solution, the appearance of the eggs at different times, the mass of the eggs, the volume of the solution, or any other evidence you find.

    *Before you work with the shell-less eggs, use the slotted spoon to remove each egg from the vinegar. **Avoid touching the eggs with your hands.** Gently blot the eggs with a paper towel. Use a coffee filter to hold each egg as you measure its mass.*

4. Enter your data in the class data table.

5. Discuss the following questions:

   a. Why is it useful to combine data from the entire class?

   b. What do the data tell you about the relationship between the internal contents of a shell-less egg and its external surroundings? Justify your response with specific evidence.

   c. How would you account for any differences you noticed in the behavior of the three eggs under different external conditions?

   d. What controls did you use in your experimental design? Why?

Part B Observing Cell Activity

To learn more about cells as containers, you can change their environments by adding different solutions and then observe the behavior of the cells. Scientists recognize the relationship between solutions and the internal environments of a cell. An **isotonic** solution, for example, provides an environment in which the concentration of solutes outside the cell equals the concentration of solutes inside the cell. The dissolved substance is a **solute**; the liquid in which it is dissolved is a **solvent**. For example, when table salt dissolves in hot water, salt is the solute and water is the solvent. A **hypertonic** solution is one in which the concentration of solutes outside a cell is greater than the concentration inside a cell. A **hypotonic** solution is one in which the concentration of solutes outside the cell is less than the concentration inside the cell. (The prefix *hyper-* refers to an excess; *hypo-* refers to a shortage.) In this part of the activity you will observe microscopic cells responding to changes in their environment.

1. View the videodisc segment *Blood Cells in Solution*, which illustrates the behavior of cells in different solutions. Before you watch this segment, create in your journal three columns (labeled *isotonic*, *hypertonic*, and *hypotonic*) to record your observations about changes in the cells.

2. To observe the responses of plant cells, prepare a wet mount of onion skin by following these steps:

    a. Remove one layer from your onion wedge.

    b. Snap the layer backward, as shown in Figure 4.1.

        This should expose the edges of several smaller layers.

    c. Use forceps to separate a piece of the transparent, paper-thin layer from the outside of the original layer.

    d. Lay the piece flat on a clean microscope slide.

    e. As necessary, use the scalpel to trim the piece so that it will fit under a cover slip.

    f. Use the dissecting needle to smooth out any bubbles or wrinkles.

        CAUTION: Scalpel blades and needles are sharp; handle with care. Replace cork on needle tip after use.

    g. Use the dropping pipet to add 1 or 2 drops of water to the slide. Then place a cover slip over the piece of onion skin.

Figure 4.1 **Preparing an onion skin specimen**

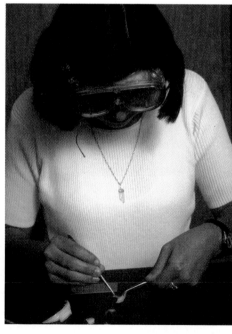

3. Examine the onion skin under the low power of your microscope.
    Take turns observing the cells.

4. Switch to high power, and focus sharply on a few cells. Make a sketch of the cells in your journal. Then place a small piece of paper towel at one edge of the cover slip (see Figure 4.2).

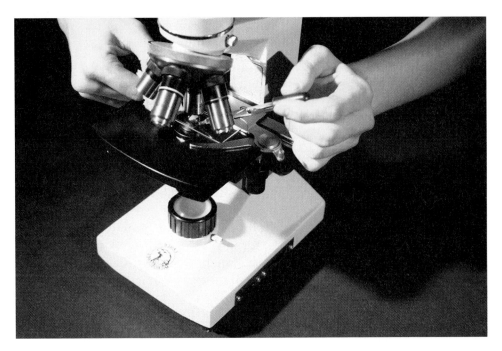

Figure 4.2 **Adding solution to a wet mount**

5. Test the effects of changing the external environment of the cells you are viewing. To do this, place several drops of 5 percent salt solution against the edge of the cover slip that is across from the paper towel. Then observe what happens, and record it in your journal.

   Take turns observing the cells.

6. Add more salt solution, if necessary, until you see changes in the cells. Record your observations in your journal, and include simple sketches that show the cells both before and after the salt solution was added.

   Be sure to record how much salt solution you added.

7. Dilute the salt solution on the slide by adding distilled water. Continue to add water until the cells return to their original condition. Make observations while you are doing this, and record them in your journal.

   To add water, use the same technique that you used to add the salt solution. Remember to record how much water you added.

Analysis

Discuss the following questions with your teammates; then record your responses in your journal.

The essay Compartments *on page E50 will be a helpful resource for this* Analysis *and for the next activity.*

1. What evidence did you collect that indicates that the external environment affects the internal environment?

2. A chicken egg is a single large cell. In this activity, how did the egg serve as a model of how cells function as containers in living organisms?

3. What do eggs, animal cells, plant cells, and the human body have in common?

4. Based on your observations of cells, what might have been happening in Ralph's body in the story *A Pause That Refreshes?* when he

- first became hot and started sweating,

- became thirsty,

- went to the house for refreshments, and

- returned to work?

Further Challenges

To learn more about the way cells function as containers, examine a fresh wet mount of an onion cell, and look for internal structures. A stain, such as Lugol's iodine solution, may make such structures more visible. Make sketches of what you observe, and label all the compartments and boundaries.

WARNING: Lugol's iodine solution is a *poison* if ingested, is a *strong irritant*, and can stain clothing. Avoid skin/eye contact; do not ingest. If contact occurs, flush affected area with water for 15 minutes; rinse mouth with water; call the teacher immediately.

A CELL MODEL

Explain

You have been exploring examples of interactions between internal and external environments. In living systems, boundaries separate these environments. These boundaries exist in every organism, from the smallest cell to large plants and animals. The cell is a good example of a compartment in which the cell membrane serves as a boundary. In this activity you will use dialysis tubing and several solutions to build a cell model that you can use to investigate how cell boundaries affect the internal cellular environment. Your investigation will help you develop an explanation for how boundaries and compartments help living systems maintain and regulate the conditions necessary for life.

Materials (per team of 2)

2 pairs of safety goggles
2 lab aprons
materials to carry out the experiment that you design

Process and Procedures

1. Read Steps 1–5. Then generate at least two *testable* questions about how membranes affect the internal environment of the cell and that meet the criteria in Step 3.

Refer to the activity Primates Exploring Primates *(Chapter 1) on page 21 for a description of a testable question.*

The essay Membranes *on page E52 also might provide useful information.*

2. With your partner, choose the testable question that you would like to investigate. Have your teacher approve your question.

3. Design an experiment to test the question you chose. Your design must be safe, and it must use

   - dialysis tubing,

   - starch suspension,

   - glucose solution, and

   - appropriate indicators.

   The Background Information *on page 88 may help you design your experiment.*

4. Have your teacher check and approve your design.

5. Create a data table in your journal in which to record a brief summary of your experimental design, your predictions, and the results of your experiment. Use the table in Figure 4.3 as a model.

Experiment: Predictions: Results:	Control 1	Control 2	Variable
5 min			
10 min			
15 min			
20 min			

Figure 4.3 Sample table A table like this one will help you to record your predictions and results.

6. Predict what will happen when you conduct the experiment, and record this information in your journal.
   SAFETY: Put on your safety goggles and lab apron.

7. Set up and conduct your experiment.
   Make observations of your setup for as long as possible, and record all observations and results in your data table.

8. Wash your hands thoroughly before leaving the laboratory.

9. During the next class session, observe your setup again. Record your final observations and results in your data table.

10. Wash your hands thoroughly.

11. With your partner, develop possible explanations for your experimental results. Record your explanations in your journal.
    The essay Molecular Movement *(page E53) may help with this step.*

CHAPTER FOUR
The Internal Environment of Organisms

Analysis

Work with your partner to write a lab report about your experiment. Your lab report should include the following:

- a statement of the question you tested and of the results you predicted;

- a description of your methods (how you performed the experiment), including the materials that you used;

- a description of the results that you obtained, perhaps presented in a well-organized table; and

- an explanation of your experimental results that clearly states how they demonstrate the selective nature of a cell membrane and how the dialysis tubing setup serves as a model of a cell.

Background Information

Dialysis tubing is a man-made membrane, made of a thin, cellophane-like material. Microscopic pores in dialysis tubing allow molecules smaller than a certain size to pass through the membrane. The tubing that you have in your lab allows the passage of small molecules such as glucose but will not permit the passage of large molecules such as starch. You can make a bag from a piece of dialysis tubing by tying one of the ends with a piece of string. You can close off the bag completely by tying both ends of the tubing with string.

Glucose is a simple sugar that dissolves readily in water.

Glucose test strips indicate the presence of glucose in solution by changing color. (Your teacher will give you information on how to interpret the color change.)

Lugol's iodine solution is an indicator that changes color in the presence of starch. Use one drop of Lugol's iodine solution for every 1 mL of starch suspension; a blue-black color indicates the presence of starch.

WARNING: Lugol's iodine solution is a *poison* if ingested, is a *strong irritant*, and can stain clothing. Avoid skin/eye contact; do not ingest. If contact occurs, flush affected area with water for 15 minutes; rinse mouth with water; call the teacher immediately.

Starch is a large, complex molecule that forms a suspension in water. Starch turns blue-black in the presence of Lugol's iodine solution.

The size of a molecule is an important characteristic that partially governs how the molecule behaves. Chemists have shown that all the molecules of a given substance are the same size and that molecules of different substances can vary significantly in size. Measurements show that both iodine and water molecules are very small, that glucose molecules are considerably larger than iodine molecules and water molecules, and that starch molecules are much larger than glucose molecules.

REGULATING THE INTERNAL ENVIRONMENT

Shrinking cells and exploding cells dramatically illustrate the internal response of cells to external conditions. By now, you have begun to understand some of the processes that allow substances to move between internal and external environments. In addition, you have seen that the exchanges between compartments are influenced by the membranes that form boundaries around the contents of cells.

You might wonder whether the same processes of exchange, diffusion, and osmosis are important in large body systems as well. To begin to answer this question, consider two important compartments within the human body—the circulatory system and the urinary system. In this activity you will see how these two systems help regulate the internal environment in humans.

Process and Procedures

1. Imagine that you are a red blood cell.

    a. Describe your journey as you travel from the left little toe through the heart and on to the right big toe.

       You will know that you have described your journey adequately if you have included

       - *capillaries,*

       - *veins,*

       - *arteries,*

       - *the four chambers of the heart, and*

       - *the lungs.*

    b. Describe how you, as a red blood cell, are involved in exchanges in each of the following places:

       - the tissues of the toe,

       - the kidneys,

       - the lungs,

       - the liver, and

       - the intestines.

       You might set up a table in your journal. Include information that you can gather at this time, and then add to your table as you learn more in this and the next chapter.

       The videodisc segment The Circulatory System *and the essay* Making Exchanges throughout the Body *on page E56 will help you with these tasks.*

       Your descriptions might be written or illustrated, or you might use a combination of written and graphic work.

2. To expand your understanding of how various systems help regulate the internal environment of the human body, watch the videodisc segment *Regulation in the Urinary System.* Respond to the video questions in your journal.

    Read the essay Disposing of Wastes *(page E58) as a background resource before you watch the videodisc segment. Continue to refer to it, as necessary, to complete this step. You may want to take notes as you watch or read.*

Analysis

Work with your team of four to complete the following task. Divide your team into two pairs, and decide which pair will develop a response to Question 1 and which pair will develop a response to Question 2. Include a supporting illustration in your response. As you answer your question, refer to the information that you developed in Steps 1 and 2 of *Process and Procedures*. When both pairs are finished, present your responses to each other.

1. How does the circulatory system help regulate the internal environment of the body? How does the urinary system influence the work of the circulatory system?

2. How does the urinary system help regulate the internal environment of the body? How does the circulatory system influence the work of the urinary system?

CAN YOU STAND THE HEAT—AGAIN?

In this activity you will return to the opening story, *A Pause That Refreshes?* and examine Ralph's situation. You then will evaluate your understanding of how internal balance is maintained when the external environment places significant stress on it.

Process and Procedures

1. Read to yourself the following conclusion to Ralph's story, *A Pause That Refreshes?*

   "You drank what?" his father exclaimed when Ralph reported that he had returned to the house and had a couple of beers. "You were seriously dehydrated. What did you learn about dehydration when you took biology in high school? Didn't you learn that alcohol is a diuretic?"

   "A *diuretic*! What is that? And how could I be seriously dehydrated?" Ralph responded. "All I know is that the beer was good and cold and that I had a glass of water as well!" Ralph felt considerably better now that he had drunk more water and cooled down a bit, but he was irritated that his father had questioned his thinking.

2. To learn more about Ralph's condition, examine the data in Figures 4.4 and 4.5, and study the *Additional Information*.

Change in Ralph's Body Mass

Figure 4.4 **Change in Ralph's Body Mass**

	Before Working (a.m.)	After Completing Work (p.m.)
Ralph's body mass (kg)	77.25	73.55*

* This measurement of Ralph's mass was taken **before** he drank more water at the end of the story.

Tracking Fluids

Fluids	While Working	During Afternoon Break	After Completing Work
In	.95 L juice and water	.71 L beer + .24 L water	.78 L water
Out	constant perspiration	.91 L urine	

Figure 4.5 **Tracking Fluids**

Additional Information

If athletes, or other people performing strenuous exercise or exposed to extreme heat, lose 3 percent to 8 percent of body mass in the form of fluids, they place themselves in a condition of dehydration. This dehydration can be quite serious if the person is unable to replace the lost fluids within a short period of time.

A **diuretic** is a substance that causes the membranes in the kidney to remove from the body more water than is taken in. Common diuretics are caffeine (found in coffee, tea, chocolate, and cola beverages) and alcohol. These diuretics increase the loss of water, resulting in an increased concentration of solutes, such as sodium ions and potassium ions. These increases can seriously disrupt the body's internal balance. For example, a high concentration of sodium ions can result in high blood pressure. Diuretic drugs that physicians prescribe to adjust blood pressure increase both the loss of water and of sodium, thus avoiding problems that result from altered solute concentrations.

Analysis

Answer the following questions to explain how Ralph could be dehydrated. Record your responses in your journal.

1. What percentage of body mass did Ralph lose in the form of fluids? Do you think this represents a serious condition? Explain your answer.

2. In what ways was Ralph's body attempting to maintain an internal balance in spite of the changing nature of his external environment?

3. How did Ralph become dehydrated even though he drank 1.9 L of liquid?

4. How could Ralph's water loss have caused him to become dizzy?

5. Do you think that Ralph would have been better off to replenish his fluids by drinking cola or iced tea rather than beer? Explain.

6. How could Ralph have achieved a better balance between his internal and his external environment?

5
MAINTAINING BALANCE IN ORGANISMS

What do swelling and shrinking eggs and a thirsty college graduate have in common? They all have the ability to communicate information about external conditions to an internal environment. Of course, in eggs this communication is simply a matter of chemistry. But communication in thirsty college students and many other organisms is generally much more complicated. This is because complex living systems rely on a broad range of behavioral responses, in addition to chemical and physiological responses, to help them regulate internal conditions.

Luckily, humans and other living systems do not have to consciously plan and execute a response to every change in the environment. Most regulatory processes occur *automatically* through a combination of physiology and behavior. Well-coordinated and rapid changes in our organ systems mean that we usually can take the internal conditions of our bodies for granted. Understanding and appreciating these conditions and the interactions that keep them within normal limits, however, can contribute to a longer, healthier life. In this chapter you will have an opportunity to consider some of the ways in which your body maintains its internal balance as you perform some of the basic activities of life.

ACTIVITIES

Engage / **Explore**	The Body Responds
Explore	What's Your Temperature Now?
Explain	Stepping Up the Pace
Elaborate	On a Scale of 0 to 14
Elaborate	How Do They Stay So Cool?
Evaluate	Homeostasis in Organism X

THE BODY RESPONDS

The bus is just starting to pull away as you turn the corner so you break into a sprint to try to catch it. After 10 or 12 seconds the driver finally sees you in the side mirror and stops to wait for you. Out of breath and red-faced, you climb onto the bus and collapse in a seat to the sarcastic cheers of your friends. After a minute or so your breathing has returned to normal, and you are talking with your friend about the party last Friday night.

Figure 5.1 Hurry, hurry, hurry
How can your body respond so quickly to changing conditions?

This brief sequence of events probably seems trivial because your body responds to environmental factors all the time. However, the changes you felt while running for the bus and recovering from the run actually required your body to coordinate a tremendous number of responses. Think for a moment about how your rapid breathing slowed to normal. For that matter, consider *why* it sped up in the first place.

Before you plunge into the details about how such *regulation* of response can occur, let's examine a range of possible responses. In this activity you will view a set of simple human activities and try to match them with their internal responses. When you think about the surprisingly complicated biological responses, you may be grateful that you usually can take your internal conditions for granted!

Process and Procedures

1. As you watch the videodisc segment *Just a Body Responding*, think about similar experiences that you have had. Your teacher will show the segment a second time, stopping the videodisc briefly after each scene to allow you to think of a short descriptive title for that scene. Record the scene number and your proposed title in your journal.

   These titles can be funny or serious, as you choose. Make sure that each title is descriptive enough to help you remember what happened in each scene.

2. Examine the *Internal Events* copymaster that your teacher has given you. Each paragraph describes a set of events that might have gone on inside someone's body in response to one of the external situations that you

have just watched on the videodisc. Make a table in your journal that allows you to record answers to each of the following questions:

a. Which videodisc scene do you think matches each description?

b. What information from each description is most helpful to you in suggesting each match?

Analysis

Participate in a class discussion of the following questions:

1. Did the internal responses benefit the people involved? Explain your answer.

2. Can you identify the *behavioral* and the *physiological* components of each response? (The word *physiological* pertains to internal biological and chemical activities.)

WHAT'S YOUR TEMPERATURE NOW?

Explore

The characters in the videodisc segment *Just a Body Responding* each experienced a different set of internal changes, which occurred in response to different external demands. Did each character's body respond in a random fashion, sensing only a change in external conditions but not a *particular* change? No, each body seemed to respond in a manner that was *appropriate* for each situation. For example, the dilated blood vessels in the stomach of the person napping increased stomach circulation, a response that speeds digestion. Likewise, the constricted vessels in the stomach of the frightened person reduced stomach circulation, a response that diverts blood to other areas of the body, in this case the muscles of the arms and legs. Understanding how the body responds appropriately to change requires an understanding of the automatic physiological processes that detect and respond to change.

In this activity you will explore the relationship between core (or internal) and surface temperatures under different conditions. This exploration will help you understand how the body automatically accommodates *specific* external changes with *specific* internal responses.

Materials (per team of 4; 2 teams per LEAP-System™)

Macintosh or IBM-compatible computer
printer
LEAP-System™
3 thermistors
graph paper
materials to carry out the experiment that you design

Process and Procedures

1. Together with your teammates, begin designing an experiment to test the following hypothesis:

   If the human body can regulate its internal temperature automatically, this internal temperature will stay fairly constant when the skin is cooled.

   Steps 2, 3, and 4 will help you to design your experiment.

2. To design an appropriate experiment, you and your teammates first must think about the following question:

   How can we measure and compare changes in internal body temperature, surface body temperature, and environmental temperature, all at the same time?

   For this experiment, assume that the temperature of the inside of the elbow joint represents the internal, or core, body temperature even though it is a few degrees cooler than actual core temperature. The temperature of the index finger accurately reflects the surface temperature of the body.

3. With your teammates develop an outline of your experimental design, and record it in your journal. You will know your experimental design is complete when you have included the following:

   - a measure of the initial temperatures of the body core, body surface, and environment;

   - a setup that cools the temperature of the body surface in a safe manner (for instance, you must **not** cool the body surface for more than 30 seconds at one time because excessive cold can damage the skin);

   - a record of the temperatures of the body core, the body surface, and the environment at regular intervals until the surface temperature warms to its initial temperature;

   - any of the materials that your teacher provides; and

   - your teacher's approval of your experimental design before you conduct the experiment.

   When the experiment has been completed, each team member should obtain a complete set of data from the recorder and the observer and then copy these results into a data table in his or her journal.

4. To conduct a precise experiment, assign the following roles and related responsibilities among your teammates:

   Test Subject: Describes his or her feelings and sensations throughout the experiment.

   This person's body temperature is monitored during the experiment.

   Timer: Keeps track of time as the experiment proceeds.

   The timer also can operate the computer.

   Observer: Records the time, observations of the test subject, and any comments the test subject makes.

   Observations might include color of the skin or changes in the test subject's behavior.

Recorder: Records temperatures.

Records temperature data in a data table in his or her journal.

Scientists often divide duties so that a large or complex project can be managed efficiently. Each scientist's individual role and responsibility, however, is crucial to the overall results. As you conduct this experiment, pay attention to your individual responsibilities as well as the goals of the team.

Figure 5.2 This team is sharing the responsibilities required to conduct its experiment.

5. Use *Protocol for Measuring Temperature* (Figure 5.3) to conduct your experiment.

6. Analyze the results of your experiment by constructing a line graph that illustrates how the test subject's core and surface temperatures changed throughout the experiment. Do this as a team.

   Plot the temperature on the y axis and the time on the x axis. Use dashed and solid lines or different colored lines for the core temperature and the surface temperature data. If you need assistance with graphing techniques refer to Appendix B, Techniques.

7. To complete your experiment, answer the following questions with your teammates, and record your answers in your journal.

   a. How do the data from the temperature readings compare with observations made by the observer?

   b. What changes did you observe in the core and surface temperatures of the test subject?

   c. Explain whether your experimental results support or disprove the initial hypothesis.

   In other words, can the human body regulate the temperature of internal and surface body regions in response to cold? Does this regulation occur automatically?

Protocol for Measuring Temperature

1. Using the procedures explained in ALPHAEXP, run experiment HUMAN01.*

   This experiment is located in the directory HUMBIO\ (IBM) and in the BSCS Human Biology Folder (Macintosh).
   IBM users, run HUMAN01T for two teams.

2. Set the experiment to SAVE the data. When asked to reinitialize, reply yes (Y).

3. Identify your team's thermistors.

   Team A, use the thermistors from Inputs 1–3; Team B, use the thermistors from Inputs 4–6. Good contact is essential for accurate readings.

4. Use the following steps to measure temperatures:

Timer/Operator:	Take the experiment off WAIT and MARK the graph.
Recorder:	Record the starting time, the MARK number, and the current temperatures from the numerical data beside the graph for core, surface, and environment.
Recorder:	Record time and temperatures at regular intervals.
Timer/Operator:	When you have collected all of your data, MARK the graph and put the experiment on WAIT.
Recorder:	Record the ending time and MARK number.

   It will greatly simplify taking measurements if both teams coordinate their steps throughout the activity. The timer/operator from one team might concentrate on timing, and the timer/operator from the other team might operate the computer or the mouse.

5. On IBM computers, the graph lines for the two teams are likely to overlap, but you can review each team's saved data separately as follows:

   - Team A: Run HUMAN01A
     Team B: Run HUMAN01B

   - This question will appear:
        Filename is HUMAN01.OUT ok?
        Press **Y**.

6. When both teams have completed their experiments, QUIT HUMAN01.

Figure 5.3 **Protocol for Measuring Temperature**

Analysis

Complete the following tasks on your own, and record your answers in your journal.

The essay Homeostasis *(page E60) may help you with these tasks.*

1. How might the changes that you observed in the core and surface temperatures benefit the test subject and help to maintain homeostasis?

2. In addition to temperature, name two other physiological processes in the human body that automatically adjust to external changes. Give examples of the types of adjustments that the body makes in each case.

3. Explain whether regulation to maintain homeostasis is a random or specific activity of the body.

4. Write three sentences that describe the importance of your role in this experiment.

STEPPING UP THE PACE

The average temperature of human beings is 37°C (98.6°F). That temperature fluctuates over a small range, but in healthy people it stays fairly constant and predictable. This balance is maintained by the automatic responses of the circulatory system. The circulatory system reponds specifically to specific changes in temperature. Perhaps you observed some of these responses during *What's Your Temperature Now?*

Did any of your observations suggest that organ systems other than the circulatory system might be involved in regulating temperature? The circulatory system is just one of several different systems that all work together. Most of those other systems, however, are difficult to observe because their responses occur inside the body and are not visible externally. The role of these other systems is quite important because homeostasis most often is maintained through the *interaction* of many systems rather than through the action of just one. To understand homeostasis fully, you must gain an appreciation for how changes in one system affect the behavior of another system.

One way to observe the interaction of systems is to consider systems for which internal changes have measurable external effects. As you have seen, you can detect change in the circulatory system by measuring temperature. Can you think of any other external methods of detecting change in circulation? Although external measures provide only an indirect view of the internal environment of the body, they are valuable in helping to illustrate the complex interactions of internal organ systems. In this activity you will examine physical exercise as a way of helping you explain how organ systems in the human body interact to maintain homeostasis.

Materials (per team of 5)

Macintosh or IBM-compatible computer
LEAP-System™
breathing rate sensor
graph paper

Process and Procedures

1. Complete the following in your journal as you work individually:

    a. How do complex, multicellular organisms interact with the external environment to satisfy their homeostatic needs?

    b. What role do the endocrine and nervous systems play in regulating homeostasis?

    c. Explain the mechanisms by which the human body determines how to adjust internal conditions in response to change.

    The essay Careful Coordination *on page E62 contains information that you may find useful.*

2. Make a prediction about how heart rate, breathing rate, and exercise are related.

3. As a team examine the predictions that each of your teammates made in Step 2, and develop one hypothesis that you think offers the best explanation of the relationship among heart rate, breathing rate, and exercise. You will conduct an experiment to evaluate your hypothesis.

A hypothesis must be testable. One way to test a hypothesis is to see how well it accounts for new observations. You also can test a hypothesis by making a prediction and then collecting data to determine whether or not the prediction was accurate. For this type of hypothesis, it is useful to state the hypothesis as an if-then statement such as "If milk gets hotter than 80°C, then it will boil."

4. Rewrite your hypothesis as an if-then statement.

5. As a team read *Protocol for Conducting a Step Test* (Figure 5.4), and discuss how you would use this exercise protocol to test your hypothesis/if-then statement.

Figure 5.4 **Protocol for Conducting a Step Test**

Protocol for Conducting a Step Test

1. In your journal construct a data table for the experiment.

   Include columns for each of the exercise rates and resting conditions that you decide to use and rows for pulse rate, breathing rate, and any other conditions. The exercise rate is the speed at which the test subject steps.

2. As a team decide who will be the test subject, who will be the timer, who will count and record the test subject's breathing rate, and who will count and record the test subject's pulse.

   You will use a step test to measure pulse rates and breathing rates for the exercise rates that you determine in your experimental design. You might begin with 1 step every 5 seconds. This activity will succeed only with the cooperation and support of all team members.

3. With the test subject sitting quietly in a chair, count for 30 seconds the test subject's resting pulse rate and breathing rate. Record the resting pulse rate and breathing rate per minute in your data table.

   To record the pulse rate and breathing rate do the following:

   a. Using the procedures explained in ALPHAEXP, run experiment HUMAN02. This experiment is located in the directory HUMBIO\ (IBM) and in the BSCS Human Biology Folder (Macintosh).

   b. Fasten the breathing rate sensor of the LEAP-System™ around the test subject's chest so that it is snug but not uncomfortably tight.

   c. Complete the following tasks:

Computer Operator:	Set the experiment to SAVE the data.
Computer Operator:	Take the experiment off WAIT, and MARK the graph.
Timer:	Start timing the pulse and breathing rate counters for 30 seconds.
Pulse Counter:	Locate the test subject's pulse as shown in Figure 5.5. Count the test subject's pulse for 30 seconds and record the rate per minute in the data table. Also record the MARK number. *To calculate the total pulse rate per minute, multiply the count by two.*
Breathing Counter:	At the end of 30 seconds, count the test subject's breathing rate from the computer screen, and record the rate per minute in the data table.
Computer Operator:	Put the experiment on WAIT.

Figure 5.5 **Checking pulse rate**

4. Perform the step test by following this process. This example assumes an exercise rate of 1 step every 5 seconds.

Timer:	Tell the test subject when to start, and then quietly call out "step" every 5 seconds for 1 minute.
Test Subject:	When the timer says "step," step up onto the platform with one foot, then with the other foot; next step down with the first foot, then with the other foot.
Timer:	Say "stop" after 1 minute.
Computer Operator:	Take the experiment off WAIT, and MARK the graph.
Timer:	Begin timing for the pulse counter for 30 seconds.
Computer Operator:	MARK the graph at the end of 30 seconds.
Pulse Counter:	Count the test subject's pulse for 30 seconds immediately after exercise, and record the rate per minute. Record the beginning and ending MARK numbers.
Breathing Counter:	Count the test subject's breathing rate from the screen, and record the rate per minute. Record the beginning and ending MARK numbers.

5. Repeat Step 4 to measure the other rates according to your experimental design.

6. Based on this protocol, identify the variables in your experiment, and decide on an appropriate control. Record the variables and the control in your journal.

Variables might include exercise rate, step height, and test subject's weight. Your control should allow you to test one variable at a time.

7. Design an experiment to test your hypothesis. Your experiment needs to meet all of the criteria listed below. Your experiment must

- be safe,

- be manageable in a classroom setting and appropriate for the length of the class period,

- use the materials available, and

- allow each team member to handle materials and record data.

8. Have your teacher approve your design; then begin your test.

   As you proceed with your test, you may need to modify the design of your experiment. If this happens, record the changes in your journal.

9. When you have completed the step test, share your data with your teammates, and then complete your own data table.

10. In your team, construct a graph that illustrates how pulse rate and breathing rate changed from resting state through increasing rates of exercise. Copy the graph into your own journal.

11. When you have finished your graphs, analyze your experiment by developing a conclusion that describes the interaction that occurs between the circulatory and gas exchange systems during exercise.

    For a strong conclusion, analyze the data that you collected, including the graph, and relate these data to the hypothesis that you tested.

Analysis

Complete the following tasks individually. Write your explanations in your journal.

The essay The Breath of Life *(page E68) may help you with your answers.*

1. Use the concept of interacting systems to explain how oxygen from the atmosphere is able to reach cells deep within the body.

2. Explain how the acidity of blood controls breathing rate.

ON A SCALE OF 0 TO 14

Elaborate

In *Stepping Up the Pace*, you learned that several organ systems can interact to maintain homeostasis. In addition, you began to explore a chemical mechanism—the amount of acid in the blood—that links the circulatory system to the gas exchange system. The amount of acid in the blood is related to the amount of carbon dioxide in the cells of the body. By sensing and responding to these levels, the body can maintain a proper balance of oxygen and carbon dioxide in its internal environment. In this case the level of acidity is a regulatory signal to which the body responds when maintaining the balance between two large organ systems. The level of acidity itself, the condition known as **pH**, is also an internal condition that must remain balanced in humans and many other organisms.

In this activity you will elaborate on your understanding of physiological regulation by addressing the question, How do living systems maintain internal conditions of pH? You will have the opportunity to compare the responses of several materials to the addition of an acid and a base. Among these materials are homogenates (mixtures that are uniform throughout) made from living cells. These homogenates act as models of the internal environment of living systems, and they will allow you to investigate how the level of acidity inside cells can be maintained within certain limits.

Materials (per team of 3; 2 or 3 teams per LEAP-System™)

2 pairs of safety goggles
2 lab aprons
50-mL beaker
50-mL graduated cylinder
petri dish half
Macintosh or IBM-compatible computer
LEAP-System™
pH probe
forceps
4 sheets of graph paper (optional)
jar of tap water for storing pH probes
tap water
dropping bottle of 0.1M HCl (acid)
dropping bottle of 0.1M NaOH (base)
50 mL of liver or potato homogenate
materials that your teacher provides

Process and Procedures
Part A   pH Is Everywhere

1. With your partner determine the pH of the available solutions by using a pH probe with the LEAP-System™. Record these pH readings in a data table in your journal.

    See Protocol for Measuring pH *(Figure 5.6).*

2. Read the following information to learn more about pH:

    Recall from the Engage Section that pH is a measure of how acidic or basic a solution is. A pH of 7 represents a neutral solution that is neither acidic nor basic. The pH scale ranges from 0 (very acidic) to 14 (very basic). The scale is logarithmic; each difference of 1 pH unit means a tenfold difference in acidity. For example, a solution with a pH of 8 is 10 times more acidic than a solution with a pH of 9. In other words, a pH of 8 is 10 times less basic than a pH of 9. Strongly acidic and basic solutions can be quite harmful to the external environment of living systems, and strong acids and bases can burn skin badly. Even minor imbalances in internal pH, however, can disrupt the normal regulation of cells.

Protocol for Measuring pH

1. Using the procedures explained in ALPHAEXP, run experiment HUMAN03.*

   This experiment is located in the directory HUMBIO\ (IBM) and in the BSCS Human Biology Folder (Macintosh).
   *IBM users, run HUMAN03T for two teams or HUMAN03X for three teams.

2. Identify your team's pH probe.

   Team A, use the probe from Input 1; Team B, from Input 2; and Team C, from Input 3.

3. Remove your team's pH probe from the storage jar of tap water.

   When not using the probes, return them to the storage jar; do not allow the tips of the probes to dry out.

4. Place your team's pH probe in the solution to be tested and determine the pH as follows:

   It will greatly simplify taking measurements if all teams coordinate their steps throughout the activity.

Computer Operator:	When you are ready to begin, take the experiment off WAIT and MARK the graph.
Recorder:	Record your team's MARK number.
Computer Operator:	Take measurements for 5 seconds to allow the readings to stabilize.
Computer Operator:	When all teams have taken measurements, put the experiment on WAIT.
Recorder:	In the data table, record the current pH from the numeric data displayed near the graph.

5. When all teams have completed the activity, QUIT HUMAN03.

Figure 5.6 **Protocol for Measuring pH**

Figure 5.7 **Household products** Do you know which products in your home have a pH low enough or high enough to hurt you?

3. Discuss the following questions with your class:

    a. Does the pH of any of the solutions that you tested surprise you?

    b. Do you think any of the solutions that you tested could be harmful to the pH balance of your organ systems?

Part B Regulating pH

SAFETY: Put on your safety goggles and lab apron.

1. To investigate how pH is regulated by living cells, you and your partner will compare how water and one type of cell homogenate respond to the addition of acids and bases. Decide who will operate the computer, who will perform the tests with the acid, and who will perform the tests with the base.

When you are not performing the experiment, you will record the data.

2. Use a table similar to the one labeled *pH Changes* in Figure 5.8 to prepare a data table in your journal.

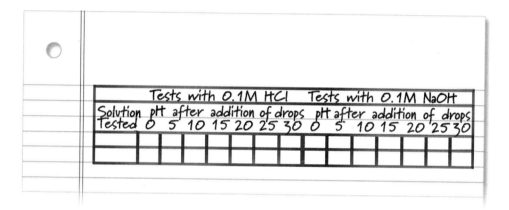

Figure 5.8 **pH changes** You will need a table like this one to record the results of your tests.

3. Then conduct your experiment by following the steps below:

    a. Pour 25 mL of tap water into a 50-mL beaker.

    b. Determine the initial pH of the solution as you did in Part A, Step 1.

    c. Record the initial pH in your data table under the column labeled *0 Drops*.

    d. Add 0.1M HCl to the beaker one drop at a time until you have added 5 drops. Gently swirl the mixture after each drop.

    You are making the water more acidic.

CAUTION: 0.1M HCl is a *mild irritant.* Avoid skin/eye contact; do not ingest. If contact occurs, flush affected area with water for 15 minutes; rinse mouth with water; call the teacher.

    e. Determine the pH of the solution and record this reading in your data table.

    f. Repeat Steps (d) and (e) until you have added a total of 30 drops of acid.

    You should have a total of seven pH readings.

CHAPTER FIVE
Maintaining Balance
in Organisms

105

g. Discard the mixture, and rinse the beaker thoroughly.

h. Repeat Steps (a) through (g), but this time add 0.1M NaOH drop by drop instead of HCl.

You are making the water more basic.

CAUTION: 0.1M NaOH is a *mild irritant.* Avoid skin/eye contact; do not ingest. If contact occurs, flush affected area with water for 15 minutes; rinse mouth with water; call the teacher.

i. Repeat Steps (a) through (h), but instead of adding tap water in Step (a), pour 25 mL of the homogenate into the beaker.

You will be observing how the homogenate responds to different pH levels.

You will use either liver homogenate or potato homogenate. Your teacher made these homogenates by blending pieces of liver or potato at high speed to break open the cells and release their contents. Remember that these homogenates will act as models of the internal environment of living systems.

j. Wash your hands thoroughly.

4. On graph paper or in your journal, draw one graph for your tests with water and a second graph for your tests with the homogenate. Each graph should show the relationship between pH and the number of drops of an acid or a base.

Plot two lines on each graph—a solid line to plot the changes with acid and a dashed line to plot the changes with base. Plot the pH on the y axis and the number of drops on the x axis.

5. Discuss the following question with your partner, and record an explanation in your journal:

Does the homogenate respond to the addition of acid and base in a manner that is similar to or different from the way water responded?

6. Join a team that tested the other homogenate, and compare data.

7. Discuss the following questions with your partner, and record the answers in your journal.

The information in Background on Buffers *may help you with your answers.*

a. Based on your results, do you think potato and liver cells are buffered when compared with water? Explain your answer.

b. Is it likely that all living systems contain buffers? Why or why not?

c. At what pH do you think the inside of liver and potato cells functions best? On what do you base this inference?

Analysis

Answer the following questions in your journal. Be prepared to contribute your ideas to a class discussion.

1. Based on your data from the experiment in Part B, how might a buffer help maintain homeostasis? Explain your answer.

2. Many manufacturers claim that their health-care and hair-care products are pH balanced or buffered. How would you test their claims?

Background on Buffers

Proper pH balance is important because different chemical and metabolic processes in living organisms work best under different pH conditions. In humans, for example, digestion in the stomach requires a low pH, but the functions of blood require a nearly neutral pH, not varying much from pH 7.2 to 7.4. In fact, each cell, tissue, and organ has a characteristic pH that is important for homeostasis. Yet many things that living organisms encounter have a pH that differs significantly from their internal pH. For example, much of the food that you eat has a pH that is different from that of your stomach. How is it, then, that living systems maintain a relatively narrow range of pH in their internal environment despite changes in the external environment?

A **buffer** maintains the pH of a solution within a narrow range of values even when external conditions threaten to change pH. In effect, this means that small additions of acid or base to a buffered solution do not cause a change in pH. Cells have physiological buffers that help maintain the cell's characteristic pH; this protects the internal environment. Different types of buffers are effective in maintaining pH within different ranges. One type of buffer might keep the pH of the inside of one cell in a slightly acidic range, and a different type might keep the pH of the inside of another cell in a neutral range.

HOW DO THEY STAY SO COOL?

Elaborate

Animals regulate their internal conditions in a variety of ways. What methods of regulation have you experienced or observed so far in class or in the world around you? Do you have a dog that pants on hot days? Have you seen a bird fluff up its feathers when the temperature drops? These are examples of temperature regulation through behavior.

During the heat of summer in the Arizona desert, a Gila monster (*Heloderma suspectum*) like the one in Figure 5.9 will wait beneath the cool shade of a rock. Other lizards dart about in the undergrowth. These reptiles survive in temperatures that would cause most humans to wilt in their tracks.

Figure 5.9 Gila monster Gila monsters (*Heloderma suspectum*) can grow up to 60 cm long. This species of lizard has adapted to the extreme temperatures of a desert environment. What behavior do you see here that demonstrates that adaptation?

(Temperatures at the surface of the soil can exceed 70°C, or 158°F.) Yet the reptiles survive even though they have no sweat glands. What are the mechanisms that allow these reptiles to survive such high temperatures? In this activity you will gather information that allows you to compare and contrast the mechanisms of temperature regulation in lizards and two other animals with those of the human body.

Process and Procedures

In this activity you will examine the question, How does the behavior of animals help them to maintain homeostasis? You will investigate this question by gathering information that will help you develop an explanation.

1. Observe the videodisc segment *Temperature Regulation in Animals*. Record your observations in a data table, and record in your journal your responses to questions in the videodisc segment.

2. In the last column of your table, record the behaviors of a living animal that you can observe directly.

    Focus your observations on the behaviors that indicate how the animal is regulating its temperature or responding to changes in the external temperature.

    Use the information in the essay Behavior and Homeostasis *(on page E70) and other available resources to describe an animal's behavioral responses to changes in external (environmental) temperature.*

Analysis

Use your notes, data, and the information in the essay *Behavior and Homeostasis* (on page E70) to develop responses to the following questions. Record your responses in your journal.

1. What behaviors related to temperature regulation do humans and lizards have in common? How do their behaviors differ?

2. What similarities and differences do you see between the way dogs regulate temperature and the way humans regulate temperature?

    Think about the physiology of dogs and of humans as each responds to temperature extremes.

3. In addition to temperature, describe two other homeostatic processes that you or other animals regulate through behavior. For each process identify the feedback and response.

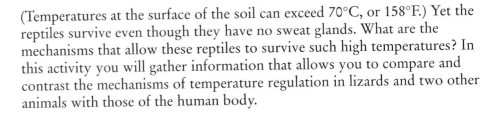

Evaluate HOMEOSTASIS IN ORGANISM X

Throughout Chapters 4 and 5 you have been exploring the concept of homeostasis and the complex systems and processes that are involved in helping living systems maintain an internal condition of dynamic balance even while external conditions change dramatically.

In this activity you will return to the description of the organism you "discovered" in *First Encounter with Organism X* from Chapter 3. In doing so you will have the opportunity to evaluate what you have learned in this chapter. By applying what you have learned about homeostasis to Organism X, you can reflect on some of the ways in which organisms maintain their internal environment when the external environment changes.

Materials (per person)

descriptions and diagrams of Organism X from Chapter 3

Process and Procedures

1. Work individually to answer the following questions in your journal:

   a. Based on what you have learned in this chapter and the preceding one, how would you describe *homeostasis*?

   b. Why is it important for organisms to maintain homeostasis?

2. Recall Organism X that you described in Chapter 3 and the environment in which your organism lives. List in your journal the specific environmental stress factors that your organism most likely would be subjected to in that environment.

3. Use the medium of your choice to demonstrate your response to this question:

   In what ways does your organism maintain its internal environment considering the external stress factors that you listed in Step 2?

   You will know that you have adequately addressed the question if your response includes an explanation, illustration, or demonstration of the following:

   • how your organism regulates internal temperature in response to heat and cold,

   • how your organism regulates water and salt balance,

   • how your organism deals with at least two other disparities between the external and internal environment, and

   • how at least two key systems in your organism interact to adjust internal conditions.

   A medium is the method or materials you use to convey your response. You could write a response in your journal, compose a song, create a poem, paint a picture, or develop a collage or mobile.

Analysis

Evaluate yourself according to the outcomes and indicators that your teacher has distributed. Your teacher will use these guidelines to evaluate you as well.

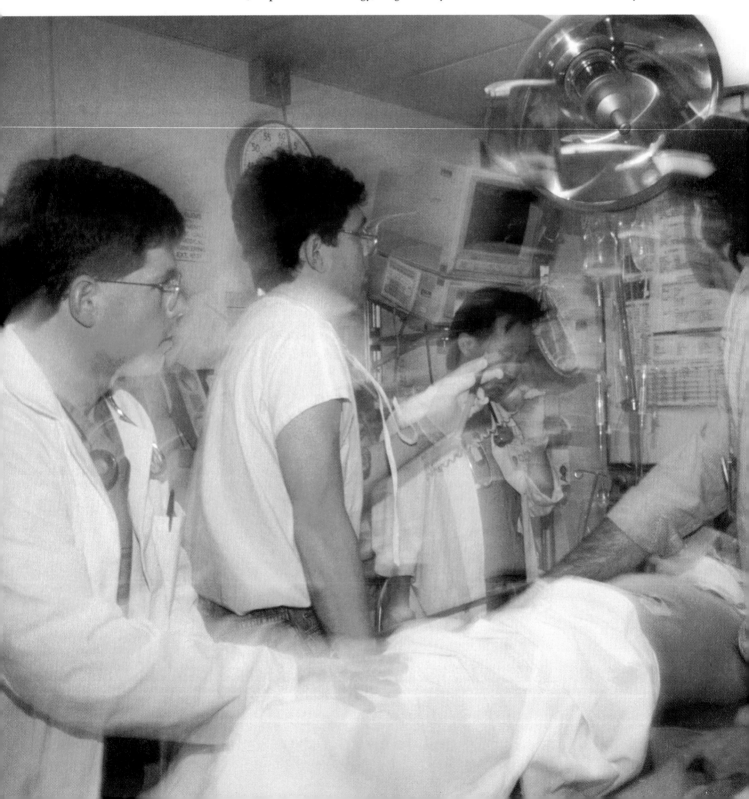

6 HUMAN HOMEOSTASIS: HEALTH AND DISEASE

The processes of homeostasis allow most people to enjoy relatively long and healthy lives. For these people the common cold and other minor illnesses are only temporary disruptions, and their bodies are quite capable of recovering through normal regulatory processes. Occasionally disruptions can affect our homeostasis and, therefore, our health in more serious and sometimes permanent ways. In these cases medical care often is required to help the body stay within the internal limits necessary for life.

In this chapter you will examine what happens when homeostasis is disrupted in various ways. You will have the opportunity to explore the effects of severe dehydration, allergic reactions, trauma, and a variety of other emergency health care situations. You also will see how medical technology can help compensate for disruptions in the body's homeostasis. A simulation will help you to understand how the immune system helps protect the body from serious disruptions and maintains homeostasis. Then you will examine how certain personal behaviors can increase or reduce your risk of disrupting your body's homeostasis. After you have completed these activities, you will evaluate your understanding of disruption by developing a mock proposal that requests funds for a worthy health care initiative of your choice.

ANDREAE VESALII BRVXELLENSIS
QVINTA
MVSCVLO.
RVM TABV.
LA.

ACTIVITIES

Engage	Pushing the Limits
Explore Explain	Hospital Triage
Explain	Self Defense!
Explain (Alternate)	Immunity Simulation
Explain Elaborate	What's the Risk?
Evaluate	Health Care Proposal

PUSHING THE LIMITS

Engage

Just how much stress can the human body endure and still function? As you work through this activity, you will examine the limits of the human body. In the activities that follow, you will use your knowledge of the mechanisms for maintaining an internal balance to develop a more complete understanding about the limits of homeostasis.

Process and Procedures

1. Read the story *A Sweltering Experience* (page 113). As you read, record in your journal details about the physical condition of the two hikers at the time of their rescue.

2. With your partner complete the following tasks. Write your responses in your journal.

    a. List the human body systems and mechanisms of regulation that would be involved in restoring internal balance in the hikers.

        Use your understanding of homeostatic body systems, which you developed in Chapters 4 and 5, to complete your list.

    b. List three situations that are not discussed in the story that also could disrupt homeostasis severely.

Analysis

Respond in your journal to the following.

Information in the essay Beyond the Limits *(page E73) may help you with your answers.*

1. List two mild stressors and two severe stressors. Describe the effects that each stressor would have on the human body.

2. From your personal experience, describe two examples of disruptions of homeostasis in nonhuman organisms.

HOSPITAL TRIAGE

Explore — Explain

You and your partners are a team of physicians in the emergency room of Desert Metropolitan Hospital. It's about 8:30 p.m. on a Friday in July. You have been busy, as you usually are on summer weekends. Unfortunately you are about to become even busier. You have just received a flurry of calls from local paramedics, and now you and your colleagues are preparing for an influx of patients who are suffering from a variety of illnesses and trauma injuries. This influx includes a flight-for-life helicopter carrying the two canyon hikers discussed in *A Sweltering Experience.*

Because there are many patients and only a handful of doctors, you cannot treat every patient at the same time. Your first task is to assess the severity of each patient's condition and establish a "priority for treatment," which is called **triage**. You will need to use everything you know about the body's normal homeostatic mechanisms to complete this triage.

A SWELTERING EXPERIENCE

Their descent into the canyon that morning had been easy and enjoyable. Many parts of the trail were steep, but the hikers had gravity to help them down. Nine miles went by quickly. All Monique and her father, Nelson, had to do on their first desert adventure was to watch their footing and enjoy the constantly changing views of the vast canyon.

They were surprised, however, at how much warmer it was deep in the canyon than it had been at the top. They soon began to feel so warm that they shed their jackets and rolled up their sleeves. Eventually, they became so hot and uncomfortable that they decided to turn around early, before they reached the bottom.

Now that gravity was their enemy, the trail seemed to stretch endlessly ahead of them, and their progress was very slow. Each mile seemed to drag on forever. The heat of the day had dried the trail, and the air was dusty. Both Monique and her father stopped frequently to mop their flushed faces and sweaty brows. They had brought only one canteen, and they regretted how freely they had drunk the water on the way down. They now were very thirsty, but they had to ration the last few mouthfuls of water for the slow trip up the canyon.

By mid-afternoon their conversation had died away, the water was gone, and both hikers realized that they were in over their heads. Their legs ached, and they were out of breath. Suddenly Monique gave a cry of pain and slapped her leg. She had been stung by a bee.

Within minutes Monique's leg began to swell near the sting, making walking very painful. Still they had to go almost a third of the distance before they reached the top of the canyon and medical help. Nelson became frantic. He decided that Monique should stay in a small shady spot at the next switchback while he tried to hurry up the trail for help.

Nearly two hours passed before Nelson reached the canyon rim, and by then he was suffering terribly from thirst. His body was so dehydrated that it could no longer cool itself adequately by perspiration. He had an excruciating headache, he felt dizzy and nauseous, and he was too confused to concentrate on what to do next.

When Nelson finally found medical help at the backcountry office a short while later, he was breathing loudly and rapidly through his mouth. A member of the medical team touched Nelson's hand and noticed that his skin was dry and hot to the touch. Yet his face was not very flushed. The paramedic knew that Monique was not the only one in trouble; Nelson was dangerously ill himself.

The rescue squad reached Monique three hours after she had been stung by the bee. She was not in good shape. Her vision was blurred, her breathing was labored, her leg was very swollen and red near the bee sting, and she had a rash over her leg and stomach. Also, her lips were cracked, and at first she had difficulty drinking the fluids her rescuers gave her.

The helicopter flight to the hospital did not take long, but Nelson's and Monique's fates still were uncertain. Now it would be up to the emergency room personnel to try to help them.

Process and Procedures

Part A Triage in the Emergency Room

1. Meet with your team of doctors, and review *Patients' Vital Signs— Preliminary Information* (Figure 6.1). Pay particular attention to the description of each patient's injury or illness.

   You may find Figure 6.2, Glossary of Vital Signs, useful for interpreting this information.

2. Discuss the vital signs information with your teammates, and record in your journal your team's speculations about the possible outcomes for each patient.

Patients' Vital Signs—Preliminary Information

Patient	Description	Vital Signs			
		Heart Rate	Temp.	Blood Pressure	Breathing Rate
Esther	85-year-old female complaining of dizziness. Has a history of heart problems and suffered a moderate stroke (blood vessel in brain ruptured) six years ago, resulting in mild left-side paralysis.	105 beats/ min, weak	36°C (96.8°F)	95/65 mm Hg	22 resp./ min, labored
Albert	Adult male, approximately 25 years old, with a head injury from a motorcycle accident. Victum was not wearing a helmet, is unconscious, and has no reflexes in his arms or legs.	45 beats/ min	37°C (98.6°F)	80/50 mm Hg	10 resp./ min
Maria	15-year-old girl with symptoms of severe diarrhea and vomiting.	80 beats/ min	40°C (104°F)	110/70 mm Hg	21 resp./ min
Mark	19-year-old male with a gunshot wound to the chest. Victim is conscious, bleeding moderately, and having great difficulty breathing	150 beats/ min	37°C (98.6°F)	100/60 mm Hg	23 resp./ min
Ed	41-year-old man, obviously overweight, called paramedics complaining of crushing chest pains. He is now unconscious.	110 beats/ min, weak	37°C (98.6°F)	variable	12 resp./ min
Monique	18-year-old woman rescued from the Canyon; stung by bee. She is dizzy, nauseous, has cracked lips, is having difficulty breathing, and has a rash on her legs and stomach.	150 beats/ min	41°C (105.8°F)	87/50 mm Hg	20 resp./ min
Nelson	51-year-old man rescued from the Canyon. He was unconscious and having seizures when brought into the emergency ward.	45 beats/ min	43°C (109.4°F)	85/50 mm Hg	10 resp./ min

Figure 6.1 Patients' Vital Signs—Preliminary Information

3. Divide the responsibility for completing each of the following tasks among your teammates:

Task 1:   Compare the vital signs of each patient with the normal ranges. (one team member)

Procedure: Use Figure 6.3, *Range of Vital Signs*, to decide whether each vital sign for each patient is within normal limits. In your journal set up a table in which to record this information. Use a check mark (✓) to indicate vital signs that are within normal limits and a minus sign (–) to indicate those outside normal limits.

Task 2: Compare the vital signs of each patient with the disrupted ranges. (one team member)

Procedure: Use Figure 6.3, *Range of Vital Signs*, to decide whether each vital sign for each patient is serious, critical, or neither. In your journal set up a table in which to record this information.

Task 3: Suggest initial triage treatments for each patient. (two team members)

Procedure:
a. Read Figure 6.4, *General Triage Guidelines*.

b. Divide the list of patients in Figure 6.1 between you and your partner.

c. Reread the description of each patient that is assigned to you.

d. For each of your patients, develop a list of emergency medical treatments that might help each patient restore homeostasis. Record this list in your journal.

Glossary of Vital Signs

Why do health care workers take measurements of vital signs in an emergency situation? Think about what systems these tests measure.

- **Pulse** is the rate (how fast or slow) at which the heart is beating. It provides clues to how well the heart is functioning and how well blood is carrying oxygen and other important substances to the tissues, including the tissues of the brain.

- **Blood pressure** is another indicator of the heart's ability to pump blood throughout the body. It is measured at an artery and is usually represented as two numbers such as 125/85. The first number is a measurement of the force of blood against the walls of the arteries, veins, and chambers of the heart as the heart contracts. The second number is a measurement of the lowest pressure reached as the heart relaxes.

- **Body temperature** indicates whether the body's thermal regulation is working. An elevated body temperature could indicate an infection. A depressed central body temperature, or *core* temperature, indicates a type of shock.

- **Rate of breathing** reflects how well oxygen is being delivered to the body; it also is associated with heart rate and circulation.

Taken together, the measurement of vital signs gives the physician a very quick view of a patient's internal state, even if the patient is unconscious and cannot explain how he or she feels. These measurements vary over a relatively narrow range in healthy people, indicating that the body has precise control of normal internal conditions. When the vital signs are far outside these normal ranges, homeostasis usually is disrupted in the patient. Under these circumstances the vital signs are direct indicators of problems with homeostatic systems. Vital signs, however, usually do not indicate the *cause* of the disruption of homeostasis.

Figure 6.2 Glossary of Vital Signs

Range of Vital Signs

Vital Sign	Normal	Serious	Critical
Blood Pressure Systolic pressure (the top number) is the pressure created when the heart is contracting and pumping blood. Diastolic pressure (the bottom number) is the pressure between contractions when the heart relaxes and fills. In general, blood pressure must be evaluated in terms of other vital signs.	110/70 to 140/90 mm Hg (systolic/diastolic)	90–100 mm Hg systolic	<90 mm Hg systolic
Resting Pulse	60–100 beats/min	<60 or >100 beats/min	<50 or >120 beats/min
Temperature	37°C (98.6°F)	39–40°C (102.2–104°F)	>40°C (>104°F)
Breathing Rate	10–20 resp./min	<10 or >20 resp./min	<10 or >30 resp./min
Cholesterol	131–200 mg/dL, normal; 200–239 mg/dL, borderline high	<131 or >250 mg/dL	>300 mg/dL
Glucose	65–120 mg/dL		
Chloride (Cl⁻)	96–109 mEq/L		
Sodium (Na⁺)	135–145 mEq/L		
Potassium (K⁺)	3.5–5.3 mEq/L		

Key to units: mm Hg = millimeters of mercury, a measure of pressure
mg/dL = milligrams per deciliter, a measure of concentration
mEq/L = milliequivalents per liter, a measure of concentration

Note: The figures quoted above are simplified guidelines, based on those used by medical personnel. More precise ranges for different age groups and sexes also are used. Occasionally healthy individuals have "normal" readings that are outside of these average ranges. For example, many young people have systolic blood pressures lower than 100 mm Hg. For the purposes of this activity, you should assume that the normal state for each patient is within the normal ranges listed in the table.

Figure 6.3 **Range of Vital Signs**

Figure 6.4 **General Triage Guidelines**

General Triage Guidelines

1. All emergency care begins with the ABCs: make sure there is an open **A**irway, that the patient is **B**reathing, and that the patient has adequate **C**irculation.

    A. Airway. Remove obstructions from the mouth, if necessary; move the tongue if it is obstructing the airway; and close any openings such as the nose that prevent the lungs from filling with air.

    B. Breathing. Restore breathing by artificial resuscitation (a technique by which another person or device can temporarily provide air to a patient) or by administering oxygen, if necessary.

C. Circulation. Stop blood loss from serious wounds. Restore heartbeat by cardiopulmonary resuscitation (CPR) if necessary. CPR is a technique by which another person can temporarily provide air and heart contractions for a patient whose heart has stopped beating or is not pumping blood effectively.

2. Look at the patient, and assess his or her injuries. Immobilize any injuries to the neck. The patient may become paralyzed if you initiate any movement. Always suspect neck injuries when there is extensive injury to the head or face.

3. **Shock** is extremely serious and life-threatening; it occurs when blood pressure drops so low that it no longer delivers adequate supplies of oxygen and nutrients to the tissues. Shock can result from failure of the heart to pump vigorously enough, from serious blood loss, or from a reduction of effective blood volume due to pooling in the capillaries or to dehydration.

Shock due to reduced blood volume can be treated by elevating the feet, by using pressure suits that force blood from the extremities back into the body core, or by infusing blood or saline solution into the circulatory system. Shock due to weakness of the heart or damage to the circulatory system may require medications or mechanical devices that assist circulation.

4. Internal body temperature normally is controlled by the hypothalamus, but if this control is lost, the core body temperature can rise to dangerously high levels, a condition known as **hyperthermia**. Extreme hyperthermia can kill cells, particularly brain cells. In these cases external measures must be taken, such as rubbing the patient with ice to bring the body temperature back within normal limits.

Conversely, the body can cool to dangerous levels, a condition known as **hypothermia**. Hypothermia can occur when people are cold and wet for a long period of time. The rapid evaporation of water can cool a person quite quickly, even if the air temperature is not extremely cold. In such cases the body must be warmed slowly to bring it back within normal limits.

---

4. Construct a journal table like the one shown in Figure 6.5. Working as a team, fill in each patient's name in the first column.

This table and the steps that follow will help you to organize your inferences about each patient's condition. Leave enough room in each column to record important information.

Data Organizer for Triage			
Patient Name	Disrupted Systems	Treatment Priority	Emergency Treatment

Figure 6.5 **Data Organizer for Triage**

5. Work with your teammates to identify which homeostatic systems most likely were disrupted in each of the patients.

List these systems in the second column.

6. Classify each patient into one of the following three categories of treatment priority:

- critical (patient requires immediate treatment):   +++

- serious (patient requires treatment very soon):   ++

- stable (patient requires treatment but can wait):   +

Put the appropriate number of plus signs (+) in the column labeled Treatment Priority.

7. Suggest some emergency care treatments that might help restore the homeostasis of each patient. Record these in the last column of your table.

Part B Let's Get More Information

While you were performing your triage evaluation, medical technicians and laboratory staff conducted some additional tests, including analyses of blood and urine samples. These new data may provide you with important information about each patient's condition that you were not able to discern from vital signs alone.

1. Review the information in the handout *Patients' Vital Signs—New Information*, and discuss the importance of this new information with your teammates.

2. Construct a journal table like the one you created in Part A, Step 4. Working as a team, decide in what order you should treat each patient. In the first column of your table, record the name of each patient in order of priority.

   Remember to place the most critical patient first and the most stable patient last. Be prepared to share your ideas and your team's triage results in a class discussion.

3. In the second column indicate any additional homeostatic systems (systems that were not identified as being disrupted in Part A) that your team thinks are disrupted, based on the new information.

   The additional information in Patients' Vital Signs—New Information *may help you with this task.*

4. With your team discuss any additional factors that may have contributed to each patient's condition. List them in the column labeled *Treatment Priority*.

   These may include nonmedical factors such as behavior and luck.

5. As a team decide what the prognosis, or long-term outlook, is for each patient, and record these ideas in the last column and label it *Prognosis*.

   For instance, you might think that some patients will recover completely, others will recover slowly and may suffer long-lasting effects, and others may never recover and even die as a result of their injury or illness.

Analysis

Work individually to respond to the following. Write your answers in your journal.

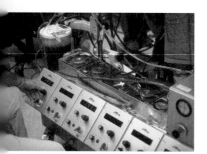

Examples of technology in medicine a. Open heart surgery bypass pump b. CT scanner c. Prosthetic devices

The essay Coping with Disruptions: The Role of Medicine in Homeostasis *(page E75) as well as the information in the various tables may help you with your task.*

1. Why are vital signs so valuable in quickly assessing a patient's condition?

2. Explain how a head injury, such as the one Albert suffered when his motorcycle crashed, could affect so many homeostatic systems.

3. How did a bacterial infection cause a higher than normal temperature in Maria?

4. Monique and Nelson both had very high temperatures when they were brought into the emergency room. Why do you think Monique's heart rate was high and Nelson's heart rate was low?

5. Nelson died because his body could not restore its homeostatic balance. Which other patients have a balance that is sufficiently disrupted that they too might die? Explain your answers.

6. What, if any, nonmedical considerations did you use to rank the patients? Explain your response.

7. In one or two paragraphs, compare and contrast an illness or injury that the body can recover from on its own with an illness or injury that requires medical intervention. Explain how the responses of homeostatic systems differ in the two situations.

8. Describe one reason why performing your assigned task was critical to your team's success in this activity.

Looking Ahead

In the Evaluate Activity for this chapter, you will be working with your teammates to develop a health care proposal. Because you will need extra time to collect information for your proposal, you will need to start now.

With your team turn to *Health Care Proposal* (page 131), read the introduction to the activity, and complete Steps 1 and 2. Then begin Step 3. You will find that developing your proposal will be more rewarding if you are well-prepared.

SELF DEFENSE!

You have seen how the body corrects minor disruptions in homeostasis and how medical technology can help when the disruptions are more serious. But would you believe that at all times, even when you are healthy, a battle is going on inside your body? This battle is being fought by homeostatic systems that work to protect your body against external conditions *before* they threaten to disrupt your body's normal balance.

As a living system your body constantly fights against foreign chemicals and invading microorganisms and viruses. From where do these invaders come? Recall the bee sting that Monique, the hiker, suffered. When a bee stings you, it injects venom into your body. This venom is composed of

Explain

molecules that are foreign to your body, and your body reacts to the foreign molecules. Another source of invading particles is the air that you breathe, air that contains many unhealthy things, including particles of pollutants, spores of fungi and bacteria, and virus particles. The number of potential invaders is even higher if you are breathing the air near a person who has a cold and who sneezes.

Figure 6.6 A single "kachoo" can release 10,000 to 100,000 virus particles.

Even though it might seem impossible for living organisms to protect themselves against the disrupting influence of so many threats, all living systems have some means of protecting their internal environment against infection. As a human you have a particularly elaborate system of natural defense, which is known as the immune system. Although it is not perfect, the immune system generally wins its battles. This activity will help you explain how your body defends itself.

Materials (per team of 4)

Test Subject card
Scenario card

Process and Procedures
Part A Natural Defenses

1. Respond to the following statements regarding how the immune system helps maintain homeostasis. Record your answers in your journal.

   a. Why is it important for the body to distinguish material that is part of itself from nonself material? Explain what happens when the body fails to make this distinction.

   b. Provide two examples of the immune system's nonspecific defense mechanisms. Explain how nonspecific immunity differs from specific immunity.

c. Explain why viruses are unaffected by antibiotics such as penicillin and tetracycline.

d. Describe how vaccination is a technological innovation that takes advantage of a basic property of the immune system.

The essays Avoiding Disruptions: The Immune System *(page E78)*, Self and Nonself *(page E81), and* Immune System Memory *(page E82) may help you with this task.*

2. Share your answers as part of a class discussion.

Part B Diagnosis: A Puzzle

Now you will use your knowledge about pathogens and the immune system to solve a puzzle. Use all available resources, including what you have learned in earlier activities. First, you will see how each Test Subject reacts to the *same, known* pathogen. Then you will use this information and similar strategies to try to determine the identity of an *unknown* pathogen.

1. Have one team member obtain a Test Subject card from the container that your teacher provides and share the information on the card with the rest of the team.

2. Assume that your subject has just been exposed to the influenza virus. In your team complete the following tasks. Record in your journal the evidence and inferences that support your conclusions.

   a. Generate a prognosis for your subject.

      A prognosis is a prediction based on evidence and inference about whether a person will become ill and, if so, how soon he or she will recover.

   b. Discuss what effect penicillin would have if it were administered the first day that symptoms occur.

3. As your teacher directs, share your team's prognosis for your Test Subject with your classmates.

4. Try a different version of this exercise. Instead of knowing the identity of the pathogen and determining the prognosis, you now will know the prognosis of your Test Subject and determine the identity of an unknown pathogen.

   a. Assume that your Test Subject has been exposed to one of the pathogens from the list below:

      • *Streptococcus* bacterium

      • common cold virus

      • rubella virus

   b. Obtain a Scenario card that describes what happened after your Test Subject was exposed to the unidentified pathogen.

   c. Work with your teammates to analyze all available clues. Use this first level of analysis to determine the identity of your Test Subject's pathogen.

You might wish to organize your reasoning by filling in a table like the one shown in Figure 6.7. The available information may not help you determine with certainty the identity of your Test Subject's pathogen. You may, however, be able to use this information to eliminate some pathogens from further consideration and to narrow the possibilities. Support your decision with reasons based on evidence. In your team practice using the skill of consensus-building.

Figure 6.7 Pathogen Identification Table Use a table like this to organize your thinking.

Pathogen Identification Table		
Level 1 Analysis	Pathogen	Test Subject
Can Eliminate	Remaining Possibilities	Reasons

5. Pick up a copy of *Complete Scenario Information* so that you can begin a second level of analysis.

6. Use this new information to try to determine your Test Subject's pathogen with greater certainty and to justify your refined answer. Include reasons and evidence that support your refined conclusions.

   The additional information should allow you to use a process of elimination to determine the unknown pathogen. For this second level of analysis, you may wish to add rows to your table to organize your reasoning.

7. Use the table that you generated as a record of your thinking, and participate in a class discussion of the results of this exercise.

Analysis

Complete the following in your journal:

1. In Part B of this activity, you used the prognoses of different patients to determine the identity of an unknown pathogen. Explain how a combination of information from different scenarios was more helpful in identifying the pathogen than the information from a single Scenario card.

2. What effect do risk factors such as fatigue, anxiety, and smoking have on a person's ability to defend himself or herself against infection and other homeostatic disruptions?

3. Assume that one healthy person had rubella as a child, a second healthy person was vaccinated against rubella as a child, and a third healthy person never encountered the rubella pathogen or vaccine as a child. If all of these people are exposed to rubella when they are 25 years old, how will their bodies respond to this pathogen?

IMMUNITY SIMULATION

ALTERNATE FOR *SELF DEFENSE!*

Explain

You have seen how the body corrects minor disruptions in homeostasis and how medical technology can help when the disruptions are more serious. Would you believe that at all times, even when you are healthy, a battle is going on inside your body? This battle is being fought by homeostatic systems that work to protect your body against external conditions *before* these conditions threaten to disrupt your body's normal balance.

As a living system you constantly must fight against foreign chemicals and invading microorganisms and viruses. From where do these invaders come? Recall the bee sting that Monique, the hiker, suffered. When a bee stings you it injects venom into your body. This venom is composed of molecules that are foreign to your body, and your body reacts to the foreign molecules. Another source of invading particles is the air that you breathe, air that contains many unhealthy things, including particles of pollutants, spores of fungi and bacteria, and virus particles. The number of these potential invaders is even higher if you are breathing the air near a person who has a cold and who sneezes.

Figure 6.8 The air you breathe can contain unhealthy particles of pollutants, fungi, bacteria, and viruses.

Even though it might seem impossible for living organisms to protect themselves against the disrupting influence of so many threats, all living systems have some means of protecting their internal environment against infection. As a human you have a particularly elaborate system of natural defense, which is known as the immune system. Although it is not perfect, the immune system generally wins its battles. This activity will help you explain how your body defends itself.

Materials (per team of 4)

Macintosh or IBM-compatible computer with Windows™
Immunity Simulation Disk

Process and Procedures

Part A Natural Defenses

1. Respond to the following statements regarding how the immune system helps maintain homeostasis. Record your answers in your journal.

   a. Why is it important for the body to distinguish material that is part of itself from nonself material? Explain what happens when the body fails to make this distinction.

   b. Provide two examples of the immune system's nonspecific defense mechanisms. Explain how nonspecific immunity differs from specific immunity.

   c. Explain why viruses are unaffected by antibiotics such as penicillin and tetracycline.

   d. Describe how vaccination is a technological innovation that takes advantage of a basic property of the immune system.

   The essays Avoiding Disruptions: The Immune System *(page E78),* Self and Nonself *(page E81), and* Immune System Memory *(page E82) may help you with this task.*

2. Share your answers as part of a class discussion.

Part B Who's Winning?

Now you can apply your knowledge of the immune system in a computer simulation that illustrates the mechanism by which this system protects you from infection. Your goal is to overcome a pathogen, which has invaded your body, before it overcomes you. You can see the effects of each of three pathogens and of each (or any combination) of five lifestyle factors that affect the immune system.

1. In your team decide who will be the computer operator, who will be the recorder, and who will be the two advisors.

   The advisors must observe the screen carefully for clues that can help the operator work effectively. Rotate responsibilities between each run.

2. To begin the program, double click on the icon labeled *Immunity.*

3. Click on the menu item for the function that you wish to perform.

   The two immunity stories explain the immune system components. (A Novel Approach is humorous but accurate; A Serious Approach is more straightforward.) You need to know the function of each of the components and their relationship to each other to run the simulation.

4. Move through either story as follows:

   • To move between pages, click on the left or right arrow.

   • To move to the next text page without showing all of the intervening animation, press the shift key while clicking on an arrow.

   • To switch from any point in one version of the story to the same point in the other, use **File** and click on the desired version.

5. Replay the stories as often as you need to gain an understanding of how immune components interact to fight infection.

   For additional information consult the essays Avoiding Disruptions: The Immune System *(page E78)* Self and Nonself *(page E81), and* Immune System Memory *(page E82) as well as the* Glossary of Immune System Components *(Figure 6.10), which provides details about the components of the immune system.*

6. As you are studying the explanations in the stories, prepare a flow chart that shows the sequence in which the components of the immune system work and how each component is activated.

7. Return to *Immunity* (use **File**), return to Main Menu, and select the simulation. This screen will appear:

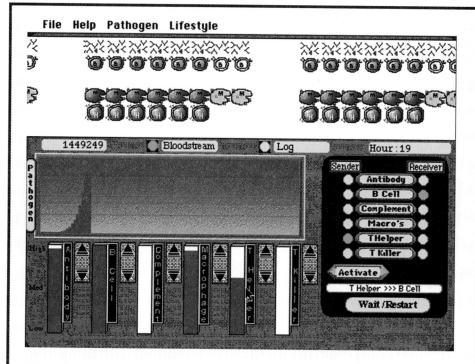

Figure 6.9 Immunity simulation screen The upper part of this screen shows the blood stream with levels of different immune system components. The lower part of this screen contains several fields including a bar graph, bars showing the level of each immunity component, and a control area for activating the immunity components.

Explanation of the menu bar

- Use **File** to REINITIALIZE (start the simulation over), return to the main menu, or QUIT.

- Use **Help** for information about immune system components.

- Use **Pathogen** to select any of three different infections.

- Use **Lifestyle** to select any combination of factors that can affect the immune system.

   When you select a factor, a check mark will appear beside it.

Explanation of the simulation screen

- The white area at the top of the screen shows a microscopic view of the blood stream or a detailed log, depending on which option you select.

   The blood stream provides information about immune system components and their status.

The log provides detailed information about what happened during the run.

- The four fields under the white area show, from left to right, the number of pathogens, two option buttons (blood stream and log), and the number of hours since the infection started.

  When you are in the wait mode, the hour alternates with the word "WAITING."

- The box below the four fields shows a graphic representation of the number of pathogens present at any time.

  The rate of growth of pathogens depends on the type of pathogen and the immune response that you have requested.

- The bars below the graph show the level of immune system components.

  You control the level of each component.

- The box to the right of the graph allows you to activate immune system components.

8. Use the following steps to run the simulation until you are familiar with how the program works.

   - To start, click on the **Wait/Restart** button.

   - To start over, use **File** to REINITIALIZE.

   - To increase or decrease the level of any component, click on the up/down arrow to the right of the component's bar. Keep pressure on the mouse button to continue to increase or decrease the level.

   - To activate a component, click on the appropriate **Sender** button, click on the appropriate **Receiver** button, and then click on the **Activate** button.

     The blood stream at the top of the screen indicates the level of each component.

     You can activate any portion you wish of any available component.

   - To put the simulation on WAIT or to RESTART the simulation from the wait mode, click on the **Wait/Restart** button.

   - To select blood stream or log, click on the button to the left of the word.

9. The objective of the simulation is to use your knowledge of the immune system to defeat the pathogen before it defeats you. As a team, discuss the best scenario to fight an infection and come to a consensus about your plan of action. Record this plan in your journal.

   There is more than one way to combat the infection without dying, but the more efficient your plan of action, the greater your chance of winning the battle. Your success depends on how well you understand the interaction of immune system components and how long it takes to conquer the pathogen. The idea is to use the lowest possible level of each component needed while conquering the pathogen as quickly as possible.

10. Now use the plan to direct your choice of steps in the simulation. In your journal, record notes about which pathogen and lifestyle factor you

selected, what happened, and how long it took to win (or lose) the battle against the pathogen.

Consult the log displayed in the program to keep track of what you have done so that you can make a record in your journal.

11. Repeat Steps 9 and 10 if time allows and try to conquer the pathogen more quickly by developing a more accurate and efficient plan of action.

   You also might select different pathogens and lifestyles.

12. In your journal record an explanation for your success (or lack of success) against the pathogen.

Glossary of Immune System Components

Antibodies are protein molecules that B cells produce. They can be located on the cell surface or circulate free in the blood stream. Surface antibodies bind to specific antigens and serve as a signal to other cells. Free antibodies are produced by B cells that have been activated in response to foreign antigens, including those present on pathogens. Antibodies, which have two binding sites, can bind two pathogens, forming clumps that inactivate the pathogens.

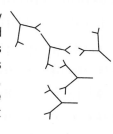

Antigens (*anti*body *gen*erators) are protein molecules that are present on the surface of most cells. Antigens that are recognized by the immune system as nonself, or foreign, can induce an immune response.

B-cells are immune cells that are made and mature in the bone marrow. When B-cells are activated, they divide and produce two groups of cells—**plasma B-cells** and **memory B-cells**.

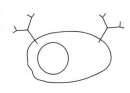

Plasma B-cells produce free antibodies (identical to their surface antibodies), which counter the infection by circulating throughout the body.

Memory B-cells are long-lived cells that ensure that future infections by the same pathogen will elicit a much more rapid response, making it unlikely that a person will become ill a second time.

B-cells can be activated in two ways. One way is by directly encountering an antigen that they bind and process. The processed antigen presented on the B-cell surface can signal other immune cells. The second way B-cells are activated is through interaction with antigen presented on a macrophage or helper T-cell. B-cells require the presence of helper T-cells to mount a full immune response.

Complement is a group of proteins found in the blood that act in a sequential manner, destroying pathogens. Antibody-antigen complexes can activate the first complement factors, which in turn activate other complement factors. After a series of step-wise activations, enough complement factors are activated to eliminate antigens in two ways. First, some of the activated complement proteins can attract macrophages, which destroy the antigen. Second, activated complement molecules can aggregate within the cell membrane of a pathogen, eventually forming holes that cause the pathogen to burst.

Lymphocytes are the white blood cells involved in the specific immune response. They include B-cells and the various types of T-cells.

Lymphokines are molecules released by helper T-cells and other lymphocytes that influence the activities of other cells.

Figure 6.10 **Glossary of Immune System Components**

Macrophages are nonspecific scavenger cells of the immune system that are present throughout the body. They engulf the foreign material, including pathogens, that they encounter. Then they degrade it and present its antigens on their surface.

Macrophages move about in a manner similar to that of one-celled amoebae and can be attracted to a site of infection. If enough macrophages are present, the infection may be stopped at this stage.

Macrophages can interact with and activate other immune cells, including B-cells and helper T-cells.

T-cells are immune cells that are formed in the bone marrow but mature in the thymus. There are at least two types: **helper T-cells** and **killer T-cells**.

Helper T-cells are activated by interacting with macrophages or B-cells that have encountered foreign antigens. Activated helper T-cells secrete chemicals known as lymphokines that activate B-cells and killer T-cells.

Killer T-cells carry out cell-to-cell combat, destroying cells infected with virus as well as cells that have become cancerous. These T-cells recognize virus-infected cells because the infected cells express viral antigens on their surface, antigens which are recognized as foreign. Killer T-cells must be activated by interaction with helper T-cells before they can multiply and attack virus-infected cells.

WHAT'S THE RISK?

Have you ever cut your finger while you were slicing vegetables or fallen and broken a bone? Perhaps you have had a serious illness. Your homeostasis can be disrupted at any time by accident or illness because you are exposed to many risks in your daily life (see Figure 6.11). You probably accept, consciously or subconsciously, some of these risks as unavoidable

Figure 6.11 Humans are exposed to many risks. Some risks are affected by behavior, and others are unavoidable despite an individual's behavior. Think about the risks involved in each of the images depicted and how those risks compare with each other.

while you strive to minimize other risks. In this activity you will explore controllable and uncontrollable risks, and you will explain what some of these risks mean in your life.

Materials (per student)

1 prepared test tube in a rack
additional materials that your teacher supplies

Process and Procedures
Part A Fluid Exchange

1. Choose one test tube from those at the station your teacher has prepared.

2. Follow your teacher's instructions.

3. When you have completed the tests, return the test tubes to the station, and wash your hands thoroughly.

4. Discuss the following questions with your classmates:

   a. What observations about the fluid exchange surprised you?

   b. Many illnesses, including the common cold, hepatitis, and AIDS, are spread by fluid transfer. What types of behaviors spread these illnesses, and what body fluids are involved?

   c. How can people eliminate completely their chances of contracting a sexually transmitted disease?

Part B Risk Assessment

1. Develop a list of at least 20 risks faced by humans.

   Use your personal experience and the essays Avoiding Disruptions: Behavior, Choice, and Risk *(page E84) and* Individual Behavior Can Affect Larger Groups *(page E86) to make your list.*

2. With your teammates look through the headings on the copymaster *Risk Assessment Data*, and decide who will be responsible for analyzing each of the following risk categories:

   • Smokeless Tobacco and Smoking

   • Alcohol, Marijuana, and Other Drugs

   • Cancer

   • Sex and Sexually Transmitted Diseases

3. Divide up the remaining categories evenly, and record in your journal the categories for which you are responsible.

4. Study the information in your risk categories.

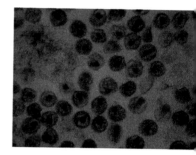

Electron micrograph of the human immunodeficiency virus (HIV) in a cell (magnified at 200,000 X)

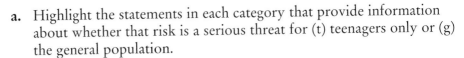

a. Highlight the statements in each category that provide information about whether that risk is a serious threat for (t) teenagers only or (g) the general population.

   Label each of these statements with a t or g to distinguish them.

b. Mark with an asterisk (*) the most shocking or surprising statistic in each of your categories.

c. For each of your categories that are affected by behavior, record in your journal one or two behaviors that might reduce the risk.

5. Briefly discuss the results of your analysis by sharing some of the more relevant or surprising statistics with your teammates.

6. As a team complete the following tasks in your journals:

a. Make a list of three risks in our society that could be reduced if people would change their behavior.

   Select risks that would have a significant impact on the entire population if those risks were reduced.

b. Make a list of three risks in our society that are unlikely to be reduced, either because the risk is not controllable or because people would be unwilling or unable to change their behavior.

c. Next to each of the items on your lists, write two or three reasons to support that choice.

   You will use this information in the Analysis and in the next activity, Health Care Proposal.

7. Complete the copymaster *Taking Risks: A Self-Evaluation* to determine your own personal risk level.

   Your teacher will collect these surveys and tally the responses for the entire class. Your individual responses will remain anonymous. You will use the summary data in the Analysis.

8. In your journal write two paragraphs that analyze behaviors that you could change to reduce your own risk level.

Part C Ethical Analysis

1. Draw a vertical line to divide a page of your journal in half. Use the left half of a page to make a simplified list of the steps in ethical analysis.

   For example, you might list the first step as: Identify the question precisely.

   The essay Ethical Analysis (page E87) will help you with this task.

2. On the right half of the page, next to the steps in the simplified list, list the related steps in scientific inquiry as you have been learning about them in previous chapters.

3. With your team discuss the following questions:

   • How does the process of ethical analysis compare with the process of scientific inquiry?

   • How similar or different are their uses?

Analysis

Respond to the following as part of a class discussion:

1. Identify several behaviors that are socially acceptable even though they have a negative impact on others. Explain why you think these behaviors are tolerated.

   Use the information in your journal and in the Risk Assessment Data.

2. Explain whether the behaviors identified in Question 1 of the *Analysis* pose any ethical dilemmas. State how the behaviors pose an ethical dilemma and whether society has made any decisions that affect those behaviors.

3. Review the class data that your teacher has compiled from the survey *Taking Risks: A Self-Evaluation.*

   a. Compare the high-risk behaviors that are most common among students in your class to those of society.

      Use the list your team made in Part B, Step 6.

   b. Which, if any, of the class risks match the risks that your team felt could be reduced if people changed their behavior?

   c. If any of these risks are the same, why do you think students in the class take these risks?

HEALTH CARE PROPOSAL

Evaluate

Entrepreneur to Fund Worthy Health Care Programs

Samantha S. Jones creates health care endowment

Phoenix, AZ—Samantha S. Jones, whose son, Nelson, died recently in a hiking mishap, revealed Thursday that she will contribute some of the profits of her software company to establish a health care endowment fund, the J. Nelson Jones Foundation. An annual award of $1 million, which may be split among as many as three different groups, will be given to worthy health care programs. The money will be distributed by a panel of health care experts, which will evaluate proposals on a competitive basis. Groups interested in obtaining funds must demonstrate that their program is biologically sound, cost effective, beneficial to a significant number of people, and sensitive to ethical concerns in society and within the health care industry. When asked why she decided to fund this type of program, Jones replied, "I'd like to create a world with better health care, better education, and a better understanding of the limits of the human body so that this kind of tragedy can be prevented."

In this activity your team will develop a health care proposal and apply for some of these funds. You will need to have a clear idea of what you want to do with the money and how your idea addresses society's needs. In addition, you must explain which of the body's homeostatic mechanisms are affected by your proposed program, and you must justify your proposed budget. To complete your task, gather information from different sources such as this program, your journal, and the library.

Your teacher will evaluate your proposal based on the criteria established by the private foundation and listed on the copymaster, *Scoring Rubric*.

An exceptional proposal may be funded in its entirety, or as many as three proposals may be partially funded. If, however, two or more proposals are submitted that deal with the same topic, only the strongest will be chosen. Remember, you may propose to spend up to $1 million, but your budget must not exceed this amount.

Process and Procedures

1. As a team review *Possible Health Care Options* (Figure 6.12), and choose an option that interests your team.

   Your teacher will keep track of which teams are selecting which options.

2. As a team develop a short description of your program, and then decide which team members will gather information for each of the following sections of the proposal:

   - Homeostasis

   - Risk Assessment

   - Ethical Issues

   - Cost and Budget

   Refer to the J. Nelson Jones Foundation's Guidelines for Proposal Development in Figure 6.13 to review the questions and issues that you must address in each of these sections. Regardless of which option your team chooses, you may use information from other options to support your position.

3. Individually, gather all of the information that you will need to develop your part of the proposal. Analyze this information to identify the evidence that will support your arguments.

   Your teacher will suggest where you can find additional material. You will have about one week to complete this step.

4. Present to your team the information that you have analyzed, and explain to them how this information will strengthen the proposal.

   Practice the skill of advocating a position.

5. As a team discuss this information, and decide what specific data to include in the proposal.

   Focus on the most important and persuasive data because proposal space is limited.

6. As a team use the steps presented in the essay *Ethical Analysis* (page E87) to conduct an ethical analysis of an issue that is involved in the health care area that you chose.

The team member who was responsible for gathering information on ethical issues should provide the team with several ethical issues to consider. Your team must choose one issue on which to focus the ethical analysis.

7. Write your proposal according to the *Guidelines for Proposal Development* (Figure 6.13).

   Divide this task evenly among your teammates. You will be evaluated according to the criteria listed in the Scoring Rubric *that your teacher distributed. You may want to review those criteria before you start writing.*

8. Submit your proposal to your teacher.

Analysis

In your journal write one paragraph that reflects on your experience of developing the proposal. Consider these questions as you write the paragraph:

- What section was hardest to write? Why do you think that was so?

- Why is ethical analysis a useful tool in science and in society?

- How can you positively or negatively influence the homeostatic systems in your body?

Possible Health Care Options

Option 1 Alcohol and Drug Treatment Program

- In the United States treatment for a drug-addicted mother costs less than $5000 for 9 months; medical care for a drug-exposed baby costs $30,000 for 20 days.

- Most people suffering from alcohol or drug abuse who cannot afford private treatment in hospitals receive their care as outpatients at clinics, or they receive no treatment at all. Because the addiction is so strong, many of these people eventually turn to prostitution or other crimes to obtain the money that they need to pay for more alcohol or drugs.

- One federally funded treatment facility in a city of 300,000 serves 700 heroin addicts of all age groups per month. An annual budget of $1 million supports 18 full-time counselors, 17 part-time employees, and a program designed to eliminate chemical dependency. This program provides heroin addicts with daily doses of methadone, a chemical narcotic that minimizes the craving for heroin and helps the addicts "stay clean." There is no limit to how long an addict can participate in this program. Approximately 25–40 percent of those on methadone stay on it for a year or more, but many find it extremely difficult to quit the drug habit completely. About 30 percent, however, are able to find work and at least partially support themselves. Methadone treatment dramatically reduces crime by addicts (from 237 crime days per person per year before treatment to 69 crime days per person per year after four months of treatment) and HIV infection rates (from 39 percent among addicts not in treatment to 18 percent among addicts in treatment for three years).

Option 2 Heart Disease Prevention and Treatment

- The dietary habits of Americans are substantially different from those of other countries. This difference has contributed to the prevalence of heart disease in the United States. For instance, the typical diet in Japan contains far less cholesterol and saturated fat than the typical American diet, and consequently, a 50-year-old

Figure 6.12 **Possible Health Care Options**

Japanese man has an average blood cholesterol level of 180 mg/dL compared with an average of 245 mg/dL for a 50-year-old American man.

- After an individual has a heart attack, there is at least a 50 percent chance that the individual will die in less than five years unless the individual takes preventive measures.

Option 3 Education Programs Focusing on the Prevention of AIDS and Other Sexually Transmitted Diseases

- Fifty-four percent of high school students have engaged in sexual intercourse.

- Sixty-nine percent of these students reported that they are currently sexually active; fewer than half of these students use condoms.

- Two percent of high school students have used intravenous drugs.

- The average lifetime cost of treating a person from the time he or she is infected with HIV until he or she dies of AIDS is $119,000.

- The average length of an HIV infection, before AIDS develops, is 10 years.

- The drug zidovudine (often mistakenly referred to as AZT), which is the drug most often used to treat those infected with HIV, costs $8000 per year per individual in 1990.

- For every dollar spent notifying sex partners of HIV-positive patients, at least $11 is saved in annual medical care costs for each case of HIV that is prevented.

Option 4 Hospital Equipment and Procedures (1994)

- Dialysis, a procedure that substitutes for the normal functioning of the kidneys, costs $16,000 to $18,000 for a typical four- to seven-day hospital stay for kidney failure. Many of these patients will never need dialysis again. By contrast, those with a serious kidney disease may require dialysis several times per week for their entire lives. Each treatment costs $35 to $40. A dialysis machine costs $13,000 to $15,000.

- A heart-lung machine, which is used as a temporary substitute for the functioning of the heart and lungs during serious heart or lung surgery, costs $21,000 to $27,000.

- An electrocardiograph (ECG), an instrument used to monitor and diagnose heart problems, costs $10,000 to $12,000.

- Pacemaker surgery, a procedure performed on patients who are suffering from certain forms of heart disease, costs $21,000 to $23,000.

- Angioplasty, a surgical procedure to open arteries in the heart that are blocked by the build-up of cholesterol and plaque, costs $8000 to $16,000.

- Appendectomy, removal of a diseased appendix, typically requires a three-day hospital stay and costs $5000 to $7000.

- Cholecystectomy, removal of a diseased gall bladder, costs $12,000 to $14,000.

- Mastectomy, removal of a cancerous breast, costs $9000 to $13,000.

- Arthrotomy, a surgical procedure to repair an injured shoulder, costs $4000 to $6000.

- Magnetic Resonance Imaging, or MRI, is a technique in which strong magnetic fields generate a picture of the inside of the body (similar to an X-ray). Physicians use it to help them locate tumors in cancer patients (about $1000 per use) or injuries and obstructions in people with difficult-to-diagnose illnesses ($950 per use). The device typically costs about $3 million.

- Tonsillectomy, removal of tonsils and adenoids, costs $1400 to $1600.

- Radiation therapy often is used to reduce the size and spread of malignant tumors. This therapy can be used in place of or in addition to surgery. It is frequently necessary if the patient is to have a chance of survival. A course of 30 treatments for lung cancer costs $8200 with an additional physician's fee of $2750 to $3000.

- The average salary for nurses and physician's assistants is about $35,000 to $45,000 per year.
- Physicians typically earn more than $100,000 per year.

Option 5 Prenatal Care

- Prenatal care for a pregnant woman for nine months (not including delivery) averages a total of $600.
- For a very low-birth-weight baby, each day in the intensive care unit costs $1720. Low birth weight is often due to prematurity, which may be a consequence of poor prenatal care. Hospital stays that last several weeks or even months are not uncommon for unhealthy babies.
- The cost of having a normal delivery is $1700 to $2300, assuming that there are no complications and that the mother is in the hospital for only one day. The cost of caring for a healthy newborn is $650 to $750 per day in the hospital.
- The cost for a typical Caesarean-section delivery, including a three-day hospital stay, is $4600 to $5300.
- For $600,000, a public health care clinic can provide prenatal care for 400 women, including education about nutrition, exercise, and avoidance of harmful behaviors. Such a clinic also could provide regular visits by nurses and handle uncomplicated deliveries. These programs are very successful at producing full-term, normal-birth-weight babies (greater than 5 lbs 8 oz). In fact, with a low birth-rate incidence of only 7 percent, they produce a higher percentage of normal-birth-weight babies than privately funded health care facilities that do not offer a prenatal care program.

   With an additional $200,000 such clinics can offer annual gynecological exams and counseling about reproduction (fertility) and pregnancy prevention for 2000 women. The Office of Technology Assessment has estimated that prenatal care that averts a low-birth-weight baby saves the U.S. health care system $14,000 to $30,000 per baby.

Option 6 Quit Smoking Program

- Cigarettes kill more Americans than do AIDS, alcohol, car accidents, murders, suicides, drugs, and fires combined (about 400,000 people per year).
- In 1991 tobacco companies spent $4.6 billion to advertise and promote cigarette consumption. This means more than $12.6 million a day, $8750 a minute.
- In 1991 the tobacco industry spent more than $100 million to advertise and promote smokeless tobacco products.
- The Office of Technology Assessment places the social cost of smoking for 1990 at $68 billion. This estimate includes $20.8 billion in direct health costs, $6.9 billion in lost productivity due to smoking-related disability, and $40.3 billion in lost productivity due to smoking-related, premature deaths.
- Every dollar that is spent on smoking-cessation programs that are successful saves $21 during a working lifetime (defined as ages 20 to 64).

Option 7 Vaccine Programs

- A measles shot costs $8; hospitalization for a child who has measles costs $5000.
- The measles-mumps-rubella vaccine led to a savings of $1.4 billion in 1983 with $14.40 saved in health care costs for every dollar spent on the program.

Disease	Cases in U.S. (year)	U.S. Cases (1991)	% Change
Measles	894,134 (1941)	9488	-98.9
Mumps	152,209 (1968)	4031	-97.4
Poliomyelitis	21,269 (1952)	0	-100

- Forty to 50 percent of the people who are at high risk for influenza or who die from influenza and pneumonia (usually people over 65) received medical attention in the previous year but failed to receive an influenza vaccine. Perhaps as many as 60,000 people die each year from influenza; the vaccine costs $3.

- Hepatitis B infects more than 200,000 people each year and kills more than 4000; the vaccine costs $130 per person.

Option 8 Programs for Women, Infants, and Children

- A nutritious diet for a young child for one year costs $842; special education for a child with a mild learning disability, which may be caused by malnutrition, costs $4000 for one year.

- Many U.S. counties have programs that try to prevent malnutrition in pregnant or nursing women and children under age 5 by providing food vouchers that families can redeem for wholesome foods such as formula, baby food, milk, cheese, eggs, cereal, dry beans, peanut butter, and tuna fish.

- One such program has a staff of 41 people and an administrative budget of $1.2 million. In the course of one year, this program disbursed more than $4 million in food to 18,200 people (an average monthly enrollment of 10,159 people) and prevented some of the malnutrition that is common in very low income families.

Figure 6.13 J. Nelson Jones Foundation, Guidelines for Proposal Development

J. Nelson Jones Foundation
Guidelines for Proposal Development

You must include each of the following sections in your proposal, and you must address each question or issue presented.

Short Description of Proposed Program

This description should be a brief overview of the program that you are proposing.

Homeostasis

- Which organs or regulatory systems does your proposal most directly affect or influence? The biology of this system will be the focus of your proposal. What is the normal function of these organ systems in maintaining homeostasis? This might include a description of the pertinent anatomy, physiology, and any involvement of the immune system. (Many options involve several organ systems, but you need to choose only one system.)

- What is the nature of the homeostatic disruption that your proposal seeks to correct? How will this correction be accomplished?

Risk Assessment

- How common is the illness or injury that your proposal seeks to treat, or how many people will take advantage of your services?

- If your proposal targets a particular population, explain how this population's needs will be met.

- How does behavior affect a person's likelihood of experiencing the risks that your proposal addresses?

- Describe the controllability or uncontrollability of the risks involved and how a person can change his or her behavior to minimize the risks.

Ethical Issues

- Identify an ethical dilemma that is associated with your proposal. Describe the concerns surrounding this dilemma.

- Use the six steps of ethical analysis to analyze this ethical dilemma, and explain your decision about what should be done.

Cost and Budget

Design a budget, not to exceed $1 million, that covers a period of one to three years. Address the following items as you justify your budget:

- Include categories such as salaries, permanent equipment, travel, materials, and supplies.

- Is the cost of the equipment or program a one-time expense or an ongoing expense?

- Is the treatment or program that you propose the only one available? (In other words, do the participants in your proposed program have any other choices?) If it is not the only choice, how does this method of treatment or prevention compare in terms of cost effectiveness with other available programs?

ENERGY, MATTER, AND ORGANIZATION:
RELATIONSHIPS IN LIVING SYSTEMS

7 **Performance and Fitness**

8 **The Cellular Basis of Activity**

9 **The Cycling of Matter and the Flow of Energy in Communities**

Energy. We all use it in varying amounts, 24 hours a day, 365 days a year. What exactly is energy, and where does it come from? How is energy related to the matter we take in each day as food? How do matter and energy help organisms like us perform? Think of a runner nearing the finish line. From where does the runner get the energy needed for that final burst of speed?

In this unit you will explore matter, energy, and the relationship between them. You will investigate how matter and energy can explain levels of human performance that allow a runner to sprint to the finish line. Then you will see how cellular processes in the body extract energy from the food consumed by this runner and where the energy present in food originates. You also will see how matter and energy link all of the organisms in a community.

By the end of Unit 3, you should understand how
- an individual's performance depends on diet and exercise,
- matter and energy are involved in maintaining fitness,
- energy is stored in the organization of matter,
- living organisms obtain and process matter and energy for activity and build and maintain body structures, and
- communities of organisms depend on the cycling of matter and the flow of energy.

You also will continue to
- collect, organize, and analyze data,
- propose explanations, and
- test hypotheses.

> *John was proud of his new energy level,*
> *which markedly increased his alertness at work. . . .*
> *On the day of the race, he thought, 'I feel ready!'*

7
PERFORMANCE AND FITNESS

What do the people in the opening photos have in common? Each is engaged in a physical activity that requires a certain level of fitness. The people out for a walk may not necessarily have the same degree of athletic fitness as a professional tennis player, but to live an active, healthy life, they still must meet a certain standard of fitness.

In this chapter you will have the opportunity to learn how matter and energy are related to human physical performance. As you do this, you will explore your understanding of the term *fitness* and why being fit should be an important priority for all of us. You will investigate how exercise and good eating habits promote fitness and how the foods you eat provide the matter and energy necessary to build your body and keep it functioning.

ACTIVITIES

Engage   Thinking about Fitness

Explore   What Determines Fitness?

Explain   What Is in the Food You Eat?

Explain   You Are What You Eat

Elaborate Structures and Functions

Evaluate Marathon

THINKING ABOUT FITNESS

Engage

What is required of the body during extreme levels of human performance? What is required to sustain even basic levels of non-athletic activities such as climbing stairs or playing catch with a friend? We begin our exploration of human performance with a look at what it means to be physically fit. Fitness often means different things to different people. Physicians may view fitness as freedom from disease, whereas physical education teachers may emphasize physical performance. In this activity you will have the opportunity to think about some possible meanings of the term *fitness* and how this concept might apply to you.

Process and Procedures

1. Carefully consider your answers to the questions below, and record your responses in your journal.

   a. What is your personal definition of fitness?

   b. What do you think are the most important factors that affect your level of fitness as you defined it?

2. Copy into your journal the fitness scale diagrams shown in Figure 7.1.

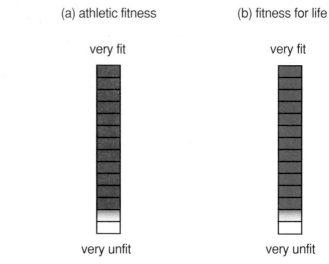

Figure 7.1 **Types of fitness** An individual's level of athletic fitness does not necessarily correspond to his or her level of fitness for life.

3. As you read the story *The Sky Awaits*, think about the fitness scales and what they mean.

4. Refer to the fitness scales as you discuss the following questions, and complete the following tasks with the members of your team:

   a. What does Scale *a* represent? What physical and behavioral characteristics would you expect to find in an individual who scores very high on such a scale?

   b. What does Scale *b* represent? What characteristics would you expect to find in an individual who scores very high on this scale?

   c. Where on each scale would you place Sullivan and Yates? Mark and label in your journal the positions on each scale at which you would place Sullivan and Yates. Below the diagrams, write a sentence explaining why you placed Sullivan and Yates at those positions.

THE SKY AWAITS

It's four o'clock in the morning. The sun has not come up yet on this day in early August, but there's work to be done. Captain Yates rolls out of bed, takes a shower, dons a flight suit and boots, and heads for the kitchen of the small apartment. Breakfast consists of a slice of grilled ham, two eggs over easy, wheat toast with strawberry jam, a glass of orange juice, and a cup of coffee.

Yates steers the bright-red sports car onto the highway and heads out to the air base. As the car approaches the main gate of the base, the sky begins to brighten in the east. It looks as though it will be a good day to fly. As an instructor pilot in the U.S. Air Force, Yates flies a T-38 Talon—a supersonic jet trainer capable of speeds up to 1450 kilometers per hour (900 mph).

After parking the car, Yates checks the schedule and learns that the first student pilot today is Lieutenant Sullivan. Yates then checks the weather forecast—no ceiling, unlimited visibility, and calm winds until about 1100 hours, when clouds will begin to build. After the preflight briefing, in which Yates explains the objectives of the flight to Sullivan, the two pilots get into their G-suits and parachutes. At the airplane they go through the preflight checklist, run up the engines, and check the equipment that will deliver 100 percent oxygen for them to breathe. They taxi onto the runway. Almost immediately, it's their turn to take off. Yates stands on the brakes and sets the throttles at full afterburner. The takeoff roll is smooth, and Sullivan retracts the wheels as the jet leaps off the runway into the brilliant sky.

This flight is an acrobatics mission in a practice area some 120 km (75 miles) west of the air base. At an altitude of 7925 m (26,000 feet), the outside air temperature is –26.2°C (–14°F). Yates executes a roll to look in every

U.S. Air Force T-38 supersonic jet trainer

direction, and then Sullivan takes over for the next phase of the mission.

The first maneuver is a "G-awareness turn" that prepares Yates and Sullivan for the rigors of the training mission. In the first half of the turn—a 70° to 80° bank—the pilots strain against the nearly instantaneous pull of up to four times the force of gravity, or four Gs. Because of the acceleration forces that result from making sharp turns at high speeds, Yates and Sullivan weigh almost four times their normal weight. Their G-suits automatically inflate around the lower portion of their bodies so blood will not pool in their hips, legs, and feet and possibly cause them to black out from a decreased flow of blood and oxygen to their brains. Sullivan tightens the turn to a 90° bank, which produces almost six Gs, before the plane rolls back to a straight and level flight.

This warm-up is the beginning of a strenuous and physically exhausting routine that lasts almost 30 minutes. The routine consists of loops, rolls, stalls, and other similar high-G maneuvers. During this routine, the pilots' bodies strain against blackouts from the added force, their heart rates soar, their muscles tense as hard as rock, and their breathing is labored.

When their maneuvers are complete, Yates and Sullivan begin to relax. The return flight is much less demanding, and the approach to the base and the landing are uneventful. As they walk off the flightline and enter the squadron building, both pilots realize that they now are very hungry and a bit tired. It's no wonder—the physical demands of this morning's flight were similar to those required of highly trained, competitive athletes. The training and level of fitness necessary for this work are a way of life for Captain Jennifer Yates and Lieutenant John Sullivan.

d. These separate scales, which represent athletic fitness and fitness for life, suggest that it may be possible for a person to lack special athletic skills and still be very fit for life. Mark and label the positions on each scale at which you would place such an individual. What benefits do you think might be associated with a high level of fitness for life?

e. Consider where you would place yourself on the fitness-for-life scale. List several ways that you and the other members of your team could modify your lifestyles to improve your positions on this scale.

Explore

WHAT DETERMINES FITNESS?

At times in our evolutionary past, *Homo sapiens* depended on strength and endurance to survive, and one's fitness was linked to one's physical ability to obtain the fundamental requirements of life. Advances in technology, however, have helped shape a lifestyle in which most people in technological societies are required to do little *strenuous* physical work, either on the job or at home. Nevertheless, medical evidence suggests that a basic level of physical fitness is essential to withstand the stresses of life and to maintain an optimal state of well-being.

A number of factors appear to determine fitness. In this activity you will explore two factors, exercise and diet, that affect your level of fitness.

Process and Procedures

Part A Looking at Physical Activity

1. Work individually to complete the *Physical Activity Analysis* that your teacher provides.

    Attach this analysis to your journal; you may wish to refer to it during subsequent activities.

2. Answer the following questions individually, and record your answers in your journal. Base your responses on your activity level that you determined in Step 1.

    a. How do you think your activity level compares with that of a typical student in your class?

    b. How do you think your activity level might compare with that of Yates and Sullivan from the story *The Sky Awaits*?

Figure 7.2 **Comparing activities** Who is using more energy?

3. On the sheet that your teacher circulates, place a check mark (✓) next to your activity level, which you determined in Step 1.

    Your teacher will use this information to create an activity level profile for the class.

4. Enter the activity level profile for the class in the appropriate spaces on the worksheet in the *Physical Activity Analysis*.

Part B Looking at Diet

1. Work individually to complete the *Dietary Analysis* that your teacher provides.

   Attach this analysis to your journal; you may wish to refer to it in subsequent activities.

2. Answer the following questions individually, and record your answers in your journal. Base your responses on the *Dietary Analysis* that you completed in Step 1.

   a. How do you think your diet compares with that of a typical student in your class?

   b. How do you think your diet might compare with that of Yates and Sullivan?

3. Work with your team to complete the following steps, which will help your teacher generate a dietary profile of a typical student in your class.

   a. Choose one of the food groups listed on the *Dietary Analysis*, and ask four students outside your team how many servings from that group they ate yesterday. Record their responses.

   Each member of your team should choose a different food group.

   b. Calculate the *average* number of daily servings from the food group you chose that you and the four students you interviewed ate yesterday.

   This will give you a rough approximation of how much a typical student in your class eats from that food group.

   c. Contribute your data to the dietary profile your teacher will compile on the chalkboard.

4. Enter the information for this typical student into the column titled *Class Profile* on the worksheet in the *Dietary Analysis*.

5. Participate in a class discussion of the following. Base your responses on the information that you collected in Parts A and B.

   a. Are you surprised at the activity profile or dietary profile of your class? Are you surprised at your own profiles in contrast to the class profiles? Explain your answers.

   b. Identify ways in which the class profiles do *not* accurately represent activity levels and dietary patterns.

   c. Why do you think many people fail to sustain an adequate level of physical activity or fail to eat an appropriate number of servings from each food group?

Analysis

1. Work individually, and use all of the information in this activity to help you answer the following questions. Record your responses in your journal.

   The essay Human Performance: A Function of Fitness *(page E88) will be helpful.*

a. What resources does your body require during extreme levels of physical performance?

b. What resources does your body require to sustain basic levels of non-athletic activity?

c. What was your personal definition of fitness from Step 1 of the previous activity, *Thinking about Fitness*? Do you think that it needs to be revised? Why or why not? If you think your definition of fitness needs to be revised, enter the revised version in your journal.

2. Work with your team to complete the following task. Then follow your teacher's instructions for posting your advertisement.

Imagine that you are the owner of a new health club in your neighborhood. Create a one-page newspaper advertisement for your club that would draw a reader's attention to the most compelling reasons you know for maintaining or improving one's fitness for life.

WHAT IS IN THE FOOD YOU EAT?

Explain

Do you ever read food labels or the nutrition panels on the boxes of the cereal you eat for breakfast? These labels list the technical names of all the ingredients, some of which are probably familiar while others may be unfamiliar. Many of these ingredients include the nutrients that supply the matter that is so essential to the normal functioning of your body. From the previous activity, *What Determines Fitness?*, you know that the source of the energy required for fitness is food. Thus, the nutrients in food supply both the matter and the energy that your body requires for human performance. Are all of these nutrients equivalent, and how do food scientists know what nutrients are present in particular types of food?

In this activity you will have the opportunity to determine the presence or absence of five specific nutrients in a set of foods that your teacher will provide. Then you will combine these test results with the dietary analysis that you completed in *What Determines Fitness?* to discover what you *really* ate last week.

Materials (per team of 5)

5 pairs of safety goggles
5 lab aprons
5 pairs of plastic gloves
dropping pipet
500-mL beaker
3 10-mL graduated cylinders
19 18-mm x 150-mm test tubes
test-tube clamp
3 test-tube racks
2 glass-marking pencils
hot plate
brown paper
Benedict's solution in dropping bottle
Biuret solution in dropping bottle

indophenol solution in dropping bottle
isopropyl alcohol (99%) in screw-cap jar
Lugol's iodine solution in dropping bottle
2 food samples

Positive Controls

5 100-mL beakers
5 10-mL graduated cylinders
50 mL of 1% ascorbic acid (vitamin C)
small tub of regular margarine or small bottle of vegetable oil
50 mL of a 6% suspension of gelatin
50 mL of a 10% solution of glucose
50 mL of a 10% solution of sucrose
50 mL of a 6% suspension of starch

Process and Procedures

To begin developing your own explanation for what is in the food you eat, set up and complete the following tests:

1. Assemble in the teams of five that your teacher assigns, and obtain the two food samples that your team is to test.

2. In your journal create a table in which to record the foods the class will test, your predictions about what nutrients each food contains, and the actual test results.

   Each team will test two foods from one food group, but your table should have space to record your team's predictions and the class's results of five tests for 12 foods. Indicators, which are chemical or physical methods used to test for the presence of certain substances, are available to test for the following nutrients: starch, sugar (glucose and sucrose), vitamin C, lipids, and protein.

3. Begin to fill in your table by entering your predictions about what nutrients you will find in each of the 12 foods the class will test. Discuss your predictions with the other members of your team.

4. Decide which nutrient each member of your team will test.

   Each team member will test both food samples for the presence of this nutrient.

   Be sure to use the correct indicator for each test and to follow carefully the directions for its use.

5. Read the following information to help you understand the role of indicators in certain types of investigations.

   When scientists use indicators, they run positive and negative controls side by side with the unknowns. Positive controls show the expected results if a given substance is present; negative controls show the expected results if a given substance is absent. For example, a known sample of glucose (a positive control) tested with a glucose indicator gives a positive result. If you see the same result after testing a food, you could conclude that glucose is part of that food. In contrast, water (a negative control) tested with a glucose indicator would give a negative test result. Recall that all controls, except for the variable under study, should be handled in exactly the same manner as the experimental materials.

SAFETY: Put on your safety goggles, lab apron, and gloves. Tie back long hair.

6. Complete your tests according to the instructions in *Protocol for Nutrient Tests*, Figure 7.3. Record a plus sign (+) in the results column of your table if the food contains a given nutrient and a minus sign (–) in the same column if the food does not contain the nutrient.

Remember to label your test tubes and to use a negative control (water) and a positive control (a sample known to contain the substance in question) at the same time that you test each of your two foods.

Pay particular attention to the warning/caution statements for each of the indicators you will use in your tests.

Figure 7.3 **Protocol for Nutrient Tests**

Protocol for Nutrient Tests

Nutrient	Test
Lipids	Rub a drop of ground-up food on a piece of brown paper. Hold up the paper to the light after the water in the sample has evaporated. Lipids make a translucent greasy spot on paper. No translucent spot appears in the absence of lipids. **Note:** When food contains only a small amount of lipids, the lipids may not be detected by this method. If no lipids are detected, do the following: a. Place the assigned food in 10 mL of a lipid solvent such as isopropyl alcohol (99%). **WARNING: Alcohol is *flammable* and is a *poison*. Do not expose the liquid or its vapors to heat, sparks, open flame, or other ignition sources. Do not ingest; avoid skin/eye contact. If contact occurs, flush affected area with water for 15 minutes; rinse mouth with water. If a spill occurs, flood spill area with water; *then* call the teacher.** b. Allow the food to dissolve in the solvent for about 5 minutes. c. Pour the solvent on brown paper. The spot should dry in about 10 minutes. d. Check the paper for a translucent spot.
Protein	Place 5 mL of ground-up food in a test tube. Add 10 drops of Biuret solution. The Biuret test gives a pink to purple reaction in the presence of protein. No color change occurs in the absence of protein. **WARNING: Biuret solution is a *strong irritant* and may damage clothing. Avoid skin/eye contact; do not ingest. If contact occurs, flush affected area with water for 15 minutes; rinse mouth with water; call the teacher immediately.**
Starch	Add 5 drops of Lugol's iodine solution to a 5-mL sample of ground-up food. Lugol's turns blue-black in the presence of starch. No color change occurs in the absence of starch. **WARNING: Lugol's iodine solution is a *poison* if ingested, a *strong irritant*, and can stain clothing. Avoid skin/eye contact; do**

not ingest. If contact occurs, flush affected area with water for 15 minutes; rinse mouth with water; call the teacher immediately.

Sugar

Add 3 mL of Benedict's solution to a 5-mL sample of ground-up food. Place the test tube in a beaker of boiling water, and heat for 5 minutes. The Benedict's test gives an orange or brick-red color in the presence of glucose and a green color in the presence of sucrose. No color change occurs in the absence of sugar.

CAUTION: Benedict's solution is an *irritant*. Avoid skin/eye contact; do not ingest. If contact occurs, flush affected area with water for 15 minutes; rinse mouth with water; call the teacher.

WARNING: Use test-tube clamps to hold hot test tubes. Always hold a hot tube in such a way that the mouth of the tube is pointed away from your own or anyone else's face. Boiling water will scald, causing second-degree burns. Do not touch the beaker or allow boiling water to contact your skin. Avoid vigorous boiling. If a burn occurs, *immediately* place the burned area under *cold* running water; *then* call the teacher.

Vitamin C

Add 8 drops of indophenol solution to a 5-mL sample of ground-up food. Blue indophenol becomes colorless in the presence of vitamin C. (Disregard the intermediate pink stage.) No color change occurs in the absence of vitamin C.

CAUTION: Indophenol solution is an *irritant*. Avoid skin/eye contact; do not ingest. If contact occurs, flush affected area with water for 15 minutes; rinse mouth with water; call the teacher.

7. Wash your hands thoroughly.

Follow your teacher's instructions for disposing of all waste materials.

8. Share with the other members of your team the results of your tests on the two foods, and enter their test results in your data table.

Be sure that you understand the results of each test.

9. In the class data table, list the foods that your team tested and enter your test results.

10. Complete your data table by entering the class data.

11. Discuss the following questions with your teammates, and record your answers in your journal.

   a. How did the predictions that you made in Step 3 compare with the test results? Did anything surprise you? If so, what?

   b. How might the natural colors of the foods affect the results?

   c. Why was it important to test each indicator using water as the negative control substance?

   d. Why was it important to test each indicator with a substance *known* to contain the nutrient in question?

Analysis

Complete the following tasks, and record your responses in your journal.

The essay Food: Our Body's Source of Energy and Structural Materials *(page E90) will help you complete these tasks.*

1. Examine the *Dietary Analysis* that you completed in the activity *What Determines Fitness?* Notice that the foods you tested in the laboratory included one from each of the six food groups in this analysis. If you assume that foods from the same food group contain many of the same nutrients, then you can use your test results to determine the actual nutrients that were present in the foods you listed on your dietary analysis. On the basis of your tests *alone*, which foods did you eat that your body could use as a source of

   a. protein?

   b. sugar?

   c. starch?

   d. vitamin C?

   e. lipid?

2. Compare your test results with the information given in the essay *Food: Our Body's Source of Energy and Structural Materials*. What does the information in this essay suggest about the sensitivity and/or the accuracy of the tests you completed?

3. Which, if any, of the foods that you ate contained all of the nutrients for which you tested? What does this mean for eating a balanced diet?

Further Challenges

Study the two graphs that are shown in Figure 7.4. These graphs represent the results of an experiment in which a cracker was placed in a test tube along with saliva. Points on the graphs were determined by using the same indicator tests for starch and sugar that you used in this activity. Use the results displayed in these graphs to explain the changes caused by the saliva.

Figure 7.4 **Changes in nutrient content across time** Graph A shows the changes in the amount of starch across time, and Graph B shows the changes in the amount of glucose across time.

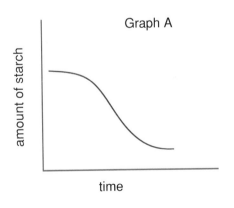

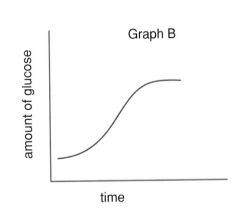

YOU ARE WHAT YOU EAT

Now you know what is in the food you eat, but once this food is inside you, how does it become useful to your body? What does your body do to this matter so that you can use the energy it contains for performance? How does your body prepare this matter so that you will have the building blocks necessary for growth and repair? In this activity you will look at digestion to understand the role it plays in preparing to release the energy stored in the molecules of food and in providing a source of building blocks for biosynthesis.

Materials (per team of 4)

materials to carry out the experiment that you design

Process and Procedures
Part A Food for Energy

1. Read the following paragraph to yourself.

    Starch is an energy storage molecule in plants, and it makes up a large part of the food of many organisms. Plants manufacture starch using energy from the sun and then break down the starch to its component sugars, releasing the energy that was stored in the starch molecules. Animals that eat plants can use starch in the same way. Starch is too large to be absorbed into the blood stream directly from the intestine, but many animals have enzymes that break down starch to small sugars. In humans the enzyme amylase, which is present in saliva, accomplishes the breakdown of starch.

2. With your team develop an outline of a controlled experiment to demonstrate that amylase breaks down starch to sugar.

    Use the following resources to help you complete this task: the essay What Happens to the Food You Eat? *(page E96), Background on Controlled Experiments (page 83), and the* Background Information *in this activity.*

3. Have your teacher check that your outline demonstrates the appropriate reaction and that you have designed a controlled experiment.

4. Choose a variable that might affect the amylase/starch reaction, and develop a hypothesis about its effect on the reaction.

    Variables that you might consider include

    *   presence of light

    *   concentration of amylase

    *   concentration of starch

    *   temperature

    *   pH

5. Have your teacher approve your hypothesis, and then write a detailed procedure for your experiment. Again, ask your teacher to approve it.

Background Information

Amylase is an enzyme found in human saliva, but it also is available in pure form from commercial suppliers.

Buffers can be added to a solution to control the pH.

Enzymes are proteins that speed up chemical reactions. A solution containing the enzyme must be added to a solution containing the substrate before the reaction can begin. The temperature, the pH of the incubation mixture, the time of incubation (the time that the enzyme and the substrate are in contact with each other), and other factors can affect the activity, or behavior, of the enzyme.

Glucose test strips indicate the presence of maltose in solution by changing color. (Your teacher will give you information about how to interpret the color change.)

Lugol's iodine solution is an indicator that changes color in the presence of starch. In a spot plate or test tube, use 1 drop of Lugol's iodine solution for 1 mL of sample solution that you are testing.

WARNING: Lugol's iodine solution is a *poison* if ingested, a *strong irritant*, and it can stain clothing. Avoid skin/eye contact; do not ingest. If contact occurs, flush affected area with water for 15 minutes; rinse mouth with water; call the teacher immediately.

Maltose is a sugar that results when starch is broken down by amylase. It can be obtained in pure form from commercial suppliers.

Starch is a macromolecule that can be obtained in pure form from commercial suppliers.

Substrates are molecules to which enzymes bind. Enzymes act to speed up reactions between substrates.

6. Carry out your experiment, and record all results in your journal.

   SAFETY: Put on your safety goggles and lab apron.

7. Wash your hands thoroughly when you are finished.

8. Participate in a class discussion of each team's results.

   Take notes on the effects of variables on enzyme reactions. Your classmates' results will be important to developing a complete lab report.

9. Prepare a lab report of your experiment.

   You will know your lab report is complete when you have included

   • *a statement of the question or hypothesis,*

   • *the procedure for conducting the experiment,*

   • *your results, and*

   • *an analysis of the data, including an explanation of the role of enzymes in digestion and the effect of different variables on enzyme reactions.*

Part B Food for the Body's Building Blocks

Food not only provides energy, but it also is the source of matter that animals use to produce new structures necessary for body maintenance and continued operation.

1. Take notes as you watch the videodisc segment *Introduction to Biosynthesis*.

2. Participate in a class discussion about the question, What happens when a foreign protein enters an animal?

STRUCTURES AND FUNCTIONS

Elaborate

Recall from the previous activity, *You Are What You Eat*, that the food you eat is broken down by the digestive system. The raw materials that result from digestion, materials such as amino acids, sugars, and fatty acids, may serve as building blocks in the synthesis of various body structures. Muscle tissue is a good example. For instance, amino acids are the building blocks your body requires for the repair and growth of muscle tissue. Once these building blocks are synthesized into muscle protein, they become part of a larger structure, a muscle. The function of muscles is to provide mobility, but not all proteins—for example, enzymes—provide mobility. What is special about how muscle proteins are arranged into structures that allow physical activity?

In this activity you will think about how matter from the building blocks produced by the digestion of food becomes organized into larger structures that have very specific functions. You also will relate the importance of the relationship between structure and function to human fitness and performance.

Materials (per team of 2)

brass brad
2 rubber bands
scissors
sheet of thin cardboard
25-cm piece of string
roll of tape

Process and Procedures

1. View the videodisc segment *Muscle Movement at the Molecular Level*, and with your partner suggest an answer to the following questions:

   a. What type of movement does this structure permit?

   b. What are the advantages and disadvantages of this structural arrangement of muscle fibers?

2. Perform the following steps to explore the function of muscle fibers at a *higher* level of organization, a level at which matter is organized in a way that allows physical motion.

   a. Bend and straighten one arm while using your other hand to feel what happens to your biceps and triceps.

   Consult Figure E7.10 in the essay The Structural Basis of Physical Mobility *(page E99) if you are not sure where the biceps and triceps are located.*

b. Repeat Step a until you develop an explanation of how your biceps and triceps generate these movements.

c. Discuss your observations and understandings with your partner.

3. Working with your partner, use the materials provided and the information in Figure 7.5 to construct a working model of your thigh and lower leg. Be sure to show the attachment sites of the quadriceps muscles located on the front of the thigh and the hamstring muscles located on the back of the thigh.

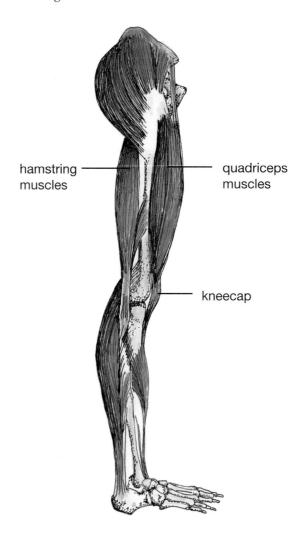

Figure 7.5 Muscles of the human leg The quadriceps and the hamstring muscles each extend across the knee to connect to the bones of the lower leg.

hamstring muscles

quadriceps muscles

kneecap

You might use the cardboard for bone, the string or rubber bands for muscle, the tape for tendons, and the brad for the knee. You may use a different combination of parts to form your model. Try to make the model as realistic as possible.

4. Place your model on the table with the leg straight. Grasp the hamstring just below the upper attachment site. Gently pull the hamstring. What do you observe? Release the muscle, but do not reposition the lower leg. Record your observations in your journal.

5. Now grasp the quadriceps just below its upper attachment site and pull gently. What do you observe? How does this movement differ from that in Step 4? Record your observations and explanation in your journal.

6. Discuss the following with your partner. Record your responses in your journal.

   a. Explain the statement, Muscles work in pairs. Why is this important?

   b. What is the role of the joint in producing movement?

   c. Recall from the videodisc segment that the molecular filaments in muscles can shorten mucles but cannot lengthen them. How do you think it is possible for us to push on anything?

   You may wish to test your answer by pushing on a wall and feeling both your biceps and triceps muscles. How are they acting to stabilize your arm? Why is this important to your ability to exert force against the wall?

Analysis

Participate in a class discussion of the following. The essays *The Structural Basis of Physical Mobility* (page E99), *The Ant That Terrorized Milwaukee* (page E103), and *Energy's Role in Making Structures Functional* (page E104) will help with this task.

1. Explain the *basic* matter and energy requirements needed for a muscle to contract.

2. What happens biologically when muscle fatigue occurs?

3. Models seek to mimic a structure or an event. Very good models mimic the actual structure or event so closely that changes in the model predict what would happen in the real world. Describe the strengths and weaknesses of your leg model.

4. Vertebrate muscles contract against the resistance of an internal skeleton made of bone. Compare some of the advantages and disadvantages of a hydrostatic skeleton, an exoskeleton, and an endoskeleton.

5. How does increased physical activity promote fitness? Answer specifically by including the effect of increased activity on the structure and function of an individual muscle.

6. Recall that your heart is a muscle. During vigorous activity your heart pumps faster and harder, delivering blood more rapidly to both the lungs and the exercising muscles. How would vigorous activity promote the increased fitness of the heart itself as well as help muscles in other parts of your body function more effectively?

MARATHON

Remember the last Summer Olympic Games? Many athletes broke world records; if you watched the games on television, you may have thought that those athletes made it look easy. Nevertheless, you were watching some of the highest levels of human performance ever recorded. A tremendous number of biological and behavioral factors had to be just right for such exceptional performances.

Evaluate

One of the most physically challenging of all of the events is the marathon. This race covers a distance of 41,920 m (26.2 miles). Athletes usually train for many years to build up to the level of endurance that is required to compete in this event. In this activity you will follow the progress of four people who entered a marathon, and you will propose explanations for the role that matter and energy played in their performance.

Process and Procedures

1. As a class watch the videodisc segment *A Good Day for Running*.

2. With your team of four, read the story *The Race*, which describes a marathon and the training and performance of four people who participated in it.

3. Decide which person in your team will study each runner in depth.

4. Review the information provided in *Physiologic Data Related to Physical Performance*, Figure 7.6.

   Think about how each set of information might help you analyze your runner's training and performance or help you suggest general strategies for a marathon runner.

5. Copy *Energy Expended in Training and Racing* (Figure 7.6A) into your journal. Use information in *Exercise and Energy Expenditure* (Figure 7.6C) as well as information in *The Race* to complete the table in your journal.

6. Use your understanding of biology along with *Physiologic Data Related to Physical Performance* and the table that you just completed to analyze your runner's training and performance on race day. Consider each of the points listed below, and record important information in your journal.

   Although each of you should analyze only your own runner, you may wish to remain in your teams as you do so. This strategy will allow you to share ideas and to begin comparing the runners as you examine their training and performance.

   a. Examine your runner's training schedule. In what ways did this schedule prepare him or her to finish the race? How did your runner's energy expenditure per week of race training compare with the amount of energy he or she expended during the marathon?

   b. Examine the diet of your runner in the weeks preceding the race. Did your runner appear to be increasing or decreasing his or her intake of any particular class of nutrients during training? Explain your answer.

   c. Summarize the strategy that you think your runner was using during training.

   d. Examine your runner's behavior on the race day before the race began. What strategies do you think he or she was using to prepare for the race?

   e. Examine your runner's performance during the marathon (for example, his or her pace, fluid intake, and apparent stamina and success). What strategies did he or she seem to be using?

f. Propose reasons why your runner's body behaved as it did.

g. Propose ways your runner could have improved his or her performance. State your proposals in an if-then format, such as "If Amy had done Y, then she would have seen Z effect on her performance." Support your if-then statements with data.

7. Meet with members of other teams who studied the same runner, and compare your findings. Modify your conclusions based on the group input.

Analysis

With your team discuss the following tasks, and record your responses in your journal.

Refer to the essay Factors Influencing Performance *(page E105) to provide greater depth to your understanding of fitness and performance.*

1. Which runner expended the most energy during training? Which runner expended the most energy during the marathon? What factors do you think were responsible for this?

2. In order to compare the training schedules and diets of the four runners before the race, complete the following tasks:

   a. List at least two training and dietary strategies that you think would be valuable for a person to consider if he or she were preparing for the same race next year.

   b. Explain the physiologic change(s) that you would expect to occur as a result of each strategy.

   c. Describe why such changes would be important to finishing the marathon. Support your answer with specific information from the *Physiologic Data Related to Physical Performance.*

3. Compare each runner's behavior on race day (during breakfast as well as during the race), and list at least two race-day strategies that you think would be valuable for each runner if he or she were running the same race next year. Explain how each strategy would be important to finishing the race. Support your answer with specific information from the *Physiologic Data Related to Physical Performance.*

4. Explain how the process of energy release from matter is more efficient in highly trained athletes than in most other people.

THE RACE

The Scenario

It is a mid-August day in a high-altitude Colorado town. Runners are gathering for an annual marathon that has been held at this site for many years. This race is interesting because several shorter races and a full marathon are held simultaneously, with all of the runners starting together. Each runner, either before or during the race, decides the exact distance he or she will run. The runners do so simply by stopping at certain measured increments. They may run 5K, 10K, 21K, or the full marathon distance of 41.9K. About 100 runners are lined up and ready to start.

Four of the Runners

Mel

Mel, a grandfather in his late forties, is a college professor who began running in his early thirties. Mel decided to begin running to control his weight and discovered that he really enjoyed this activity. As his running program progressed, he went from recreational running to competitive running because of the many positive changes he saw in his body and his lifestyle.

Amy

Amy was a member of her cross-country team in college. Now, at age 33, she is an attorney and maintains a high level of competitive fitness, continuing her running program as part of her lifestyle. She trains regularly and enters several races each year.

Neal

Neal is an exercise physiologist, currently employed as a scientist in a government laboratory. At the lab he manages a wellness program for employees, putting together exercise prescriptions for people who want to attain various levels of fitness. Neal was a track star and a classmate of Amy's in college. He also decided to make competitive running a major part of his life. He and Amy have run together many times.

John

John is in his early forties. He is an engineer who always had enjoyed watching runners and had wished that he could run. One evening after supper John announced to his wife that he was going to train for and complete a marathon. In contrast to the other three runners listed, John was a smoker, slightly overweight, had done no running except for occasional short-distance jogging, and drank alcohol "slightly more than moderately" (as he explained it).

How the Runners Trained

Mel

This was Mel's first full marathon, but he had participated in many shorter-distance races and fun runs. The farthest he had run competitively was a half-marathon. He had a practice course that he especially liked, and he ran the course on each practice day, a distance of about 8K (5 miles). He ran the course four days a week. About one month before the marathon, he increased his commitment to 13K (8 miles), four days a week. Mel lived and trained at high altitude in Colorado. Leading up to the race, Mel ate regular meals with his family. These balanced meals included carbohydrates, proteins, low amounts of fats, and plenty of vegetables. He had a meal of French toast and juice the morning of the race.

Amy and Neal

Their approaches to training and their lifestyles were nearly identical. They worked out at a moderate pace, consistently drinking sufficient fluids, eating diets that emphasized carbohydrates, and participating in a regular training program in which they ran a variety of distances at different speeds. About six weeks before the marathon, they finally settled on a steady workout regimen of running 16K, four days a week. They included rest periods in their weekly schedules to allow recovery from mild stiffness, soreness, and tired muscles. For several days before the race, they ate large amounts of whole-grain bread, cereals, and pasta. On the morning of the race, their breakfast consisted of oatmeal with a little milk and several glasses of juice.

John

John described his first month of training as "terrible." He vowed to quit smoking for the year of training that led to the marathon, but he struggled with shortness of breath during the early weeks. During the first month, his knees hurt enough to make him reconsider his decision to train and compete. Due to his increased activity, he lost 7 kg (15 lb) in two months. He ran mostly at lunch and occasionally again after work. He limited his running to about 5K per workout, five days a week. After about six weeks his knees stopped hurting. John was proud of his new energy level, which markedly increased his alertness at work. He also found that he required less sleep and that he slept very well. In the six weeks before the race, John increased his training distance to 10K, four days a week. He did not have time to work out more than this. On the day of the race, he thought,

"I feel ready!" Like Mel, during training John ate meals that consisted of whatever his family was eating. About two hours before the race, he ate ham and eggs, grits, three pieces of toast, several glasses of juice, and one cup of coffee.

The Race

The race began at 8:00 a.m. sharp. The coolness of the high-altitude summer morning was invigorating and added to the sense of excitement that all of the runners felt. Some of the participants had arrived nearly an hour before start time and were slowly stretching both upper and lower body muscles, concentrating on their leg muscles to prevent pulling and cramping. Others were slowly jogging and drinking fluids. Friends and spectators were gathering for the start. Runners prominently displayed their numbers. The weather promised to stay clear and dry.

Following some brief instructions and last minute information about the condition of the course from the starter, the runners lined up. How many would go the whole distance? The starter's pistol cracked loudly, and the mass of runners moved forward.

During the first 3K (1.8 miles), the line of runners gradually spread out. Five runners ran in a small pack a good distance out in front of the others and established a quick pace. Amy and Neal, running together for the moment, were in the front third of the main pack and were running at a respectable but comfortable pace. John was slightly behind them. Mel was at the beginning of the final third of the pack. A few runners straggled well behind.

Because the first quarter of the race was a gentle downhill stretch, most of the runners felt good. Each established his or her desired pace and settled down for the long haul. At the 5K mark about 10 runners decided to call it a race. It had been fun! The others could go farther if they wanted. At the 10K point Amy, Neal, John, and Mel were in the same respective positions, all running steadily without tiredness, soreness, or fatigue. All felt that their respective training regimens were serving them well. They watched as several more runners, including a couple of the front runners, decided to stop.

The runners now were spread out over a 1.5K length. As the 21K marker came into view on an uphill segment, some runners obviously were struggling to continue, having established a plodding gait, and were ready to call 21K their distance. Several of the front runners stopped here as well. Amy and Neal were now about one-fourth back in the remaining pack and still were running together. John was

among the last 10 runners left. Mel was in the middle, now 400 m ahead of John. John was feeling a slight pulling sensation in his right calf muscle. He had altered his stride slightly to see whether he could "work it out."

The next 10K segment was quiet and uneventful. All four of our racers and their fellow runners settled into an automatic pace. Amy and Neal were running in relative comfort, pushing themselves slightly, but doing well overall and still maintaining their positions. Mel maintained his middle-of-the-pack position and was beginning to experience some leg muscle fatigue. In fact, the race was becoming a serious effort, but he was still all right and willing to go the entire route. John noticed the beginning of a blister on his right foot as his shoe rubbed the same spot over and over. Both of his calf muscles were very tight and beginning to hurt, especially now that he was running on hard pavement. Like Mel he was experiencing leg muscle fatigue, and he considered stopping where he was. Still he ran on. Several other runners were dropping out, some limping, a few holding their tightened or pulled leg muscles. Some were holding their cramping abdominal muscles. Most just suffered severe fatigue.

The last 12K produced the greatest change in the positions and welfare of the remaining runners. About 30 of the original 100 were left. Neal was running sixth. Amy was about 500 m back, but she still was running smoothly and steadily. Mel's stride was short, and he felt as though a brick was at the bottom of his chest. The race had become very hard work. His leg muscles were beginning to cramp, and the only way he could relieve these effects was to reduce his pace somewhat and to run with an exaggerated heel-toe gait. With 8K to go, John "hit the wall." His legs became so tired, heavy, and cramped that he could do little more than make slow and laborious forward progress. His pace was only slightly faster than a walk. His chest muscles ached severely, and he began to feel somewhat nauseated. He was in last place, hurting all over, but he still was determined to finish.

Neal was now in fourth place. As he entered the last 2K, however, he, too, experienced a "wall" effect. He was pushing himself hard, seeing not only the end of the race, but the possibility of improving his position as well. He crossed the finish line in third place at 3 hours, 5 minutes, 5 seconds, edging out the next male competitor by 50 m. His body went limp, and he had difficulty standing upright. Amy finished at 3 hours, 29 minutes and was the fifth female competitor to cross the line. She experienced the same final effects as Neal. At 3 hours,

55 minutes, 10 seconds, Mel finished. His "wall" experience in the last 5K had been quite dramatic. He had no energy left for a final sprint to the finish.

At 4 hours, 22 minutes, 20 seconds, John completed his first marathon. His finishing pace was a slightly elongated walk. He held his middle. His legs would no longer support him, and he went first to his knees, then over on his back in total collapse. His chest heaved with exaggerated breathing for several minutes before he was able to sit upright. As the runners were recovering from the race and congratulating one another, John's thought was, "Maybe one is enough!"

Figure 7.6 Physiologic Data Related to Physical Performance

Physiologic Data Related to Physical Performance

The following collection of data provides various types of evidence related to diet and physical performance. Use the data to help you analyze your runner's preparation and performance and to help you suggest strategies for improving them.

Figure 7.6A Energy Expended in Training and Racing

Energy Expended in Training and Racing

Runner	Weight	Kcals Used/Week Normal Workout (assume 7 min/ mile pace)	Kcals Used/Week Race Training (assume 7 min/ mile pace)	Kcals Used for Marathon (see finish time)
Neal	68 kg (150 lbs)	N/A		
Amy	50 kg (110 lbs)	N/A		
Mel	82 kg (180 lbs)			
John	75 kg (165 lbs)			

Figure 7.6B Muscle Glycogen Levels in Relation to Perceived Effort during 3 Hours of Treadmill Training

Muscle Glycogen Levels in Relation to Perceived Effort during 3 Hours of Treadmill Training

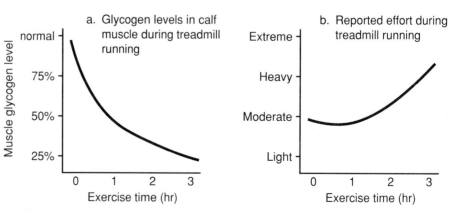

Source: From *Physiology of Sport and Exercise,* by Jack H. Wilmore and David L. Costill. ©1994 by Human Kinetics.

Exercise and Energy Expenditure

Exercise	Kcals Used/ Pound Every 10 Minutes	Exercise	Kcals Used/ Pound Every 10 Minutes
Bicycling		Skiing	
slow 8 km/h (5 mph)	0.25	Downhill	0.59
moderate 16 km/h (10 mph)	0.50	Cross-country (noncompetitive)	0.78
fast 21 km/h (13 mph)	0.72		
		Soccer	0.63
Golf	0.29		
		Stationary Running	0.78
Hiking	0.42	(70–80 counts/minute)	
Running		Swimming (crawl)	
9.6 km/h (10 min/mile)	0.79	20 meters/minute	0.32
10.7 km/h (9 min/mile)	0.84	50 meters/minute	0.71
12 km/h (8 min/mile)	0.89		
13.6 km/h (7 min/mile)	0.95	Walking	
16 km/h (6 min/mile)	1.00	3 km/h (2 mph)	0.22
		8 km/h (5 mph)	0.64
Racquetball	0.63		

Figure 7.6C Exercise and Energy Expenditure

Energy and Nutrients

Food	Percentage of Fat	Percentage of Protein	Percentage of Carbohydrate	Kcals (Food Value) per 100 g
Apples	0.4	0.3	14.9	64
Bacon, fat	76.0	6.2	0.7	712
broiled	55.0	25.0	1.0	599
Beef, medium lean	22.0	17.5	1.0	268
Bread, white	3.6	9.0	49.8	268
Butter	81.0	0.6	0.4	733
Cabbage	0.2	1.4	5.3	29
Carrots	0.3	1.2	9.3	45
Cheese, Cheddar	32.3	23.9	1.7	393
Chicken	2.7	21.6	1.0	111
Corn (maize)	4.3	10.0	73.4	372
Haddock (fish)	0.3	17.2	0.5	72
Lamb, leg	17.5	18.0	1.0	230
Milk, whole	3.9	3.5	4.9	69
Oatmeal, dry uncooked	7.4	14.2	68.2	396
Oranges	0.2	0.9	11.2	50
Peanuts	44.2	26.9	23.6	600
Peas, fresh	0.4	6.7	17.7	101
Pork, ham	31.0	15.2	1.0	340
Potatoes	0.1	2.0	19.1	85
Spinach		2.3	3.2	25
Strawberries		0.8	8.1	41
Tomatoes	0.3	1.0	4.0	23

Figure 7.6D Energy and Nutrients

Figure 7.6E **Effect of Diet on Muscle Glycogen and Muscle Endurance**

Effect of Diet on Muscle Glycogen and Muscle Endurance

Diet	Amount of Glycogen in Muscles (g/kg)	Average Endurance Running at Speeds Characteristic of a Marathon (min to Exhaustion)
High carbohydrate diet	40	240
Mixed carbohydrate and fat diet	20	120
High fat diet	6	85

Source: From *Textbook of Medical Physiology*, 8th edition, by Arthur C. Guyton. ©1991 by W.B. Saunders Co.

Figure 7.6F **Effect of Exercise on Muscle Structure**

Effect of Exercise on Muscle Structure

1. Mild to moderate increase in number of muscle fibers.
2. Increased capacity to transport oxygen from the blood to the mitochondria.*
3. Increased numbers of mitochondria.
4. Increased growth of capillaries serving the muscle.

* Mitochondria are the parts of the cells that are primarily responsible for the oxygen-requiring release of energy from glucose.

Source: From *Textbook of Medical Physiology*, 8th edition, by Arthur C. Guyton. ©1991 by W.B. Saunders Co.

Figure 7.6G **Comparing Cardiac Outputs of Marathoner and Non-Athlete**

Comparing Cardiac Outputs of Marathoner and Non-Athlete

	Stroke Volume* (mL)	Heart Rate (beats/min)
Marathoner		
Resting	105	50
Maximum	162	185
Non-Athlete		
Resting	75	75
Maximum	110	195
* Volume of blood moved in one heartbeat		

Source: From *Textbook of Medical Physiology*, 8th edition, by Arthur C. Guyton. ©1991 by W.B. Saunders Co.

> *We have entered the cell, the mansion of our birth, and have started the inventory of acquired wealth.*

Albert Claude, 1898–1983, Belgian-born American cytologist and Nobel Prize winner

8 THE CELLULAR BASIS OF ACTIVITY

Gasoline, nuclear power, and electricity all are obvious forms of energy, but do you also think of energy when you see sunlight or green plants? In fact, energy is found wherever matter is organized, from the molecules of rocks to cells to entire living organisms. Even though most people are unaware of the tremendous importance of energy in living systems, without energy life would not be possible.

In this chapter you will begin to investigate the important relationship between matter and energy by considering examples of where and how energy is stored and released. From Chapter 7 you realize that food contains energy that humans and other organisms require for daily activities as well as for less common physical performances. Eating and digestion, however, do not explain exactly how food molecules become useful as energy. Now you will have the opportunity to learn how energy in food is converted into a form that fuels the chemical reactions that keep cells, tissues, and organisms alive. In addition, you will learn about the original source of the energy that is present in food. Once you understand these basic connections between energy and matter, you will explore some of the specific cellular reactions for which energy is needed, in particular, the reactions that create new molecules and maintain the organization that is characteristic of living systems.

ACTIVITIES

Engage   Releasing Energy

Explore
Explain   Energy in Matter

Explain   Keep on Running!

Explain
Elaborate   Using Light Energy to Build Matter

Elaborate   Building Living Systems

Evaluate   Tracing Matter and Energy

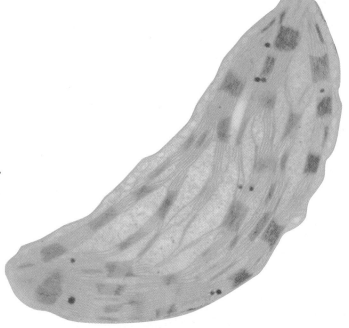

RELEASING ENERGY

A runner in a marathon is pushing hard several miles before the finish line, but then she suddenly slows to a walk, clutches her side, and sways dizzily, about to fall. This athlete is said to have "hit the wall." She did not actually slam into a brick barrier, but she may feel as if she has done so. In this case "hitting the wall" means that the runner has exhausted the energy supplies necessary to keep running. After eating food and resting, those energy supplies will be replenished, but how does this additional food become the energy needed for physical activity? In this activity you will begin to examine the relationship between matter, such as food, and energy.

Process and Procedures

1. Read the story *A Matter of Explosions* (page 167), and think about the following questions:

   - What provided the energy in the grain explosion?

   - How can energy be stored in grain?

   - What started the explosion that released the energy?

   - Why do you think you do not explode when you eat grain products?

2. Contribute your thoughts to a class discussion.

ENERGY IN MATTER

Explosions are a dramatic example of the release of a tremendous amount of energy, but what is the source of all this energy? In the case of the grain elevator explosion, the energy came from grain dust and air, two simple forms of matter that often are not considered energy sources. Grain and air, however, are not exceptional forms of matter; all matter contains energy. All matter does not, however, contain the same *amount* of energy. The precise amount of energy in any substance is determined by the particular organization of **atoms**, the building blocks of matter, in that substance. In this activity you will investigate some of the links among matter, energy, and organization and attempt to describe the close relationship among them.

Materials (per team of 2)

2 pairs of safety goggles
2 lab aprons
125-mL flask
Pyrex test tube
graduated cylinder
2 microscope slides
test-tube rack
dropping pipet

spatula or spoon
materials for molecular model
hand lens or microscope (optional)
2-cm piece of magnesium ribbon
ammonium nitrate
hydrochloric acid solution
saturated urea solution

A MATTER OF EXPLOSIONS

Figure 8.1 Grain storage elevators

Hank celebrated his 20th birthday by starting a new job at the town's grain storage facility. He was eager to make a good impression. Middletown Grain Storage was the largest employer in town, offering good wages, great benefits, and opportunities for advancement. The facility consisted of three huge grain elevators that stored wheat. Each storage elevator was shaped like a giant cylinder and rose 100 feet above ground and descended 20 feet underground. A series of tunnels connected the underground portions of the cylinders. The tunnels contained machinery that dropped the grain onto concave conveyer belts and then moved the grain from one storage tower to another, where giant elevating platforms lifted the grain to the top of the tower.

Hank was in a small underground storage room next to one of the towers, waiting for his supervisor to arrive and give him instructions. The floor was several inches deep with dust from the grain, and he was making a mess as he paced impatiently. Noticing a metal shovel in a corner, Hank decided to get busy and clean up the place a bit. He took the shovel and was scraping it along the cement floor when his boss and another worker appeared at the door. In an instant they jumped at him, one grabbing his arms while the other grabbed the shovel. Hank was dumbfounded.

"Sorry to startle you," the boss said, "but if you're going to work here, you've got to learn some rules. First and most important: Don't do anything that could cause sparks!"

"But there's no gasoline around, just wheat. What difference would a few sparks make?" Hank responded, feeling embarrassed.

"Here, read this," the boss stated grimly as he removed a worn letter that had been posted on a bulletin board. This is what Hank read:

10 August

Dear Shannon,

I am amazed to be alive and able to write this letter to you—or actually to dictate it; my hands are burned too badly to write it myself.

As you heard on the news, the worst happened: an explosion in elevator number two. We all had been warned, but I never could have imagined the horrible experience. It happened amazingly fast. First there was an odd "whoosh" noise, and my right side was seared with heat. Before I could think about what was happening, there was a deafening noise. I still have ringing in my ears.

I guess I was lucky to be a bit removed from the center of the explosion. Four guys here did not make it. It's hard to believe that the whole thing probably started with a spark from the gears of the conveyer belt while we were shifting grain—we'll never know for sure. The explosion was like dynamite. It blew the top right off the number two elevator, and then the fire took over.

I know your facility is similar to ours. Whatever you do, don't let this happen there.

Yours truly,

Bill

Hank looked silently at the others in the room. He had no idea that the flourlike dust from simple wheat could produce such an explosion and fire, and he certainly couldn't explain *why* such a thing could happen or *how*.

Process and Procedures
Part A Energy in Reactions

Work with your partner during Part A.

SAFETY: Put on your safety goggles and lab apron.

1. Discuss how you would define matter and energy and how they are related. Based on these definitions, identify two or three examples of evidence that support the idea that matter and energy are related.

    Record your team's definitions and examples in your journal. As you work, encourage your partner to generate ideas.

2. Try this exercise.

    a. Use a graduated cylinder to measure 5 mL of 0.7M hydrochloric acid. Pour this into a test tube that is firmly seated in a test-tube rack. Do **not** remove the test tube from the rack.

    WARNING: 0.7M HCl is a *strong irritant*. Avoid skin/eye contact; do not ingest. If contact occurs, flush affected area with water for 15 minutes; rinse mouth with water; call the teacher.

    b. Feel the test tube to note the relative temperature (Figure 8.2).

    c. Place 2 cm of magnesium ribbon into the acid in the test tube, and observe what happens as the solid dissolves.

    d. Feel the test tube, and record your observations in your journal.

    e. Discuss the energy changes that you observed, and compare these changes with the events in the opening story.

3. Now try another exercise.

    a. Put approximately 40 mL of room-temperature water in a flask.

    b. Feel the flask to note the relative temperature.

    c. Put about two spatulas of dry ammonium nitrate into the water in the flask, and swirl the water gently to make the powder dissolve.

    CAUTION: Ammonium nitrate is an *irritant*. Avoid creating dust; avoid skin/eye contact; do not ingest. If contact occurs, flush affected area with water for 15 minutes; rinse mouth with water; call the teacher.

    d. Feel the flask, and record your observations in your journal.

    e. Discuss the energy changes that you observed.

4. Finally, try this exercise:

    a. Use a dropping pipet to place one drop of saturated urea solution on a microscope slide.

    CAUTION: Urea solution is an *irritant*. Avoid skin/eye contact; do not ingest. If contact occurs, flush affected area with water for 15 minutes; rinse mouth with water; call the teacher.

    b. Use the edge of the second slide to spread the drop in a thin layer across the surface of the first slide.

c. Watch the layer closely for changes, which may take several minutes to develop. Use hand lenses or microscopes, if available.

d. In your journal answer the following question:

Solute molecules in a solution have the freedom to move about according to the forces of diffusion; in other words, these molecules have a great deal of entropy. But what appears to happen to the organization of urea molecules as the water in the urea solution evaporates?

5. Wash your hands thoroughly.

6. Complete the following tasks in your journal. Be prepared to participate in a class discussion about the relationship between energy and the organization of matter.

a. Is the reaction in Step 2 exothermic or endothermic? What about the reaction in Step 3? State your evidence in each case.

b. For the endothermic reaction, describe where the energy needed for the reaction came from. For the exothermic reaction, describe where the energy produced by the reaction went.

c. As the ammonium nitrate dissolved in the water, what do you think happened to the organization of the molecules in this solid? How did this differ from what happened to urea?

Answer the questions thoroughly; you will refer to your notes throughout this chapter.

The essay Matter and Energy Are Related *(page E108) contains information that you will find useful.*

Part B Molecular Models

Complete Part B individually.

1. Look at Figure 8.3 to see several ways that molecules can be represented. Which representation would you use to demonstrate molecular structure

Figure 8.3 **Four different representations of the molecule methane**

(atoms bonded together) to someone who did not understand that matter is organized? Explain your choice.

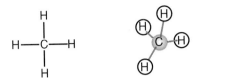

a. chemical
formula

b. structural
formula

c. ball-and-stick
model

d. space-filling
model

2. Watch the videodisc segment *Molecular Models* so you can answer the following question.

    How do the atoms of simple molecules serve as building blocks for much larger and more complex molecules?

3. Use the materials that your teacher provides to construct a model of a molecule of urea. Base your model on information from the videodisc, the essays, the following rule of molecular bonding, and the following information about molecular bonds.

    The number of bonds that an atom usually has is based on the number of electrons in its outer shell.

    - Urea is represented by the chemical formula $CO(NH_2)_2$.

    - Carbon usually has four bonds.

    - Oxygen usually has two bonds.

    - Nitrogen usually has three bonds.

    - Hydrogen usually has one bond.

Analysis

Respond to the following tasks in your journal. Be prepared to share your responses in a class discussion.

You might find the essay Energy Is Converted and Conserved *(page E110) helpful.*

1. If an exothermic reaction releases heat, why is it inaccurate to say that an endothermic reaction releases cold?

2. What is the difference between kinetic energy and potential energy? Provide at least two examples of each to help with your explanation.

3. Explain how energy can be stored in a molecule yet not be destroyed when that molecule is broken down into smaller molecules.

4. Define ATP's role as a link between matter and energy.

5. How would you change your definitions of matter, energy, and their relationship from those that you developed in Step 1 of Part A? Feel free to use diagrams or other visual aids to help explain your definitions.

Further Challenges

1. In your journal draw a diagram of your own design to show that starch is a macromolecule composed of many molecules of glucose. Then explain how the arrangement of matter determines its energy content.

2. A saturated solution is a solution in which the concentration of solute is so high that no more solute can dissolve. With this knowledge explain what happened during the crystallization of urea in Step 4, Part A.

   Do you think a dilute solution of urea (one in which the concentration of solute is very low), if left uncovered for a long period of time, would eventually form crystals as well? Explain your answer.

KEEP ON RUNNING!

The connections among food, the energy stored in matter, the release of potential energy from molecules, and the importance of exercise to human performance was so interesting to one of your classmates that she decided to investigate the career opportunities for nutritionists and dietitians. She began by subscribing to several health and nutrition magazines. In last month's issue of *Athlete's World*, she ran across an interesting ad:

Explain

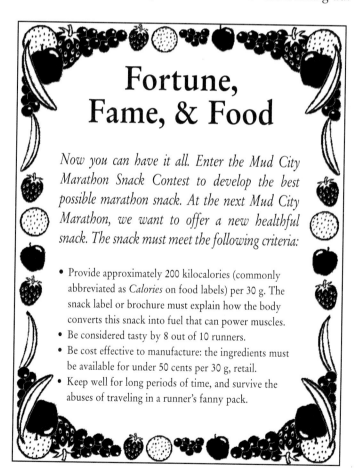

Fortune, Fame, & Food

Now you can have it all. Enter the Mud City Marathon Snack Contest to develop the best possible marathon snack. At the next Mud City Marathon, we want to offer a new healthful snack. The snack must meet the following criteria:

- Provide approximately 200 kilocalories (commonly abbreviated as *Calories* on food labels) per 30 g. The snack label or brochure must explain how the body converts this snack into fuel that can power muscles.
- Be considered tasty by 8 out of 10 runners.
- Be cost effective to manufacture: the ingredients must be available for under 50 cents per 30 g, retail.
- Keep well for long periods of time, and survive the abuses of traveling in a runner's fanny pack.

This activity provides the tools for you and your teammates to compete in the Mud City Marathon Snack Contest for a prize that your teacher will announce.

Materials (2 or 3 teams per LEAP-System™)

3 pairs of safety goggles
3 lab aprons
100-mL graduated cylinder
250-mL Erlenmeyer flask
Macintosh or IBM-compatible computer
LEAP-System™
thermistor
balance
forceps
tin can with cutout air and viewing holes
cork with sample holder
kitchen matches
20-cm x 30-cm piece of extra heavy aluminum foil
2 pot holders
small container of water
food samples
materials to design a food label

Process and Procedures

1. Your challenge is to select three different foods from the available
    choices, measure the calories in each according to a calorimetry protocol,
    and use at least two of these foods (plus your understanding of energy,
    matter, and nutrition) in a combination that provides the best chance of
    winning the Mud City Marathon Snack Contest. With your team discuss
    and develop your strategy, and record it in your journal.

    A good strategy will include the following:

    *   a brief written explanation about why you would like to test the three
        foods you have chosen,

        *You might look back at Chapter 7 to remind yourself about the role of the
        different components of food.*

    *   the use of the *Calorimetry Protocol* (Figure 8.4),

    *   calorimetry tests of three samples of each food, and

    *   a data table for recording your results.

    *Decide who in your team will be the experimenter, who will be the recorder/computer
    operator, and who will be the safety monitor who assures that the correct safety
    procedures are followed.*

2. Have your teacher approve your strategy.

3. Conduct your tests.

    Remember to follow the required safety cautions.

    SAFETY: Put on your safety goggles and lab apron. Tie back long hair and roll up
    long, loose sleeves.

Calorimetry Protocol

A calorimeter is an instrument that measures in calories the amount of energy in foods. A calorie is the amount of heat required to raise the temperature of 1 g (1 mL) of water 1°C. The caloric values of foods in diet charts are given in kilocalories (1000 calories), or kcals.

Using a simple calorimeter (Figure 8.5) with the LEAP-System™ and a thermistor, you can measure the change in temperature of a known volume of water. The temperature change is caused by the absorption of the heat given off by burning a known mass of food. Based on the change in temperature, you can calculate the amount of energy in the food.

1. Using the balance, determine the mass to the nearest 0.1 g of each food sample. Record each mass in a data table.

2. Obtain a 250-mL Erlenmeyer flask, a tin can, a cork with sample holder, and a piece of extra heavy aluminum foil.

3. Assemble a calorimeter like the one shown in Figure 8.5. Practice assembling and disassembling the equipment.

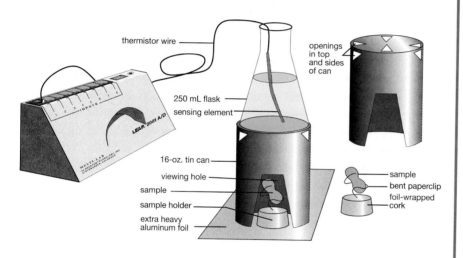

4. With the calorimeter disassembled, measure 100 mL of tap water and pour it into the flask.

5. Using the procedures explained in ALPHAEXP, run experiment HUMAN04.*

    This experiment is located in the directory HUMBIO\ (IBM) and in the BSCS Human Biology Folder (Macintosh).

    *IBM users, run HUMAN04T for two teams or HUMAN04X for three teams.

6. Identify your team's thermistor.

    Team A, use the thermistor in Input 1; Team B, the thermistor in Input 2; and Team C, the thermistor in Input 3. It will greatly simplify taking measurements if all teams coordinate their steps throughout the experiment.

7. Prepare your thermistor by following these steps:

    a. Bend the thermistor wire at a right angle about 3 cm from the tip of the temperature sensing element.

    b. Insert the thermistor into the flask. The sensing element should be submerged in the water but should not touch the glass.

    c. Bend the thermistor wire at the top of the flask to hold the sensing element in position.

Figure 8.4 **Calorimetry Protocol**

Figure 8.5 **The calorimeter setup**

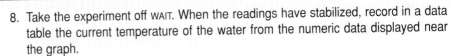

8. Take the experiment off WAIT. When the readings have stabilized, record in a data table the current temperature of the water from the numeric data displayed near the graph.

9. Place a food sample in the wire holder anchored in the cork. Then place the cork on the piece of aluminum foil.

   For runny foods, you may need to use extra heavy aluminum foil to construct a sample holder that is shaped like a boat or platform.

10. Carefully set fire to the food sample. This may require several matches. Discard burned matches in the container of water.

    WARNING: Matches are flammable solids. In case of burns, place burned area under *cold* running water; call your teacher immediately.

11. Place the tin can over the burning sample with the viewing hole facing you. Place the flask of water on top of the tin can.

12. MARK the graph to indicate that burning has begun. Record your team's MARK number. Continue recording the water temperature, even when the sample has burned completely, until the water temperature begins to decrease.

    The temperature will continue to rise after the sample has burned completely as the water absorbs heat from the tin can.

13. Record the maximum water temperature from the numeric data displayed near the graph.

14. Allow the calorimeter to cool about 2 minutes before disassembling.

    CAUTION: The flask and tin can will be hot; use potholders to handle these items, and place them *only* on the aluminum foil. The sample holder also will be hot; use forceps to remove the burned sample. In case of burns, call your teacher immediately; place burned area under *cold* running water.

15. Repeat the procedure until you have data for three samples of each food. Change the water in the flask each time. REINITIALIZE the experiment before burning each sample.

    If several teams are using one LEAP-System™, do not REINITIALIZE until all teams have finished their current samples.

16. When all teams have finished, QUIT HUMAN04.

17. Wash your hands thoroughly before leaving the laboratory.

18. Analyze your data using the following steps:

    a. Determine the average change in temperature for each of your three foods.

    b. Calculate the number of calories produced per gram for each of your three foods. To do this, multiply the increase in water temperature (average change) by 100 (the number of mL of water used).

    c. Convert the number of calories to kcals by dividing by 1000.

    d. Calculate the kcals produced per gram of food. To do this, divide the number of kcals by the number of grams of food burned.

    e. Enter all data in your table.

Analysis

You will find the essays *Controlling the Release of Energy from Matter* (page E116) and *Cellular Respiration: Converting Food Energy into Cell Energy* (page E117) helpful in completing this task.

1. Develop a package label or brochure that is informative and is an effective marketing tool for your Mud City Marathon Snack. Your package label or brochure should answer the following questions:

   - What combination of two or three foods did you decide to use in your snack?

   - How did your calorimetry data and your understanding of nutrition influence your decision to develop your snack using these two or three particular foods?

   - How much does your snack cost per 30-g serving?

     You may find Figure 8.6 helpful in calculating costs.

   - What test results demonstrate that this is a good snack for marathon runners?

   - How does the matter in the food become usable energy for the body? In particular, how does it help a marathon runner keep on running?

Cost of Common Foods

Food	Cost per 30 g (¢)	Food	Cost per 30 g (¢)
Almonds	55	Oatmeal	10
Cashews	56	Potato chips	20
Corn flakes	16	Peanut butter	12
Dried coconut	13	Pretzels	15
Corn chips	18	Raisins	12
Cheese snacks	19	Roasted peanuts	16
Honey	13	Walnuts (English)	38

Figure 8.6 **Cost of Common Foods**

2. Submit your package label or brochure to the Mud City Marathon Snack Contest judges, and wait for a decision on the winner.

Further Challenges

Complex processes for extracting energy from matter and using that energy to fuel cellular reactions have evolved in living organisms. As a result, when a cheetah begins accelerating toward a gazelle that it desires as its next meal, the cheetah can produce enough ATP in its muscles to fuel the rapid contractions necessary for an explosive sprint. If the cheetah had too little energy, it would have no chance of catching its prey. You might guess from your understanding of homeostasis that the processes of energy production probably are regulated in some way to maintain an appropriate supply of energy for the activities of living organisms. To understand how organisms balance their energy needs with their energy supplies, read the essay *Regulation and Energy Production* (page E120), and draw a feedback loop that demonstrates how energy is regulated.

Explain
Elaborate

USING LIGHT ENERGY TO BUILD MATTER

In the last activity you studied cellular respiration, a method for obtaining energy from matter in a form that is useful for cell activity. This process is dependent on glucose molecules, which can be derived from glycogen, starch, fats, and other macromolecules. In humans this glucose, and indeed all of the carbon in human macromolecules, originally came from molecules produced outside our bodies. Many of these molecules come from the plants we eat; the rest come from animals or other organisms that are dependent on plants. Humans, and all organisms that depend on previously assembled molecules for their carbon and energy, are called **heterotrophs** (*hetero* = other, *troph* = to feed). This dependence on other sources raises some interesting questions about the *original* source of energy and matter. For instance, how did energy first come to be stored in these molecules on which we depend, and what is the original source of the energy contained in their structures?

The answer is green plants, which are quite different from heterotrophs. They are called **autotrophs** (*auto* = self, *troph* = to feed), which means that they do not have to rely on other organisms for the complex molecules necessary for life. Autotrophs are able to make all of their own macromolecules. Plants need only light, water, air, and a few essential elements that are available from the soil to grow. In other words, plants are able to make their own food—the carbon-containing molecules such as simple sugars and starch that are used to support their cellular activities. The process of making these carbon-containing molecules is called **photosynthesis**, a term that comes from two words: *photo*, meaning light, and *synthesis*, meaning to make.

A number of variables can influence the efficiency, or rate, of photosynthesis. In this activity you will identify this range of variables and test one. This process will help you put together the entire cycle of energy, matter, and organization by examining the processes that capture energy and trap it in matter.

Materials (per team of 4, 2 teams per LEAP-System™)

2-L beaker
100-mL beaker or small jar
250-mL flask
2 25-mm x 200-mm test tubes
250-mL graduated cylinder
wrapped drinking straw
lamp with 100-watt spotlight
Macintosh or IBM-compatible computer
LEAP-System™
2 pH probes
thermistor
printer
red, blue, and green cellophane
tape
ruler
glass-marking pencil

distilled water
tap water at 25°C
ice water (10°C)
jar of tap water for storing pH probes
2 15-cm sprigs of young, healthy *Anacharis* (elodea)

Process and Procedures

1. With your teammates develop a list of at least three variables that could affect the rate of photosynthesis.

   The information in the introduction to the essay Getting Energy and Matter into Biological Systems *(page E123) will help you generate your list.*

2. Compare your list with one other team's list. Add any new ideas to your list.

3. Participate in a class discussion to identify two variables that you could test easily in your classroom.

4. Use the *Photosynthesis Protocol*, Figure 8.7, and the materials available to you to outline an experiment that tests the influence of one of these variables on the rate of photosynthesis. Your experiment should:

   • answer some variation of the question, What affects the rate of photosynthesis? and

   • include an if-then statement that relates your question to your experiment.

5. Write a plan for your experiment, and have your teacher approve it.

6. Conduct your experiment.

   Record your data in your journal.

7. For the variable that your team tested, draw a graph that plots the rate of photosynthesis, as indicated by pH, on the *y* axis (vertical) against elapsed time on the *x* axis (horizontal).

   Remember to include a line that represents your control.

8. Record your team's data in a class data table, and present your team's conclusions to the class.

   Note the variables that the class tested and the range of values recorded.

9. Review the graphs from each team so that you can discuss with your team your conclusions about the effects of each variable tested on the rate of photosynthesis.

Figure 8.7 **Photosynthesis Protocol**

Photosynthesis Protocol

There are several ways to measure the rate of photosynthesis. This protocol outlines a model system that allows you to determine how quickly carbon dioxide is being used in photosynthetic reactions. You will do this by measuring the decrease in carbon dioxide concentration in the surroundings of a photosynthetic system. You can measure carbon dioxide (CO_2) concentration in a water environment indirectly by measuring pH because carbon dioxide gas dissolves in water, producing carbonic acid (H_2CO_3). If CO_2 is removed from the water, there is less carbonic acid in the water, and the pH increases.

This model system uses two sprigs of *Anacharis* (a common aquatic plant called elodea) as the photosynthetic organism. A complete system for monitoring photosynthesis also should include a light source to provide energy and a LEAP-System™ with a thermistor to monitor the temperature and two pH probes to measure pH changes associated with carbon dioxide use.

The following variables can be tested with the equipment you have available:

Light intensity: The intensity of light reaching the chloroplasts of *Anacharis* can be changed by moving the light source. Place the light source at a distance not greater than 50 cm from the *Anacharis*.

Wavelength: The wavelength (or color) of light reaching the chloroplasts of the *Anacharis* can be changed by using red, blue, or green cellophane; for example, red cellophane isolates red light. Because light intensity drops as it passes through colored cellophane, place the light source at a distance not greater than 10 cm from the *Anacharis*.

   Do **not** attach the cellophane to the light source. The heat from the light source may cause the cellophane to melt.

1.   Put 125 mL of distilled water in the flask.

2.   Put the unwrapped straw in the flask, and blow gently through the straw into the water for 3 minutes. This adds CO_2 to the water.

   Be careful not to suck any liquid into your mouth. Discard the straw after use.

3.   Place two sprigs of *Anacharis,* cut end up, into one of the test tubes.

   This is your experimental tube. Mark it as such.

4.   Mark a second test tube as a control, and then fill both test tubes three-fourths full with the water that you blew into.

5.   Stand the test tubes in a 2-L beaker and add 25°C water to the beaker until it is about two-thirds full.

6.   Perform the following steps:

   a.   Using the procedures explained in ALPHAEXP, run experiment HUMAN05.*

      This experiment is located in the directory HUMBIO\ (IBM) and in the BSCS Human Biology Folder (Macintosh).

      *IBM users, run HUMAN05T for two teams.

   b.   Set the experiment to SAVE the data.

   c.   Team A, use Inputs 1–3; Team B, use Inputs 4–6.

      It will greatly simplify taking measurements if both teams coordinate their steps throughout the experiment.

   d.   Insert the thermistor into the water in the beaker.

   e.   Remove the pH probes from the storage jar of tap water.

   f.   Insert the pH probe from Input 1 (or Team B, insert the probe from Input 4) into the experimental test tube and the pH probe from Input 2 (Team B, Input 5) into the control test tube, as shown in Figure 8.8.

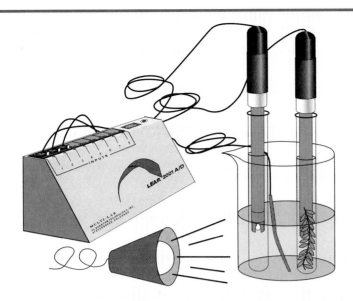

7. Let the entire assembly stand for about 5 minutes to permit the temperature to become uniform throughout the system.

8. Take the experiment off WAIT. When the initial readings have stabilized, record the temperature in the beaker and the pH in each test tube from the numeric data displayed near the graph. Put the experiment on WAIT.

9. Maintain the initial temperature throughout the experiment.

   Use the small beaker to add ice water and/or to remove water from the 2-L beaker.

10. Arrange your setup to test your experimental question.

11. Perform the following steps:

    a. Take the experiment off WAIT.

    b. Illuminate the system.

    c. MARK the graph to indicate when you illuminated the system.

    d. Record the MARK number for your team.

12. Take pH and temperature readings for 30 minutes. In your journal record the current readings every 5 minutes.

13. Perform the following steps.

    a. Put the experiment on WAIT.

    b. Using the procedures explained in ALPHAEXP, print a stripchart or a screen print of the saved data for your team.

       IBM users, print HUMAN05A for Team A and HUMAN05B for Team B.

    c. If you have not already done so, QUIT HUMAN05.

    d. Remove the pH probes and rinse them with tap water. Replace the probes in the storage jar of tap water.

    e. Clean up and return the other equipment to storage.

14. Wash your hands thoroughly before leaving the laboratory.

Analysis

Complete the following tasks and questions individually:

1. Write three questions that test your understanding of your experimental results and the concepts in the essay *Getting Energy and Matter into*

Biological Systems (page E123). Your teacher will collect these questions and develop a quiz from these and other questions.

2. The earth's early atmosphere had no oxygen. Use your understanding of plant photosynthesis to explain how photosynthetic organisms made possible the evolution of aerobic organisms.

3. Do plants carry out cellular respiration? Explain your response.

4. How does the trapping of light energy provide energy for carbon fixation?

BUILDING LIVING SYSTEMS

Elaborate

The cells of your body must receive a constant supply of usable energy and matter if you are to grow into an adult. Even after growth has stopped, your body must be able to make new cells to replace damaged or infected ones, a process that occurs wherever healing is necessary, such as at the site of a skin cut. Cells also must build small simple molecules, such as amino acids and simple sugars, into the more complex biological molecules, such as proteins and glycogen, that are important for your daily activities. This building process is called **biosynthesis**.

At the same time that biosynthesis is taking place, many breakdown activities also are occurring, including the digestion of the food you eat. Food molecules can come from a variety of sources: plants, animals, or fungi. Not surprisingly, the molecules that make up these other organisms are not always the same molecules needed by your body. In this activity you will see how your body is capable of taking matter from an outside source, converting it into usable energy and matter, and rearranging the molecular structure of that matter to provide building materials for new molecules, cells, and tissues.

Materials (per person)

descriptions and diagrams of Organism X from Chapter 5

Process and Procedures
Part A Metabolism

1. Use the following short story and your understanding of energy, matter, and cellular respiration to answer the question, How can my body take in materials from a cow and make it a part of me?

    Your high school girl's basketball team has just beaten its cross-town rivals and has qualified for the state basketball tournament. To celebrate, you and two friends go to your favorite fast food restaurant and order the usual burger, fries, and soft drink. You notice that a new menu is on display, and you see that the word hamburger has been replaced by the word beefburger. After a brief discussion with your friends, you see the logic of the change. Your burger is a beefburger. This term accurately describes what you are about to eat. That realization leads to another discussion. The burger you are about to eat is rich in protein; with luck, it is not too rich in fat! This cow protein and fat will be used by your body as a source of energy and as building material for biosynthesis. The

discussion centers around the question of how your body can take in materials from a cow and make it a part of you.

2. With your partner list three biological processes involving biosynthesis and/or breakdown that you think are necessary for maintaining the human body.

3. Choose one of the processes your team identified in Step 2, and write an explanation in your journal of how energy and matter are organized during this process. Consider the following questions as you write your explanation:

    •   Why is this process necessary for survival?

    •   What is a source of energy for this process?

    •   What is a source of matter for this process?

    The essay Metabolism Includes Synthesis and Breakdown, *page E127, will help you with your explanation.*

Part B Energy and Matter for Organism X

Use an explanation, diagram, model, or demonstration of your choice to construct a response to the question, How does your organism use matter and energy to maintain its organization?

You will know that you have addressed adequately the preceding question when your response

•   indicates the organism's source of energy and matter and how these are obtained from its surroundings,

•   demonstrates how energy is stored and made available for its activities, and

•   distinguishes the macromolecules that it is able to synthesize from the macromolecules that this organism must obtain from its external environment (through its diet or other means).

Your teacher will use these criteria to evaluate your project.

Evaluate

TRACING MATTER AND ENERGY

Scientists can investigate the steps in metabolic processes (synthesis and breakdown) by feeding extremely small (or trace) amounts of radioactively labeled compounds to laboratory organisms. These labeled compounds undergo changes in molecular structure during chemical reactions and transfers within and between the organisms. As these changes and transfers occur, the scientists can collect samples and trace the course of events by following the radioactivity.

Figure 8.9 **Scientist working with radioactivity**

In this activity you will trace the path of an imaginary, radioactively labeled carbon atom through the various molecules in which it is organized. The atom begins its journey as part of a carbon dioxide molecule and ends up as part of a muscle protein in a human arm. Your task is to use the knowledge that you have gained in this chapter to draw a diagram of what happens to the atom during its journey. Then you will explain the source of energy for these events.

Materials (per person)

materials needed to complete your project

Process and Procedures

1. Construct a diagram or other visual aid to show a plausible set of events that could explain how a labeled carbon atom in a molecule of atmospheric carbon dioxide ends up in a human muscle protein.

    There is more than one possible scenario, but you must show a sequence that actually occurs in nature, and you must be able to justify and explain the sequence that you choose.

2. Write at least one paragraph that explains the sequence of events that you have diagrammed. You should include both the flow of matter and the energy sources that make these events possible.

Analysis

1. Join the class in answering the discussion questions that your teacher raises.

2. After participating in the discussion, note in your journal appropriate changes to steps in your sequence from *Process and Procedures*, Step 2.

> *The ecstatic upland plover, hovering overhead, poured praises on something perfect: perhaps the eggs, perhaps the shadows, or perhaps the haze of pink phlox that lay on the prairie.*

Aldo Leopold, *A Sand County Almanac*, 1949

9 THE CYCLING OF MATTER & THE FLOW OF ENERGY IN COMMUNITIES

Think about your school as a community—the organization and interaction of people, books, paper, furniture, and food. People move through the hallways in repeated patterns throughout the day; books and papers are moved from lockers, to class, and back to lockers; desks are rearranged to accommodate various activities and events; and food is moved from the kitchen, to lunch trays, to hungry students.

In much the same way as these things move through the organized community of your school, energy and matter move through organized communities of living organisms of all sizes. In biology we speak of a **community** as the group of living organisms that live and interact with each other in a specific area. An **ecosystem** is a community of organisms interacting with their environment. Consider the upland sandpiper (or plover) in the opening photograph; now think about the insects on which it feeds and the many other organisms that inhabit this prairie. Together these organisms make up a community, and matter and energy move through this community in different ways. In this chapter you will use your experiences from a variety of activities and related essays to develop an understanding of how matter and energy are organized within communities.

ACTIVITIES

Engage
Explore   A Matter of Trash

Explore   Exploring the Cycling of Matter in Communities

Explain   Spinning the Web of Life

Elaborate   Generating Some Heat

Evaluate   What Have I Learned about Matter and Energy in Communities?

Engage
Explore

A MATTER OF TRASH

You probably have chores around your home. Washing dishes, mowing the lawn, feeding pets, baby-sitting for a younger brother or sister, and taking out the trash are activities that might be your responsibility. Do you remember the last time you took out the trash? Did you notice the contents of what you were about to throw away? Where did the trash go after you put it out on the curb or tossed it into a Dumpster™? What happened to the matter in the material you just discarded? Was any energy stored in that matter? Where did that energy go? This activity will engage you in thinking about various forms of matter that you may consider to be waste and about what happens to this matter after you throw it out.

Figure 9.1 Trash Where will the matter and energy in this trash end up?

Process and Procedures

1. Examine the discarded items in the trash demonstration that your teacher presents, and create a table or other visual diagram that includes the following information:
   - a list of the trash in the demonstration,
   - a list of the origin of the trash,
   - an indication of which items of trash in the demonstration match your household trash, and
   - a list of possible fates for the trash.

2. In your journal write a short description of what you think might happen to the matter and energy in this trash after it is thrown out.

Analysis

Working individually, write your responses to these questions in your journal, and be prepared to share them in a class discussion. The essay *Garbage among Us—from Then until Now* (page E128) will provide you with information about how matter cycles in other communities.

1. In what ways might the waste of one organism be useful to another organism? Give examples to support your answer.

2. How does your answer to Question 1 support the idea that organisms in communities depend on one another for matter and energy?

3. As you compare how matter cycles in different communities, what do you notice about the type of matter and the length of the cycle? What problems have these differences caused for modern human societies?

Further Challenges

On a separate piece of paper, write 10 to 15 things you've thrown away in the past week. Exchange lists with your partner, and write four or five things that you might infer about your partner from his or her trash. Support your inferences.

EXPLORING THE CYCLING OF MATTER IN COMMUNITIES

Have you ever watched ants in an ant farm? They always appear to be busy modifying their environment in some way. In this activity you will complete the observations of the earthworm habitats that you set up some weeks ago and think about what your observations tell you about how these organisms interact with their environment. You also will reflect on the design of your experiment and then use your understanding to design another experiment to explore how other organisms interact with each other and with their environment.

Materials (per team of 4)
Part A

6 slides
hand lens
stereomicroscope
2 spoons

2 paper towels
small tray
earthworm habitat
control habitat

Part B (2 teams per LEAP-System™)

4 25-mm × 200-mm test tubes
test-tube rack
aluminum foil
light source
Macintosh or IBM-compatible computer
LEAP-System™
4 pH probes
printer
jar of tap water for storing pH probes
dechlorinated water
1- to 1½-cm freshwater snails
15-cm sprigs of *Anacharis* (elodea)

CHAPTER NINE
The Cycling of Matter and the
Flow of Energy in Communities

187

Process and Procedures

Part A Reflections on the Earthworm Habitats

1. With your teammates look at the earthworm habitats that you set up several weeks ago, and study the observations that you have recorded in your journal.

   Remember that the purpose of these habitats was to provide evidence of interactions between living systems and the physical environment as well as evidence about the nature of these interactions.

Figure 9.2 In what ways do you think earthworms interact with their environment?

2. Discuss the following with your teammates, and record your ideas in your journal.

   a. What evidence did you collect that supports the idea that earthworms modify (interact with) their environment?

   b. Describe this interaction. What do you think happened in the containers of earthworms?

   c. What was the specific purpose of each container in helping you identify and describe the interaction of earthworms and their environment? That is, what did your observations of *each* container tell you that helped you answer Questions 2a and 2b?

   Compare the containers that had earthworms with those that did not, and record your comparisons in your journal. Why was it important to observe both the containers with earthworms and the containers without earthworms?

   Compare the containers that had earthworms (one with soil and organic matter and one with only soil) with each other. Why was it important to observe both types of containers? What did your comparisons of these containers tell you about the interactions of earthworms with their environment?

3. To generate a context for your observations and prepare you to design your own experiment that explores the transformation of matter, read the essays *Matter in Nature Is Going Around in Cycles . . . What Next?* (page E132) and *Worms, Insects, Bacteria, and Fungi—Who Needs Them?* (page E135).

   Check to see that your teammates understand the concepts in the readings and how they relate to your findings.

Part B Snails and *Anacharis*: What Can I Learn from Them?

1. Devise an experiment to provide evidence of the cyclical movement of matter in a community. See the *Materials* section for the materials and equipment that are available.

   a. Think back to previous chapters, and remember what you learned about photosynthesis and respiration. Also, read the following information, which provides some important background.

      Plants and animals interact in a variety of ways. By setting up closed systems with an aquatic plant and an aquatic animal, you can study an interaction that is related to the carbon cycle. As you remember from Chapter 8, carbon dioxide dissolves in water and forms a weak acid, which lowers the water's pH. A decrease in pH indicates an increase in the concentration of carbon dioxide. Conversely, an increase in pH indicates a decrease in the concentration of carbon dioxide.

   b. Develop a hypothesis about the cyclical movement of matter in a community. You must be able to test the hypothesis using the materials and information listed above.

   c. Design an experiment to test your hypothesis, and outline your experiment in your journal.

      Look at the design of the earthworm habitats, and apply your understanding of that design to the one you develop here.

   d. Have your teacher approve your design.

2. Use the *Protocol for Monitoring Change in pH* (Figure 9.3) to conduct your experiment during the next two hours.

3. Prepare a lab report of your findings.

   You may want to refer to the guidelines for a lab report that were presented in Part A of the activity You Are What You Eat, *Chapter 7. Be sure to answer the following questions as part of your report.*

   a. What was the specific purpose of each test tube that you set up? That is, what evidence did each test tube provide? What did you learn from each test tube?

   b. How was each test tube important in helping you develop your conclusion?

   c. Do your data support your hypothesis? Explain.

   d. Based on what you have learned from this experiment, what question would you like to ask and answer next?

Analysis

1. Take turns presenting your team's experimental design and results to the rest of the class. Also, share ideas from your lab report.

2. Participate in a class discussion of the various experiments conducted and the results that emerged.

CHAPTER NINE
The Cycling of Matter and the
Flow of Energy in Communities

189

3. With your teammates create a visual diagram, such as a concept map, that represents your current understanding of the cycling of matter through a community. Base your diagram on the two populations that you studied in this activity.

Figure 9.3 **Protocol for Monitoring Change in pH**

Protocol for Monitoring Change in pH

1. Using the procedures explained in ALPHAEXP, run experiment HUMAN06.*

   This experiment is located in the directory HUMBIO\ (IBM) and in the BSCS Human Biology Folder (Macintosh).

   IBM users, run HUMAN06T for two teams.

2. Identify which pH probes Team A will use and which pH probes Team B will use.

3. Remove the pH probes from the storage jar of tap water, and insert a pH probe in each of your team's test tubes.

4. Set the experiment to SAVE the data.

5. Take the experiment off WAIT.

6. Record the current pH readings for each tube from the numeric data displayed near the graph.

7. Illuminate your setups.

8. Leave the LEAP-System™ running (off WAIT); the system automatically will take readings every 15 minutes.

   Allow your experimental setups to run for at least 2 hours.

9. Using the procedures explained in ALPHAEXP, PRINT HUMAN06.

   IBM users, print HUMAN06A for Team A and HUMAN06B for Team B.

10. If you have not already done so, QUIT HUMAN06.

11. Remove the pH probes and rinse them with tap water. Replace the probes in the storage jar of tap water.

12. Wash your hands thoroughly before leaving the laboratory.

SPINNING THE WEB OF LIFE

The next time you eat a hamburger, think about what it takes to make a pound of beef.

Recipe for One Pound of Beef

Begin with one calf and add the following ingredients over a period of about two years:

43 square yards of grazing land	0.8 ounces of phosphorus
8 square yards of farmland	1.6 ounces of potassium
13 pounds of forage	antibiotics
3 pounds of grain	hormones
6.4 ounces of soybeans	pesticides
18 ounces of petroleum products	herbicides
3 ounces of nitrogen	1300 gallons of water, added regularly

You already have explored how matter cycles through various communities, and you are aware that energy is stored in matter. When you eat various forms of matter, some of that energy is transformed and is available to you so that you can carry out the daily activities of life and the distinctive activities that make you who you are. Not all foods contain the same amount of energy, and not all organisms require the same amount of energy for growth. So why does it take so much energy to make one pound of beef, and what is the difference between eating plants and eating animals?

Let's examine the larger picture of matter and energy, of which you are a part. By relating the food that you eat for one day to the plants and animals from which it came, you can begin to develop a greater understanding of how the energy that is stored in matter flows through a community and fuels the activities of its organisms.

Materials (per person)

set of colored pens or pencils, 3 colors

Process and Procedures

1. To begin to develop a more detailed understanding of how energy flows through a community as matter cycles, try to generate a **food web**, which is a visual diagram of the interactions of matter and the consequent flow of energy.

    a. In your journal list all of the foods that you ate yesterday, including snacks as well as meals. (Alternatively, use the list that you generated for the *Dietary Analysis* that you completed in the activity *What Determines Fitness?*, Chapter 7.)

    Remember that many foods are combinations of different plants, animals, fungi, or bacteria. Record the ingredients of each food separately. For example, if you had a piece of cake for dessert, you should list oil, flour, sugar, butter, eggs, and milk.

    b. Next to each item that you listed, record the type of organism from which it came.

    For example, write wheat next to flour, sugar cane or sugar beet next to sugar, and chicken next to egg.

a.

b.

Figure 9.4 Common sources of sugar are (a.) sugar cane and (b.) sugar beet.

CHAPTER NINE
The Cycling of Matter and the
Flow of Energy in Communities

191

c. For every animal that you have listed, do the following:

    1. Next to it, list several foods that it eats.

    For example, next to hamburger, you already would have listed cattle. Now list the grass and corn the cattle eat.

    2. List animals that might eat it.

    For example, next to cattle you might list wolf, bear, mountain lion, or human. Next to fish, you might list raccoon, otter, or sea gull, depending on the type of fish it is.

d. Reorganize the information you analyzed in Step 1c in the following way.

    Using one of the colored pencils, record the names of organisms, not foods.

- List all the names of the plants across the bottom of the page.

- Write the names of all the herbivores (plant eaters) 4-6 cm above the plants.

- Write all the names of the omnivores (animals that eat plants and animals) 4-6 cm above the herbivore row. Write your name at the end of this row.

- Write all the names of the carnivores (animals that eat only meat) 4-6 cm above the omnivore row.

e. After you have all the names of the plants and animals organized, draw arrows from the organisms that are providing energy to the organisms that are receiving energy.

    For example, if your plant row includes grass and your herbivore row includes cattle, then draw an arrow from grass to cattle with the point of the arrow aimed at the cattle.

2. Expand your food web in the following manner:

a. Think of other organisms that might compete with you for your food. Make a list in your journal of organisms that eat some of the same foods that you eat.

b. Using a pen or pencil of a different color, add the organisms that you listed in Step 2a to the appropriate level of your food web; add the appropriate arrows.

3. Work with a classmate to briefly discuss the following questions:

This strategy provides you with an opportunity to develop the skill of using your classmates as a resource.

a. What is the key difference between how producers and consumers build biomass?

b. What sorts of matter do you think archaeologists might uncover years from now from the suburban family in Maineville, Ohio, that you read about? How would this compare to what archaeologists have uncovered from the Anasazi Indians?

c. Why is there no predator in Africa that lives by eating only lions and leopards?

d. What effect are cattle and other domesticated livestock having on certain ecosystems?

To help you with this task, read the letters in Let's Ask Drs. Ricardo and Rita *(page E136) and the essay* Losing Heat *(page E141).*

4. To demonstrate your more complete understanding of food webs, use a pen or pencil of a third color to add decomposers to your food web. Use arrows to indicate the relationship of these decomposers to the other organisms present.

Analysis

Participate in a class discussion of the following. You may want to record some of your ideas in your journal.

1. Describe how the energy flows through the community, as shown in your food web.

2. Where is the most energy available in your food web? Explain your answer.

3. Compare the food web of a vegetarian in your class with the food web of someone who is not a vegetarian.

   • What differences are evident?

   • What do you think is significant about these differences?

   • How does this relate to the recipe for one pound of beef?

4. How have we made our task of acquiring food easier than it was for the Anasazi Indians?

GENERATING SOME HEAT

Elaborate

It's Saturday. You have a soccer game at noon. You promised your friends that you'll be there a half hour early to practice, but you know that your Saturday chores have to be finished before you leave. This is your week to cut the grass, so you get up early to begin the task. Because the morning is cool, you find yourself making excellent progress, but time is passing quickly, and the practice session is about to begin. In an effort to shave some time off the grass-cutting job, you pile all of the clippings on some newly turned soil in a corner of the yard. You say to yourself, "I'll bag the clippings later." Saturday and Sunday come and go.

School begins again on Monday, and the clippings still are sitting there on the soil in the yard. Another Saturday rolls around, and you have another game. It's your sister's turn to cut the grass. Good! No grass today! Your game is early in the morning. On the way out of the house, your dad reminds you that last week's clippings are still sitting in the yard and that it's your responsibility to bag them before you head out. You run to the yard with some lawn bags and begin scooping up grass clippings. Again, the morning is very cool. As you dig into the pile of clippings, you notice something interesting. Steam is rising from the clippings. In fact, the center of the clippings is very warm, especially down near the soil. You wonder why the inside of the pile of clippings is so warm on such a cool morning.

Figure 9.5 Wow! This stuff is hot! I wonder why.

From your experiences in this chapter, you might guess that microorganisms were beginning to break down the clippings, especially those next to the soil, and that their action was generating heat. In this activity you will elaborate on your knowledge of the cycling of matter and the flow of energy in communities. You will look especially at the role of decomposers. You and your teammates are about to participate in a compost design competition. Your goal is to design a compost system that generates the most change in temperature.

Materials (per team of 4)

4 pairs of plastic gloves
nonmercury thermometer
pan balance
1-gal plastic milk container with 4 cm of top cut off
230-250 mL measuring cup
masking tape
4-gal plastic trash bag
sheet of plain newsprint
trowel or spatula for mixing compost
foam insulation material
packet of compost starter inoculum
bag of potting soil
bag of grass clippings
bag of shredded leaf mulch or bark mulch
water

Process and Procedures

1. As a class make the following decisions about your compost designs:

   - how much compost starter inoculum all teams will use,

   - how much water all teams will use, and

   - how much organic matter (the mass) all teams will use.

   Why is it important that all teams keep certain features constant?

2. With your team consider some recipes that you might use for your composting system. To help you decide on a recipe, discuss the following questions with your team. Record your answers and your recipe in your journal. Be sure to justify your decisions.

   a. What organic matter should you use?

   b. Should you use one source of organic matter or a combination of sources?

   c. If you use a combination, what proportions should you use?

   Composting is not an exact science. Gardeners frequently have their own personal compost recipes. Now it's your turn to create a compost recipe. You have all of the materials necessary for decomposers to work effectively. The grass clippings, shredded leaf mulch, and shredded bark mulch provide excellent carbon sources (organic matter to be consumed as food) for the microorganisms (starter inoculum). Because these microorganisms thrive in soil, the potting soil should provide a suitable environment for initial growth. Remember, your entire compost needs to fit in the plastic milk container.

   As you develop your recipe, practice the working relationship skill of reaching consensus.

3. Carefully create your composting system.

   SAFETY: Put on your plastic gloves.

   a. According to the recipe that your team agreed on, measure your soil and your food sources one at a time from materials you have available, and pour them into the plastic bag.

   b. Mix the ingredients thoroughly with a trowel.

   c. Add the amount of compost starter inoculum that your class decided to include.

   d. Add the amount of water that your class decided to include.

   e. Again, use the trowel to thoroughly mix the compost ingredients. Then place your mixed compost into the plastic milk container.

   f. Insulate your container of compost.

   Cut enough insulation material to surround the container. Tape the insulation in place. Cut a small piece of insulation for the top. For the composting inoculum to work most efficiently, the system must retain the heat that is generated by the microorganisms. If not insulated, the compost systems will lose this heat to the environment.

4. Keep your system going for two more days, and record the temperature in the center of the compost twice a day.

   Create a data table in your journal in which to record the temperature of your compost.

CHAPTER NINE
The Cycling of Matter and the
Flow of Energy in Communities

195

5. Graph the temperature changes that you observed in your composting system.

Analysis

Complete the following tasks as a team:

1. Present your team's compost recipe and results to the rest of the class. Be sure to share the graph of your results.

2. Based on all of the teams' reports, determine which compost recipes generated the most change in temperature.

   You and your classmates will judge the effectiveness of each composting recipe by examining each team's data.

   a. Why do you think the most effective recipes worked better than others?

   b. What do your results tell you about the cycling of matter and the flow of energy?

   c. What role did the microorganisms play?

3. The heat that you noticed is a form of energy. In what form was this energy before it was released as heat? What happens to this energy after it is released as heat?

4. What does your answer to Question 3 reveal about the flow of energy in a community?

WHAT HAVE I LEARNED ABOUT ENERGY AND MATTER IN COMMUNITIES?

Evaluate

By now you are aware of how closely connected the flow of energy is to the cycling of matter in communities. In this activity you will have an opportunity to evaluate what you have learned about these concepts. You will work individually to think about the impact that a natural disaster would have on various communities on earth, and then you will work with a classmate to respond to some questions about survival in different communities.

Process and Procedures
Part A A Natural Disaster

1. Read the following scenario:

   The earth is entering a phase of instability that no one had predicted. Throughout both hemispheres, hundreds of volcanoes are erupting with great force. The atmosphere of the earth is thick with minute volcanic debris and dust. As much as 75 percent of the sunlight now is blocked from reaching the earth's surface. This period of eruptions is expected to continue indefinitely, and it is likely that soon virtually all sunlight will be blocked from reaching the earth's surface.

2. Answer the following questions, and record your answers in your journal. Be sure to answer each part of each question. Your teacher will collect your journal and use the criteria listed in the scoring rubric to assess your understanding.

   a. What might be the effect if only 80-85 percent of the sunlight were blocked from the earth? What might be the effect on the following organisms: an earthworm, a shark, a maple tree, a saguaro cactus, and a teenager?

   b. Imagine that the trend described in the scenario continues and eventually all sunlight is blocked from reaching the earth's surface.

      • What might be the effect on the following organisms: the producers, the consumers, and the decomposers?

      • Describe how the cycling of matter through a community would be affected.

Part B Strategies for Survival

Work with a partner to discuss the following questions; then record your responses in your journal.

1. From the thousands that sprout, why will only one or two healthy trees grow into the available space between other existing trees?

2. Are the fish that live 2 km deep in the ocean likely to be herbivores or predators? Explain.

3. During the years of heavy DDT (a pesticide) use in the United States, the decline in the population of small songbirds was far less than the decline in the population of birds of prey such as ospreys, peregrine falcons, and bald eagles. Why do you think this was so?

CHAPTER NINE
The Cycling of Matter and the
Flow of Energy in Communities

197

4. Human societies that live by hunting and gathering usually have much smaller populations than groups in a similar setting that live primarily by growing crops. Why do you think this is so?

5. Suppose you found yourself snowed in for the winter in a remote mountain cabin with no way of contacting the outside world. You must survive for several months with only what is on hand to eat. Aside from a small supply of canned peaches, your only resources are two 100-lb sacks of wheat and a flock of eight hens. Discuss the relative merits of the following strategies:

   a. Feed the grain to the hens and eat their eggs until the wheat is gone, and then eat the hens.

   b. Kill the hens at once, freeze their carcasses in the snow, and live on a diet of wheat porridge and chicken.

   c. Eat a mixture of wheat porridge, eggs, and one hen a week, feeding the hens well in order to keep the eggs coming until all of the hens are killed.

6. Every breeding pair of bullfrogs produces hundreds of eggs each spring. During the time they are growing up in the pond, the small tadpoles feed entirely on microscopic water plants. Predators living in the pond eat a large fraction of the tadpoles before they transform into frogs. As adults, however, bullfrogs themselves are predators. Discuss the logic behind this scenario compared with one in which the tadpoles would be predators and the adults would be herbivores.

Figure 9.7 Bullfrog and tadpoles These bullfrogs (*Rana catesbiana*) are predators, but they begin life as prey.

Further Challenges

1. There are several hypotheses about how dinosaurs became extinct. One of these hypotheses involves a climatic catastrophe that has some similarities to the one presented in Part A. See what you can find out about this hypothesis, and report your findings to the class.

2. Write and perform a skit that depicts organisms defending their role in a community. In it describe the advantages, the disadvantages, and the importance to the community of a producer, a predator, and a decomposer.

CHAPTER NINE
The Cycling of Matter and the
Flow of Energy in Communities

199

Explain
CONDUCTING YOUR OWN INQUIRY

Science can be a career, as it is for the people pictured here. Science also can be one way of studying and knowing the world around you. In this section you will build on the inquiry skills that you began developing in the Engage Section. Now that your are half finished with this biology program, you should have enough experience with these skills to realize that thinking scientifically is a valuable way of answering many questions. In addition, you should be able to apply your critical-thinking skills to evaluate new scientific information.

In the activities of this section, you first will look at examples of science in the popular press and critique these examples for the adequacy of their coverage. Is the science accurate? Was the coverage complete? Do the scientists or investigating agencies have an unbiased perspective? You next will begin investigating a scientific question of your own. What would you like to study? Where will you find background information? How will you conduct your experiment? The choices are yours.

The goals of this section are

- to give you the opportunity to evaluate new scientific information using your inquiry skills, and
- to allow you to conceive, design, and conduct a scientific investigation of your own choosing.

ACTIVITIES

Engage
Explore   Science All Around You

Explain
Elaborate
Evaluate   Being an Experimental Scientist

Engage
Explore

SCIENCE ALL AROUND YOU

One reason to study biology is to learn to use the methods of science to study the organisms, interactions, and processes that surround you. Another reason is to understand the events that influence your life. The story *Glasses for the Ears May Help Treat Dyslexia* is an example of how science and technology are reported in the newspaper. It describes experiments related to a new technology that may represent a breakthrough in helping young people deal with language and learning difficulties. What can you tell from the article about *how* the scientists who did the work got their results? Can you tell what questions the scientists asked?

Materials (per person)

newspaper or news magazine
scissors
tape, stapler, or glue

Process and Procedures

1. Scan the newspaper or news magazine for articles about science.

2. Cut out the article that most interests you, and attach it to a page in your journal.

3. To analyze this news article, answer the following questions in your journal.
   Base your answers on the information in the article itself and what you infer.
   - What question did the scientists ask?
   - What background information informed the scientists?
   - What type of investigation did the scientists conduct?
   - What tools did the scientists use?
   - What results did the scientists get?
   - What conclusions did the scientists draw?
   - What new questions did the scientists ask?

Analysis

1. Discuss your article and analysis with a partner.

2. Record in your journal at least two scientific questions you would like to research. These may be related to the article you analyzed or to another area of science.

"GLASSES FOR THE EARS" MAY HELP TREAT DYSLEXIA

Scientists have developed a radically different treatment for children with severe language and reading difficulties, one that may have applications for children with dyslexia.

They call it "glasses for the ears."

The treatment uses a special form of computer-generated speech sounds. Just as glasses correct faulty vision, these changes in the auditory cortex sharply improve the children's ability to perceive spoken sounds and to decode written words.

The new treatment exploits what most parents already know—namely, that children will sit for hours in front of a video game. Scientists developed four colorful computer games with processed speech. The games drill children in hearing pairs of tones and phonemes—the basic units of language—at faster and faster rates of speed.

Recent experiments show that after just four weeks of treatment, language-disabled children advanced two full years in their verbal comprehension skills, researchers say. They said the improvements endured after training had stopped.

The two scientists spearheading the research, Dr. Paula Tallal of Rutgers University in Newark and Dr. Michael Merzenich of the University of California School of Medicine in San Francisco, said they believed that the treatment would help many children and adults with milder forms of language and reading disability—the condition widely known as dyslexia.

But Dr. Tallal, who is director of the Center for Molecular and Behavioral Neuroscience at the Newark campus, and Merzenich cautioned that dyslexia had many causes and that not everyone would respond to the treatment.

BEING AN EXPERIMENTAL SCIENTIST

In the Engage Section at the beginning of this course, you investigated the way that scientists think when they do their work, and you conducted experiments to test a hypothesis. Although the Tri-Lakes activity was not an actual situation, it was based on several very real scenarios. Now you will be a scientist as you carry out a *full inquiry* of your own design. Remember that each of the *thinking* steps, such as asking a good question, deciding how to test it, and analyzing the meaning of the data you collect, is just as important as the hands-on step of *doing* an experiment. Your performance in this activity will demonstrate both your understanding of the particular area of biology that you investigate and your ability to explain and use scientific processes.

Materials (per person)

Materials will depend on the experiment you design. You will need your teacher's approval before you assemble materials.

Explain
Elaborate
Evaluate

Process and Procedures
Part A Preparation

1. Use Figure Ex.1 to answer the following questions in your journal:

    a. Which of these steps were evident in the article you analyzed in the previous activity? Explain how they were evident.

    b. Identify specific times or activities in this course when you have used these steps.

    Looking through your journal may help you answer this question.

- Asking questions

- Gathering information

- Proposing explanations

Figure Ex.1 The processes of science When have you used these processes of science this year?

2. Examine the criteria for the "excellent" categories on the scoring rubric in Figure Ex.2. Tell your partner what you think an "excellent" project would look like when it is finished and what it would be like to complete an "excellent" project.

Part B Conducting a Full Inquiry

Now you will carry out your own full inquiry by following these steps:

I. Asking the Question

1. Choose an area of biology that interests you, and identify a testable question.

    If you are having difficulty thinking of a question, look back at your response to the Analysis questions for the previous activity, or do some library research about a topic that interests you. The new information will provide useful background and may give you an idea for a testable question.

    a. Record in your journal the question in one or two sentences.

    b. Explain why your question is significant.

    To do this, you will need to write several sentences describing what already is known about the topic that you wish to investigate.

Figure Ex.2 (page 205)
Scoring Rubric for Being an Experimental Scientist Read these criteria and decide what an excellent project would look like.

Scoring Rubric for Being an Experimental Scientist

Level of Achievement	Asking the Question	Gathering Information	Analyzing Your Data and Drawing Conclusions	Communicating Your Results and Listening to Others
Excellent-Good (full–3/4 credit)	• Proposes a significant question (e.g., asking whether plants need sunlight is not significant, but asking why some plant species require more sunlight than others is significant) • Proposes a biologically interesting question (e.g., asking whether plants are killed by alcohol is not biologically interesting because alcohol is not relevant to a plant's life) • Proposes a question that is focused enough that it can be tested	• Uses diverse resources • Designs an experiment that is straightforward and efficient • Designs an experiment that appropriately uses controls • Designs an experiment that specifically addresses the question • Can explain the design of the experiment • Shows creativity in experimental design • Carries out an experiment carefully, paying attention to details • Meets safety requirements • Keeps accurate and complete records • Demonstrates repeatability (if time permits)	• Organizes data in a manner that is easy to understand • Draws logical conclusions • Uses data to support conclusions • Uses data to answer original question • Connects results to most appropriate unifying principle(s) • Makes predictions about new hypotheses or future research possibilities	• Presents background material • Displays clearly the – question – design – results – conclusions • Uses appropriate means to display data (e.g., uses graphs, tables, etc. when necessary) • Explains connections to unifying principle(s) • Makes connections to technology, culture, ethics, and history or explains lack of connection • Displays material creatively and neatly • Is able to find evidence of scientific processes in other's presentations
Adequate-Needs Improvement (3/4–1/2 credit)	Meets about 1/2 of the criteria listed above	Completes about 1/2 of above criteria well or all of it with less thoroughness	Completes about 1/2 of above criteria well or all of it with less thoroughness	Satisfies above criteria but with less thoroughness
Inadequate (<1/2 credit)	• Proposes a trivial question that has no biological significance • Proposes a biologically uninteresting question • Proposes a vague question that cannot be tested	• Does not do outside research • Designs an experiment that does not address the question • Does not complete the experiment • Does not keep accurate records • Conducts test carelessly • Does not meet safety requirements • Designs an inefficient experiment • Does not use controls	• Fails to organize data in a manner that is understandable • Draws illogical conclusions • Does not use data to support conclusions (or uses it incorrectly) • Does not accurately address original question • Does not connect material logically to unifying principle(s) • Fails to make predictions about possible directions for future research	• Does not present background material • Lacks clear display of – question – design – results – conclusions • Presents messy or confusing display • Displays data in a misleading or inaccurate manner • Does not clearly show connection to unifying principle(s), technology, culture, ethics, or history • Unable to find evidence of scientific processes in other's presentations
Relative Percentage	20 percent	40 percent	20 percent	20 percent

c. Restate your question as a hypothesis that can be tested.

d. Record which of the six unifying principles of biology (Figure Ex.3) is most related to your hypothesis.

BSCS

Evolution: Patterns and Products of Change in Living Systems

Homeostasis: Maintaining Dynamic Equilibrium in Living Systems

Energy, Matter, and Organization: Relationships in Living Systems

Continuity: Reproduction and Inheritance in Living Systems

Development: Growth and Differentiation in Living Systems

Ecology: Interaction and Interdependence in Living Systems

2. Show your question to your teacher for approval before you proceed.

II. Gathering Information

1. Use the library, local scientists, the Internet, or other available resources to gather information related to your question.

In the Tri-Lakes activity you answered a scientific question by using data collected by others as well as data you gathered through experimental investigations. You should use a similar process at this time.

2. Design an experiment to test or answer your question by doing the following:

a. Write in your journal a description of your experimental design that includes

- rationale, that is, how this experiment will test your question (include a description of the role your controls will play);

- procedure (include the materials you will need);

- data collection (explain how you will collect data); and

- data analysis (explain how you will analyze the data).

Your teacher may have specific suggestions about the length of time you will have or the equipment that is available.

b. Write in your journal a safety plan for your experiment. In your procedures record the precautions that you will follow when you

- use chemicals,

- handle equipment, and

- handle biological hazards such as bacteria or yeast.

Ask your teacher to explain any hazards that you do not understand and to help you identify the precautions necessary to prevent harm from an accident.

Review Appendix A, Laboratory Safety, on pages 406–413 and be sure that you understand all the safety considerations involved in your experimental design.

Make sure that you have read and understood the hazards and precautions described on the labels and Material Safety Data Sheets for all the chemicals you plan to use in your experiment. Report all accidents, no matter how small, to your teacher.

Figure Ex.4 Scientists protect themselves by following safety procedures while working in the laboratory. What safety precautions has this scientist taken?

3. Discuss your library research, experimental design, and safety plan with your teacher before you continue. If your plans are reasonable and safe, your teacher will approve further work.

4. When you have your teacher's approval, carry out the experiment you have designed to test your hypothesis.

   Remember to record data carefully in your journal and to use the proper controls to make it a valid test.

III. Analyzing Your Data

1. Organize your data in a way that makes it easier to see patterns or understand what the data show you (see Figure Ex.5).

   This step will help you when you present your work in Part V.

2. Decide what your data tell you, and record your preliminary conclusions. Include a description of any limitations of your experimental design and any unexpected results that you may have found.

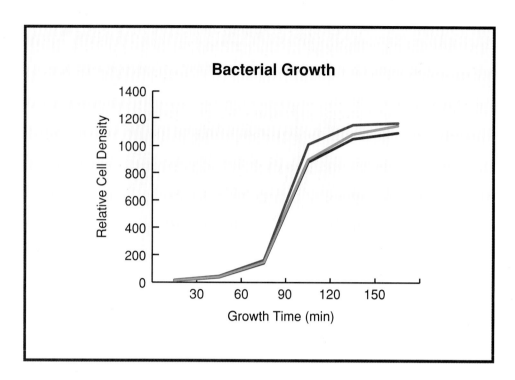

IV. Drawing Conclusions

1. Write an explanation of what your conclusions indicate about the question you asked.

   Support your conclusions by making specific references to your data.

2. Describe how your work connects to the unifying principle most related to your inquiry.

V. Communicating Your Results

1. Assemble a presentation of your full inquiry that makes it possible for someone else to understand what you did, why you did it, and what you found out.

   A poster, a written or verbal report, or a videotape are some examples of how you can communicate your results.

2. Be sure to identify the connections between your inquiry and the following aspects of biology:

   - the unifying principles of biology
   - technology
   - culture
   - history
   - ethics

   All inquiries will have connections to at least one of the unifying principles of biology, but your inquiry may not have connections to all of the other aspects of biology. If you cannot identify a technological, historical, cultural, or ethical connection relevant to your inquiry, explain this in your presentation.

3. As you listen to other students present their results, look for evidence or examples that illustrate why the approach they took to answering their question was a *scientific* approach.

CONTINUITY: REPRODUCTION AND INHERITANCE IN LIVING SYSTEMS

10 Reproduction in Humans and Other Organisms
11 Continuity of Information through Inheritance
12 Gene Action

What do the things depicted in this collage have in common? If you think about it carefully, you may realize that each thing is capable of storing or transferring information, or both. *How* each stores or transfers information varies a great deal, however. In this unit you will explore the idea of continuity by examining the complex mechanisms that make the transfer of genetic information possible.

By the end of this unit, you should

- understand that the continuity of a species depends on the transfer of genetic information,
- understand how information is transferred and preserved through reproduction and the structure of the genetic material,
- understand the processes by which genetic information is expressed,
- understand how sexual reproduction and mutation increase genetic variation and why this is important for the evolution of a species,
- appreciate that human reproduction takes place within a cultural setting and involves ethical issues, and
- become aware of the impact of genetic engineering technology.

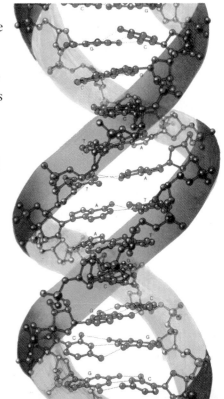

> *If a single cell, under appropriate conditions, becomes a man in the space of a few years, there can surely be no difficulty in understanding how, under appropriate conditions, a cell may, in the course of untold millions of years, give origin to the human race.*

Herbert Spencer, 1820–1903, English philosopher

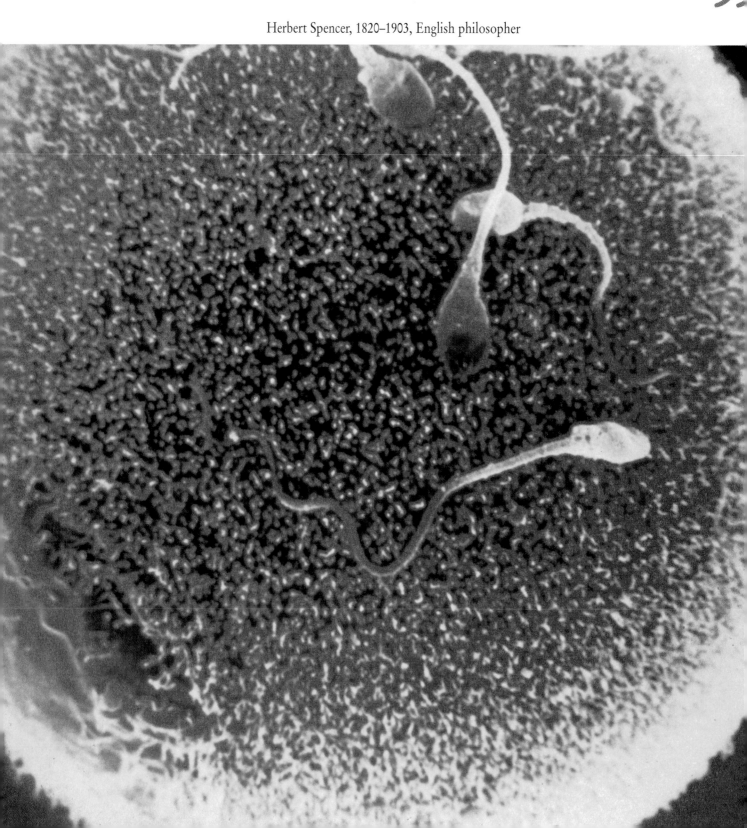

10
REPRODUCTION IN HUMANS AND OTHER ORGANISMS

These opening photographs capture the moment of fertilization as a human sperm makes its way into an ovum. It is in this way that the development of every human begins. Not all organisms begin this way, however. Among the diversity of organisms, there are many ways to make more offspring. Certain organisms are even capable of reproducing themselves, without the contributions of another organism. In this chapter you will explore concepts about reproduction in general and about human reproduction specifically. As you do so, you will begin to develop an understanding of how reproduction contributes not only to the continuation of life on earth but to the continuity of species.

ACTIVITIES

Engage   A Zillion Ways to Make More

Explore   Making Sense of Reproductive Strategies

Explain   Making Sense of Human Reproduction

Elaborate Observing Reproductive Behavior in Nonhuman Animals

Elaborate Cultural Influences on Human Mating Behavior

Evaluate   A Reproductive Strategy for Organism X

A ZILLION WAYS TO MAKE MORE

You probably know how humans reproduce, but humans are just one of millions of organisms on earth. Have you ever thought about how organisms that are very different from you (such as bacteria or evergreen trees) reproduce? This activity will give you an opportunity to think about what you really know and will introduce you to some of the many reproductive strategies that exist.

Process and Procedures

1. How does a small cluster of trees become a forest? To begin thinking about this question, read *The Aspen Story*.

2. What other ways are there to make more? To give you some idea of the range of reproductive strategies and behaviors that exist, watch the videodisc segment *What's Going on Here?* and think about what is happening in each of the images.

Analysis

Now that you have begun to think about how organisms reproduce to create more organisms, write a statement in your journal that shows what the term *biological continuity* means to you at this point in your study of biology.

This statement is a way for you to record your early ideas on the subject so that you can refer back to it and observe how your view changes as you work your way through the chapter and unit.

MAKING SENSE OF REPRODUCTIVE STRATEGIES

Reproduction, the making of offspring, is an essential process for the continuation of a species. But is it essential for an individual? In humans reproduction requires the interaction of a male and female. Is that true of all species? To begin to answer these questions and to develop your understanding of reproduction, you will look at similarities and differences in the reproductive strategies of a number of species.

Materials (per team of 2)

2 reproduction cards
felt-tipped marking pen
large sheet of paper or poster board

THE ASPEN STORY

The pleasant fragrance of evergreen trees surrounded the small group of students and their teacher as they stopped by a stream for a rest before continuing their field trip.

"How tall is that tree?" a student named Kim asked, pointing at a ponderosa pine.

"Oh, I'd say around 100 feet tall," estimated the teacher. "Is that the largest tree you see here?"

"Yeah, but it's not all that big," observed another student, Randy. "Last summer I went to Yosemite Park in California, and they've got *enormous* trees there, called sequoias. They are much taller than this."

"How do you know?" the teacher asked Randy.

"Well, there were pine trees like this nearby, and the sequoia just towered over them. And the trunk was much thicker, like five or 10 of these other pines put together. And the big sequoias are *old*. I read that some are close to 2000 years old," Randy replied.

"Wow. Think of how many cones and new trees it must have made during that length of time," commented Maria. "It's amazing the whole forest wasn't just sequoias!"

"Look at this cluster of trees with the heart-shaped leaves. Does anyone know what they are?" asked the teacher.

"Aspens," several students said simultaneously.

"Well, here's a puzzle: if I tell you this aspen is even bigger than the sequoia, how could that be true?"

"I know! I read about that in a magazine," Phil said excitedly. "All these aspen trees we see are really like branches coming off one giant tree trunk that is underground. Some parts die, but others grow up from the underground part that connects them all. So this whole group of aspens is really just one big tree."

"Then how do you explain that lone one over there?" Kim asked, pointing to a tree on a slope far beyond the stream.

"Aspens also can grow new individuals from seeds in the same way oaks and pines reproduce," explained the teacher. "That one is most likely a separate tree, not part of the cluster here."

"So if that lone tree grows some more tree-looking sprouts from its roots, has it reproduced?" Randy asked.

"What do you think?" his teacher inquired in return.

Process and Procedures

1. To begin your study of the variety of reproductive strategies, with your partner choose two reproduction cards, and read through the material on each one.

   Your teacher will provide these cards. The glossary on page 217 will help with unfamiliar terms.

2. With your team look for similarities and differences in the reproductive strategies described on your cards. Record this information in your journal.

3. Prepare a short (3- to 5-minute) presentation that compares and contrasts the reproductive strategies of the two organisms that you and your partner chose. In your presentation answer the following questions:

   a. What are some of the characteristics of the reproductive strategies of each organism?

You might include characteristics such as

- *the number of offspring produced during each reproductive cycle,*
- *the frequency of the reproductive cycle,*
- *the structures that are involved in reproduction,*
- *the age of sexual maturity,*
- *the life span of individuals, and*
- *the mating behavior.*

b. How are the reproductive strategies similar and how are they different?

c. What are the advantages and disadvantages of the two organisms' reproductive strategies?

d. How does each strategy ensure survival of the species?

e. What are the biological costs of these strategies in terms of time and energy?

The essay Continuity through Reproduction *on page E142 may help you with this task. Your teacher may have other resources available for you as well.*

You may want to create a visual diagram to go along with your presentation.

4. As a team make your presentation to the rest of your classmates.

   As you listen to the other presentations, take notes in your journal that will help you begin to see similarities, differences, and patterns in reproductive strategies.

5. When the presentations are finished, participate in a class discussion of the patterns that are emerging.

6. In your journal make a list of three concepts related to reproduction and continuity that you think are important to the study of biology.

   These would be ideas about reproduction rather than reproductive structures. For example, you might list the idea that reproduction can be a sexual or an asexual process.

7. Work with your partner to create a visual representation of your understanding of reproduction and continuity. Use a large sheet of paper or poster board, and be sure to include your ideas from Step 6.

Analysis

Complete the first item as a class, and then complete the remaining items individually.

1. Display your visual representations around the room, and be prepared to explain and defend the relationships that you have depicted.

   Your teacher will give you time to study the posters after they are displayed and to participate in a short poster session.

2. Think about the following questions, and record your responses in your journal.

   a. Is reproduction necessary for the survival of an individual? Explain.

   b. Is reproduction necessary for the survival of a species? Explain.

c. Do you think that the reproductive strategies that are effective for organisms today will be effective 100,000 years from now? Why or why not?

3. Explain the connection between natural selection and reproduction in terms of biological continuity.

4. Has your view of biological continuity changed? If so, revise the statement that you recorded in your journal during the activity *A Zillion Ways to Make More*.

Background Information: A Glossary of Reproduction Terms

Asexual reproduction involves the growth of a new organism without the fusion of nuclei (which occurs in sexual reproduction). Asexual reproduction usually involves one parent and leads to offspring that are genetically identical to each other. Growing a new plant from a cutting is an example of asexual reproduction.

Binary fission is the division of a cell into two cells of equivalent size. It involves a replication of genetic material (DNA) in the parent cell prior to cell division. Binary fission is asexual. It is the chief means of asexual reproduction in prokaryotic cells.

Budding is a type of asexual reproduction in which new individuals begin as outgrowths on the body of a single parent. The offspring eventually separates from the parent and becomes independent.

Fertilization is the fusion of nuclei from two gametes during sexual reproduction. The result is one nucleus in a zygote.

Fragmentation is a type of asexual reproduction in which a piece broken from a parent organism grows into a new individual.

Gametes are special reproductive cells produced in eukaryotes as a result of a type of cell division called *meiosis*. Each cell contains half the genetic complement of a body cell. During sexual reproduction, the nuclei of gametes fuse and become the first cell of the offspring (called a *zygote*).

Gestation is the internal incubation of embryos or the carrying of young, usually in a uterus, from conception until delivery.

Meiosis is a special process of cell division in eukaryotes that produces reproductive cells known as gametes. Gametes contain half of the genetic complement of the parent.

Mitosis is the production of two identical nuclei in one cell, usually followed by division of the cell into two cells. Each new cell resulting from a mitotic division has the same genetic makeup as the original cell. Asexual reproduction in most unicellular protists occurs by mitosis. Cell growth in multicellular organisms also occurs by mitosis.

Ova (singular, **ovum**) are the gametes produced by females. Also known as eggs, ova are larger than sperm and contain cytoplasmic substances that will influence the early development of the zygote. Eggs also contain subcellular compartments that will be passed on to the offspring and nutrients needed to sustain the earliest stages of a developing organism.

Sexual reproduction involves the fusion of the nuclei of two gametes. Most often the gametes are produced from two parents. Because each parent contributes information, the offspring of sexual reproduction are not genetically identical to either parent.

Sperm are the gametes produced by males. They consist of a compact nucleus surrounded by a membrane, a flagellum for propulsion, mitochondria that provide energy, and a small sac at the tip containing enzymes that help them penetrate the ovum.

Spores are asexual reproductive cells that can develop directly into a complete organism. By contrast, most gametes must join with another gamete before development can occur.

Sporulation is the process of reproducing asexually by producing spores that grow directly into new individuals.

Vegetative reproduction is reproduction that involves parts of an organism that are not specialized for sexual reproduction. Examples include fragmentation, budding, and sprouting from roots as aspen trees do or sprouting from runners as strawberries do.

Zygotes are the cells that result from the fusion of gamete nuclei. A zygote is produced when an ovum and a sperm unite to form a fertilized egg. Zygotes contain a full set of genetic information, half from each parent's gamete.

MAKING SENSE OF HUMAN REPRODUCTION

Explain

In most mammals other than humans, the females are not receptive to sexual activity during the times that their bodies are not ready to conceive. In humans the female may be receptive to sexual activity at any time, even though her body is prepared to conceive only for a short time during each menstrual cycle. Humans have improved their ability to promote and prevent pregnancies through modern technologies.

Your task in this activity is to explain the *biological basis* for birth control methods. In other words, you will explain how birth control technologies work and why they differ in their effectiveness. You also will explain how the chances for conception might be improved. To complete this task, you will have a variety of resources.

Materials (per person)

miscellaneous art supplies

Process and Procedures

As you work, use the following resources as well as others that you have available to help you complete this activity:

- the NOVA video, *Miracle of Life* (The first half shows events related to fertilization.)

- the essays *Making More People* on page E144, *Hormones and Sexual Reproduction* on page E146, and *Sexual Activity and Health Hazards* on page E149

- the videodisc segments *Human Menstrual Cycle* and *Conception*

- additional materials (Your teacher may supply these, or you may find them on your own.)

1. Sort the birth control methods listed in the table in Figure 10.1 into the following categories: Physical Barriers, Chemical Methods, and Behavioral Methods.

   Record in your journal the decisions of your sorting task. Remember that you have a variety of resources to help you decide how to categorize these methods.

Examples of Birth Control Methods

Device	Description	Failure Rate* (Pregnancies per 100 Women per Year)
Abstinence	Abstaining from sexual contact (completely prevents AIDS via sexual route).	0
Cervical Cap	Small dome-shaped rubber cap that is inserted via vagina to closely cover the opening entrance to uterus (cervix).	22.0
Condom (for males)	Thin sheath made of latex or animal skin that is placed over penis. Often coated with spermicide. **Only latex** will help protect against AIDS.**	7.3 (if spermicide is not used)***
Diaphragm	Flexible wire circle covered with latex that is placed over entrance to uterus (cervix). Generally used with spermicide.	10.0 (if spermicide is not used)***
Douche	Vagina is rinsed after sexual activity.	40.0
Injectable Progesterone-like Substance	An injection that lasts up to three months. May cause menstrual irregularities.	1.0
Implant Devices	Capsules containing synthetic progesterone that are inserted surgically under a female's skin. They slowly release hormone for three to five years.	1.0
Intrauterine Device (IUD)	Small plastic or copper device placed (by a physician) inside the uterus. Effective for 10 years. Probably causes mild inflammation of uterine wall.	1.0
Natural Family Planning (Rhythm Method)	Abstaining from sexual intercourse for several days around the time of ovulation.	26.0
Oral Contraceptives (Birth Control Pills)	Synthetic estrogen and synthetic progesterone (or progesterone only) that a female takes in tablet form throughout menstrual cycle.	0.3
Spermicide	Foam, cream, jelly, or suppository preparations that contain chemicals that kill sperm. Spermicides are placed in the vagina prior to sexual activity.	15.0 (if used alone)
Tubal Ligation (Sterilization)	Oviducts in female are surgically severed and tied off.	0.5
Vasectomy (Sterilization)	Vas deferens in male is surgically severed and tied off.	0.6

*Failure rate can be *much* higher if the method is not used properly. For example, if the condom is used *after* sexual intercourse has begun, some sperm already may have been released.

**Natural membrane condoms, as compared with latex condoms, have larger micropores and thus may permit passage of pathogens such as HIV.

***Combined use of spermicide with condom or diaphragm is more effective than either of these methods used alone.

Figure 10.1 **Examples of Birth Control Methods**

2. Using the available resources, prepare a brochure about one contraceptive method from each category. Your brochure should explain the *biological basis* for the contraceptive methods in question.

    You will know you have developed a good brochure if you do each of the following:

    • Describe how each birth control method you have chosen interferes with reproduction.

    • Explain how each method affects the reproductive systems of males and females. (For example, does the method alter the hormonal levels of the male? the female?)

    • Explain why each method needs to be used at a particular time.

    • Explain the differences in the effectiveness of the methods that you chose in each category.

    You can develop one brochure that describes all of the methods, or you can develop one brochure for each method.

3. Write an explanation in your journal that describes how reproductive behavior can be regulated to increase the chances of conception.

Analysis

Join the class in a discussion of the following:

1. The word *progesterone* means "to promote gestation," and *estrogen* means "to generate estrus." Why are these good names? What do you think the prefix *contra-*, as in *contraception*, means?

2. What is the significance of gametes in sexual reproduction in relation to the continuity of information?

3. Imagine that a couple wants to have children, but tests show that the man has a fairly low sperm count and the female has blocked oviducts. In your journal explain whether or not the use of pharmaceutical drugs alone will improve their chances of beginning a pregnancy.

Further Challenges

1. Use what you know about human reproduction and about current contraceptive technologies to suggest one or two possibilities for new contraceptives.

2. After the development of oral contraceptives for women, many people predicted the development of birth control pills for men. No such pills are yet available in the United States, and the area is not a major focus of research. Discuss the reasons for and against development of oral contraceptives for men, both culturally and physiologically.

OBSERVING REPRODUCTIVE BEHAVIOR IN NONHUMAN ANIMALS

Throughout the animal kingdom, the males and females of many species put on a show of behavior aimed at attracting a mate. When a peacock wants to attract a peahen, he displays his extravagant plumage, he struts, and he shakes his colorful tail. Similarly, a male elk (bull elk) calls out loudly (bugles) during mating season. The bugling of a bull elk has at least two effects: it attracts potential mates, and it announces to the other bulls that he is in the area and that they should stay away. Such animal displays often are showy and complicated. Ethologists, biologists who develop and test hypotheses about animal behavior, are interested in learning more about the connections between various behaviors and mating. In this activity you will observe some animals' mating behaviors and develop a question that you would like to answer about the mating behaviors that you observe.

Materials (per team of 4)

cricket cages (clear plastic shoe boxes work well)
cricket food
watering device (a wet sponge in a small plastic container works well)
live crickets

Process and Procedures

1. Observe the crickets in their habitats. As you make your observations, do the following and record the information in your journal:

   a. Observe the general structure of the crickets' bodies.

   b. Observe how they move.

   c. Determine their sex.

   Study Figure 10.2 to learn how to distinguish male and female crickets.

   d. Observe their behavior (both mating and nonmating).

2. With your team develop a question about some aspect of the mating behavior of crickets that you would like to answer (and think you could answer with the time and resources available).

3. Now develop a set of hypotheses that answer your question. Record your hypotheses in your journal, and circle the one your team intends to test.

   The essay Mating Behaviors of Nonhuman Animals *on page E151 may give you some ideas for your hypothesis. You also may use outside sources of information.*

4. Design an experiment to test your hypothesis. Record your design and detailed procedures in your journal.

   Have your teacher approve your design and your procedures when you have completed them.

CHAPTER TEN
Reproduction in Humans and Other Organisms

Gryllus domesticus

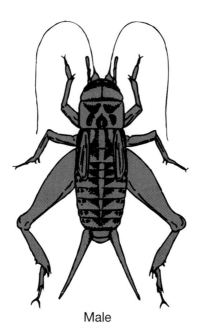 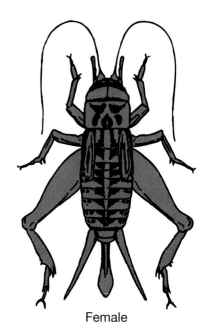

Male   Female

Figure 10.2 Notice the differences between the male cricket and the female cricket (*Gryllus domesticus*). The female has a long ovipositor that is externally visible. It is through the ovipositor that the female lays her eggs.

5. As a team carry out the experiment that you designed. Record your results in your journal, and be prepared to share them with the class.

6. Wash your hands thoroughly after you have completed your experiment.

7. Share the results of your experiment with the class.

8. Discuss the range of hypotheses that the class tested and the range of results that the class obtained.

9. As a class develop the following:

- a summary of what your results from these experiments indicate about cricket mating behavior,

- a list of some of the things you still do not know but would like to know about cricket mating behavior, and

- a list of practical and conceptual difficulties that ethologists face as they make observations and conduct experiments to investigate animal behavior.

Analysis

Study the mating behavior of the animals in the videodisc segment *Animal Mating Behaviors*; in your journal record the observations that you make of each animal. Write a short report describing what you have learned about mating behaviors in nonhuman animals. Be sure to do the following:

a. Compare and contrast what you have learned about cricket mating behavior with what you have learned about the mating behaviors of other animals that you watched on the videodisc.

b. Provide specific evidence from the videodisc segments and from your observations of the crickets.

c. Suggest one or two additional questions that you now have about cricket mating behavior specifically or reproductive strategies in general.

Your teacher will review your journal entry to assess your level of understanding.

Further Challenges

Design an experiment to answer one of the additional questions that you thought of in the *Analysis* and, with the approval of your teacher, carry out this new experiment.

CULTURAL INFLUENCES ON HUMAN MATING BEHAVIOR

Elaborate

In the activity *Making Sense of Human Reproduction*, you explained the physical aspects of human reproduction and how it is regulated. In this activity you will extend that understanding to include cultural behaviors that are associated with human reproduction.

Materials (per team of 4)

scissors
poster board or large sheets of paper
materials collected earlier
tape or glue

Process and Procedures

1. Use the materials that you have collected to create a display that demonstrates various American cultural behaviors that are associated with reproduction. Include characteristics of human behavior that you think are similar to those of other animals and those that you think are distinctive.

   You can use the essay Cultures and Mating Patterns *found on page E153 or any other reference source to help you support your claims. Reading about cultures other than your own helps you realize the variety of patterns that exists and usually helps you observe your own culture more objectively.*

2. Set up your display in your classroom.

3. Study the displays of other teams, and look for patterns in the cultural influences that lead to human mating behaviors.

   Remember that you are looking for patterns about behaviors that lead to mating, not about the mechanisms of mating.

Analysis

Work in your team to respond to the following. Record your responses in your journal, and be prepared to discuss your responses with the class.

1. List the various aspects of American culture that you think most influence human mating behaviors today. For example, you may think American music has an important influence.

2. Compare and contrast human mate selection in the United States 200 years ago with mate selection today, and think about the changes in American culture that have contributed to the differences that exist.

Figure 10.3 What might these images convey about mate selection in American culture?

Further Challenges

1. An ongoing debate in the scientific community is whether most human sexual behavior is biologically based. Research the scientific literature, and report the conclusions your evidence supports.

2. Historically in the United States, families use the surname of the father. Perhaps for this reason, Americans may think automatically of the male as the head of the household, although this view is changing as more women work outside the home. This view of the significance of males is not universal. For instance, among some Native American tribes in the Pacific Northwest, each tribe is composed of clans that are linked through the mothers' line.

   Consider the following observations, and then discuss how societies draw their conclusions about the importance of men and women.

   • A human embryo automatically develops with female characteristics unless it receives male hormones such as testosterone very soon after conception.

   • The ancient Greek philosopher Aristotle suggested that the female is a defective male because she does not have a penis.

A REPRODUCTIVE STRATEGY FOR ORGANISM X

One egg sac of a spider hatches, and swarms of tiny spiders emerge. Although there are huge numbers of spiders born, each one lives a relatively short time. In contrast, an elephant gives birth to one baby, but this young elephant has the potential to live for almost a century. In both cases mechanisms exist that provide for a continuation of the species, even though the individual organisms eventually will die. In this activity you will evaluate what you have learned about reproduction.

Figure 10.4 Many spiders are born at one time, but each spider has a short life span.

Materials (per person)

descriptions and diagrams of Organism X from Chapter 8

Process and Procedures

1. Revisit Organism X that you discovered, and think about how it might reproduce.

   Use your journal notes, the essays, and videodisc segments to help you.

2. Write a detailed description of the organism's method of reproduction. Be sure to address the following:

   - the structures involved,

   - how reproduction is regulated,

   - how the organism behaves to ensure the production of offspring,

   - the number of offspring and their approximate life span,

- how the organism nurtures its offspring, and

- how the organism's overall method of reproduction compares and contrasts with human reproduction.

You may want to include a drawing or diagram with your description.

3. Share your descriptions and drawing with the rest of the class.

4. Ask questions of your classmates to ensure that you have a clear understanding of the reproductive strategies that they are describing.

Analysis

Return to the visual representation of reproduction and continuity that you created in Step 7 of *Making Sense of Reproductive Strategies*, and study the ideas and relationships that you have represented. Think about how your understanding has increased, and add three or four more ideas or relationships to your visual representation. You may want to simply add these ideas to your existing diagram or you may want to create a new one.

"Why is it that children look like their parents? Think of your own family or the families of your friends. In many cases you can see physical resemblances among family members even when they are different ages."

11

CONTINUITY OF INFORMATION THROUGH INHERITANCE

You probably have noticed how certain traits are shared by some members of a family. Even in your own family you might notice some characteristics that seem to be passed from generation to generation. Look at the opening photographs and see if you can identify any similarities between members of the different generations. In this chapter you will begin to explore the processes involved in the transfer of genetic information and what the results mean for various populations of organisms, including humans.

For instance, you will learn how a disease was inherited by several generations of royal families in Europe and was especially tragic for the family of Nicholas and Alexandra Romanov, the last czar and czarina of Russia. You also will learn how the laws of probability allow you to make predictions about inheritance, how cellular processes account for patterns of inheritance, and how genetic concepts explain much of human variation.

ACTIVITIES

Engage	Gifts from Your Parents
Explore	Game of Chance
Explore Explain	Patterns of Inheritance
Explain	Understanding Inherited Patterns
Explain Elaborate	Can You Sort It Out?
Elaborate	The Genetic Basis of Human Variation
Evaluate	Continuity and Change

GIFTS FROM YOUR PARENTS

"You have your mother's nose." "You smile just like your grandfather." "You'll be as handsome as your uncle." Comments such as these often are heard when families get together, and they remind us of the biological link from one generation to the next. The processes of reproduction ensure that species can survive through many generations. What is it, exactly, that survives? It is not the organism itself because regardless of species all organisms eventually die. Yet the survival of a species or a family line shows that something is handed down from parents to offspring.

Throughout this chapter you will investigate the mechanisms of inheritance that underlie reproduction and that allow families to continue and species to survive even though individual members of a particular family or species have finite life spans. In this activity you will begin to think about how genetic information is transferred from one generation to the next as you examine the biological inheritance of one trait in the family of the last czar of Russia.

Process and Procedures

1. Read the story *A Royal Tragedy*, which describes a disease that was inherited through several famous families.

2. Discuss the following questions as a class:

   a. How did the young czarevitch come to have hemophilia?

   b. Why did his father and mother not suffer from the disease?

   c. Suggest some traits that people acquire during their lifetime. What is the difference between acquired traits and traits that are inherited?

   d. How might cultural practices have influenced the frequency of occurrence of hemophilia among the czar's family?

GAME OF CHANCE

If someone flips a coin, you know that there are two possible outcomes: heads or tails. You can predict how likely either is to occur. Can you make similar predictions about the outcomes of genetic events? In this activity you will use biological data to explore the concept of probability, which is a mathematical tool that enables us to make predictions. We will provide you with information about the offspring of two rabbits, and you will look for patterns in the results. In addition, you will begin a genetic experiment that you will conduct throughout this chapter. The experiment will help you see the relationship between genetic information, sexual reproduction, and inherited traits.

Materials (per team of 2)

coin
calculator

A ROYAL TRAGEDY

In the year 1904 the vast Russian Empire was swept by a great wave of celebration and public enthusiasm. At long last, after four daughters in succession, the czar and czarina produced a son—an heir to the throne. This little boy, who was given the name Alexis, seemed destined for a great future. In addition to having been born heir to the throne of the world's largest country, he was connected by ties of kinship to many of the royal and aristocratic families then existing in Europe. His great grandmother Victoria, who died only three years before, reigned as Queen of England for more than half a century, the period in which the British Empire reached its greatest heights of power and influence. The kings of England and of Spain, the prince of Prussia, and the kaiser of Germany all were cousins of Alexis on different family branches, and his own mother, Czarina Alexandra, was the daughter of one grand duke of Hesse and the sister of another. Even though the huge empire of his father Nicholas II was troubled by unrest, everyone expected that this new little heir would provide a sense of stability as he and the new century grew together. In time he would preside over a more modern and progressive Russia.

Unfortunately, the little boy was not well. He had been born with hemophilia A, a disease in which the blood clots so slowly that the victim can bleed to death from a minor injury. Today treatments are available that allow victims of hemophilia to lead fairly normal lives, but this was not the case in little Alexis's time. For him even the most minor childhood accidents meant bouts of painful and potentially fatal illness.

Czar Nicholas was so obsessed with the poor health of his only son that he failed to devote adequate attention to the many problems of Russia. The boy's mother, Czarina Alexandra, became even more preoccupied than the czar. She may have been tormented by the possibility that her son's illness came to him from her side of the family. A number of male relatives in different branches of her family were afflicted or had been afflicted, including one of her brothers (see the pedigree shown in Figure 11.1). Desperate, the czarina became an easy victim for a succession of quacks and mystics, whose undue influence at court was viewed with suspicion and alarm by regular government officials and the nation at large. Indeed, the preoccupation of the czar's family with Alexis's hemophilia may have contributed to the downward spiral of social disorder that triggered the Russian Revolution of 1917. Ultimately this revolution led to the murder of the entire imperial family by agents of the Bolsheviks in July of 1918.

The story of young Alexis and his family is filled with drama and tragedy, but it also illustrates a number of features about the inheritance of biological traits in both humans and other complex organisms. In this chapter we take a closer look at the basic systems that govern both the continuity and change of genetic information.

Figure 11.1 Hemophilia in Czarevitch Alexis's family A pedigree is a family tree that enables us to trace the transfer of a specific genetic trait through several generations.

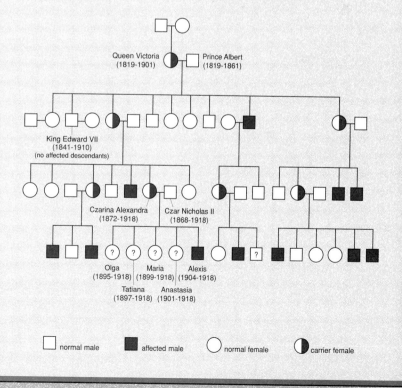

Process and Procedures

1. Work with your teammate to solve the following problems, and record your answers in your journal.

    a. If a pair of rabbits mated and produced 10 offspring, how many males and how many females would you predict in those offspring?

    Explain how you made your prediction and whether you need additional information.

    b. Even if you are reasonably confident that your prediction is correct, can you guarantee how many males and how many females will be born in the litter? Explain your answer.

2. Test your prediction by using a coin to simulate the genders of the 10 offspring.

    a. In your journal prepare a table in which to record your team's predicted and actual results.

    b. Flip the coin 10 times, and record your results.

    c. Explain the role of chance in determining your results.

3. Look at the data under the heading *Small Sample Size* in Figure 11.2. These data show the results of three rabbit matings. Discuss the following with your teammate:

    a. Do these results match your predictions from Step 1 or your test results from Step 2? Explain your answer.

    b. The result of each of these three crosses obviously is not the same. Explain why the actual outcomes vary from 50 percent males and 50 percent females.

Figure 11.2 **Results of Rabbit Matings**

Results of Rabbit Matings

Small Sample Size			
Trial	**# of Offspring**	**Males**	**Females**
1	10	4	6
2	10	6	4
3	10	6	4
Large Sample Size			
Trial	**# of Offspring**	**Males**	**Females**
1	600	279	321
2	600	296	304
3	600	316	284

4. To investigate the relationship between probable outcomes, actual results, and sample size, examine the data under the heading *Large Sample Size* in Figure 11.2. Note that the number of offspring is 600. Calculate the percentage of rabbits that are male for each group of 600 offspring, and record these results in your journal.

5. Answer the following questions with your teammate, and record the answers in your journal.

   a. Are these results generally closer to 50 percent than the results you observed in the small sample size (Step 3)? Explain your answer.

   b. Based on your observation, what effect does sample size have on the match between probable outcomes and actual results?

6. You can test the accuracy of large sample sizes by generating your own data with coins and combining the data of the entire class.

   a. This time, instead of flipping the coin 10 times, flip it 20 times, and record the results in your table from Step 2.

   b. What is the percentage of heads for this sample size of 20 flips?

   c. Contribute your data to a class data table that your teacher develops on the chalkboard.

   d. What is the percentage of heads for this large sample size?

   e. What do these results suggest about the effect of sample size on the match between probable outcomes and actual results?

Analysis

1. Use what you have learned about the importance of sample size to evaluate the following medical study reported in a local newspaper.

   "A study reported in the medical journal *Acta Artifacta* appears to link ownership of fast cars with premature balding. The study, consisting of 17 men who own sports cars, found that nearly 60 percent suffered from premature balding. The authors of the study conclude that because this percentage of balding is much higher than in the general population, there is an increased chance of suffering from premature baldness if one owns a fast car."

2. If you flip a coin five times and get heads every time, what is the probability that you will get tails on the next flip?

   When your teacher indicates, you will begin a genetic cross experiment that will continue throughout the chapter. The procedures for this experiment, which uses baker's yeast, *Saccharomyces cerevisiae*, are set forth in Figure 11.3, *Yeast Genetics Protocol*.

Yeast Genetics Protocol

Baker's yeast (*Saccharomyces cerevisiae*) is a unicellular organism that reproduces both sexually and asexually. Because the cells have a characteristic shape at each stage, it is possible to distinguish all the major stages of the yeast life cycle under the microscope. A brief description will help you understand the stages that you observe.

Certain types of yeast cells, referred to as haploid cells because they contain only one chromosome from each chromosome pair, can occur in two mating types (or "sexes"): mating type **a** and mating type α (alpha). When cells of opposite mating types (**a** and α) come in contact, they secrete hormone-like substances called pheromones. These pheromones cause haploid cells of the opposite mating type to develop into gametes,

Figure 11.3 **Yeast Genetics Protocol**

which are pear-shaped. The gametes then can fuse and produce zygotes, which appear as peanut-shaped cells. When cultured on a solid growth medium, the fused peanut-shaped cells can reproduce asexually by budding (they appear as cloverleafs) and may grow into a visible colony that contains up to 100 million cells. In times of stress, however, these cells cease to bud and undergo a process that produces reproductive cells called spores. Spores can survive stressful environmental conditions, and when growth conditions become favorable, the spores can germinate and reenter life stages that lead once again to reproduction by budding.

In this experiment you will use two yeast strains of opposite mating types so that you can complete an entire yeast life cycle. You will mate the yeast and let them grow through a complete life cycle until you regenerate the original mating types. As you progress through this sequence, you will learn some important principles, such as how genetic material interacts when it is combined, how it is sorted during preparations for reproduction, and how it is transferred through inheritance.

Materials (per team of 2)

microscope slide
cover slip
dropping pipet
glass-marking pencil
compound microscope
container of sterile flat toothpicks
bottle of water
6 zip-type plastic bags labeled *waste* (1 for each day)
MV medium agar plate
3 YED medium agar plates
"unknown" medium agar plate
YED medium agar plate with pregrown yeast strains of **a** and α mating types
(labeled *Plate I*)

Procedures

Remember to record all of your results in your journal.

Day 0

1. Observe the techniques in the videodisc segment *Yeast Monohybrid Cross*.

    These techniques will help you perform the steps in this experiment.

2. Obtain Plate I from your teacher, and use a microscope to examine some yeast cells of either mating type. Use the following procedure.

    a. Put a small drop of water on a microscope slide.

    b. Gently touch the flat end of a sterile toothpick to the streak of either the **a** mating type (strain that has a red or pink phenotype) or the α mating type (strain that has a cream phenotype) on Plate I.

    c. Mix the material on the toothpick with the small drop of water on the slide.

    d. Place a cover slip over the drop.

    e. Discard the toothpick in the plastic waste bag.

    f. Examine the cells with the high-power lens of the microscope.

    g. In your journal sketch the cells and note the phenotype of each mating type.

    Use a new sterile toothpick for each step that requires a toothpick. Be careful not to touch the ends of the toothpicks to anything except yeast or the sterile agar. Discard used toothpicks in the plastic waste bag. **Keep the lid on the plate except when transferring yeast.**

3. Prepare fresh cultures of both mating types on Plate II as follows:

   a. Touch the flat end of a sterile toothpick to the **a** mating type.

   b. Gently drag the toothpick across the surface of a YED medium plate to make a vertical streak about 1 cm long and 1 cm from the edge at a position of 12 o'clock (see Figure 11.4, *Yeast Culture Plates*).

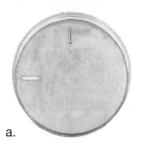 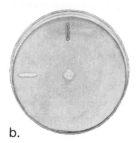

      a.   b.

Figure 11.4 Yeast culture plates a. Appearance of Plate I on Day 0 b. Appearance of Plate II on Day 1 after the mating mixture has grown

   c. Discard the toothpick in the waste bag.

   d. Use a glass-marking pencil to label the plate with an *a* near the streak.

   e. Use a new sterile toothpick to put a horizontal streak of α mating type at 9 o'clock on the same plate.

   f. Label this streak α.

   g. Label this culture *Plate II*, and add your names.

4. Prepare a mating mixture on Plate II as follows:

   a. Use a new sterile toothpick to transfer a small amount of the **a** mating type from the streak in Plate I to the middle of the agar in Plate II.

   b. Use another sterile toothpick to transfer an equal amount of the α mating type to a point just next to the first spot.

   c. Use a third sterile toothpick to thoroughly mix the two dots of yeast to make a mating mixture.

   Be careful not to tear the agar surface. Discard used toothpicks in the waste bag.

5. Invert Plate II and incubate at room temperature for 3 hours so that the cells can begin mating. Then refrigerate the plate until the next lab period.

   If your class meets in the afternoon, you can refrigerate the plate immediately and then incubate it at room temperature for 4 to 5 hours before Step 7.

6. Wash your hands thoroughly before leaving the laboratory.

Day 1

7. Examine Plate II, and discuss the following with your partner:

   • What is the phenotype of the mating mixture?

   • If you compare this result with the results of your bean cross, which color bean is equivalent to the yeast α mating type?

8. Use the procedure in Step 2 to examine the mating mixture through the microscope.

   a. Sketch what you see, and compare it with your earlier drawings.

   b. Describe any differences in the types of cells you see, and explain why these changes may have occurred.

9. Make a subculture of mating type **a**, some of mating type α, and some of the mating mixture. Use sterile toothpicks to transfer the yeast to an MV medium agar growth plate.

Use the same pattern that you used on Plate II with the two different mating type streaks and the circular mating mixture colony in the middle. Discard used toothpicks in the waste bag.

10. Label this subculture *Plate III*, and add your names.

11. Invert the plate and incubate overnight.

12. Wash your hands thoroughly before leaving the laboratory.

Day 2

13. Examine Plate III, and discuss the following with your partner:

    • What do you observe about the colonies on Plate III after incubation?

    • Explain how the information present in the fused cells differs from the information in either one of the original mating types.

14. Use the procedure in Step 2 to examine the freshly grown mating mixture in Plate III through the microscope.

15. With your partner discuss the following:

    • What types of cells are present? Sketch each type.

    • If any of the types seen in Step 8 have disappeared, explain what happened to them.

16. Make a subculture by transferring some of the mating mixture with a sterile toothpick to a fresh YED medium agar plate.

    a. Streak the cells in one horizontal line across the middle of the plate.

       Discard used toothpicks in the waste bag.

    b. Label this subculture *Plate IV* and add your names.

    c. Invert, and incubate overnight.

17. Wash your hands thoroughly before leaving the laboratory.

Day 3

18. On a plate of "unknown" medium, make several thick streaks of the freshly grown subculture from Plate IV.

    Discard used toothpicks in the waste bag.

19. Label this culture *Plate V*, and add your names.

20. Invert the plate and incubate at room temperature for three or more days.

21. Wash your hands thoroughly before leaving the laboratory.

Day 7

22. Use the procedure from Step 2 to examine yeast from Plate V through the microscope.

    You may need to use the fine adjustment on the microscope to distinguish cells at different levels.

23. Discuss the following with your partner:

    • What cell types are present now that were not present before? Sketch these cell types.

    • Compare these cell types with the cell types you saw at other stages.

    • How do you think the "unknown" medium differs from the growth medium?

       Refer to the introduction.

    • If the cells in the sacs are most frequently found in groups of four, what process do you think produced them? Explain.

- Are the cells in the sacs haploid or diploid? Explain.

24. Transfer some yeast from Plate V to a fresh YED medium agar plate.

- Use sterile toothpicks to streak the cells as shown in Figure 11.5.

Discard used toothpicks in the waste bag.

25. Label this culture *Plate VI*, and add your names.

26. Invert the plate and incubate at room temperature for about 5 hours, and then refrigerate until the next lab period.

If your class meets in the afternoon, refrigerate immediately, and then incubate at room temperature for 5 hours before Step 28.

27. Wash your hands thoroughly before leaving the laboratory.

Day 8

28. With your partner discuss and explain what you observe about the colonies on Plate VI.

29. Use the procedure in Step 2 to examine the growth from Plate VI through the microscope.

30. Sketch the cells, and compare them with the stages you observed before.

31. Discard used culture plates as your teacher directs.

32. Wash your hands thoroughly before leaving the laboratory.

Figure 11.5 **Pattern used to streak the spores** First streak one line of cells across the top of the plate, and then use a second sterile toothpick to streak some of these cells in a zigzag pattern across the rest of the plate. When this is done properly, individual yeast cells at the end of the second streak will grow into isolated colonies.

PATTERNS OF INHERITANCE

Why is it that children look like their parents? Think of your own family or the families of your friends. In many cases you can see physical resemblances among family members even when they are different ages. Even more striking is the similarity of physical characteristics in domestic animals that are purebred, such as breeds of dogs. This similarity of characteristics occurs only if matings occur between members of the same breed such as the collies illustrated in Figure 11.6a. In the case of asexual reproduction, for instance tulips that reproduce through bulbs, the similarity of inherited traits among individuals is even greater.

Natural populations generally are not as uniform as purebred lines of domesticated organisms. For example, dogs running free in a city can choose their mates naturally, and they do not always mate with dogs that look very similar to themselves. Such matings produce offspring with a mixture of traits. Consider the mixture of traits that results when a Great Dane mates with a collie, as shown in Figure 11.6b. A mixture of traits is common among human offspring as well.

a.   parent   offspring   parent

b.   parent   offspring   parent

Figure 11.6 Continuity of form in animals a. Purebred dogs such as collies pass their characteristic traits to offspring almost unchanged. b. Mixed mating, however, produces a mixture of traits.

In this activity you will begin to model how genetic information is passed from one generation to the next.

Process and Procedures
Part A Inheritance of One Trait

In this model you will use two colors of beans to represent the genetic information that contributes to the trait for straight or floppy ears in rabbits.

1. Select two beans randomly from the container that your teacher provides.

2. In your journal record the color of your beans and whether your pair is homozygous or heterozygous.

   *These beans represent genetic information in one of the parents. If the beans are the same color, the pair is **homozygous** (homo = same). If the beans are different colors, the pair is **heterozygous** (hetero = different).*

3. Choose one bean randomly by shaking both in your hands and then selecting one without looking.

   Each parent contributes only half of his or her own genetic information to each offspring, and parents do not know which half they contribute.

4. Add this bean to the one your partner selected.

5. Record the color of the two beans in the new combination, and indicate whether the pair is homozygous or heterozygous.

The new bean combination represents the genetic information that will determine the condition of the ears in the rabbit offspring.

6. Use the color key that your teacher displays, and list in your journal the ear trait to which your beans corresponded initially and finally.

   a. To which condition did your first bean combination correspond?

   b. To which condition did your partner's first bean combination correspond?

   c. To which condition did the new combination correspond?

7. As part of a class discussion, use the key once again and the shared results of the class to consider the following:

   a. Did one bean color (which corresponds to one piece of inherited genetic information) have a greater influence in determining the ear trait than the other bean color?

   b. Using genetic terminology, explain any patterns that you detected in the homozygous and heterozygous bean combinations.

   The essay Phenotype and Genotype *on page E155 may help you with your explanation.*

Part B Inheritance of Two Traits

What will you observe if you follow the inheritance of two traits at once? Continue working with your partner as you study the inheritance of ear type and gender in rabbits.

1. Examine Figure 11.7, which shows the results of a cross that follows two traits in rabbits, ear type (floppy ears or straight ears) and gender (female or male).

   Male rabbits that had floppy ears were mated with female rabbits that had floppy ears.

   The results for 100 offspring are shown.

Figure 11.7a Following the inheritance of two traits

| # of Offspring | Male rabbits, floppy ears | | Female rabbits, floppy ears | |
| | Males | | Females | |
	Floppy	Straight	Floppy	Straight
100	52	0	48	0

Figure 11.7b

2. Based solely on the results of this cross, what conclusions can you draw about these two traits?

3. Next male rabbits that had straight ears were mated with female rabbits that had floppy ears. The results for 100 offspring are shown. Calculate the data for gender (female/male) and ear type (floppy ears/straight ears) as percentages.

| # of Offspring | Male rabbits, straight ears | | Female rabbits, floppy ears | |
| | Males | | Females | |
	Floppy	Straight	Floppy	Straight
100	0	47	0	53

4. Discuss the following questions with your teammate:

a. What overall conclusions can you draw about the inheritance of straight ears versus floppy ears?

b. Do you think the type of ear a rabbit is born with is affected by whether a rabbit is female or male? Explain your answer.

Analysis

Work individually to develop responses to the following questions in your journal.

Base your responses on your experiences so far in this chapter and on the information in the essays Phenotype and Genotype *(page E155) and* Case Studies of Two Tragic Genetic Disorders *(page E158).*

1. Although Huntington disease is a dominant trait, symptoms do not appear until late adult years. What are the implications for young adult children of a parent who has Huntington disease?

2. Restate the following accurately: One out of every two offspring resulting from a cross between parents with the genotypes *Hh* and *hh* definitely will have Huntington disease.

3. Two healthy individuals marry and produce three children. The first two are healthy, but the third is born with cystic fibrosis, indicating that she is homozygous *cc* for the cystic fibrosis alleles. What can you conclude about the genotypes of the other people in the family?

4. What is the probability for the couple in Question 3 that their future children will have cystic fibrosis? What are the implications for their healthy children?

5. In cocker spaniels black (*B*) is dominant over red (*b*) and solid color (*S*) is dominant over white spotting (*s*). A red male was mated to a black-and-white female. They had five puppies, as follows: one black, one red, one black-and-white, and two red-and-white.

   a. What genetic principle accounts for the phenotypes of these offspring?

   b. Explain how that principle worked in this cocker spaniel family. Show the genotypes of the parents and their offspring in your explanation.

6. How do your genes and the environment interact to produce you as a person?

   Modify the diagram (Figure 11.8) below to illustrate your answer.

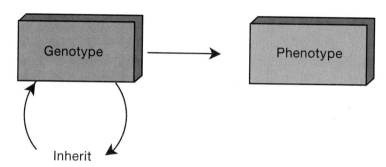

Figure 11.8 **Relationship between genotype and phenotype**

UNDERSTANDING INHERITED PATTERNS

In the activity *Patterns of Inheritance*, you learned that we can use phenotypic patterns of inheritance to derive information about the genotype of an organism. You also know from Chapter 10 that genetic information is passed from generation to generation by cells called gametes. Gametes such as egg and sperm have only half the genetic information of other body cells. If they had more, each zygote would start life with more genetic information than its parents, a situation that would continue to build up generation after generation. At some step in the life cycle of reproducing organisms, the number of chromosomes must be reduced.

How and when does this reduction occur? Most of the solution to this question was worked out just before the beginning of the twentieth century. The key is a process called **meiosis**. In this activity you will use information presented on the videodisc and in an essay to work out the mechanism through which meiosis produces gametes.

Explain

Materials (per team of 2)

red and blue modeling clay
large sheet of paper
small sheet of scrap paper
scissors

Process and Procedures

Part A Meiosis

Meiosis is the mechanism by which the amount of genetic information in a cell is reduced by one-half before gametes are produced. Meiosis also explains why patterns of inheritance often are predictable.

Work with your team to complete the following tasks:

1. Write a description in your journal of how the major events in meiosis lead to a reduced amount of genetic information in gametes.

2. Explain why there are four meiotic products.

3. Record an explanation of how individual chromosomes behave during the process of meiosis.

   You may use the essay Meiosis: The Mechanism Behind Patterns of Inheritance *(page E160) and the videodisc segment* Meiosis *as resources.*

Part B Tracking Genes through Meiosis

Consider a cell from a male diploid animal that has two pairs of chromosomes. Use Figure 11.9, your journal notes, and modeling clay to track how meiosis affects the distribution of the four chromosomes by identifying a specific gene associated with each chromosome.

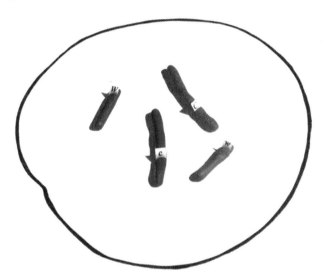

Figure 11.9 Clay models of chromosomes in a diploid animal cell

1. On a large sheet of paper, draw circles to represent a cell undergoing meiosis.

   Draw a large circle at the top of the page, two smaller circles below that, four circles below the two, and four circles with tails (to represent sperm) at the bottom of the page.

2. Use the clay to form two pairs of chromosomes about as thick as a pencil. Make one pair about 6 cm long and the other pair about 3 cm long.

   Use red clay to represent the chromosome of each pair that came from the female parent and blue clay to represent the chromosome of each pair that came from the male parent.

3. From the sheet of scrap paper, cut enough 0.5-cm squares to make gene labels for each chromosome. Mark the labels C = cream eye, c = tan eye,

W = white wing, and w = spotted wing. Press the labels into the clay models.

Put the labels for eye color on the long chromosomes and the labels for wing color on the short chromosomes (refer to Figure 11.9).

4. Place your chromosome models in the large circle at the top of your sheet of paper.

5. Prior to meiosis, the chromosomes replicate. Make more clay models with labels to represent this step.

6. Using your cell-diagram circles, move your chromosome models through the process of meiosis and into the sperm.

   Sketch or record how you line up the chromosome models during each meiotic division.

7. Answer the following questions:

   a. What is the genotype of each sperm for eye color and wing color?

   b. What other genotypes are possible?

      Check the results of other teams.

   c. How would you change the lineup of your chromosome models during each meiotic division to obtain the other possible genotypes?

Analysis

1. Explain how your cell diagrams and your answers to the questions in Part B illustrate the connection between the segregation of alleles and meiosis.

2. If an organism has the genotype AA, what can you say about the alleles present in the gametes that gave rise to that organism?

3. Use the principle of segregation to explain how the alleles of different genes are mixed during mating.

CAN YOU SORT IT OUT?

In the activity *Understanding Inherited Patterns*, you saw that the key to inherited patterns is a process called meiosis. Meiosis results in the production of gametes, with each gamete carrying genetic information from one of the parents. In this activity you will examine the results from the mating of different strains of garden peas, much as Gregor Mendel did in his monastery garden over 150 years ago. You will learn how to use a Punnett square as a tool to help you predict simple crosses and then apply your understanding of genetics to determine the relationships between the phenotypes and genotypes of the parents and offspring. Finally, you will investigate the patterns of inheritance in linked traits and in X-linked traits.

Materials (per team of 2)

calculator

Process and Procedures

Part A A Mendelian Trait: Pea Pod Color

View the videodisc segment *Mendel's Peas*. Work through the steps of this **monohybrid** (one trait) **cross** with your partner, and answer the questions from the video.

Part B Two Mendelian Traits: Pea Pod Color and Shape

When you follow the inheritance of *two* traits (a **dihybrid cross**), more complex patterns result. In garden peas the genes for the traits pod color and pod shape are on different chromosomes. As a result their inheritance conforms to Mendel's law of independent assortment.

Green pods (G) are dominant over yellow pods (g), and expanded pods (E) are dominant over constricted pods (e) (see Figure 11.10). Peas that are homozygous for both green and expanded pods are crossed with peas that have yellow, constricted pods.

Figure 11.10 **Pod color and shape in garden peas**

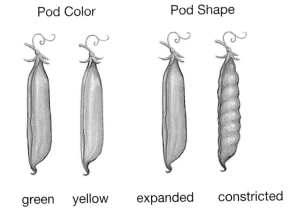

Pod Color   Pod Shape

green   yellow   expanded   constricted

1. With regard to pod color and pod shape, record in your journal the genotypes of gametes that can be formed by these parental types. List the genotypes and phenotypes of the F_1 (first generation) offspring.

2. The F_1 individuals are crossed with each other. Use a Punnett square to show the possible genotypes of the offspring.

3. Based on the data in your Punnett square, what fraction of the offspring would you predict to have each of the following phenotypes?

    a. the dominant phenotype for pod color and pod shape

    b. the dominant phenotype for pod color and the recessive phenotype for pod shape

    c. the recessive phenotype for pod color and the dominant phenotype for pod shape

    d. the recessive phenotype for pod color and pod shape

4. If there were 288 offspring, how many would you predict to have green, expanded pods? green, constricted pods? yellow, expanded pods? yellow, constricted pods?

5. Compare the actual results that your teacher provides with your predictions and answer the following questions:

   a. Do the results match your predictions?

   b. How can you explain any discrepancy between the result and your predictions?

Part C Linked Traits

Because there are many more genes than chromosomes, it is not surprising that many different genes are located on the same chromosome. When gametes are formed, alleles of genes that are located on the same chromosome tend to end up together in the same gamete. The term **linkage** is used to describe this tendency. Linkage can be disrupted during meiosis, however, when chromosome segments are broken and exchanged between matching chromosomes during **crossing over** (see Figure E11.8 on page E162).

1. Repeat Steps 1–3 in Part B, but this time suppose that the genes for pod color and pod shape are on the same chromosome (that is, *G* and *E* are linked, and *g* and *e* are linked).

2. Describe the effect that crossing over might have on the results.

   The videodisc segment Crossing Over *may help you with your description.*

Part D X-Linked Traits

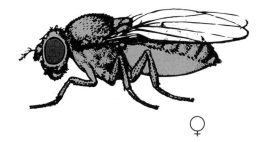 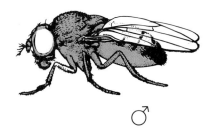

♀   ♂

Figure 11.11 **Eye color in fruit flies**

1. In the fruit fly, *Drosophila melanogaster*, red eye color is dominant over white eye color (see Figure 11.11). Predict the genotypes and phenotypes you would expect in the F_1 when a homozygous red-eyed female fly is mated with a white-eyed male fly.

   First you might want to list the gametes you would expect the parents to produce. Then combine the gametes to show the F_1 offspring.

2. Your teacher will list the actual results of such a cross on the chalkboard. Compare them with your predictions.

3. If you mated red-eyed *male* flies with white-eyed *female* flies (instead of red-eyed females with white-eyed males), would you expect similar F_1 results to those obtained in Step 2? (When males and females are switched with respect to the traits used in a cross, it is called a *reciprocal* cross.)

4. In the early 1900s biologist Thomas Hunt Morgan performed a similar cross, but instead of getting all red-eyed flies, he got half red-eyed flies and half white-eyed flies. However, all the red-eyed flies were females and all the white-eyed flies were males. How might you explain these seemingly unexpected results?

The following information may help you explain this phenomenon.

One special chromosome pair in humans and many other organisms does not always occur as a matching pair: the sex chromosomes. As their name indicates, these specialized chromosomes carry genes that control the sex phenotype of the individual. Whereas nonsex chromosomes, which are called autosomes, are designated by numbers, the sex chromosomes are designated by the letters X and Y. In humans and other mammals, as well as in fruit flies, females have a pair of equivalent X chromosomes. The cells of males contain only a single X chromosome; its pairing partner is a Y. Figure 11.12 shows a diploid set of human chromosomes arranged in matching pairs. Are these chromosomes from a male or a female?

Figure 11.12 A diploid set of human chromosomes Except for gametes, nearly every human cell contains a matching pair of each chromosome. One member of each pair is from the mother and the other member of each pair is from the father. At the beginning of meiosis, the chromosomes condense and thicken. When pictures of the condensed chromosomes are cut out of a photograph and arranged in pairs, as shown, the resulting display is called a karyotype. Karyotypes are used in genetic testing.

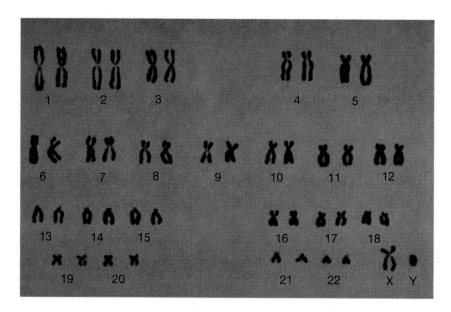

Very few genes have been detected on the Y chromosome, although it has at least one gene that causes maleness. The X chromosome, on the other hand, carries hundreds or thousands of different genes; these genes are said to be X-linked. Most of the genes carried by the X chromosome control phenotypes that have nothing at all to do with gender, for example, Alexis's hemophilia in the story *A Royal Tragedy*.

5. Use all of the information you have acquired to solve the following problems:

   a. A superscript letter, which designates the allele, often is used in connection with the chromosome symbol to diagram a cross of a trait that is carried on the X chromosome. For example, the allele for white eyes might be represented by X^r, and the allele for red eyes by X^R. The symbol Y can represent the Y chromosome and does not carry a gene for eye color. Using these symbols, diagram the cross of Step 3, and show the expected genotype and phenotype fractions.

b. Diagram the cross that would result if the F_1 offspring were mated, and show the genotype and phenotype fractions you would expect in the F_2.

Analysis

Be prepared to respond to the following as part of a class discussion:

1. Predict the possible phenotypes and genotypes of the offspring from a mating of two beagles: a heterozygous female with the dominant phenotype droopy ears (D) and a male with the recessive phenotype upright ears (d).

2. Describe the role of a Punnett square in allowing you to predict the potential ratio of genotypes that may result from a particular cross.

3. You already have learned that crossing over causes some genes on the same chromosome to assort independently. What relationship might you expect there to be between the spacing of genes on the chromosome and their frequency of independent assortment?

4. Revisit Step 1 of Part D. Now that you know that the gene for eye color in fruit flies is linked to the X chromosome, predict the F_2 results of that mating.

5. Explain why males are more likely than females to display the phenotypes associated with X-linked recessive traits.

6. Use your knowledge of X-linked traits to explain the inheritance of hemophilia in the story *A Royal Tragedy*.

THE GENETIC BASIS OF HUMAN VARIATION

Elaborate

Look around at your classmates and consider how different they are from one another. You can use any of a tremendous variety of physical characteristics, such as hair type, skin color, body type, and facial features, to distinguish one human from another. The degree of variation is limited, of course; despite the physical distinctions of humans, you easily can recognize a stranger from a distant country as a member of *Homo sapiens*. What might be the genetic basis for such variation? How might natural selection affect the variety of characteristics exhibited by humans? In this activity you will begin to answer these questions by experiencing some of the nature and range of variation in human populations and by acquiring additional information about how genotype affects phenotype.

Materials (per person)

white 3 × 5-inch card
colored 3 × 5-inch card
pencil
masking tape
small squares of paper

Process and Procedures
Part A Measuring Variation

1. Work with your classmates and teacher to construct the axes for the two histograms shown in Figure 11.13, Histogram A and Histogram B. Use the masking tape and small pieces of paper to make and label the axes on a large wall in your classroom, in the hall, or another large open area.

2. Next, write XX on a white 3 x 5-inch card if you are a female, or XY if you are a male. This is your chromosomal sex. On a colored card, write your height in centimeters; round off to the nearest 5 cm.

3. Use masking tape to affix your cards to the proper places on the histograms.

   It will be interesting to watch the histograms develop, but do not crowd in front of them or it will be difficult for other students to add their data.

4. When the histograms are complete, work with your partner to answer these questions.

   Use the data in the histograms and your understanding of genetics to help you.

   a. How many different types of individuals are represented on Histogram A?

   b. How many different types of individuals are represented on Histogram B?

   c. Describe how the shapes of the two histograms differ.

   d. What is responsible for the shape of Histogram A?

   e. What is responsible for the shape of Histogram B?

   f. What would happen to the shapes of the histograms if you added data from the rest of the students in your school?

   g. List other human traits that would result in the same types of patterns shown in the histograms.

Part B Variation and Evolution

Imagine that a terrible disaster has befallen humankind. The survivors of the catastrophe are forced underground, where they live in cramped tunnels and caves. Because it is impossible to survive above ground, these tunnel-dwellers never venture outside. The hard earth and the scarcity of digging tools force the inhabitants to dig passageways that are as small as possible. The resourceful inhabitants live primitive lives eating roots, insects, and

a.

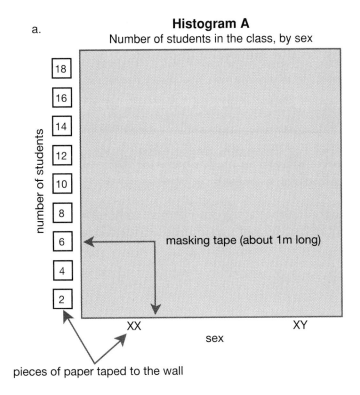

Histogram A
Number of students in the class, by sex

number of students

18
16
14
12
10
8
6
4
2

← masking tape (about 1m long)

XX   XY

sex

pieces of paper taped to the wall

b.

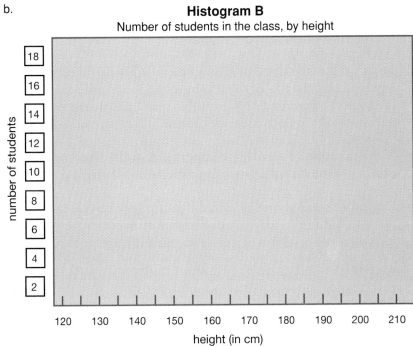

Histogram B
Number of students in the class, by height

number of students

18
16
14
12
10
8
6
4
2

120 130 140 150 160 170 180 190 200 210

height (in cm)

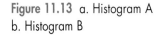

Figure 11.13 a. Histogram A
b. Histogram B

rodents; sleeping huddled together; and reproducing within these constricted surroundings. For generations and generations, through thousands of years, no human ever experiences wide open spaces.

Complete the following with your teammate. The information in the essays *The Role of Variation in Evolution* (page E164) and *Genetic Complexity* (page E168) may help you.

1. Describe the change in Histogram B that you might expect to see if the inhabitants of this underground world were to participate in the activity of Part A 5000 years after moving underground.

    Feel free to be creative by illustrating your description or expanding the story to explain your response.

2. How do we define *normal* for traits such as height? What, for example, is "normal" height, and where does "short" end and "tall" begin? How do we define other complex traits such as behavior? Where does "intelligent" end and "really intelligent" begin?

3. What evolutionary interpretation can you give to human traits that exhibit a continuous pattern of variation? Consider the effects of natural selection as you develop your answer.

Analysis

Respond to the following tasks individually, and contribute your answers to a class discussion.

1. One of the most important lessons of Charles Darwin's work is that natural selection acts on naturally occurring variation in populations. How might traits that vary continuously, such as height, arise from the mechanisms of evolution?

2. Look again at the shape of Histogram B. What part of the histogram represents inhabitants of the underground world who are likely to fare well in this new environment? Why are so many people represented by the measurements in the middle and so few at the extremes—the "outliers"?

3. Is the variation observed in Histogram B of the same type as that which causes genetic disorders such as cystic fibrosis? Explain.

4. Explain the underlying genetic difference between complex traits (that is, those exhibiting continuous patterns of distribution) and simple traits (that is, those exhibiting discrete patterns of distribution).

5. What does human variation—at the level of an individual—have to do with genetic disorders, natural selection, and evolution? Before you try to answer this question, consider the following quote written by two physicians who study disease in its evolutionary context:

    > Some genes are very disadaptive. They result in reduced life expectancy, limited reproduction, and physical and social handicaps. Most of these genes express themselves early in life. They are the first to be removed from the population by natural selection. In fact, more than 90 percent of single-gene disorders—such as cystic fibrosis, sickle cell disease, and muscular dystrophy—express themselves by the end of puberty.[1]

---

[1] *Source:* Excerpted from "Age at Onset and Causes of Disease," by B. Childs and C. R. Scriver, *Perspectives in Biology and Medicine,* Volume 29, 1986.

Further Challenges

Biologists have discovered that there are genes that make certain people susceptible to heart disease. Other people who have these genes, however, do not develop heart disease. What does this information tell you about the role of natural selection in common disorders such as heart disease? What does it indicate about our ability to identify people at risk and about the likely effectiveness of various treatments?

CONTINUITY AND CHANGE

In this activity you will apply your understanding of the basic concepts in genetics by using genetic terminology to summarize the results of your yeast experiment, by comparing the patterns of inheritance in yeast and humans, and by analyzing data from human genetic disorders.

Materials (per team of 2)

results from yeast experiment

Process and Procedures
Part A Genetic Mechanisms in Yeast

1. Study your yeast experiment results in enough detail to respond in your journal to the following:

    You may find it helpful to use the introduction to the yeast protocol (Figure 11.3) as a resource.

    a. Compare the final colonies to the original mating type colonies. Describe how they are alike or how they are different.

    b. What life-cycle stages are present in these final colonies? How do you know they are present?

    c. What evidence is there that a new life cycle has started?

    d. Use the results of your yeast experiment to explain the statement, The genetic plan of yeast cells can be transmitted through reproduction.

2. Apply your understanding of genetics to the yeast reproduction diagrams on the copymaster *Yeast Reproduction*, and complete the following tasks:

    a. At each position where new cells are represented in the sequence, describe whether the cells are haploid or diploid; justify your responses.

    b. Label in the appropriate diagram the position where meiosis occurs.

3. Working individually, write one or two paragraphs in which you compare the mechanisms that yeast and humans use to maintain genetic continuity. Include in your comparison a description or illustration with labels of the following:

   • the haploid and diploid cells in each species,

   • the meiotic process each organism uses that leads to gamete formation,

   • an obvious phenotype that can be inherited from one generation to the next, and

   • the life cycle and method of reproduction of each species.

Part B Genetic Variation and Human Genetic Disorders

Analyze the following questions using what you know, the essay *Incomplete Dominance* (page E170), and the material in Figure 11.14.

1. Which disorders show X-linkage? Explain whether you would expect to see them in males or females or possibly both.

2. Explain how young Czarevitch Alexis, from the activity *Gifts from Your Parents*, could be affected by hemophilia even though neither parent was affected. Use genetic terminology.

3. Although Huntington disease is very rare in the total population, it appears at a rate of 50 percent in affected families. Explain this statement in genetic terms.

4. Alleles that cause CF occur at a rate of about 1 in 25 individuals of European descent. Alleles that cause red-green color blindness occur at a rate of about 1 in 20 males in the United States. In a school with 500 students (50 percent male and 50 percent female), would these statistics guarantee that 20 students are heterozygous carriers of the CF gene or that at least 12 males are colorblind? Explain your response.

5. Of the disorders listed, how does the genetic abnormality of Trisomy 21 differ from the others?

6. Use the concept of natural selection to help explain why the frequency of hemophilia A is so different from the frequency of color blindness.

7. Charles Darwin did not have an understanding of genetics when he proposed his theory of evolution. Now that you have some understanding of genetic mechanisms, how would you use that knowledge to help explain evolution and the diversity seen in populations of living organisms?

8. How might the mechanisms of evolution explain how harmful alleles, such as those that lead to sickle cell disease, are retained in a population?

Human Genetic Disorders

Disorder	Genotype	Phenotype	Genetic Incidence
Cystic Fibrosis (CF)	autosomal recessive	• Life-threatening respiratory illness and blockage • Digestive problems • Salty sweat • Often lethal in childhood or early adulthood	European American Heterozygous 1:25 General Population Homozygous 1:2000 African American Homozygous 1:17,000 Hawaiian Asian Homozygous 1:99,000
Tay Sachs	autosomal recessive	• Accumulation of lipids in the brain and other organs and tissues • Progressive retardation in development • Paralysis • Dementia • Blindness • Death by age 3 or 4	Jewish-American of Eastern European Descent Heterozygous 1:30 Non-Jewish Heterozygous 1:400,000
Sickle Cell Disease	autosomal recessive	• Sickle-shaped red blood cells • Reduced oxygen-carrying capacity • Under certain conditions, some sickling in heterozygotes	African American Heterozygous 1:10 Homozygous 1:400
Familial Hypercholesterolemia (FH)	autosomal dominant	Heterozygotes • Heart disease • Blockage of blood vessels Homozygotes • Large buildup of fat deposits in arteries • Fat deposits in tendons and under skin • Heart attacks in childhood	American Heterozygous 1:500 Homozygous 1:1,000,000
Huntington Disease	autosomal dominant	• Neurological disease, fatal in middle age, characterized by uncontrollable twitching	In general, very rare Heterozygous 1:20,000 In affected families: 50%
Hemophilia A	X-linked recessive	• Lack of blood clotting, can result in fatal hemorrhage	Males 1:7000 Females almost nonexistent
Red-green Color Blindness	X-linked recessive	• Inability to distinguish the colors red and green • Reduced night vision	European-American Males 1:20 Homozygous Females 1:400
Trisomy 21 (Down Syndrome)	3 copies of chromosome 21	• Mental retardation • Poor skeletal development • Weak muscle tone • Altered facial features • Heart and organ abnormalities • Males are sterile; females may be fertile • Life span of 30 to 50 years	Incidence increases with mother's age

Figure 11.14 Human Genetic Disorders

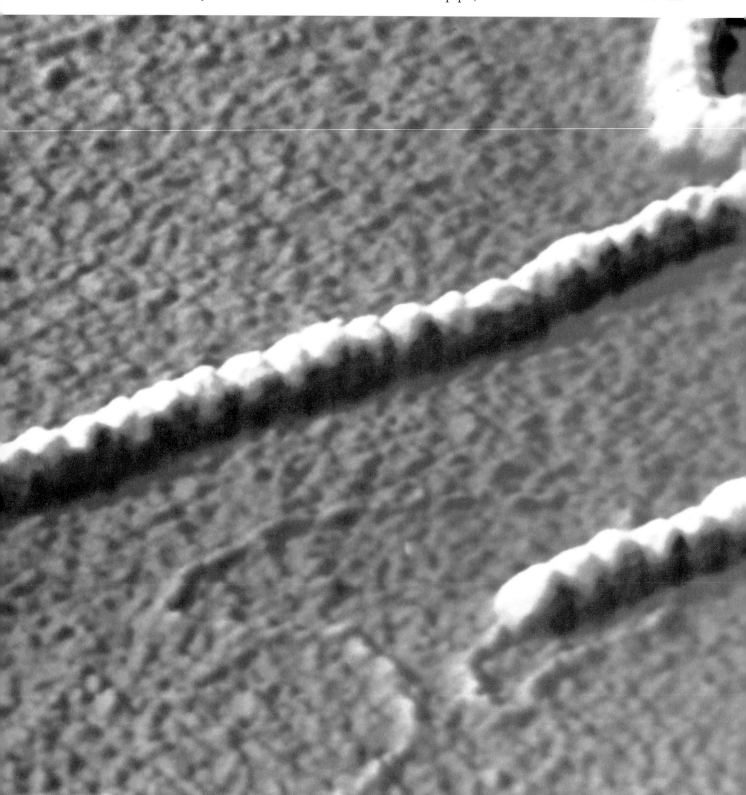

> **"** *It has not escaped our notice that the specific pairing . . . immediately suggests a possible copying mechanism for genetic material.* **"**

Last sentence of James Watson and Francis Crick's 1953 *Nature* paper, which described the structure of DNA.

12
GENE ACTION

Do you have any idea what the linguistic sequence below means? How might you begin to understand the information that is stored in an unfamiliar language? During World War II, the United States Marine Corps used this language, Navajo, to code and transfer secret information. They chose Navajo because it is a complex language that is not widely known. Consequently, if a message were intercepted by people outside the United States, it was unlikely that they would be able to make sense of it. Translated into English, this Navajo sequence means, "What does this sentence really mean when you think about it?"

Just as the symbols of linguistic sequences function as a code to store and transfer information, the molecules of genetic material such as those shown at left also store and transfer coded information, in this case from one generation to the next. In addition, the information stored in the molecular code of genetic material can be expressed in a form that is useful to living systems. The importance of this code has motivated scientists for decades to study the genetic code in more and more detail to uncover the rules and mechanisms that govern it.

In this chapter you will have the opportunity to learn about and to model the genetic code, as well as to understand other aspects of molecular genetics. You also will learn about the field of genetic engineering and consider the consequences and implications of advances in this field.

ACTIVITIES

Engage
Explore **The Stuff of Life**

Explore **Transferring Information**

Explore
Explain **Modeling DNA**

Explain **Gene Expression**

Elaborate **A Closer Look at Protein Synthesis**

Elaborate **Genetic Technology**

Evaluate **Prime Time**

Ákohgo ts' ídá haiit' éego ááhyił ní nahalingo, éí baa nitsídzíkees dooleeł?

THE STUFF OF LIFE

In Chapter 11 you learned about the importance of inheritance for the continuity of species, the way natural selection acts on the variation in populations, and the relationship between genotype and phenotype. The key to all of these processes is deoxyribonucleic acid, DNA. In this activity you will begin to investigate DNA, and later you will see how an understanding of DNA has led scientists to develop technologies that are having far-reaching effects on biological research, medicine, public policy, and the judicial system. Because these technologies have the potential to affect the lives of all citizens, an understanding of DNA and its molecular function is no longer only the realm of the scientist. But first, what is this stuff we call DNA?

Materials (per team of 2)

2 pairs of safety goggles
2 lab aprons
2 15-mL culture tubes
500-mL beaker
calibrated pipet or graduated cylinder
glass stirring rod
thermometer
test-tube rack
hot plate
glass-marking pencil
ice bucket with ice
1 mL Woolite® cold water wash or Dawn® dishwashing detergent
6 mL 95% ethanol
5 mL bacterial suspension from lima beans

Process and Procedures

1. Answer the following questions as part of a class discussion:

    a. When you hear the term "DNA," what images come to your mind?

    b. If you could see DNA, what do you think it would look like?

    SAFETY: Put on your lab apron and safety goggles.

2. Investigate some of your ideas about DNA by using the following procedure to isolate real DNA from bacteria, a process that is called extraction.

    a. With your partner collect the materials that your team will need to complete this extraction.

    Your teacher will tell you which materials you will need in your team of two and which materials you will be sharing with several other teams at stations.

    b. Use the glass-marking pencil to label one culture tube as your bacterial suspension culture tube, so that you and your partner will be able to identify it later.

    c. Carefully add 5 mL of bacterial suspension to this culture tube.

d. To release the DNA from the cells, you must break open the cells with a detergent. To accomplish this, do the following:

   1. Add 1 mL of Woolite® or Dawn® to the bacterial suspension that is in your culture tube.

   2. Cap the tube tightly, and invert it *very* slowly two times.

     It is important to do this step carefully and extremely slowly (as if in slow motion) so that the DNA does not break into small pieces. If the DNA breaks, the rest of the procedure will not work.

   3. Let the tube stand undisturbed for 5 minutes.

e. According to your teacher's instructions, place your tube in a hot-water bath that has been preheated to 65°–70°C. Let your tube stand in the bath (incubate) for 15 to 20 minutes; monitor the temperature of the bath carefully to keep it within this range.

   WARNING: Hot water will scald, causing second-degree burns. Do not touch the hot-water-bath beaker or allow the hot water to contact your skin. If a burn occurs, *immediately* place the burned area under *cold* running water; *then* call the teacher.

   Exposure to this high temperature range and the detergent will help separate the DNA from the other macromolecules that make up the cell. A temperature greater than 80°C will break apart the DNA.

   While the suspension is incubating, continue with Steps f and g.

f. Pour 6 mL of ethanol that has been stored in a freezer into the second culture tube, cap it, and place it in a bucket of ice.

   WARNING: Ethanol is *flammable* and is a *poison*. Do not expose the liquid or its vapors to heat, sparks, open flame, or other ignition sources. Do not ingest; avoid skin/eye contact. If contact occurs, flush affected area with water for 15 minutes; rinse mouth with water. If a spill occurs, flood spill area with water; *then* call the teacher.

   For this extraction to work, the ethanol must be ice cold.

g. As the bacterial suspension warms up and the ethanol cools down, contribute to a team list of questions that come to mind as you read the story *Where Were You on the Night of August 23?*

h. After your bacterial suspension has been in the hot-water bath for 15 to 20 minutes, remove it and allow it to cool to room temperature in a test-tube rack.

   This will take about 5 minutes. Handle the tube carefully; it should remain stationary. Do not put it on ice or wave it around to cool it.

i. Add the cold ethanol to the suspension by tipping the tube containing the bacterial suspension at an angle and allowing the ethanol to flow slowly down the side of the tube as you pour.

   This method will prevent the two layers from mixing. The ethanol causes the DNA previously dissolved in the detergent solution to become insoluble (no longer able to be dissolved); the insoluble, white precipitate in the ethanol layer (upper layer) is DNA.

WHERE WERE YOU ON THE NIGHT OF AUGUST 23?

"Can you state, under oath, the whereabouts of the defendant on the night of the 23rd of August, between the hours of 7 p.m. and 11 p.m.?" the defense attorney asks, looking directly at the witness. The witness does not hesitate before answering.

"Yes, I can. He was with Tom and me, at Tom's house, watching TV," she states.

You are a juror, seated in the jurors' box. As you listen carefully to every word of the trial proceedings, you are keenly aware that your decision, along with those of 11 other jurors, will determine a man's fate.

The prosecuting attorneys have not produced the murder weapon, a knife. Nor have they called an eye witness. In fact, their case so far rests almost entirely on a description of a car that was seen leaving the murder scene. The car could have belonged to the defendant. Yet the defense attorney has just finished questioning two witnesses who both swore under oath that the defendant could not have murdered the victim because he was at home with them at the time of the murder.

Cross-examination by the prosecutor revealed that the female witness was emotionally involved with the accused man, and the male witness has been friends with him for 10 years. Still, even though both witnesses have a close personal relationship with the accused, neither has a criminal record, and the prosecutor has not been able to discredit the alibi. The witnesses' testimony seems truthful to you, and you wonder whether the other jurors also are thinking of setting the accused man free.

Now, late in the trial, the prosecution introduces a new type of evidence. Suddenly you find yourself doubting your first impressions. A forensic specialist steps up to the stand and testifies that he was able to recover several small samples of skin and blood from under the victim's fingernails. Apparently this type of evidence often is available as the result of a struggle in which a desperate victim scratches the attacker in an effort to escape.

Then the next witness, an expert in the relatively new technique of DNA fingerprinting, states that she analyzed the DNA that was extracted from cells derived from these samples. When the prosecuting attorney asks her to explain the details of the DNA analysis, you learn that the DNA in each person's cells is unique and that this fact allows scientists to use it as a molecular fingerprint for each person. Although the technique involved is complex and hard to follow, you understand that the results of the test may have an important bearing on this case.

The courtroom becomes very quiet when the scientist finally displays a pattern of lines on a large piece of film (Figure 12.1). She states that the visible pattern of lines represents the DNA fingerprint of the attacker because this person's skin and blood were under the victim's fingernails. She also states that this DNA fingerprint does *not* match the DNA fingerprint of the victim, whose blood also was tested. You glance at the defendant's impassive face and wonder what he is thinking.

"And did you use the same technique to analyze a sample of DNA from the defendant?" asks the prosecutor.

"Yes. Yes, I did."

The prosecutor pauses for a moment and then, very quietly, asks the obvious question. "And what did your analysis of the defendant's DNA indicate?"

"The patterns derived from the defendant's blood match exactly the patterns derived from the materials under the victim's fingernails," the scientist replies. "In my expert opinion, the samples belong to the same person."

When the excited buzz in the room finally dies down, the prosecutor asks what seems to you to be a crucial

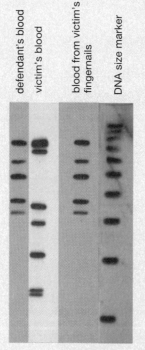

Figure 12.1 DNA fingerprinting
Scientists cut DNA with specific enzymes and then separate the small DNA fragments by size in an electric field. Each line represents a DNA fragment. DNA fragments of known size are included for size comparison. What can you conclude by comparing and contrasting the pattern of lines in lanes for the defendant's blood, the victim's blood, and the blood from under the victim's fingernails?

defendant's blood | victim's blood | blood from victim's fingernails | DNA size marker

question, "And is it possible that the match between these two patterns is just a coincidence?"

The scientist's answer is clear and direct. "The odds of a coincidental match are one in 15 million. This technique is extremely sensitive at detecting differences between individuals. We can detect no differences in this case."

As you retire to the deliberation room, you are struck by the confidence the scientist placed in this technique. She seemed quite sure that the DNA evidence pointed clearly to the defendant's guilt, but you are not quite so sure. You recall the witnesses who placed him somewhere else at the time of the crime, and you are uncertain about how to proceed. One problem is that you really do not understand this new procedure. In particular, you are curious about how any technique could be so accurate that the odds of a random match are only one in 15 million.

Although you still have questions, the defense and the prosecution have rested their cases, and the verdict now depends on you and the other jurors. Under the law, you must convict or acquit the accused. How will you decide?

j.   With your partner take turns spooling the DNA as follows:

  1. Hold the tube at a 45° angle.

  2. Place the glass stirring rod into the tube, and rotate the rod slowly.

      Fibers consisting of many molecules of DNA should come out of the solution and attach to the glass rod as you rotate it. This process, called spooling, is possible because the DNA is soluble in water and insoluble in alcohol. Remember to rotate the rod slowly and to look carefully for fine white threads accumulating around the rod.

  3. Continue this process until no more DNA comes out of the solution.

  4. Describe in your journal whatever you can observe about the DNA you have extracted.

k.   Return the materials as your teacher directs, and wash your hands thoroughly before leaving the laboratory.

3. Participate in a class discussion that your teacher guides.

Analysis

Complete the following tasks in your journal:

1. Make a list of your observations of the process of isolating DNA and of what the DNA looked like. Compare these observations with your previous impressions from Step 1 of *Process and Procedures*.

2. Is it likely that the cells of living organisms other than bacteria could serve as a source of DNA? Explain your answer.

3. The story *Where Were You on the Night of August 23?* illustrates one example of how the legal system uses DNA technology. Describe another application of DNA technology of which you are aware.

4. To begin learning more about DNA and to develop a context for the next activity, read the essay *Genetic Information Is Stored in Molecular Form* (page E172).

TRANSFERRING INFORMATION

How often does someone in your school use a photocopier to make a duplicate of an important document? As recently as the 1970s, photocopiers were uncommon, and people used manual typewriters and carbon paper or they used mimeograph machines to make copies of original documents. Compared with modern technology, such processes seem inefficient. Consider, however, how difficult duplication was before the widespread use of movable-type printing presses in the late-1400s, when expert scribes made handwritten copies of entire books. Errors often occurred during hand copying, a fact that is verified by reviewing historical documents.

Written language is a powerful technology for storing and transmitting information. How does it compare with the challenge faced by a cell every time a new cell or a new organism is produced? Cellular mechanisms must copy (or replicate) enormous amounts of DNA, the genetic material. In this activity you will explore the general process of information transfer and think about the importance of accuracy. See what you can discover during this exercise about how living systems carry out the transfer of information.

Materials (per team of 8)

1 or 2 large sheets of paper
pen or pencil

Process and Procedures

1. To begin thinking about information transfer in living systems, join your teammate to discuss the following questions. Record your answers in your journal.

   At this point in the chapter your answers may be brief and simple.

   a. All molecules store energy, but not all molecules store the information that affects inheritance. Which molecules contain the information that results in an organism's genotype and phenotype? What ideas do you have about how information is stored in molecules?

   b. What is the role of genetic information in the life of an individual organism?

   c. Why is replication of DNA important to the continuity of a species?

   d. Why is it important that genetic information be transmitted accurately?

2. Join with another team, and briefly discuss your answers to the questions in Step 1. Revise your answers if you wish.

3. According to your teacher's instructions, create a team of eight, and follow your teacher's directions for carrying out two tests of information transfer.

Analysis

In your team of two, complete the following tasks, and record your responses in your journal. Your teacher may collect your journal to assess your understanding of the ideas that you explored in this activity.

1. Examine the results of each test you conducted in Step 3. Explain how each set of observations might apply to the genetic mechanisms that are responsible for storing and transferring information.

2. What effect might inaccurate transfer (or replication) of DNA have on an organism? on its offspring?

MODELING DNA

In the activity *Transferring Information*, you experienced some of the challenge of transferring information accurately and you began to think about how cells manage this task. Now you will build a model of DNA to help explain how this molecule can store genetic information and transfer it to new cells.

As you build your model, keep in mind that modeling is a tool for understanding structures and processes that may be difficult to observe directly. James Watson and Francis Crick, the scientists credited with proposing the first detailed description of the structure of DNA, used model building extensively in their work. When they began to develop their DNA model, they were unable to explain all of the physical data collected from DNA, and their first attempts did not match the actual structure of DNA very well. Eventually, by modeling arrangements of atoms suggested by the X-ray photographs of Rosalind Franklin and Maurice Wilkins and by using a great deal of chemical insight, their model began to mimic the natural situation more closely.[1]

Remember also that even the best model only partially imitates the phenomenon it is constructed to illustrate. Models are useful for showing some aspects of an actual structure or process, but they do not portray every characteristic perfectly. As noted above, the power of models rests in their ability to help us visualize structures and processes that we cannot actually see. If there is a danger in modeling, it lies in our tendency to assume that every feature of a particular model is accurate. This activity will help you learn not only to develop but also to critique a biological model.

[1]*Source: The Double Helix*, by James D. Watson, ©1968 by Atheneum Publishers.

Materials (per team of 4)

pop-it beads or colored paper clips
twist ties
rubber bands
double-sided tape
wire

Process and Procedures

Part A Modeling DNA Structure

Imagine that you are on a team of research scientists involved in an effort to describe the likely structure of the DNA molecule. You have decided to tackle this task by modeling, that is, by building physical representations of a variety of possible structures and by changing these models to reflect new information as it becomes available.

Although you will work as a team to share information, each of you should build your own model. This will allow you to compare different ways of representing the same physical characteristics of the DNA molecule.

Model 1

1. Build your first model using both the pop-it beads and the following observations about the structure of DNA:

    - DNA is a polymer, a very long, chainlike molecule composed of small subunit molecules. Subunit molecules are like the links in a chain and are attached to each other by covalent bonds.

    - Four different types of subunit molecules exist.

2. Compare your model with the models built by the other members of your team. Discuss any differences that you notice.

3. Analyze your model by responding to the following in your journal:

    a. What features of your model represent the properties of DNA described above?

    b. Look closely at your model. How do you think that the structure of DNA might allow it to store information? How might DNA store different information along different parts of its length?

Model 2

4. Modify your DNA model to reflect the following additional information:

    - DNA consists of two long chains of subunits twisted around each other to form a double helix. (A helix is the shape a pipe cleaner takes when you wrap it around a pencil.)

    - The two helical chains are bonded together weakly, with subunits on one chain or strand bonding to subunits on the other strand.

    - The diameter of the DNA molecule is uniform along its length.

5. Compare your new model with the new models built by the other members of your team. Discuss any differences that you notice.

---

6. Analyze your second model by responding to the following in your journal:

   a. How well does your pop-it bead model represent each of the five structural characteristics of DNA listed so far? Briefly list these five characteristics, and then explain how each is represented by your model.

   b. Examine the figures that your teacher presents, and compare the superficial resemblances with a spiral staircase and with a zipper. List the strengths and weaknesses of using a spiral staircase or a zipper to illustrate DNA structure. How do these models compare with your pop-it bead model?

Model 3

7. Now add another layer of detail to your model of DNA. Consider the following observation, and modify your design accordingly.

   - The order of subunits in one strand of DNA determines the order of subunits in the other strand of the DNA.

     As you try to solve this portion of the model, consider that the aspect of DNA's structure that you now are modeling is the key to how DNA is replicated.

8. Compare your new model with the new models built by the other members of your team. Discuss any differences that you notice.

9. Analyze your third model by responding to the following in your journal:

   a. Use your modeling results to describe the relationship between subunits bonded to each other on opposite strands of the DNA double helix.

   b. As you think about this last characteristic of DNA, consider how the relationship between the subunits on each strand might suggest a means of replicating the molecule. Describe your ideas.

Analyzing a Video Model

10. With your class view the videodisc segment *DNA Structure.* This segment will give you more information about DNA structure (the four subunits and the interactions between subunits on each strand) and the importance of this structure to the replication of DNA molecules.

    As you watch the videodisc segment, look for information that will help you answer the following questions. Take notes in your journal. The new information that you gather will help you better understand the structure of DNA.

    a. What characteristic of the subunits allows for a uniform diameter of the double helix? What type of interaction takes place between the subunits of each strand that encourages the formation of a double helix?

    b. How does the sequence of subunits on one strand provide a template (a pattern or guide) for the sequence of subunits on the other strand? Is this important to replication? Explain your answer.

11. Use the new information that you gathered from the videodisc segment and the tasks below to re-evaluate your third model.

    - Write a short paragraph in your journal that describes how you would have to modify your third model to reflect this new information.

    - List three ways in which pop-it beads limit your ability to create a more detailed or more accurate model of a DNA molecule.

    - State what materials (other than pop-it beads) you would use and how you would assemble them to model DNA more accurately. (Alternatively, you may use these materials to construct another model on your own.)

Part B Modeling DNA Replication

As you saw in the first part of this activity, one of the reasons that scientists build models is to help them visualize and better understand *structures* that are difficult to observe directly. Models also can help scientists visualize and better understand a variety of complex and hard-to-observe *processes*.

Imagine that you now have determined the structure of DNA. Your research team has turned its attention to the task of explaining how the replication of DNA occurs. In particular, your team has decided to try to answer the two research questions that follow.

Complete the following steps to begin to develop answers to these research questions.

The essays DNA Structure and Replication *(page E173) and* Replication Errors and Mutation *(page E178) and the videodisc segment* DNA Replication *contain information that is necessary to answer these questions. Be sure that your responses to the questions and tasks reflect the information in these resources.*

Research Question 1

What are the critical characteristics of DNA that allow both the lasting storage of information and the transfer of information through copying (replication)?

1. Use the pop-it beads and the following key to build a model of a molecule of DNA.

   Key: black = adenine (A)
   white = thymine (T)
   red = cytosine (C)
   green = guanine (G)

   Although you will work as a team to share information, each of you should build and manipulate your own model. This will ensure that you will be able to demonstrate to your teacher your own understanding of replication.

2. Manipulate your pop-it bead model to illustrate the process of replication.

3. Discuss the following statements and questions with your team and with your teacher when he or she visits your group. Record your conclusions in your journal. It is important that you understand these points before you attempt to complete Step 4.

   a. Scientists use the phrase *complementary base pairing* to refer to the pairing of G with C and A with T in the DNA molecule. How does complementary base pairing make accurate replication possible?

   b. The bonds that attach adjacent nucleotide subunits in a strand of DNA are covalent bonds, which are strong. How would strong bonds favor the capacity of the DNA molecule to store linear information in an accurate and lasting manner?

   c. The bonds that attach the two DNA strands to each other in a double helix are hydrogen bonds, which are much weaker than covalent bonds. How would weaker bonds favor the capacity of the DNA molecule to replicate?

d. How does your model reflect the difference between the strength of the bonds between nucleotide subunits and the strength of the bonds between the two strands of a DNA molecule? Illustrate your answer with specific references to your pop-it bead model.

4. Analyze your work in Steps 1–3 above, and then develop an answer to Research Question 1. Record your answer in your journal.

Research Question 2

What are the advantages and disadvantages of an information transfer system that involves the use of a physical template?

---

5. Use your pop-it bead model to demonstrate a mutation in your DNA molecule.

6. Use your pop-it bead model to demonstrate what would happen to this mutation as your DNA molecule replicates.

7. Discuss the following questions with your team and with your teacher when he or she visits your group. It is important that you understand the answers to these questions before you attempt to complete Step 8.

   a. Think back to the tests of information transfer that you conducted in the previous activity, *Transferring Information*. How did the accuracy of information transfer when the message was passed orally compare with the accuracy of transfer when the message was passed using a written template?

   b. Under what conditions might the extremely high accuracy of replication be advantageous to a species?

   c. Under what conditions might such accuracy not be advantageous?

8. Review your work in Steps 5-7, and then develop an answer to Research Question 2. Record your answer in your journal.

Analysis

During this activity, you used scientific modeling to explore two important characteristics of DNA: (1) its ability to *store* information, and (2) its ability to *transmit* that information to subsequent generations. Fit these characteristics into the big picture of genetic continuity and reproduction by completing the following tasks:

1. Copy Figure E12.1 (page E172 in the essays) into your journal. Label the portion of the figure that specifically illustrates the *storage* of genetic information. Label the portion of the figure that specifically illustrates the *transmission* of information to subsequent generations. Label the portion of the figure that specifically illustrates the *use* of genetic information to maintain life.

2. In your journal explain the relationship between DNA structure and

   a. information storage.

   b. accurate information transfer during the reproduction of organisms (meiosis). Include a diagram to illustrate your explanation.

3. Imagine that a mutation occurs in an organism's genotype.

   a. Explain in molecular terms how this change might have occurred and how this change can be passed along to offspring and to the next generation of offspring after that.

   b. What effect might this mutation have on an organism carrying it? Why? Refer to Figure E12.1 in your answer.

Further Challenges

Briefly review the tests of information transfer that you completed in the activity *Transferring Information*. Although those tests used written and spoken language instead of the language of DNA, each test was a rough model of the process by which information is transmitted from one cell to another (or one organism to another) following DNA replication. As a model each test had strengths and weaknesses; each test portrayed the processes of information transfer with varying degrees of accuracy.

Ask yourself how accurately each test represented the replication of DNA as you understand the process now, and then answer the two questions below.

1. How is the copying of coded language messages inadequate as a model of replication, the biological mechanism by which one DNA molecule produces a second, identical molecule?

2. How is the transfer of coded language messages inadequate as a model of reproduction, the biological mechanism by which genetic information maintains the continuity of a species?

GENE EXPRESSION

Explain

You have learned how a DNA molecule is able to act as a template for its own replication, and in so doing, you have discovered how genetic information maintains the continuity of a species from one generation to the next. You also know that genetic information is used to build and maintain the phenotype of individual organisms. How is the information stored in DNA used by each organism that possesses it? How do your cells use the information that is stored in the sequence of nucleotides in your DNA to build and maintain the physical being that is you?

In this activity you will develop your own explanation of the relationship between genotype and phenotype by tracing the series of events that leads to gene expression. The gene you will study is the gene for sickle cell disease, a potentially fatal condition about which you read in Chapter 11. As you and your partner work through this activity, you will create a poster that illustrates the molecular basis of sickle cell disease. When you have completed your poster, you will be able to use your understanding of gene expression to suggest explanations for how other genes may exert their effects.

Materials (per team of 2)

poster board
assorted construction paper
felt-tipped markers or crayons
scissors
tape or glue

Process and Procedures
Part A Looking at Sickle Cell Disease

1. Begin your study of the molecular basis of sickle cell disease by reading the following information about this inherited disorder and about the associated gene.

Hemoglobin and Red Blood Cell Abnormalities in Sickle Cell Disease

Each year about one in 625 African-American children in the United States is born with sickle cell disease. This disease is caused by an abnormality in hemoglobin, the protein in red blood cells that carries oxygen to the cells of the body. When the oxygen supply in the blood is low, these abnormal hemoglobin molecules clump together instead of remaining separate as normal hemoglobin molecules do. Figure 12.2a shows this difference between the behavior of sickle cell hemoglobin and normal hemoglobin.

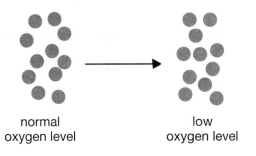

normal
oxygen level   low
                 oxygen level

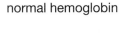

normal hemoglobin

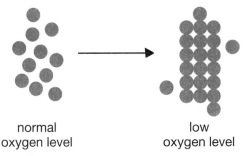

normal
oxygen level   low
                 oxygen level

a.   sickle hemoglobin

The clumping of the hemoglobin molecules at low oxygen levels causes the red blood cells in a person with sickle cell disease to become long and rigid like a sickle instead of remaining round and flexible (Figure 12.2b). This change in cell shape causes a variety of problems in the body. For example, as cells become sickled, they tend to block small blood vessels, causing pain and damage to the areas that are not receiving an adequate blood supply. The long-term effect of repeated blockages may permanently damage a person's internal organs, including the heart, lungs, kidneys, brain, and liver. For some people the damage is so severe that they die in childhood. With good medical care, however, many people with sickle cell disease can live reasonably normal lives.

Figure 12.2b Hemoglobin and Red Blood Cell Abnormalities in Sickle Cell Disease
Comparison of the shapes of normal and sickle cell red blood cells under conditions of low oxygen

b.   normal   sickle-shaped
          red cells   red cells

As you may recall from Chapter 11, sickle cell disease is associated with the genotype Hb^SHb^S. People who have this condition have two abnormal genes, one inherited from each parent.

2. Discuss the following questions with your partner, and contribute to a class discussion as your teacher directs.

You may wish to review the essay Incomplete Dominance *on page E170.*

a. What medical symptoms might a person with sickle cell disease experience?

b. What is the physiological cause of these symptoms? What problem in the red blood cells causes these symptoms to occur? What problem in the behavior of the hemoglobin molecules is associated with these changes in an individual's red blood cells?

c. Think back to your knowledge of DNA structure. What might be the *molecular basis* for the phenotype of sickle cell disease?

3. Working with your partner, use the materials that your teacher provides and the information in Figure 12.3 to begin creating a poster that will illustrate the molecular basis of sickle cell disease. Your poster should have a place for a title (you will add this later), and it should have each of the numbered sections that you see in Figure 12.3.

4. Use the information that you gathered in Steps 1 and 2 to complete areas one, six, seven, and eight of your poster. Refer to Figure 12.3 to determine the information required in each of these sections.

Include on your poster pictures and words that you think would be appropriate and helpful. Add a descriptive label to each area so that a viewer will understand what each section is displaying.

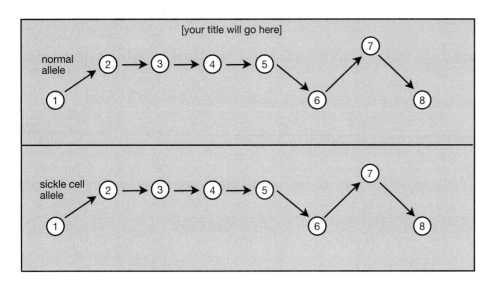

Information that eventually will be included in each section:
1. Genotype
2. Structure of the DNA
3. Structure of the mRNA
4. Structure of the polypeptide
5. Shape of the hemoglobin molecule
6. Behavior of the hemoglobin molecule
7. Shape of the red blood cell under conditions of low oxygen
8. Medical implications

Figure 12.3 **Sample poster design** Use this outline as a model for the poster that you will create to explain the molecular basis of sickle cell disease.

Part B Looking at the Structure of the Gene Involved in Sickle Cell Disease

1. To understand in more detail how the information present in the hemoglobin gene is related to sickle cell disease, refer to the DNA sequences on the copymaster that your teacher provides. Use these sequences as paper models of the same portion of two different alleles of the hemoglobin gene.

2. Compare the two nucleotide sequences.

    a. Draw an arrow or a circle on your poster to indicate the nucleotides in the sickle cell sequence that differ from those in the normal sequence.

    b. What type of mutations exist in the sickle cell allele?

3. Attach your DNA sequences to the appropriate places on your team's poster.

Part C Looking at the Expression of the Gene Involved in Sickle Cell Disease

1. To understand how the difference in sequence between the normal and sickle cell alleles of the hemoglobin gene results in the symptoms associated with the disease, determine the messenger RNA (mRNA) sequence that corresponds to the DNA sequence that you examined in Part B.

    One member of your team should generate an mRNA based on the DNA sequence that represents the allele for normal hemoglobin. The other member of your team should generate an mRNA based on the DNA sequence that represents the sickle cell allele.

    You will find information about messenger RNA and about the process by which mRNA is synthesized in the essay The Expression of Genetic Information *(page E180) and in the videodisc segment* Transcription.

2. Attach your mRNA models onto the appropriate places in your team's poster.

3. Compare the mRNA that results from the transcription of the normal allele of the hemoglobin gene to the mRNA that results from transcription of the sickle cell allele. Use an arrow or a circle to indicate on your poster the nucleotides in the sickle cell mRNA that differ from those in the normal sequence.

4. Use the genetic code table (Figure E12.12, page E183) to determine the sequence of amino acids that would result from translating the mRNA that you built from your original DNA sequence.

    Each member of the team should translate one of the mRNA molecules.

    You will find information about the genetic code and about how mRNA is translated into protein in the essays Translating the Message in mRNA *(page E182) and* News Flash! White-coated Sleuths Decipher Genetic Code *(page E184).*

5. Post your amino acid sequences in the appropriate places on your team's poster.

6. Compare the amino acid sequence that results from transcription and translation of the normal allele for the hemoglobin gene with the amino acid sequence that results from transcription and translation of the sickle cell allele. Use an arrow or a circle to indicate on your poster the amino acids in the sickle cell protein sequence that differ from those in the normal sequence.

7. Read the following information about the relationship between the sequence of amino acids in a hemoglobin molecule and the molecule's shape.

    Inside the environment of the red blood cell, a molecule of normal hemoglobin consists of four protein chains folded into a globular shape. The molecule remains folded in this manner due to attractive forces that occur between amino acids in different parts of the protein chains that make up the molecule.

    The change in the amino acid sequence that occurs as a result of the single nucleotide mutation in the hemoglobin gene has no effect on the overall shape of the molecule when oxygen levels are normal, so sickle cell hemoglobin behaves just like normal hemoglobin under these conditions. When oxygen levels are low, however, the amino acid change alters the attractive forces inside the molecule, causing molecules of sickle cell hemoglobin to assume a different shape from those of normal hemoglobin. As Figure 12.4 shows, it is this change in molecular shape under low oxygen levels—a change in shape that results from only one change in the amino acid sequence—that causes sickle hemoglobin to form the rigid rods characteristic of the condition.

8. Add the information presented in Step 7 to the appropriate places in your team's poster.

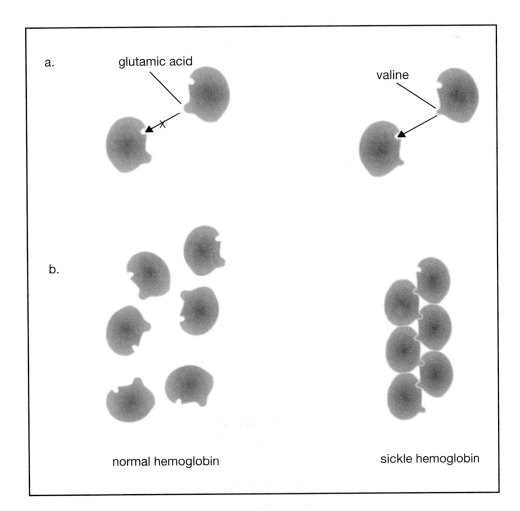

a.

glutamic acid

valine

b.

normal hemoglobin

sickle hemoglobin

Figure 12.4 **Normal and sickle hemoglobin** The difference in behavior of sickle cell hemoglobin is related to a shape change that occurs at low oxygen levels. This shape change results from the substitution of the amino acid valine for a glutamic acid. a. Molecules of normal hemoglobin will not associate with each other because the bulge created by the glutamic acid is too large to fit into a pocket that occurs in another part of the hemoglobin molecule. Molecules of sickle hemoglobin, however, *will* associate with each other because the bulge created when a valine is substituted for the glutamic acid is small enough to fit into the pocket. (The size of the pocket does not change.) b. Molecules of normal hemoglobin remain in solution, even under conditions of low oxygen. In contrast, molecules of sickle hemoglobin associate together to form rigid cells under conditions of low oxygen.

9. Complete your team's poster by adding a descriptive title and any other details that you think would help someone else understand the information that it presents.

Analysis

Use the information available on your poster and in the essays to respond to the following. Record your responses in your journal. Your teacher may collect your journal and evaluate your responses as a way to assess your understanding of gene expression.

1. List and describe the steps involved when a gene in a person's genotype is expressed to create a biochemical or physical characteristic that is part of that person's phenotype.

    You may want to organize and to illustrate your response by referring to the specific example of the steps involved in expression of the genotype for sickle cell disease.

2. In Chapter 11 you read about cystic fibrosis, a recessive condition that is quite common among Caucasians. The gene involved in cystic fibrosis normally codes for a transport protein that affects the flow of ions across membranes. When mutated, the now faulty version of this protein limits the body's ability to normally secrete fluids into the respiratory and

digestive systems. Use your understanding of gene expression to create a flowchart that illustrates how the genotype *cc* might lead to the medical problems associated with the phenotype of cystic fibrosis.

The poster that you completed on sickle cell disease may suggest a general scheme for your outline.

3. Refer back to the copy of Figure E12.1 in your journal. Into what area on that figure do the processes of transcription and translation fit? Explain.

A CLOSER LOOK AT PROTEIN SYNTHESIS

Elaborate

When scientists describe protein synthesis conceptually, they generally focus on the process of translation as the final step in the transfer of information from DNA to RNA to protein. As the final products of gene expression, proteins are used by the cell in many different ways. Scientific inquiry into protein synthesis does not end, however, with this general appreciation of the important role of translation. The details of this process are the focus of much current research. To understand exactly how the final transfer of information from nucleic acid to protein takes place, scientists investigate the cellular mechanisms that control how the language of nucleic acids is translated into the language of proteins. In this activity you will elaborate on your understanding of gene expression by examining the intricate and elegant details of translation.

Process and Procedures
Part A   Understanding Translation

1. Copy into your journal the symbols from Figure 12.5, and label them.

   You will need to know these symbols to explain the events in the videodisc segment.

Figure 12.5 **Symbols used in the videodisc segment on protein synthesis**

= ribosome   = release factor

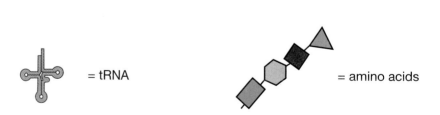

= tRNA   = amino acids

2. Watch the animated videodisc segment *Protein Synthesis* according to your teacher's instructions.

   Record notes in your journal that describe the steps involved.

   Watch the video as many times as necessary.

3. With your partner write an explanation of the translation process. Your teacher may collect your explanation at the end of the class. A thorough explanation will include answers to the following questions:

   a. What structures are necessary for protein synthesis?

   b. How is the ribosome involved in protein synthesis?

   c. How does the nucleotide sequence of each mRNA codon help position each tRNA? How many nucleotides are involved in this positioning?

   d. How is the mechanism of positioning tRNA on mRNA similar to the mechanism that holds DNA together as a double strand?

   e. What happens to adjacent amino acids once they are positioned on the ribosome?

   f. How is the sequence of mRNA nucleotides related to the sequence of amino acids in the protein?

   g. What events cause protein synthesis to stop? Is there a special mRNA nucleotide sequence or another factor that contributes to stopping translation?

   h. What happens to the protein after translation?

4. Watch the second part of the videodisc segment *Protein Synthesis*, and discuss with your team how this translation sequence differs from the first.

5. Participate in a class discussion of the explanations that you developed in Step 3 and of the significance of what you observed in Step 4.

   The essay Cellular Components in Protein Synthesis *(page E184) will help clarify the process of protein translation.*

Part B Predicting the Effects of Mutations

As you might guess, the importance of DNA to life means that significant numbers of copying errors would result in serious consequences. Complete the following steps to develop a deeper understanding of the possible effects of mutation on an individual's phenotype.

1. Write the mRNA sequence that would be produced by transcription from the following DNA strand:

   TACTTCGATCAGTAAGCTATGGACGACCAGAGCACGATCGACT

   The single strand of DNA nucleotides shown above is the strand that is transcribed into mRNA and that ultimately gives rise to a protein.

2. Use the genetic code table (Figure E12.12) to determine the sequence of amino acids that would result from translation of the mRNA from Step 1.

3. Use the same process you used in Steps 1 and 2 to determine the effect of the following mutation:

   TACTTCTATCAGTAAGCTATGGACGACCAGAGCACGATCGACT

4. Identify the mutation in the following strand of DNA and specify its effect:

TACTTCGATCGTAAGCTATGGACGACCAGAGCACGATCGACT

Use the DNA strand from Step 1 as your reference.

5. Identify the effect of the following DNA mutation:

TACTTCGATCAGTAAGCTATGGACGA<u>G</u>CAGAGCACGATCGACT

Analysis

Respond to the following with your partner, and record your responses in your journal.

1. Which of the three mutations in Part B most likely would give rise to a functional protein?

2. Why is it that many mutations do not result in observable changes in phenotype?

3. Describe why protein synthesis is important to living systems.

GENETIC TECHNOLOGY

In this activity you will apply your understanding of the basic principles of gene action by studying how scientists manipulate genetic material and by considering the potential effects of such manipulations. Then you will simulate these genetic manipulations by conducting a DNA-based genetic screening test. When you have completed this activity, you will be able to describe the molecular basis of the fingerprinting evidence presented at the trial in the opening story *Where Were You on the Night of August 23?* You also will understand how it is that in recent years geneticists have started to identify and to document the molecular similarities that make us all human as well as the molecular variability that uniquely identifies each of us.

Materials (per team of 2)

pop-it beads or colored paper clips

Process and Procedures
Part A Gene Manipulation: Science and Society

Working individually, answer the following questions in your journal. Be prepared to participate in a class discussion.

The essays News Flash! Extraordinary New Technique Changes Biology Forever *(page E186),* Manipulating Genetic Material *(page E187), and* Ethical and Public Policy Issues Related to Genetic Engineering *(page E192) will be very helpful in answering these questions.*

1. What is a recombinant DNA molecule and how (in general) do scientists build such molecules?

2. How can scientists use recombinant DNA molecules in their research?

3. What practical applications do recombinant DNA technologies have?

4. How does recombining DNA in genetic engineering differ from selective breeding?

5. Under what circumstances might it be permissible to alter the genes of future generations? Should such manipulation be forbidden forever?

6. Why might governments choose to regulate recombinant DNA research?

7. What are the risks in releasing genetically engineered organisms into the environment?

Part B Getting a Handle on Molecular Variability

1. To deepen your understanding of variability at the molecular level, read the following information:

> You had a basic understanding of human variability long before you ever started this course. Even as an infant, although you likely did not know what you were doing, you used differences in smell and sound to distinguish some humans from others. As you got older, you noticed other differences among people as well, such as differences in appearance, differences in attitude and behavior, and even differences in health. Gradually you began to understand this variability intellectually. You began to understand that although all of us are sufficiently similar that we easily can identify each other as human, we also differ from each other in a number of important ways.

> Now that you have completed Chapters 10 and 11, you also understand that both the similarities among us and the differences that we see reflect underlying genotypic similarities and differences—similarities and differences in the structures of our DNAs. Although we are more similar than we are different, the study of genetics tends to focus on differences. Until about 1980 it was only through tracing such phenotypic differences (many of them health-related) that scientists were able to recognize and describe either the ways in which we are similar *or* the ways in which we are different.

> Do all genotypic differences result in visible phenotypic differences? A hint to the answer to this question was provided in the opening story about a trial in which lawyers used DNA fingerprint evidence to try to link a suspect with biological evidence obtained at a crime scene. Why do you think that such a procedure is possible? What is it about the structure of DNA that makes one person's genetic material different *at the molecular level* from that of another person? Are all such molecular differences associated with some physical effect? What tools do we have available today that can help us study these differences?

2. Sequences A and B in Figure 12.6 represent two slightly different versions (two different alleles) of a nucleotide sequence within the same portion of a human autosomal chromosome. Some people carry Sequence A; other people carry Sequence B. Because each of us carries two chromosomes of each type, some people may carry both sequences.

Figure 12.6 Hypothetical DNA sequences

sequence A:

CCTGCAGAATTCGTTGAATTC

sequence B:

CCTGCAGAATTCGTTGATTTT

3. Use pop-it beads and the key listed below to make a model of each sequence.

   One team member should make a model of Sequence A while the other team member makes a model of Sequence B.

   Key: black = adenine (A)
             white = thymine (T)
             red = cytosine (C)
             green = guanine (G)

4. Examine the sequences, and identify any differences.

5. Discuss the following questions with your partner, and record your answers in your journal.

   a. What is the molecular difference between Sequence A and Sequence B? Answer specifically.

   b. Would you expect that the polypeptide produced by transcription and translation of Sequence A would differ from that of the polypeptide produced by transcription and translation of Sequence B? Why or why not? Support your answer with specific evidence.

      You may wish to review the essay Translating the Message in mRNA *(page E182) before you answer this question. Be sure to support your answer with specific evidence.*

   c. Based on your answer to Question 5b, do you think you could detect a difference in phenotype between a person with Sequence A and a person with Sequence B? Why or why not?

6. The restriction enzyme EcoR1 recognizes the DNA sequence GAATTC and cuts the DNA between the G and the A each time it encounters that sequence as it reads from left to right. Locate the places on your model of Sequence A that EcoR1 would recognize, and break the strand at the appropriate points. Do the same for Sequence B.

7. Discuss the following questions with your partner, and record your answers in your journal.

a. How many bases long is each fragment that is produced when EcoR1 cuts Sequence A? You can determine this by counting the beads in each fragment.

b. How long is each fragment that is produced when EcoR1 cuts Sequence B?

c. Special laboratory tests exist that can distinguish pieces of DNA of one length from pieces of DNA of another length. How could you use EcoR1 and these lab tests to distinguish a person carrying Sequence A from another person carrying Sequence B?

d. How does this procedure give you a way to identify molecular variability?

Analysis

You have just used RFLPs (pronounced riflips) to test for the presence of a particular genotype difference that occurs among members of the human population. RFLP stands for **R**estriction **F**ragment **L**ength **P**olymorphism. In the laboratory scientists determine the lengths of the fragments produced after treatment with a restriction enzyme by separating the DNA fragments in a gel by means of an electrical field. As you have seen, differences in the lengths of the fragments produced reveal differences in the nucleotide sequence of the DNA sample. These sequence differences are examples of variability *at the molecular level*. Some of these differences appear to be associated with detectable phenotypic differences among humans; others do not appear to be associated with any differences.

1. Read the following information about how a similar test can help scientists distinguish chromosomes that carry the sickle cell allele from those that carry the normal allele for human hemoglobin. Discuss with your partner the questions that follow, and record your answers in your journal. Your teacher may collect your journal to assess your understanding of the concepts presented in this activity.

    As you saw in the activity *Gene Expression*, sickle cell disease is caused by a mutation in the DNA that codes for hemoglobin. The mutation changes the amino acid glutamate in normal hemoglobin to the amino acid valine in sickle hemoglobin. The mutation also eliminates a restriction enzyme recognition site. The presence or absence of this restriction site is the basis for an accurate prenatal DNA test that determines whether a developing fetus has sickle cell disease, sickle cell trait, or normal hemoglobin.

    a. Sequences A and B in the copymaster that your teacher provides represent the same sections of the nucleotide sequences of the normal and the sickle cell alleles of the human hemoglobin gene. What is the difference between Sequence A and Sequence B?

    b. Based on this difference and on your knowledge of the genetics and the molecular basis of sickle cell disease, what can you predict about an individual who has only Sequence A in his or her genetic makeup? What if the individual has only Sequence B?

c. The restriction enzyme MstII recognizes the DNA sequence GGTCTCC and cuts the DNA between the first T and the first C each time it encounters that sequence. How could you use MstII to distinguish the gene for sickle hemoglobin from the gene for normal hemoglobin? Answer specifically in terms of the lengths of the fragments you would expect to see in each case.

d. Assume that a woman undergoes prenatal diagnosis to determine whether or not her fetus has sickle cell disease. Below are two possible test results for the developing fetus. What is the diagnosis for each result?

(i) fragment sizes = 4, 14, 10, 5

(ii) fragment sizes = 4, 24, 5

PRIME TIME

Evaluate

News Flash! "Scientists at a research institute in Boston report success in cloning a gene from human chromosome 4 . . ."

When you hear a short account of a new scientific discovery on the radio or TV or when you read about one in a newspaper, do you ever think about *why* a team of scientists would have worked so hard, often for years, to accomplish the task? How well do you understand the fundamental concepts of gene action? Has your work in this chapter helped you understand better the interactions among genotype, environment, and phenotype? In this activity you will evaluate your current understanding of the molecular basis of genetics by reporting on an actual discovery in molecular genetics or genetic engineering for your own News Flash! story. Then you will return to the court case described in the opening story, analyze the evidence presented, and decide on the defendant's guilt or innocence.

Materials (per team of 2)

reference materials

Process and Procedures

Working with your partner, select a genetic engineering topic from the list of suggestions that your teacher provides.

1. Begin to design an interview that will provide the information you need for your news report. Your news flash should be more extensive than those contained in the essays for this chapter. Make sure that both of you participate in all of the steps involved in developing your interview. (For example, you and your teammate will share the responsibility for researching, designing, and presenting your interview.)

2. Select a news medium for your interview: radio, TV, newspaper, or magazine.

3. Do the background reading that is necessary to collect sufficient information on your topic.

Your teacher may have some materials available as resources.

4. Write a script for your interview that includes a role for each team member.

   Keep in mind that your audience is the general public. To catch their attention and educate them about the scientific discovery, you will need to explain to them the molecular basis for the storage, transfer, and expression of genetic information. Use the criteria presented in Figure 12.7 as a guide for your preparation and interview.

5. Prepare to present your interview to the class.

   Again, use the criteria in Figure 12.7 as a guide. Your presentation should last 5 to 7 minutes.

6. Follow your teacher's instructions for the presentations.

Scoring Criteria

Score	Definition of Problem	Biological Concepts	Ethical Issues	Presentation
Excellent 3/4–full credit	• Clearly and completely outlines the problem and the need for research	• Demonstrates a very good to excellent understanding of the underlying biological concepts • Makes specific connections between the concepts in the chapter and the topic • Answers correctly appropriate questions from classmates or teacher	• As appropriate to the topic, analyzes the ethical issues in a manner that is well-reasoned and thorough	• Is clear and easy to follow • Demonstrates shared responsibilities • Is engaging and creative
Could Improve 1/2–3/4 credit	• Presents a fair outline of the problem and the need for research	• Demonstrates an acceptable understanding of the underlying biological concepts • Makes general and few connections between the concepts in the chapter and the topic • Answers correctly a few appropriate questions from classmates or teachers	• As appropriate to the topic, analyzes the ethical issues in a manner that is less than well-reasoned and somewhat incomplete	• Is fairly clear and easy to follow • Demonstrates less than equal responsibilities • Is somewhat engaging and creative
Needs Help 1/4–1/2 credit	• Presents a partial and vague outline of the problem and the need for research	• Demonstrates little understanding of the underlying biological concepts • Makes few or no connections between the concepts in the chapter and the topic	• Mentions ethical issues but includes no serious analysis	• Is unclear and hard to follow • Demonstrates that one person did most of the work • Shows little creative effort

Figure 12.7 **Scoring Criteria**

Analysis

Complete the following individually:

DNA fingerprinting such as that used by the expert witness in the opening story *Where Were You on the Night of August 23?* is based on a form of RFLP analysis. As you saw in the previous activity, *Genetic Technology*, this type of analysis uses the natural sequence variations that are present in every human genome. These sequence variations mean that the locations of restriction sites vary from one individual to another and, therefore, that the lengths of the fragments produced by cutting with a particular restriction enzyme also will vary. The pattern of lines that was shown to the jury in the story (Figure 12.1) corresponded to a number of DNA fragments of varying lengths. These fragments were produced by cutting DNA samples obtained from the blood evidence with the same enzyme and separating the resulting fragments on a gel.

Imagine that you are a member of the jury for this case. Your task is to analyze the evidence presented and decide how you will vote (guilty or not guilty) when the jury foreman polls you. Remember that as a properly selected and sworn member of the jury, you are one of the people charged with the responsibility of judging the innocence or guilt of the person on trial. You must vote when the foreman polls you.

Answer the questions below to help you make a decision, and explain the reasons for your decision.

1. What are the facts of the case?

   You may wish to review the opening story and make a short list of the facts.

2. How is DNA fingerprinting accomplished? Your answer may be brief.

   You may wish to review what you learned about DNA fingerprinting in the previous activity.

3. Based on what you know about inheritance, DNA replication, and how DNA fingerprinting is done, do you think that you would be more likely to obtain fragments of the same size from people who are related than from strangers? Why?

4. Do you think you would be more likely to obtain DNA fragments of the same size from two people who are members of the same ethnic group than from people from different ethnic groups? Why?

5. If the DNA fingerprint expert had said that the odds of a random match were one in 100, how would that affect your decision? How about one in 10,000? How about one in 1 billion?

6. What three questions concerning the methods, the results, or the reliability of the technique would you want to ask the expert?

7. Notice that the expert witness did not give any explanation for how she determined the odds of a coincidental match between the defendant's DNA fingerprint and the DNA fingerprint derived from the biological evidence left on the victim's body. This means that you must decide the case without this information. Would this information be important to know? Why or why not?

   One way to think about this question is to ask yourself what type of answer the expert might have given had she been asked this question. For example, at the end of the trial, suppose the defense attorney questioned the DNA expert witness and obtained the following testimony. How would this type of answer have affected your decision?

   Defense attorney: "What data have you used as the basis for arriving at the statistic of one in 15 million?"

   DNA expert: "The data are derived from an identical RFLP study of 10,000 men in Ireland."

   Defense attorney: "But the defendant is Brazilian. I have data from an RFLP study of Brazilians, and it indicates that these particular fragment sizes are quite common in this population. In fact, the likelihood of a random match may be as frequent as 1 in 200."

8. Based on your analysis of the case, how do you vote?

   Be sure that your answers to Questions 1-7 support your decision.

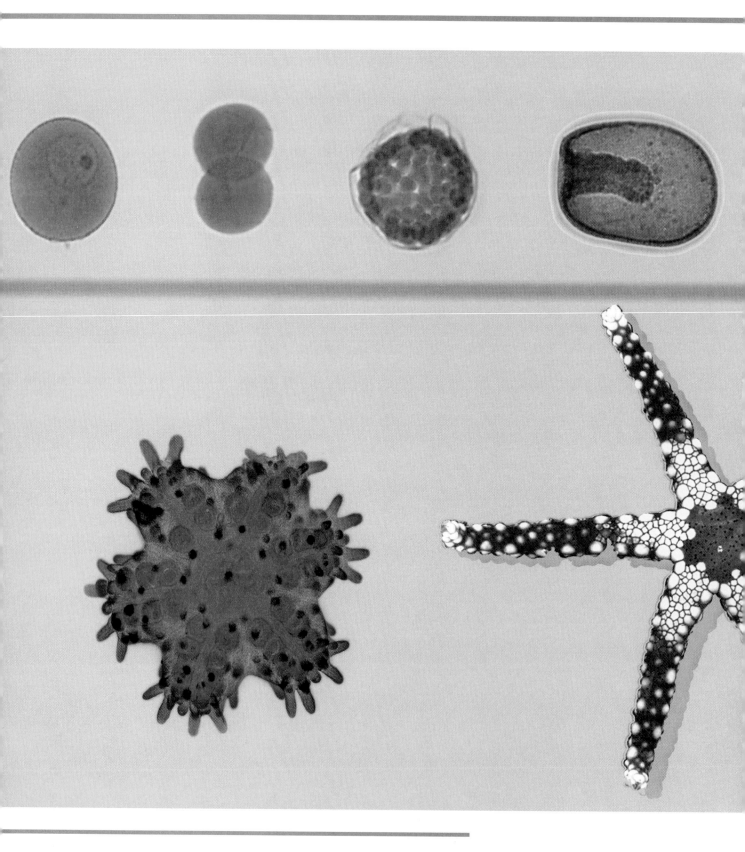

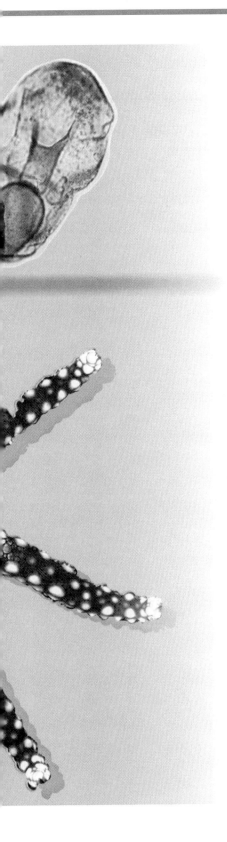

DEVELOPMENT:
GROWTH AND DIFFERENTIATION IN LIVING SYSTEMS

13 Processes and Patterns of Development

14 The Human Life Span

In Unit 4 you learned about genetic processes that allow change in species from one generation to the next. In this unit you will explore developmental processes that allow change within the lifetime of individual organisms. For instance, you will construct an understanding of how changes in cells lead to changes in tissues, organs, and organisms. Then in Chapter 14, you will focus on changes in human growth and development from birth through old age. You also will explore human life stages in other cultures to learn how culture influences the expression of life stages.

In this unit you also will use the unit assessment in a new way: as an additional guide for your learning. By the end of Unit 5, you should understand how

- development occurs from fertilization to death;
- physical development involves processes of growth and differentiation;
- humans grow and develop through various stages of life, including physical, cognitive, emotional, and social stages; and
- culture influences the expression of these life stages.

You also will continue to make observations; collect, organize, and analyze data; interpret and synthesize data; and apply knowledge to a new setting.

> *That a cell can carry with it the sum total of the heritage of the species, that it can in the course of a few days or weeks give rise to a mollusk or a man, is the greatest marvel of biological science.*

E.B. Wilson, *The Cell in Development and Inheritance*, 1990

13
PROCESSES AND PATTERNS OF DEVELOPMENT

Occasionally, scientists as well as nonscientists describe the processes of science as a simple sequence of steps that one can follow to generate scientific knowledge (the so-called "scientific method"). Such descriptions often are not very accurate or helpful. Even though practicing scientists really *do* generate questions, test hypotheses, and form conclusions, they do not necessarily complete exactly these tasks in exactly this order. Instead, "doing science" is more like a highly creative, long-term search for answers to complicated and fascinating puzzles. Searching for answers is a dynamic process, a process that requires careful questioning, hypothesis-making, and concluding, but the search also involves imagination, persistence, humor, and sometimes some good, old-fashioned luck.

One of the fascinating puzzles of life is development: How is it that organisms change in precise and predictable ways from fertilized eggs to adults? In this chapter you will have the opportunity to consider what we know about the processes and patterns of development. You also will be invited to think about how scientists have gathered this knowledge and how they might extend their understanding to answer the questions about development that continue to puzzle us today.

ACTIVITIES

Engage   One Hundred Years of Questions

Explore Explain   A Start in Development

Explore Explain   Processes that Generate Complexity

Elaborate   Development Gone Awry

Evaluate   Evaluating Where We Stand

Engage

ONE HUNDRED YEARS OF QUESTIONS

One of the questions that people often ask about science is, How do scientists think of the questions that they would like to investigate? The fact is that finding questions to ask is the easy part of science. The hard part is expressing our questions in such a way that we can answer them.

Process and Procedures

1. View the videodisc segment *From Egg to Adult*. What questions do these pairs of images raise in your mind? Record these questions in your journal.

2. View the videodisc segment *A Collection of Eggs; An Assortment of Adults*. What questions do these images raise in your mind? Record these questions in your journal.

Analysis

1. To see how some scientists have approached the study of development, read the story *Changes All Around*, and then join the class in a discussion of the following:

   a. Compare Roux's hot-needle experiment on a frog embryo to Driesch's experiment on sea urchin embryos. Indicate similarities and differences in each of the following aspects of their experiments:

      • experimental design

      • results

      • conclusions

   b. What was the specific question that Roux tried to answer? How was his question different from the question, How do organisms develop? Why was this difference important?

   c. What was Roux's contribution to the science of developmental biology?

2. Work individually to compare the questions that you recorded in your journal in *Process and Procedures* with the questions that scientists in the *Science* survey named as most interesting to them. How are your questions similar to the questions the scientists asked? How are they different? Record your answers in your journal.

3. Scientists still are asking and answering questions about development. What does this suggest about the processes of development? What does this suggest about the processes of science? Record your answers in your journal.

CHANGES ALL AROUND

The fertilized frog egg has just started to develop; it has undergone one cell division. Under the magnification of a microscope, it is clear that the two cells are in close contact with one another. The scientist picks up a hot needle and, with excruciating care, pierces just one of the two cells. The other cell remains untouched. The pricked cell dies, and the scientist continues his observations to see if the remaining cell will develop further or if it also will die. The year is 1888, and the scientist, a German named Wilhelm Roux (pictured above), is in the process of making a dramatic step forward in the study of living systems.

Why was the killing of one cell such an important experiment? How did this affect biology? Scientists in the late nineteenth century knew that vertebrates, such as frogs or humans, start life as a single cell, the fertilized egg. Scientists had proposed a number of ideas to try to explain *how* an embryo develops after fertilization. Some thought that a fertilized egg contained a tiny, fully formed—or preformed—organism that simply grew larger during development. Others thought that the structure of an organism formed as the embryo developed. Roux, however, took a big step forward by doing more than just thinking about how development might take place. He asked the large question—How do organisms develop?—and identified a simpler, related question *that he could test*. This question was, Does each of the first two cells in an embryo contain all of the structures and information needed to grow into an organism, or does each cell contain just half of the structures and information needed? (The second half of this question would be true if a fertilized ovum contained a tiny, fully formed or preformed organism.) Roux tested this question by performing the hot-needle experiment described above.

The results of Roux's hot-needle experiment were, in the words of one scientist, "spectacular." As Roux described them: "[An] amazing thing happened; the one cell developed in many cases into a half-embryo generally normal in structure, with small variations occurring only in the region of the immediate neighborhood of the treated half of the egg."[1]

In other words the cell that survived gave rise only to the portions of the embryo that it would have produced if the experiment had not been done. Roux interpreted his results as evidence for preformed embryos.

As new evidence came to light, however, Roux's conclusion was challenged. In 1892 another scientist, Hans Driesch, conducted a similar experiment, but his experimental design was different. First, Driesch used a different organism, a sea urchin. Second, Driesch actually separated the two cells and watched to see what would develop from each one. (Roux had killed one cell and left them both in place.) Driesch's results supported a view of development that was opposite to Roux's. Driesch observed that whole embryos developed from each of the separate cells. This evidence supported the idea that the organism forms *during* development, *not* that it is preformed.

Driesch was astonished with this observation. How could he account for his results? His first response was that sea urchin eggs are not frog eggs and perhaps he simply was seeing a difference between the two types of organisms. Did his results actually mean that frogs and urchins develop in completely different ways? Because that answer did not seem satisfactory, he suggested that perhaps Roux had not really "isolated" the frog cells. Roux had killed one cell, but the dead cell remained in contact with the live one. Possibly the dead cell was exerting an influence on the development of the live cell. In fact, in 1910 a scientist named J.F. McClendon removed one cell of a two-cell frog embryo by sucking it up into a tiny eye dropper. Like Driesch, he found that when he completely *isolated* one cell of a two-cell embryo, the remaining cell developed into a normal, although small, embryo. This result also suggested that the structure of an embryo is formed *as* the embryo develops.

Figure 13.1a illustrates Roux's historic experiment. Why do we remember it, despite Roux's incorrect conclusions? We remember it largely because it helped mark the beginning of the *science* of developmental biology. Roux's work and the work of other scientists of his time highlighted the importance of asking questions about development that we can answer *by doing experiments*. Sometimes the questions that we can answer are only parts of larger questions that we wonder about. We gain scientific knowledge from piecing together the answers to small questions to shed light on larger, more complicated questions.

---

[1]Wilhelm Roux wrote these words in 1888; they were translated by Hans Laufer and appeared in *Embryology: An Introduction to Developmental Biology*, by Stanley Shostak, ©1991. HarperCollins Publishers, Inc., New York.

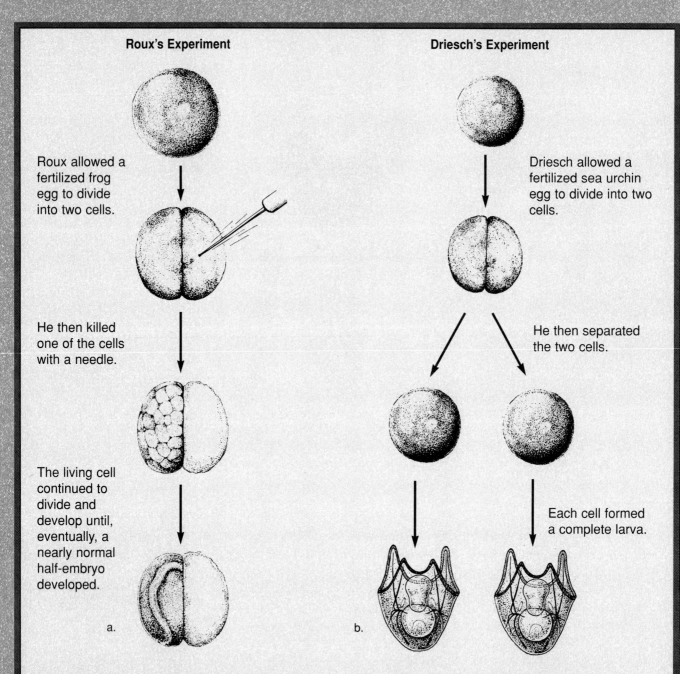

Roux's Experiment

Roux allowed a fertilized frog egg to divide into two cells.

He then killed one of the cells with a needle.

The living cell continued to divide and develop until, eventually, a nearly normal half-embryo developed.

a.

Driesch's Experiment

Driesch allowed a fertilized sea urchin egg to divide into two cells.

He then separated the two cells.

Each cell formed a complete larva.

b.

Figure 13.1 Early experiments in developmental biology a. After Roux killed one of the first two cells of a frog embryo, the remaining cell developed tissues corresponding to half of a normal frog embryo. The results would have been different if Roux had removed the dead cell. b. Driesch's experiments with sea urchin eggs demonstrated that the presence of the dead cell influenced the development of the live cell.

In 1894 Roux helped start a scientific journal to communicate new discoveries and to air new discussions about development. Today, after more than 100 years of asking questions, performing experiments, and building answers, developmental biologists have a much better understanding of how a fertilized egg develops. We also have learned how many other aspects of development occur as well. Yet we still have unanswered questions. In fact, in 1994, 100 years after the establishment of Roux's journal, the publishers of *Science* magazine surveyed scientists to find out what questions about development are most interesting to them and what questions are most likely to be answered from experiments conducted during this decade. The top two questions that scientists named were

1. How are the body's tissues and organs formed?
2. What clues does development reveal about the process of evolution?

Even as you read this chapter, scientists are devising new experiments and using new technologies to study these complex questions. New data emerge daily to help biologists find answers to the puzzles of development.

A START IN DEVELOPMENT

Simple observation reveals changes in size and shape as a human baby grows to an adult (see Figure 13.2). The nineteenth-century scientists Roux and Driesch were able to watch the early development of a fertilized egg because the technology of microscopes improved their view. Today more advanced microscopes and other technological tools, such as fiber optics and ultrasound, give scientists an even closer look at development.

In this activity you will use a series of video images to begin your study of development. As you do so, keep in mind the two questions posed by scientists in the 1994 *Science* magazine survey, and see whether you can find clues to the answers.

Figure 13.2 **Life stages** What changes occur during the development that takes place before birth? What changes occur as organisms develop to maturity?

Materials (per team of 2)

modeling clay
large sheet of paper

Process and Procedures
Part A How Does Development Occur?

The videodisc images that you saw in the previous activity provided some basic information about development, but the images did not reveal how much time was required for each developmental change. In fact, the time span that elapses between the stages of life illustrated in the images (from fertilized ovum to adult) may be from weeks to years, depending on the organism. What happens during that time? The videodisc segment *A Closer Look* will provide some clues. The images on this segment were filmed with a combination of technologies, including fiber optics (which allows tiny cameras to record images of hard-to-view areas, such as the uterus), high-resolution microscopes, and video recording equipment.

1. As you view the videodisc segment *A Closer Look*, think about the questions below. Then record your observations and your answers to the questions in your journal.

   a. What types of changes do you observe in these images?

   b. How has technology expanded our ability to observe development? Compare the images in the videodisc segment *A Closer Look* with the images you saw in the videodisc segment *From Egg to Adult* in the previous activity.

2. Developmental biologists use the terms *growth* and *differentiation* to describe the fundamental processes that occur during development. Complete the following tasks in your journal:

   a. Describe what is meant by each of these terms.

   b. List specific examples from the videodisc segment *A Closer Look* that show evidence of each of these processes taking place.

   The information in the essay The Long and Short of Development *on page E194 will help you with this task.*

Part B How Does Growth Occur?

In Part A you observed two basic processes involved in development. This part of the activity offers you an opportunity to explore one of those processes—growth.

1. Observe the cellular activity in the videodisc segment *Cell Division* that your teacher presents.

   As you watch the segment, try to determine what is happening to the cells involved, and think about how the events that you see relate to development.

2. Discuss the following with your class:

   a. Describe what you saw taking place in the videodisc segment. What appeared to be happening to the cells?

   b. How do you think this cellular activity relates to the changes that occur as humans develop?

   c. Do you think that cell division is an important aspect of the development of other organisms? Illustrate your answer with two examples.

3. Scientists refer to the cellular activity that you just observed as mitosis. As you watch the videodisc segment *Cell Division* again, answer the following questions in your journal:

   a. What major cellular structures are most active during mitosis?

   b. What is the importance of these structures to the cell?

   Think about the types of molecules of which these structures are made and what they contain.

c. What happens to these structures as a result of mitosis? (Answer as specifically as you can.) Does this make sense to you in the light of what you understand about cell structure and function? Explain your answer.

4. Working with your partner, use the materials available, including the videodisc, to construct a model of mitosis that shows the following:

   • a starting cell with two pairs of chromosomes,

   • DNA synthesis,

   • the stages of mitosis, and

   • the cells formed as a result of mitosis.

   Use the information provided in the essay The Cell Cycle and Growth Control *(page E195) to help you complete this task. You also may wish to view the videodisc segment again. Comparing the videodisc segment with the diagrams in Figures E13.3 (mitosis) and E13.4 (cell cycle) may help you follow the specific events of the cell cycle more easily.*

Figure 13.3 **Model of a cell with clay chromosomes**

5. At the anaphase stage of mitosis, how do the chromosomes in each group compare with each other (and with the original group of chromosomes) in terms of the number and types of chromosomes? How do they compare with each other in terms of genetic information?

   Answer each question in your journal.

6. Think carefully about the specific manipulations you just completed with your clay models. Why is it important that

   • chromosomes *duplicate* during mitosis?

   • chromosomes *line up in single file* during metaphase of mitosis?

   • the duplicated chromosomes *separate* during anaphase of mitosis?

   In your journal, answer each question in terms of the movement and distribution of the genetic material.

Analysis

One of the important skills that most successful scientists develop is the skill of recognizing contradictions between observations and conclusions or between the results of an experiment and an established scientific principle. Sometimes a contradiction indicates that either the observation or the conclusion is wrong. In other cases, however, the contradiction is not real. As scientists investigate the problem more thoroughly, they discover that there only appeared to be a contradiction because they failed to understand the phenomenon or the experimental system thoroughly.

Work individually to answer the following questions in your journal. Then exchange journals with your partner, and let him or her write comments about your answers. After you have read and commented on each other's answers, retrieve your journal, and rewrite your answers as necessary. Your teacher will collect your journal to assess your current understanding of development.

Examine the statements below.

Statement 1: Growth occurs during development as a result of the mitotic division of cells.

Statement 2: Mitosis results in daughter cells that are genetically identical to each other and to the parent cell that divided.

Statement 3: During development cells become both structurally and functionally different from one another.

1. To the best of your knowledge, is each of the above statements correct as written? Explain your answer.

2. What contradiction do you see between Statements 2 and 3? What is it about Statement 2 that does not appear to be consistent with Statement 3? Explain your answer.

3. If you were a scientist who recognized this contradiction, what steps would you take to try to resolve it? Answer specifically.

PROCESSES THAT GENERATE COMPLEXITY

Explore
Explain

You have seen that the development of a multicellular organism involves the process of growth. But growth is only half of the story of development. Differentiation, the change in cells that occurs during development, is the other half.

Your own body, for example, consists of trillions of cells, but these cells include hundreds of different types. Hold your hands up in front of you and think about their complexity for a moment. Your fingers and thumbs are different lengths, yet their lengths match from one hand to the other; the skin, bone, and muscles that they are made of are packaged together in essentially identical and highly functional ways. Hundreds of nerves, both in your hands and extending from your hands to your brain, form an extensive communications network that offers not only enormous sensitivity and precision, but also tremendous coordination. Now consider the rest of your body: your heart, your brain, your kidneys, and even your big toes. Along with your hands, all of these complicated structures arose from the same single cell.

a.

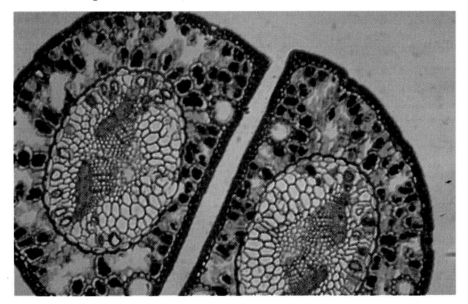

b.

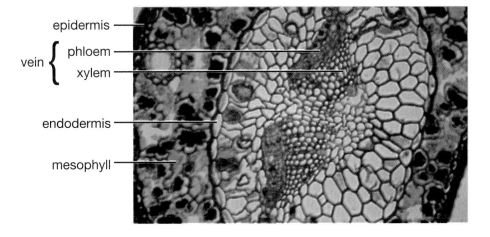

Figure 13.4 **Differentiated tissues in plants** a. How many different tissues can you see in this cross section of pine needles? b. At higher magnification, even more detail is visible. Xylem and phloem function as support and transport tissues. Xylem permits water to move from the roots to the leaves, whereas phloem conducts the carbohydrates produced as a result of photosynthesis from the leaves to the stem and roots.

In this activity you will continue your examination of the processes of development. In so doing, you will continue to consider the first question of the 1994 *Science* magazine survey: How are the body's tissues and organs formed? You will see that at first glance, the answer is surprisingly straightforward.

Process and Procedures

1. Observe the videodisc segment *Cells, Cells, and More Cells*. How do these images relate to the changes that occur as humans develop from a single-celled zygote to a mature person?

    As you watch the segment, compare the physical appearance of the cells within specific organisms, and think about what this tells you about development.

2. Examine closely the series of frames that shows the low- and high-magnification views of the cross section through the trachea. The trachea is the tube that connects your nose and mouth to your lungs. Discuss the following questions with your class:

   a. Describe some of the specialized cells that you see. How are they structurally different from each other?

   b. How do these cells compare with each other genetically?

   c. What questions do your answers in (a) and (b) raise about the process of differentiation?

3. Working with your partner, examine the table in Figure 13.5 to identify similarities and differences among the cell types listed. At what level of gene expression do genetically identical cells become structurally different? Be prepared to support your answer with specific examples from the table and to share your answer in a class discussion.

   As you examine the information in the table, ask yourself whether the structural differences that you saw among these cells correlate with any molecular *differences among them. The understanding that you gained in Chapter 12 about the expression of genetic information and the information in the essay* Differentiation and the Expression of Genetic Information *(page E199) also may help you answer this question.*

Figure 13.5 Cell types that make up the human trachea

Type of Cell	Function	DNA	mRNA	Major Proteins Produced
goblet cell	produces mucus	identical to zygote	mRNA coding for proteins required for basic cell functioning; also much mRNA coding for proteoglycans and glycoproteins, which are complex protein-carbohydrate molecules	proteoglycans and glycoproteins, the major types of protein in mucus
cartilage cell	produces cartilage	identical to zygote	mRNA coding for proteins required for basic cell functioning; also much mRNA coding for collagen	collagen, the major protein in cartilage
muscle cell	contracts	identical to zygote	mRNA coding for proteins required for basic cell functioning; also much mRNA coding for actin and myosin	actin and myosin, the major proteins in muscle

Analysis

Review your work in this and the previous activity and participate in a class discussion of the following:

1. Explain at a molecular level how genetically identical cells can differentiate into structurally and functionally diverse cells.

    Feel free to use examples from the table in Figure 13.5 to support your answer.

2. How does a human develop from a fertilized egg into an adult?

3. To what extent do you feel that you understand the answer to Question 2? What do you think you would have to know to answer this question completely?

4. What was the first question that developmental biologists identified in the 1994 *Science* survey as being of great interest and importance to them? What does the fact that scientists are still asking that question today indicate about its complexity and the level of detail with which scientists would like to answer it? Use specific examples to illustrate your answer.

Further Challenges Putting Growth and Differentiation Together

A novel called *The Boys from Brazil*[2] was published in 1976. The book told the story of a man who discovers a Nazi plot to achieve ultimate victory over the world by producing another Adolf Hitler, even after the death of the first. How did the conspirators hope to accomplish this feat?

1. Read the following excerpt from *The Boys from Brazil*. The scene opens as a man leaves a house after meeting a young boy who was astonishingly similar to another boy he met a few days earlier.

    Astounding, such a sameness. Peas in a pod . . .

    Was it possible that Mrs. Curry and Frau Doring both had affairs with the same gaunt, sharp-nosed man nine months before their sons were born? Even in that unlikely event, the boys wouldn't be twins. And that's what they were, absolutely identical.

    Twins . . . But these boys weren't twins; they only looked like twins . . .

    Leibermann wrestled with the problem . . . It had to be a coincidence. He'd seen plenty of look-alikes in his lifetime . . . But look-*alikes*, not look-*the-sames* . . .

    He took a breath, let it out. "They have the same son."

    "The same what?"

    "Son! The same son! I saw him here and in Gladbeck . . . And he's in Goteburg, Sweden; and in Brammings, Denmark. The exact same boy! He plays an instrument, or draws . . . Four different mothers, four different sons; but the son is the same, in different places . . .

---

[2]Written by Ira Levin, © 1976 by Random House, Inc.

Adolf Hitler, Berlin, 1939 Adolf Hitler, the fascist dictator of Germany from 1933–1945, provoked World War II and was responsible for extermination camps that killed more than 5 million people, mostly Jews.

2. In general, how do you think the conspirators were trying to create another Hitler, and how does their scheme relate to the topic of this activity? Discuss these questions with your partner, and be prepared to share your answers in a class discussion.

3. Work with your partner to complete the following tasks about Levin's book. Record your responses in your journal, and be prepared to share them in a class discussion.

   a. Levin has produced a terrifying fantasy by starting with a little bit of scientific knowledge and using his imagination to spin that knowledge out to its theoretical limits. Use your own knowledge of science, and let your imagination go in a similar fashion for a moment. Hitler had no children of his own, but as the story is written, one of the conspirators had collected a sample of Hitler's cells before his death. If a diabolical scientist had access to such cells, how could he or she use them to produce many young Hitlers or, at least, many little boys with the same genetic inheritance as Hitler?

   You may wish to consult the essay Cloning People? *(page E203) for help in answering this question.*

   b. What do you know about the genetic composition of most of the cells in the human body that makes it at least *theoretically* possible to produce another complete Hitler from one of his cells?

   c. What developmental processes would have to be successful to produce a complete Hitler from one of his cells? Express your answer in terms of cell division and gene expression.

   d. In the story the babies produced from Hitler's cells were placed in carefully selected adoptive homes so that they would have as similar an upbringing to Hitler's as possible. Use your understanding from

Chapter 11 of the interaction between genotype and environment to explain why this would be important if the conspirators wanted to produce another Hitler.

e. How realistic is the scenario in *The Boys from Brazil*? Support your answer with specific information from the essays you have read.

f. What ethical questions does the story raise? Other than the problems associated with *who* the conspirators cloned, what ethical problems may be associated with cloning people?

DEVELOPMENT GONE AWRY

You have seen that the growth of cells and their differentiation into specialized tissues are highly regulated events. In this activity you will discover how sensitive this regulation is and what can happen if errors occur during development. A human embryo, for example, can develop improperly if harmful environmental influences disrupt the regulation of growth or differentiation or if the genetic plan that is directing the embryo's development contains mutations.

Developmental errors are not limited to the growing embryo. They can occur even after birth. For example, sometimes harmful changes in hormonal conditions alter the pattern of growth that normally would occur as an organism matures. Likewise, sometimes body cells no longer respond properly to regulatory signals and grow unrestrained, causing cancerous tumors. As you begin this activity, apply your understanding of the processes of human development as you consider situations in which development in humans has gone awry.

Materials (per person)

resource materials that include news articles, essays, and videodisc material
video or audio equipment for recording and playing (optional)

Process and Procedures

Imagine that you are a doctor who is handling a case that involves an error in development. These errors fall into two general categories: birth defects and cancer. Follow the steps below to write a conversation in which you explain the disorder to your patient and describe what may be done about it. In this conversation you (the doctor) will need to show how normal developmental processes have been altered to produce the disorder.

1. Working individually, choose a disorder to be the topic of your script. Good ways to identify possible topics include the following:

    • Find a news event that relates to either birth defects or cancer, and use this news coverage as the basis for your script. You may find interesting news bulletins in publications such as newspapers, *Science News*, *Discover*, and *Harvard Health Letter* as well as in other sources that cover medical news.

- Interview a person who has experienced a birth defect or cancer or who has had a family member with a birth defect or cancer.

- Use an idea that occurs to you as you review the resources listed in Step 2.

2. Discuss your idea with your teacher, and have him or her approve your topic. Then collect information about the biology involved in the disorder from the available resources.

It may help you to read through the criteria for a good script, which are listed in the next step. These criteria will help you decide what information you will need.

The following resources are available in this project: the essays Development and Birth Defects *(page E203) and* Cancer: Unregulated Growth *(page E206) and the videodisc segment* Cancer in Humans. *The videodisc images show cancerous growths in a variety of human tissues, microscopic images of cancerous blood cells (leukemia), X-ray images of lung cancer before and after treatment, and MRI (Magnetic Resonance Imaging) of brain tumors. You may, of course, use any other available resources.*

normal liver tissue   cancerous tissue

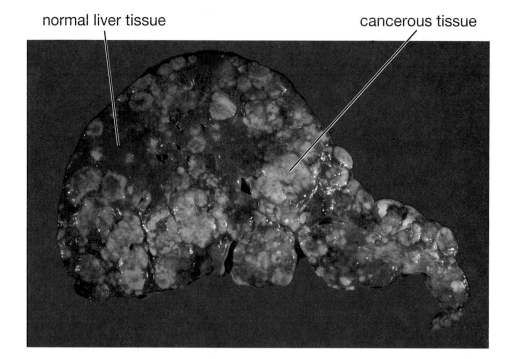

Figure 13.6 Liver cancer The extensive cancerous growths in this liver are the result of metastasis. The cancer began in the lungs.

3. Develop your script of a conversation between a doctor (you) and a patient suffering from the disorder you chose in Step 1. (The conversation may take place between the doctor and family members as well.)

You will know you have a good script when it:

- describes in detail the disorder you chose,

- explains the probable role (if any) of external (environmental) factors in producing the disorder,

- explains the probable role (if any) of genetic factors in producing the disorder,

- describes how and when (that is, in what way and, if known, through what specific events or mechanisms) growth and/or differentiation have been affected,

- describes the long-term effects of the disorder, and
- mentions any treatments and their likelihood of working.

Assume that the patient or family members of the patient are just learning of the diagnosis and are struggling to deal with it. Your goal is to explain what the disorder is, how it came about, and what can be done about it, if anything. To present the information in a way that will meet the criteria listed above, you will have to imagine what questions the patient or family members might ask and how the doctor would respond.

Your teacher will indicate if you are to submit your script in writing or if you are to enlist the help of other students and perform it for your class. Another option may be to record your script on videotape or audiotape.

Analysis

Your study of development should have convinced you that development involves the expression of a genetic plan within a set of environmental conditions that can influence and/or modify the plan's execution. Use your knowledge of development and the information about developmental errors that you gained in this activity to write a response in your journal to the following questions. After you have written your answers, join the class in a discussion of this topic.

What choices do humans make that influence the risk of birth defects and cancer? To what extent do these choices play a role in determining the risk of cancer as compared with the risk of birth defects?

Include answers to the following in your response:

a. What are some examples of developmental problems that scientists believe are largely or entirely genetic in origin?

b. What are some examples of developmental problems that scientists believe are largely or entirely environmental in origin?

c. What are some examples of developmental problems that scientists believe result from both genetic and environmental causes?

d. Are there any obvious relationships between individual choice and developmental errors? Explain.

EVALUATING WHERE WE STAND

How many arms and legs do humans have? How many limbs do oak trees have? You can see from this simple comparison that different species show qualitative differences in their developmental patterns. The fundamental processes of development are essentially the same in all multicellular species, but the results vary greatly.

In this part of the activity, you will apply your knowledge of developmental processes and patterns to a new challenge—the description of a developmental scheme for the organism that you discovered in Chapter 3. This task gives you a way to use and display your learning. To help you evaluate your progress, you will use a set of outcomes, or goals, that represents the important concepts in the chapter and a list of indicators of success that will help you (and your teacher) determine how much you have accomplished. Then you will extend your evaluation by reflecting once again on the current status of the two complex questions that have motivated developmental biologists through the years.

Materials (per person)

descriptions and diagrams of Organism X from Chapter 10

Process and Procedures

Part A Development in Organism X

1. Look through the *Outcomes and Indicators of Success* in Figure 13.7 to organize your ideas about how to evaluate your progress in this chapter.

    You may see that you need to review the essays, particular activities, or notes in your journal to meet these goals.

Figure 13.7 **Outcomes and Indicators of Success**

Outcomes and Indicators of Success

As a fully successful learner, you should be able to

1. describe the basic processes of development.

    You could indicate this ability by

    a. identifying growth and differentiation as key processes of development,

    b. describing what these processes do and how they work by using specific examples, and

    c. applying these concepts to a description of the development of Organism X.

2. describe patterns in developmental biology.

    You could indicate success by

    a. identifying key developmental steps in the life span of a variety of species and

    b. identifying some of these steps in the development of Organism X.

3. explain how developmental processes can malfunction.

    You could indicate your understanding by

    a. identifying the major developmental problems that affect humans and

    b. explaining the developmental causes of these problems.

4. creatively apply your knowledge of development to a new situation.

    You could indicate your creativity in this area by

    a. thinking of an interesting and logical way that Organism X undergoes development and

    b. presenting your ideas in a clear and interesting manner.

5. describe how scientific study is done.

    You could indicate your understanding by

    a. explaining why certain questions may be of interest to scientists for many years and

    b. describing the process scientists use to find answers to complex questions.

The numbered statements represent *Outcomes* and the lettered statements represent *Indicators*.

---

2. Describe in your journal what you have learned that helps answer the first big question that scientists asked about development, How are the body's tissues and organs formed?

Keep in mind that development involves more than just getting bigger.

3. If you took a snapshot of an outdoor scene, you would have recorded a moment in the lifetime of a variety of organisms, some of which may be in different stages of development. For example, you might see trees with leaves of bright red and orange, adult animals foraging for food, and a caterpillar consuming a leaf, building energy stores before it changes into a moth. In the diversity captured in such a snapshot, you would see many different developmental stages and evidence of many different behaviors.

Giant water bugs These male water bugs (*Abedus indentatus*) are caring for the eggs that will produce the next generation. What developmental stages will the offspring pass through before becoming adults?

Identify as many significant developmental events or stages as you can that might occur in the lifetime of a multicellular organism, and record them in your journal. Provide specific examples from at least two different species to illustrate your list.

The essay Patterns of Development *on page E210 will help you with this step. As you generate your list, ask yourself when each event typically occurs and what each event or stage accomplishes for the organism involved. For example, what does puberty accomplish in the developmental pattern of humans? What does flowering accomplish in the developmental pattern of flowering plants?*

4. Use your answers from Steps 2 and 3 to help you create a description of a developmental pattern for Organism X. The following three steps will help you organize your work.

Note: If your organism is single-celled, consult your teacher to find out how to do this step.

a. Draw your organism in its various developmental stages, and label these stages with enough detail so that someone looking at the illustration will understand what is happening.

b. Write a paragraph to explain how your organism is influenced by the developmental processes that you identified in Step 2 and the developmental stages that you listed in Step 3. As appropriate, consider how the organism's genetic plan interacts with environmental conditions to direct and regulate development.

c. Explain whether or not your organism occupies different habitats in its different developmental stages.

Part B Studies of Development As a Scientific Process

At the beginning of this chapter, you read about two questions regarding development that were of particular interest to scientists in 1994. The first question (How are the body's tissues and organs formed?) is fundamentally the same question that was asked by earlier developmental biologists. It is very possible that early scientists also wondered about the second question (What clues does development reveal about the process of evolution?). Throughout this chapter you have learned about many aspects of development that are related to these two questions. Think about what you know about the processes of science, and answer the following questions in your journal. Your answers will provide a second way to evaluate your progress in this chapter.

1. Years of experimentation have produced much data relating to the events and processes of development. Why, then, do scientists continue to ask these two questions:

   • How are the body's tissues and organs formed?

   • What clues does development reveal about the process of evolution?

2. Most developmental biologists likely would say that we understand the answers to the first question far better than we understand the answers to the second question.

   a. Does this mean that the second question is invalid or unanswerable and that we should not continue to pursue answers to it? Explain your reasoning.

   b. What does the history of our attempts to answer the first question suggest about how we should go about answering the second? How long might it take to answer this question? Illustrate your answer with specific references to the history of developmental biology.

Analysis

Use your knowledge of development, including the evidence you used to support your answers in Parts A and B, to evaluate your progress in achieving the outcomes listed for the chapter.

1. Compare your progress in this activity with the goals for learning suggested by the *Outcomes and Indicators of Success* in Figure 13.7.

   a. In your journal list the indicators that best describe your ability to talk and think about development.

b. In your journal list the indicators (if any) that you feel do not describe you.

c. Based on these lists, write at least one sentence for each outcome that evaluates your success.

2. If you had other goals for yourself that are not listed in Figure 13.7, list them and add a sentence for each that evaluates your success in reaching it.

" *Whatever begins, also ends.* **"**

Seneca, 4 B.C.–A.D. 65, Roman philosopher and writer

14 THE HUMAN LIFE SPAN

The photographs you see on these two pages are of the same individual taken 90 years apart. As you look at them, think of the immense changes this woman has experienced. During her childhood there were no automobiles. As an adult, however, she not only saw the development of cars, airplanes, and computers, but she watched as a man walked on the surface of the moon. Imagine the different stages of her life, the people who have influenced her, and those she has influenced. In this chapter you will have the opportunity to study the stages of life that a human experiences. Essentially, all humans develop physically in much the same way, but life stages also include emotional, cognitive, and social development. A person's culture influences this development in a number of ways.

To engage you in this chapter, we have put a family's photograph album into motion. During this activity you will have an opportunity to think about the phases of life through which an individual has progressed. You will begin to explore human growth and development by joining with your teammates to study a specific life stage. You will do this by observing and interviewing people. You will continue to develop your understanding of the human life span by participating in a debate about how much your culture, environment, and genetic plan contribute to your development. Finally, you will prepare to participate in a multicultural fair in which your team will present information about certain life stages in a different culture.

ACTIVITIES

Engage   A Century of Photographs

Explore   Growing Up—What Does That Mean?

Explore   A View of Life

Explain   Life-Span Development: Examining the Contexts

Elaborate Evaluate   Cultural Diversity in the Human Life Span

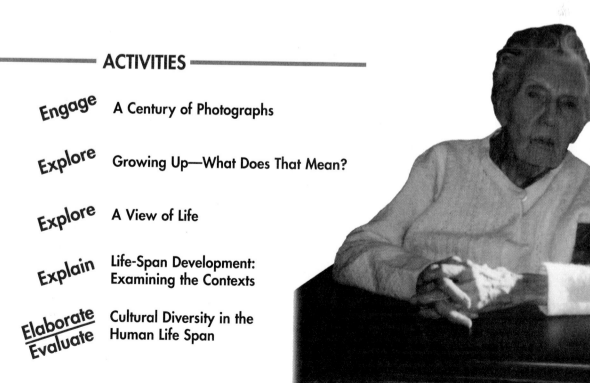

Engage

A CENTURY OF PHOTOGRAPHS

When you look through a photograph album, you are likely to see images of friends and relatives caught in a moment of time. What would you see if you could put those individual snapshots into motion and watch someone's life progress? What happens to a person during the process of aging? This activity will give you a sense of what that might look like.

Process and Procedures

1. Read through the questions in *Analysis*, and keep them in mind as you view the videodisc segment *Age Progression*.

2. Note in your journal two or three thoughts or additional questions that you had as you viewed the images.

3. Discuss with your classmates your thoughts and questions.

Analysis

Now that you have had the opportunity to discuss some of the changes associated with aging, answer the following questions in your journal:

1. What are the stages in a human life?

2. What do you think identifies particular life stages?

3. Is each stage clearly distinctive from the next? Explain your answer.

Figure 14.1 Many physical, mental, emotional, and social changes occur during the human life span.

GROWING UP—WHAT DOES THAT MEAN?

In the activity *A Century of Photographs*, you watched a baby grow into a little girl, then into a mature woman, and finally into a elderly woman. As you watched, did you wonder exactly what happens to people as they grow and develop? In this activity you will have the opportunity to make observations of humans in a specific stage of life and to focus on physical, cognitive, emotional, and social aspects of their development.

Explore

Process and Procedures

1. Join your team of four, and decide for which one of the following life stages your team will make observations.

    -   infancy (birth through 1 year)

    -   early childhood (2 years through 6 years)

    -   middle childhood (7 years through 11 years)

    -   adolescence (12 years through 18 years)

    -   young adult (19 years through 30 years)

    -   prime adult (31 years through 55 years)

    -   middle age (56 years through 70 years)

    -   old age (more than 71 years)

Figure 14.2 Developmental psychologists learn a lot about patterns of child development by observing children in a variety of settings.

2. Based on your current knowledge, record in your journal your ideas about the following for the age group that you will observe:

    a. three physical characteristics,

    b. two cognitive characteristics,

c. two emotional characteristics, and

d. two social characteristics.

Do this task individually.

3. Share with the rest of your team your journal entries for each category listed in Step 2, and note similarities and differences in your current understandings.

4. With your team of four, develop a plan for making observations that all team members will follow.

Your teacher will provide you with information about whom you will be observing and how you will accomplish this task. Human Development 101 (page E214) in the essays will help you plan and prepare for the observations that you will be making. This essay is a series of journal entries from a fictitious college student who, at the time of the entries, was just beginning her studies in human development. The information in her journal will provide you with some background about aspects of physical, cognitive, emotional, and social development.

You will form pairs within your team of four and make your observations in these pairs.

a. Have a brainstorming session, and create a list of ways to observe and obtain information about people in the age group that you chose to study.

b. Refine your list, and develop an outline with specific strategies and potential questions and tasks that you will use during your observations.

Agree on roles for you and your partner so that you share the responsibilities of recording observations and interacting as necessary with the participants.

Follow the Guidelines for Observations *in Figure 14.3.*

c. When you think your plan is complete, ask your teacher to approve it. Make revisions, if necessary, and have your teacher look at it again.

d. Based on your plan, create observation forms that will help you keep track of the information that you collect.

Review the Sample Observation Form, *Figure 14.4, for some ideas.*

5. Conduct your observations in pairs.

You likely will conduct these observations after school or on the weekends according to the structure that your teacher has set up and the arrangements he or she has made. When the observations are complete, each pair should have observed three people.

6. Meet with the rest of your team to share and summarize your observations. Include information about each type of development (physical, cognitive, emotional, and social), and discuss the following questions:

a. How do your actual observations compare with your journal entries from Step 2?

b. Between the two pairs of teammates, what were the similarities or differences in your observations?

c. How might you explain the differences in what you observed?

7. With your class, share each team's summary, beginning with the earliest life stage.

Guidelines for Observations

Introduction

1. Introduce yourself to the people you will be observing and to any teachers or supervisors present.

2. Describe the project on which you are working in a way that is appropriate to the age of your subjects.

3. Reach an agreement with the participants or supervisors on the approximate length of your observation session.

4. Answer any questions that your subjects might have, and make certain that each of them is willing to participate.

5. Remember that the participants have the right to decide not to continue at any time; you must respect such a request.

Making Observations

6. You must make specific observations in each of the following areas: physical, cognitive, emotional, and social development. In general, you should focus your observations on three different people for about 15 minutes each.

   Some of the things that you observe will cover more than one area. Remember to record your observations, not your interpretations of the observations.

7. Provide a task for the participants to complete as you observe them.

8. Ask questions as needed to obtain the information that you want.

9. Ask the participants to perform tasks that will help you obtain the information that you want.

10. If you are observing 2- through 6-year-olds , ask the participants to draw something that you can take back to your classroom with you. (You may want to refer to these as you read the essay *Physical Growth Influences Mental Growth* later in the chapter.)

11. Record your observations clearly and completely.

12. Before you conclude your observations, look over your observation forms to see that they are as complete as they can be.

Conclusion

13. Thank the participants and supervisors for their time and cooperation.

Figure 14.3 Guidelines for Observations

Figure 14.4 **Sample Observation Form**

Sample Observation Form

Participant #1: _____

Age: _____

Description of setting: _____

_____

_____

_____

_____

Observation of physical growth and development:

Description of task, if any: _____

_____

_____

1. _____

2. _____

3. _____

4. _____

5. _____

Observation of cognitive growth and development:

Questions or description of task, if any: _____

_____

_____

1. _____

2. _____

3. _____

4. _____

5. _____

Observation of emotional growth and development:

Questions or description of situation, if any: _____

_____

_____

1. _____

2. _____

3. _____

Observation of social growth and development:

Questions or description of situation, if any: _____

_____

_____

1. _____

2. _____

3. _____

Analysis

Complete Step 1 as a class, and then complete Step 2 individually. Your teacher will collect your journal after you complete Step 2.

1. Create a visual representation of human growth and development based on the aspects of development that you and your classmates observed. Your representation should

   - be chronological,

   - include aspects of physical, cognitive, emotional, and social development, and

   - include a written component.

   Use any supplies that your teacher provides; be creative.

2. In your journal write a short reflective entry that describes your current understanding of development during the human life span.

   - Which aspects of development do you think are specific to the cultural setting?

   - Which aspects of development do you think are found in most cultures; that is, which do you consider to be "universal"?

   - What other questions come to mind that you hope to answer as you continue with this chapter?

   The essay Growing Up through Life's Phases *(page E216) should help you with this entry.*

A VIEW OF LIFE

What is it like to have experienced all of the stages of life? With a healthy lifestyle, the chances of a young person living an active life extending seven or more decades are much greater today than they were several generations ago. In this activity you will have an opportunity to have an elderly guest visit your class and answer some interview questions that you and your teammates will develop. This will give you a retrospective view from someone who has experienced most of life's stages. The type of interview you will be doing is an autobiographical oral history. This is the most personal type of interview, and it should be relaxed and free-ranging.

Process and Procedures
Part A Interview Preparation

1. Assemble into your team of four from the activity *Growing Up—What Does That Mean?*

2. Refer to the class project you completed in the previous activity, and make a list of additional information you would like in order to construct a more complete understanding of the life stages.

Think about the types of insights an interview could provide, that is, insights that were not apparent from the observations that you made in the previous activity. For example, you collected information about various phases of life but little about experiencing all of them.

3. Use your list from Step 2, along with the *Guidelines for Interviews* from Figure 14.5, to develop at least five interview questions for your guest.

   You also might find the questions in Step 2 of the Analysis *helpful as you develop your interview questions.*

Figure 14.5 **Guidelines for Interviews**

Guidelines for Interviews

A primary goal of an interviewer is to create a suitable environment in which to carry out the interview. You should make the guest feel comfortable, and the interviewers should be active listeners. The following tips might help you conduct a successful interview.

1. A good interviewer is not the star of the show; the guest is. The objective is to get the guest to tell his or her story.

2. Ask questions that elicit more than just a yes or no response.

   For example, What was the most exciting time of your life? is a better question than, Did you have an exciting life?

3. Ask only one question at a time, and keep the questions brief.

4. Do not begin with controversial or sensitive questions. Save those for later when the guest warms up to the class.

5. Allow plenty of time for your guest to reflect and to respond to your questions; do not rush ahead impatiently if there is a pause.

6. Allow time at the end of the interview for the guest to add whatever he or she would like.

---

4. Order your questions with respect to importance (1 = what you feel is most important).

5. Have a representative from your team record your first question on the chalkboard or flip chart.

If there is already a question that is very similar to yours, go to the next question on your list.

6. Continue to record questions in this manner until your class has about 15 questions listed.

With your class suggest improvements for clarifying the wording or eliminating repetitive questions.

Part B The Interview

1. Choose a spokesperson from your team, and as your team is called on, have this person pose one of the prepared questions to the guest.

2. You may pose other questions that the interviewee's responses might raise. Your teacher will invite these questions periodically throughout the interview.

3. Record in your journal the major ideas from your guest's responses.

You will have access to a tape recording of the interview, which will allow you to review the responses and to add detail to your notes.

Analysis

Revisit the project your class developed for the activity *Growing Up— What Does That Mean?* Complete Step 1 as a team and Step 2 individually.

1. Determine what new information emerged from the interview, and add it to your class project.

2. Use the information you obtained during the interview to respond to the following in your journal:

   a. How did the guest's culture influence his or her growth and development?

   Explain how a different cultural setting might have had a different impact on this person.

   b. Provide examples of how technological change during the individual's lifetime influenced his or her development at each life stage.

LIFE-SPAN DEVELOPMENT: EXAMINING THE CONTEXTS

You have investigated how people develop through the major phases of life on their dramatic journey from birth to old age. Where we happen to be along that journey frequently determines how we perceive those stages. By completing the previous activities in this chapter, you have had an opportunity to broaden your perspective through observation, an interview, and some analysis. In this activity you will develop further your understanding of human life stages by preparing for and participating in a debate about the relative contributions that a genetic plan and the environment make to each individual.

Process and Procedures

1. With your teammates discuss the following two questions, and decide which question your team would like to explore.

   a. How much of the variation that we see in individuals can be attributed to a specific genetic plan?

   b. How much of the variation that we see in individuals can be attributed to the greater environment?

2. In your journal begin to develop an answer to your question by outlining relevant information. Also, record related questions that come to mind.

   In addition to what you already know and what you have learned in this program, use the following essays as resources to help you complete this step:

   Physical Growth Influences Mental Growth *(page E220)*

   Physical Growth Influences Social and Emotional Growth *(page E223)*

   All Phases of Life Require Self-Maintenance *(page E224)*

3. Join with the other teams that selected the same question that your team selected, and prepare for a class discussion of the two questions from Step 1.

   a. Choose a recorder for your new team.

   b. Summarize what you already know.

   Take turns adding new information to the team's collection of information, and make sure you can support each statement. You may want to record your ideas on a flip chart.

c. Make a list of other information that you would like to have, and explore the resources that your teacher has available to determine whether you can answer your questions.

d. Think about what the other group might say and how you might respond.

4. Participate in a class debate.

As the debate progresses, record notes and questions in your journal.

Analysis

With your classmates discuss the findings from the debate. Then, individually, write a short essay in your journal that synthesizes the two sets of responses. Briefly summarize the responses to each question, and relate how heredity and environment *together* interact to make us who we are.

CULTURAL DIVERSITY IN THE HUMAN LIFE SPAN

In this activity you will have the opportunity to explore life stages in another culture and, in doing so, to elaborate on what you have learned in this chapter. Each team will explore a different culture and create a display and a presentation for a multicultural fair that will take place as part of the Unit 5 Assessment. At this fair you and your classmates will display what you have learned about growing up and living in another culture, and you will present some aspect of that culture for the rest of the class. After you have studied a culture, you will have an opportunity to evaluate your understanding of human development over a life span by reflecting on the diversity as well as the commonality of expression in different cultures. You also will reflect on the aspects of biological development that allow for these similarities and differences.

What do you think these teenagers have in common?

Process and Procedures

1. To prepare for the multicultural fair in which you will be participating, do the following:

    a. Think about the cultural setting in which you might want to explore human life stages.

    Your teacher will have a list of suggestions as well as some resources for you to look through to help you choose.

    b. Join with other classmates who have chosen to explore the same culture, and share with them your reasons for being interested in this particular group of people.

    Use the information in the essay Culture: The Great Shaper of Life *on page E226 as a resource to help set the stage for your study.*

2. Begin your study of another culture.

    a. Look through the resources that your teacher has as well as additional resources that you might want to obtain.

    b. Divide the responsibility of reading and reviewing the resources in order to develop a general overview of the culture.

    You may want to find additional resources of your own. Remember, you are not limited to books and magazines; you may want to look for films, videos, music, or art. You even may know of someone from the culture you have chosen who might be willing to share some ideas with you.

    c. Share what you have learned so far. Exchange resources if you wish.

3. Develop a specific design for your entry in the multicultural fair. Refer to *Guidelines for Creating Your Entry for the Multicultural Fair* in Figure 14.6.

    Remember, your job is not just to learn about the life stages in another culture but to teach your classmates about it as well.

4. Design and prepare your entry for the fair.

    Be as creative as you want, but remember to follow the guidelines provided. You will be assessed according to the criteria presented in the scoring rubric that accompanies the Unit 5 Assessment.

5. Practice the presentation portion of your entry.

    You may want to schedule additional practice time outside of class. You will participate in the fair as both visitor and exhibitor during the Unit 5 Assessment.

Analysis

After you have completed Step 5, record in your journal your reflections on the following questions. Work individually.

1. Imagine that you are a young woman or man living in the culture that you have just studied. Write two or three paragraphs describing how she or he might view your American culture. Which aspects of your life

experience would seem similar to her or his own experiences, and which aspects would seem different?

2. What have you learned about biological development in humans that may help explain both these similarities and these differences?

Guidelines for Creating Your Entry for the Multicultural Fair

Your entry for the fair should include a combination of visual displays and written support material. You may want to include an audio portion as well. These displays may be arranged in your booth in any way that you want within the boundaries that your teacher establishes. You also will need to create or develop some type of presentation that includes all team members.

As you create your entry, you should do the following:

1. Provide an overview of the culture that you studied, including where this group of people lives or lived and a description of their way of life.

2. Present information about one of the following life stages:
   - infancy and childhood
   - adolescence
   - adulthood
   - old age

   a. Describe the physical and social setting that is predominant at this life stage.
      - Who is around the individuals at this stage, and what do the individuals do at this stage?

   b. Describe the cultural practices for individuals at this stage.
      - What is expected of the individual at this stage?

   c. Describe the cultural values surrounding this stage.
      - How are individuals in this stage perceived?

3. Explore some aspect of cognitive development that seems particularly significant in the culture that you are studying. (For example, in American culture the point at which a child begins to speak is considered significant.) Explain why you think this particular cognitive development is highly valued, and describe the underlying biology that allows for this aspect of development.

4. Choose and complete one of the following:

   a. Describe the different forms of cultural expression that seem to be significant during any life stage—for example, music, art, dance, mythology, religion, or dress.

   b. Describe the celebration of at least one of the following rites of passage:
      - birth
      - puberty
      - marriage
      - death

   c. Describe the differences in growing up male from growing up female in the culture you are studying.

5. Some part of your entry should be a presentation. You may, for example, decide to portray in detail one life stage that you found most interesting. You may want to incorporate music, literature, art, dress, or dance. This presentation does not need to take place in your booth.

ECOLOGY:
INTERACTION AND INTERDEPENDENCE IN LIVING SYSTEMS

15 Interdependence among Organisms in the Biosphere

16 Decision Making in a Complex World

What happens when developers cut down the trees along a river bank? The trees are gone, you say. But what effect does the loss of these trees have on the ecosystem of the river? How has a web of dependence been altered? In this unit you will have an opportunity to explore the concepts of interaction and interdependence in living systems and the issue of human influence in the biosphere.

By the end of this unit, you should understand that

- a community of organisms interacts with the abiotic environment to form ecosystems,
- ecosystems change across time,
- ecosystems are complex, but it is possible to analyze them,
- ecosystems can be modified by human actions,
- population size is affected by carrying capacity, and
- human actions follow from decisions.

You also will continue to design and conduct experiments, evaluate explanations, and explore the relationship between public policy and scientific investigation.

> *Every part of this earth is sacred to my people. Every shining pine needle, every sandy shore, every mist in the dark woods, every clearing, every humming insect is holy in the memory and experience of my people.*

Attributed to Chief Seathl (Seattle), 1786–1866, of the Suwamish Tribe

15
INTERDEPENDENCE AMONG ORGANISMS IN THE BIOSPHERE

Life in the mountains. Life in a forest. Life on a bustling city street. Life in the oceans. Life in a refugee camp in Africa. Life on a farm in rural Ohio. How are the organisms in these settings dependent on one another for survival? How can this interdependence be described? What factors influence the interdependence?

In this chapter you will learn about the interdependence of organisms in various ecosystems within the biosphere. First, you will seek evidence of interactions in your own schoolyard or through visual images provided by a videodisc. Next, you will study the patterns of population growth in a real group of reindeer that live on an island off the coast of Alaska. You will begin to develop an explanation about interdependence among organisms as you investigate interactions that occur in an aquarium. You will have an opportunity to elaborate on these concepts by studying different ecosystems to learn about how they change across time. Finally, you will evaluate your understanding by describing how your Organism X will interact with you and other organisms in a particular ecosystem.

ACTIVITIES

Engage/Explore Interactions in the World around Us

Explore Reindeer on St. Paul Island

Explain On the Double

Explain/Elaborate A Jar Full of Interactions

Elaborate Changing Ecosystems

Evaluate Organism X and Global Interdependence

INTERACTIONS IN THE WORLD AROUND US

You crawl out of bed in the morning, flip on a light and your CD player, and stroll to the bathroom as your CD player hums in the background. For breakfast you eat a bowl of corn flakes. It's an especially cold morning, so you fix yourself some hot chocolate, too. The television is blaring with news of an overnight fire in the nearby national forest. It's getting late, so you pack your lunch, fill the bird feeder that is outside your apartment window, and rush off to catch the bus.

This scenario is typical for some teenagers. In this activity you will have the opportunity to reflect on a series of images, some that may be quite familiar to you and others that may be less familiar to you. What interactions do you see in the natural world?

Process and Procedures

1. Watch the videodisc segment *Images from around the World*, and think about the question posed at the beginning of the segment, What do these images have in common?

2. Have your team of three join with another team and discuss your responses to this question.

3. Read the task card that your teacher distributes, and in your journal prepare a table similar to the one in Figure 15.1 that addresses the task described on your card.

Examples of the Positive Influences of Humans on Other Organisms

Human Action	Positive Influence on Other Organisms

4. According to your teacher's directions, look for three examples of the relationship that is described on your task card, and record each example in your table.

5. When all of the teams have finished, share your examples with the rest of the class.

Analysis

As you respond to this *Analysis* in your journal, think about the examples that you and your classmates have found and about the images that you viewed on the videodisc.

Use the opening story Early Morning Reflections *on page 324 as an additional resource to complete this task.*

1. What do these examples indicate about interactions that take place in the world?

2. Do you think that humans have more or less influence on the biosphere than other organisms? Explain your response.

3. Do you think that humans have a responsibility to monitor how they influence the biosphere? Explain your response.

REINDEER ON ST. PAUL ISLAND

Islands are intriguing ecosystems to study because they are somewhat isolated. It is especially interesting to study the patterns of population growth for certain land animals that are confined to an island. This is the

EARLY MORNING REFLECTIONS

Dear Senator Wilks,

I have just returned from a fascinating visit to a ranch in northern Nevada, and before I become enveloped by daily demands, I thought I would share with you some reflections about my visit to your part of the state.

The landscape of northern Nevada is spacious and quite barren, and yet there is still something awe inspiring about the place. I had the good fortune of spending the weekend with my friend and his family—an exceptional and thoughtful family—at their ranch on the Marys River floodplain. Not only was I inspired by their unpretentious manner, but also by their genuine stewardship of the fragile river ecosystem that is part of their vast 7000 acres.

A river meandered through the landscape, and as my friend showed me around, I noticed a group of willow trees lining the riverbank. But farther upriver and downriver, there were almost no willows. I wondered why this was. When I asked, I saw a faint smirk cross my friend's face, followed by a touch of sadness. He told me that back in the '70s, many of the farmers along the river began cutting down the willows because they thought the willows sucked up too much water. They thought that by cutting down the willows, they would increase their crop yield. My friend explained that it may have worked that way for a while, but actions like that have unintended consequences. During the floods of 1983 and 1984, my friend's ranch sustained the least damage of any along the river. The willows, along with the native hay meadow vegetation, helped stabilize the riverbanks and the floodplain soils; this reduced erosion and other damage from flooding. But my friend said there was even more to the story. To show me what he meant, he took me down to the riverbank.

I was amazed by the variety of lush vegetation that I saw in addition to the willows. My friend seemed to enjoy watching me try to survey the situation and figure it all out. I saw what I suspected were beaver dams built across the river at various places and, although I did not see any beaver, I did see two otters.

Soon my friend began to fill in between the lines. He explained that the willows and the other vegetation along the river provide material with which the beavers can build their dams. With the dams in place, the water becomes somewhat deeper just upriver from the dam, and the amount of groundwater increases in these areas as well. This additional water is what allows for and sustains the lush vegetation. As he talked, I made mental notes of all the evidence I saw. My friend continued to explain that the willows and other overhanging plants shade the river so that it stays cooler, which keeps the level of oxygen higher.

I started nodding my head. I was beginning to get the picture. With willows for shade, more oxygen in the water, and deeper water levels, the river could sustain more life—including aquatic plants and fish and, from there, continuing up the food chain to the level of the beaver (a herbivore) and the otter (an omnivore that loves fish).

Without the willows, not only did the ranchers lose out during the spring floods, but the beaver lost out because now there was no material with which to build dams. Without the dams and the shade provided by the vegetation, the river was too shallow and too warm to support much aquatic life, so the fish lost out. And without the fish, a primary source of its food, the otter lost out too.

Even though surrounding this ranch, especially to the west, the river environment is quite barren, my friend and his family have been stewards of the river and of the land—and the rewards have been immeasurable.

case for a population of deer (see Figure 15.2) on St. Paul Island in the Bering Sea, off the coast of Alaska. This activity will allow you to peek into the history of a population of reindeer that inhabit the island and reflect on some of the interactions that take place in that ecosystem.

Materials (per person)

calculator

Process and Procedures
Part A Exploring a Population of Reindeer

St. Paul Island is one of the Pribilof Islands in the Bering Sea near Alaska. It is approximately 106 km² (41 square miles) in size and is more than 323 km (200 miles) from the mainland. In 1911, 25 reindeer (4 males and 21 females) were introduced onto the island. At that time there were no predators of the reindeer on the island and hunting of the reindeer was not permitted.

The Pribilof Islands (see Figure 15.3) are windy and rainy in fall and windy and snowy in winter. During the spring, summer, and early fall there

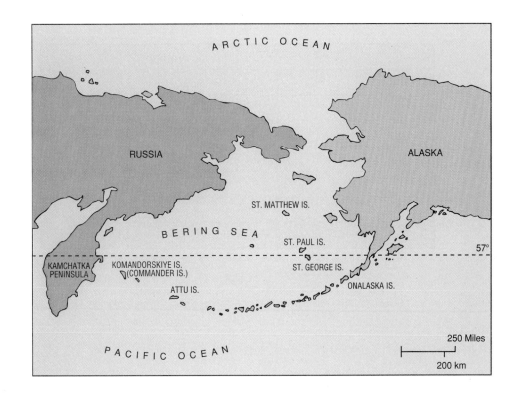

Figure 15.3 St. Paul Island is one of the Pribilof Islands located in the Bering Sea, southwest of Alaska.

is a great deal of fog. In addition to the reindeer, there are a variety of fish and fur seals along the coast. The islands are covered with low-lying shrubs and grasses as well as lichens and mosses.

1. With your teammates study the graph in Figure 15.4, and answer the question that your teacher assigns you.

   a. What was the size of the reindeer population at the beginning of the study?

   b. What was the size of the reindeer population in 1920?

   c. What was the difference in the number of reindeer between 1911 and 1920?

   d. What was the average annual increase in the number of reindeer between 1911 and 1920?

   e. What was the difference in population size between the years 1920 and 1930?

   f. What was the average annual increase in the number of reindeer between 1920 and 1930?

   g. What was the average annual increase in the number of reindeer between 1930 and 1938?

   h. During which one of the following three periods was the increase in the population of reindeer greatest: 1911–1920, 1920–1930, or 1930–1938?

   i. What was the greatest number of reindeer found on St. Paul Island between 1911 and 1950? In what year did this occur?

   j. In 1950 only eight reindeer were still alive. What is the average annual *decrease* in the number of reindeer between 1938 and 1950?

Figure 15.4 **Changes in the population size of the reindeer on St. Paul Island** This graph depicts the changes in the size of a population of reindeer from 1911 to 1950. No data were collected during 1941 and 1942; this gap in the data is represented by a dashed line.

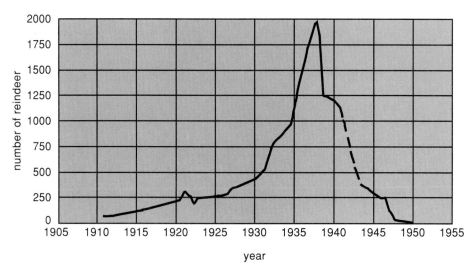

Changes in the Population Size of the Reindeer on St. Paul Island

Source: From V.C. Scheffer (1951) The rise and fall of a reindeer herd. *Science Monthly* 73: 356–362.

2. Participate in a class discussion of the answers to the questions in Step 1.

   Record the answers in your journal; you will need to use them as you complete the Analysis.

Analysis

Consider the following questions, and record your responses in your journal.

Use the essay Interdependence Involves Limiting Factors and Carrying Capacity *(page E232) as a resource.*

1. What might account for the tremendous increase in the population of the reindeer between 1930 and 1938, as compared with the rate of growth during the first years the reindeer were on the island?

2. What factor or factors finally limited the growth of the reindeer population? What other factors might limit the growth of a population? Give examples.

3. What evidence is there that the number of reindeer on the island exceeded the carrying capacity of the island?

4. What does this study tell you about unchecked population growth?

5. How might interactions between the reindeer and predators have affected population growth? How might interactions between the reindeer and hunters have affected population growth?

Part B So What about Resources?

In general, in order for individuals in a population to sustain themselves, the individuals must have food for energy. The amount of food available to each individual would depend both on the total amount of food available and on the total population size at that time. If we know both of these things, then we can calculate (by simple division) the amount of food available per individual in the different populations.

1. Copy the data table in Figure 15.5 into your journal, and with your teammates complete the table by calculating for each population the number of units of food available to each individual.

   Note that the total amount of food available is always 100 units.

   The first calculation has been completed for you. For Population A, the population size is 2, and the number of units of food is always 100: 100 divided by 2 = 50.

2. Use the data in your table to create a graph of the relationship between the size of a population (*x*-axis) and the number of units of food available per individual (*y*-axis).

Figure 15.5 In different-sized populations, how much food will each person receive if the amount of food available is always limited to 100 units?

Population	Population Size	Food Allocation per Individual (# units)
A	2	50
B	4	
C	8	
D	16	
E	32	
F	64	
G	128	

Analysis

With your team consider the following questions, and record your answers in your journal. Use your data table and your graph to help you answer these questions:

1. How does the amount of food available to each individual change as the population size increases?

2. Imagine that in order to survive individuals need a minimum daily requirement of 3.0 units of food. How would this minimum requirement affect the survival rates of individuals in Population E? What about Populations F and G?

3. How might the relationship between population size and amount of food available per individual apply to the population of reindeer on St. Paul Island that you explored in Part A of this activity?

ON THE DOUBLE

Explain

Would you rather receive an allowance of $20 a week for a year or begin with one penny and double that amount every week for a year? Before you answer this question, you might want to learn more about exponential growth and doubling time.

Materials (per team of 2)

1-lb. bag of sunflower seeds
clock or wristwatch
1 die
watercolor brushes for manipulating sunflower seeds
graph paper
calculator

Process and Procedures

Part A Covering the Earth with Sunflowers

In this part of the activity, each sunflower seed represents a sunflower plant.

1. Read through Part A of this activity, and create a data table that will help you organize the data you will be collecting during this part of the activity.

2. Roll the die to determine the starting size of your sunflower population. (The population must be 4 or less.) Count out the number of sunflower seeds that are indicated by the die (1–4), and put them in the center of your table, which represents an open farm field.

   If you roll a 5 or a 6, roll again until you roll a 1, 2, 3, or 4. Record this information in your data table.

3. Roll the die again to determine how many offspring (1–4) each sunflower in the population produces each time it reproduces, and record this information.

   For example, if you roll a 2, then for each flower present in the population, you will add two offspring to the population when it reproduces. Again, roll the die until you roll a 1, 2, 3, or 4.

4. Simulate the reproduction of sunflower seeds through 5 to 10 generations in the following manner:

   a. Assume that the sunflower seeds reproduce every few minutes and that all the seeds in the population reproduce simultaneously.

   b. Record the starting time of the experiment, and keep track of the passage of time.

   c. Use your current population, the number of offspring that you determined in Step 3, and your calculator to determine the total number of offspring that will result each time your sunflower seed population reproduces.

   d. Withdraw the offspring from the bag of seeds, add them to the population on the table, and determine the new population total.

   You may want to use a watercolor brush to seperate and count the sunflower seeds.

   e. Record the information in your data table, and begin the next reproduction event.

   f. Continue the experiment for about 5 to 10 reproduction events, depending on time and seed supply.

   If you discontinue using the seeds, continue calculating.

5. Graph your data, and observe the shape of your growth curve.

   Put the number of generations on the x-axis and the population size on the y-axis.

6. Post your graph in the classroom, and present your results to the class.

   Be sure to label your graph so that the rest of the class can determine the starting size of your population and the number of offspring that each seed produced during reproduction.

7. Compare the growth curves that different teams obtained. What factors influenced the shape of the graphs?

8. Complete the following questions as a team, and use the essays *Growing, Growing, Grown* (page E234) and *Inquiring Minds* (page E237) as resources. Record your answers in your journal.

   a. How would you describe the general shape of your population growth curve? How was your growth curve similar to or different from the curves other teams obtained?

   b. How do differences in the starting size of the population affect population growth? How do differences in the number of offspring produced per parent seed affect population growth?

   c. How would increasing the time interval between reproduction events to 6 minutes affect population growth?

   How would decreasing the time interval between reproduction events to 1 minute affect population growth?

   d. In what ways is this simulation of population growth realistic? unrealistic?

   e. If exponential growth is one of the basic principles of population growth, then why isn't the earth covered with flies, elephants, or sunflowers?

Part B Doubling Time

We can approximate the doubling time for a population by dividing the number 70 by the growth rate of the population expressed as a percentage. The growth rate is calculated by taking into account the population's birthrate, death rate, and migration rate, which is the combined effect of immigration and emigration.

$$\text{Doubling time} = \frac{70}{\text{population growth rate (\%)}}$$

This formula tells us that a population growing at a rate of 2 percent will double in size in 35 years. Even a population growing at half that rate (1 percent) will double in size during an average human lifetime (70 years). Some of the fastest-growing human populations in the world have had a population growth rate of nearly 4 percent, which means that those populations double almost every 17 years. In contrast, some slow-growing human populations are increasing by less than 1 percent per year.

The number 70 in the numerator of the doubling time formula is derived from a mathematical formula that describes exponential growth and is approximately the natural logarithm of 2 times 100. (The natural logarithm of 2 = 0.693. If you round up to 0.7 and multiply by 100 you get 70.) The denominator of the formula is the growth rate of the population expressed as a percentage.

a.

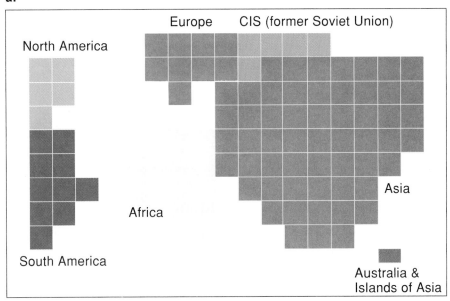

North America
Europe   CIS (former Soviet Union)
Africa
Asia
South America
Australia & Islands of Asia

Each square = 1% of world population (1990)

Source: Population data from United Nations (1990).

Figure 15.6 a. In this schematic map of the world, each square represents 1 percent of the world's population. As you see, only 5 percent of the world's population lives in North America, and 56 percent of the world's population lives in Asia. (CIS stands for the Commonwealth of Independent States.)

Figure 15.6 continued
b. These profiles indicate the percentage of males and females in each age group. Developed countries like the United States (1) generally have about the same percentage of people in each age group. The slight bulge in the middle represents the post-World War II baby boom. Also notice that at +75, there are more females than males. Emerging nations like Kenya (2) tend to demonstrate a pattern of rapid population growth with the majority of people in the younger age groups. When you average together all of the nations' profiles, the world pattern emerges (3).

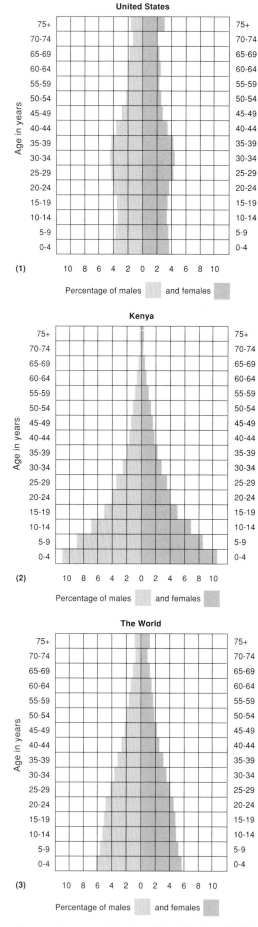

Source: Population data from United Nations (1990).

1. Use the population growth rates for the early 1990s, provided below, to calculate the approximate doubling time for human populations in each of these areas. Record these in your journal.

North America	0.8 percent
Africa	3.0 percent
Europe	0.2 percent

2. Use the information provided in the copymaster *Growth Rates and Doubling Times in Nations around the World* to determine the following:

    a. the doubling time for each population,

    b. the year that each country will double its 1995 population size if it maintains its current growth rate, and

    c. the size the population will be after doubling if the population growth rate remains constant.

    Record this information on the copymaster.

Analysis

1. Use what you have learned in this activity along with the information that you have recorded to explain whether you think there is the potential for a human population problem. Explain your answer.

2. If half of the nations in the world reduced their growth rate to zero, would that eliminate the problem of overpopulation on a global level? Why or why not?

3. In what ways is the earth as a whole similar to an island such as St. Paul Island? Is the earth capable of supporting only a certain number of humans? Explain your answer.

4. Now that you understand more about doubling time and exponential growth, would you rather receive an allowance of $20 per week for a year or begin with one penny and double the amount every week for a year? Explain why.

Further Challenges

Use your bag of sunflower seeds to investigate one of your own ideas about the effect on population growth of (a) starting size, (b) number of offspring produced, or (c) amount of time between reproduction.

A JAR FULL OF INTERACTIONS

Recall the last time you were bitten by a mosquito. You probably swatted at it, scratched the area, and went about your business. Now, let's look at this event in a different way. Where had that mosquito hatched just weeks earlier? What other creatures had the mosquito bitten? How dependent was the mosquito on its interaction with you? This activity provides you with an opportunity to develop a better understanding of

another type of interaction—that of predator/prey—and of some of the limiting factors that influence this interaction. You also will have an opportunity to reflect on how interactions of this nature influence the greater community in which they occur.

Materials (per team of 6)

2 clear containers to be used as aquaria
2 dropping pipets
thermometer
clock or wristwatch with second hand
small fish net
rinsed sandbox sand
dechlorinated water, room temperature
8 large guppies
food source: *Daphnia*
water plant(s) such as *Anacharis*
snails

Process and Procedures

1. With your teammates consider the following task:

    Your task during this investigation is to determine the effect of one limiting factor on the predator/prey interactions between a specific predator fish and a specific food source in an aquarium.

2. Work with your classmates to list possible variables that you could test that might affect these predator/prey interactions.

    Study Figure 15.7 and the control aquarium that your teacher has set up to help you generate ideas.

Figure 15.7 Your teacher's control aquarium might look something like this.

3. Develop a hypothesis that clearly indicates which variable you will test.

    You will have two aquaria similar to your teacher's; one will serve as a control, and in the other you will change one variable. For example, you may decide to vary the size of the container to test the effect of limited space on the predator/prey interactions.

Record your hypothesis in your journal.

4. Develop an experimental protocol that will help you test your hypothesis about a limiting factor. In your journal

   a. describe your experimental setup, and include the change that you will make to your experimental aquarium.

      For instance, if you are going to test the limiting factor of space on the predator/prey interactions, you will want to indicate the difference in size between your control aquarium and your experimental aquarium.

   b. record your procedural steps, and note any necessary safety precautions.

   c. prepare a data table in which to record the following data:

      • behavior of predator, every 30 seconds for at least 5 minutes

      • behavior of prey, every 30 seconds for at least 5 minutes

   d. determine what roles you will need as you make observations and record data, and assign those roles now.

      Three team members should observe the control aquarium, and three should observe the test aquarium.

5. Have your teacher approve your protocol and make any changes that are necessary before setting up the experiment.

6. Set up your aquaria by using your teacher's aquarium as a model.

   Three team members should set up the control, and three members should set up the test.

   a. Add sand to the container to a depth of about 3 cm.

   b. Pour in enough dechlorinated water to reach a depth of 30 cm.

   c. Add a sprig or two of *Anacharis* and two or three snails.

7. Add the predator fish to both aquaria, and leave them undisturbed for at least two days. Do not feed the fish during this time.

8. Wash your hands thoroughly.

9. From the list of variables that you generated in Step 2, identify three that are limiting factors. Explain how these factors are present in an aquarium and how they might influence the interactions among the living organisms present. Record this information in a table in your journal.

   Use the essay Endless Interactions (page E238) as a resource to help you do this. One way to record this information would be a two-column table with one column listing the limiting factors and the other column listing the effect each might have on the interactions.

10. After two days resume the experiment by introducing the prey (*Daphnia*). Be ready to observe and record the interactions the instant the prey enters the container.

    Remember your roles. Notice the amount of time that elapses before the predator finds the prey as well as the amount of food per time that the predator consumes. If you are

using Daphnia *as the prey, transfer them carefully; if air gets under their shell they will die. Record your observations in your data table.*

Analysis

Share your data with your classmates. Analyze the data, and develop a conclusion by answering the following questions in a class discussion. Following this class discussion, summarize the class results, and write a conclusion in your journal that addresses your hypothesis.

1. What generalizations can you make about the effect of your variable on the interactions that occurred in your experimental aquarium? Use your results to support your generalizations.

2. What do the interactions that you observed suggest about the interdependence of these organisms?

3. How are the interactions that you observed similar to and different from the ones your classmates observed?

4. What other questions about interactions, interdependence, and limiting factors could you investigate using the same organisms?

Further Challenges

If you have enough time, ask your teacher for permission to perform another experiment to retest the same variable or explore some other variable from the list that you created in Step 2 or from ideas that you generated in Question 4 of the *Analysis*.

CHANGING ECOSYSTEMS

You have been exploring and investigating how various populations of organisms interact with each other in a variety of ecosystems (refer to Figure 15.8). You also have thought about how limiting factors present in the ecosystem affect its carrying capacity. But what happens in these ecosystems across time? This activity provides you with an opportunity to elaborate on your understanding of interactions and interdependence as you investigate how a variety of ecosystems change across time.

Process and Procedures

1. With your team of four, decide who will study information about each of the following ecosystems:

    a. jack pine forests in Michigan

    b. Yellowstone National Park

    c. forests of the past

    d. ponds along the shore of Lake Erie

Figure 15.8 In this activity you will study forest and pond ecosystems, but the earth is made up of a great variety of ecosystems, some of them as different as a desert and a coral reef.

2. Study the information for which you are responsible, and look for data about how that ecosystem has changed across time.

The following essays provide the foundation for this task, and your teacher may have other resources for you to use.

The Ecology of Kirtland's Warbler *(jack pine forests in Michigan) (page E241)*

Yellowstone Burning: Is Fire Natural to Yellowstone National Park? *(page E242)*

Forests of the Past *(page E244)*

From Puddle to Forest *(page E247)*

3. Find a classmate who is responsible for the same ecosystem that you are. Discuss your ideas about the questions related to your ecosystem.

Figure 15.9 **Forest fires** a. In 1988 fires devastated much of Yellowstone National Park. b. By the summer of 1995, there was considerable evidence of regrowth.

Jack Pine Forests in Michigan

a. Why didn't the population of the warblers decline after the logging of the forests?

b. How would suppressing fires in a jack pine forest for more than 50 years change the makeup of the jack pine population?

c. What conditions create the type of jack pine forest most suitable for the Kirtland's warbler? Why?

Yellowstone National Park

a.

b.

a. What do the data tell you about the pattern of fires in Yellowstone in the past?

b. How does the extent of burning during the last 50 years compare with the previous 300 years? Why do you think this is so?

c. What is the role of fire in the Yellowstone ecosystem? Would you consider this role natural or unnatural? Why?

Forests of the Past

a. Use the information in the fossil pollen profile (Figure E15.20) to describe the changes in the types of trees that grew in this area of the northeastern United States from 15,000 years ago to the present.

b. Use the information in Figure E15.21 to describe the changes in the forests of the northern part of the midwestern United States during the past 20,000 years.

c. Compare the changes in the forests in these two locations. What similarities and differences do you see?

Ponds along the Shore of Lake Erie

a. How does the depth, surface area, and volume of each pond compare? What changes would you expect to see if we were able to check the oldest pond 50 years from now?

b. Describe specifically how the plant community of the youngest pond (A) was different from the plant community of the oldest pond (D).

c. How did the number of mollusks change from the youngest pond to the oldest pond?

d. Calculate a community similarity index for mollusks using the pond comparisons listed below:

- A and D
- B and C
- C and D

4. Describe your ecosystem, and share your findings with the rest of your team.

Be sure that your teammates understand the information that you have been studying and how your ecosystem has changed across time.

Analysis

With your team prepare a poster for a poster session that demonstrates what your team has learned about changing ecosystems. You will know that you have a successful poster when it provides answers to these questions and addresses these issues:

1. What evidence have you accumulated that ecosystems change across time?

2. What types of changes occur in ecosystems across time?

3. How long do these types of changes take?

4. How do changes in ecosystems reflect the processes of ecological succession?

5. How do changes in ecosystems reflect the processes of evolution?

6. How do changes in ecosystems relate to the other unifying principles of biology?

Evaluate

ORGANISM X AND GLOBAL INTERDEPENDENCE

In this activity you will evaluate what you have learned about global interdependence among living organisms. You will work in teams and create a story that describes interactions among various organisms in a particular environment. These organisms will include your Organism X, other classmates' Organism X, humans, and other native organisms.

Materials (per team, size will vary)

pencil and paper
colored pencils and markers
descriptions and diagrams of Organism X from Chapter 13

Process and Procedures

1. Participate in a class discussion that your teacher guides to summarize the main ideas from the chapter.

2. Join with classmates who had the same habitat card that you had in the activity *First Encounter with Organism X* in Chapter 3, and introduce your new teammates to your Organism X.

   You may need to refer to your journal to see what habitat you had. Be sure to describe all the features of your organism in detail, and respond to any questions that your teammates might have.

3. Imagine that you, your teammates, and each teammate's Organism X are living in your habitat from Chapter 3. Consider what interactions might occur as you observe a videodisc segment that presents images of your habitat.

4. With your team have a brainstorming session to generate ideas about the following:

   • possible interactions among the organisms that inhabit this ecosystem,

   • the interdependence that might exist,

   • adaptations that might evolve,

   • changes that might occur in the ecosystem, and

   • how humans from a variety of cultures might interact with other organisms in this environment.

5. Individually, write a story that features some of the ideas that your team suggested.

   Use Criteria for a Good Story *(Figure 15.10) and the scoring rubric that your teacher provides to guide the development of your story.*

Criteria for a Good Story

You will know that you have written a good story if you do the following things:

1. Describe the interactions and changes that occur within the environment.
2. Describe the interdependence among organisms.
3. Identify some of the limiting factors present.
4. Discuss how the populations of organisms are likely to change as a result of the interactions.
5. Provide information about the carrying capacity of the environment and how it might change in response to changes in the environment, for example, in response to limiting factors.
6. Include information about your classmates' organisms that are living in the same ecosystem.
7. Write a story that is at least 750 words.
8. If you think your organism might not be able to survive in the ecosystem, suggest adaptations for your organism that might allow it to survive. Provide justification for the adaptations you suggest.
9. Illustrate your story.

Figure 15.10 **Criteria for a Good Story**

Analysis

Reflect on the process you used to think about and write your story by participating in a class discussion of the following questions:

1. What was the most challenging part of writing your story? Explain.
2. Which of the major concepts of this chapter did you have the most difficulty incorporating into the story? Explain.
3. Was it easier to write about the interactions and interdependence of your organism with other organisms or about the limiting factors at work and the carrying capacity of the environment? Explain why.
4. What adaptations did you consider adding to your organism? Explain why you did or did not add them.

> *Man shapes himself through decisions that shape his environment.*

René Dubos, 1901–1982, French-born American bacteriologist

16
DECISION MAKING IN A COMPLEX WORLD

What do you think about when you look at these pictures? From space, the earth might seem so large that human influence on the planet is minuscule, but don't be fooled. In this course you have explored the tremendous diversity of life on earth, and by now you probably appreciate that a vast number of interactions connect living organisms. The interactions and interdependence among living organisms and between these organisms and the environment provide opportunities for the evolutionary processes that characterize the earth's biodiversity. Unfortunately, the complexity of these interactions makes it difficult for humans to predict how their actions affect other organisms and the environment.

In this chapter you will investigate decision making in a complex world by analyzing some past consequences of human actions and by planning some potential courses of action for humans in the future.

ACTIVITIES

Engage   Calling the Question

Explore   The Sun and Life (or Death?)

Explain   Where Do We Go from Here?

Elaborate   The Ozone Layer: A Disappearing Act?

Evaluate   The Limits of Abundance

Engage

CALLING THE QUESTION

Should you eat lunch or skip it? Buy the white pants or the gray ones? Shoot some hoops before you head home or bypass the gym?

Most of us probably make more decisions in one day than we realize. Many of these decisions are straightforward, even trivial, and we make them so easily that we hardly notice them. But how do we handle the tough decisions?

In this activity you will think about the process of making decisions by analyzing a town council meeting. When the town council "calls the question" (that is, when they ask for a vote) about building a new shopping mall in the neighborhood, how easy will it be to make the decision? How easy would the decision be for you, if you had to make it for the whole town?

Process and Procedures

1. Imagine that the cartoon in Figure 16.1 illustrates the members of the town council listening to an important proposal about a development project.

2. Listen as your teacher reads to you a short news item about the proposal.

3. Work with your partner to identify the viewpoint of each council member. What would each of these individuals likely be thinking? On the copymaster that your teacher provides, fill in each bubble in the cartoon with the corresponding viewpoint.

Figure 16.1 The individuals who serve on town councils usually represent many diverse interest groups. Who are the stakeholders at this meeting? What is each one thinking? How does a person's role in the community influence his or her thinking about the issue?

Analysis

Participate in a class discussion about the decision facing the council. Would it be easy for you to decide how to vote? Will the decision about whether to approve the mall be a simple decision for the town council? Why or why not?

THE SUN AND LIFE (OR DEATH?)

In the activity *Calling the Question*, you may have found that not all situations are as simple as they first appear. Often, when you consider a situation from a perspective other than your own, or when you look at a phenomenon a second or third time, you discover complex relationships that you did not see at first glance. Consider the sun, for example (Figure 16.2). A simple description of the sun might represent it as a glowing ball of fire in space. A more complete description of the sun also might mention the important relationships that the sun has with living systems. In this activity you will investigate some of the interactions between the sun and life. As you might expect, this relationship is more complex than many people realize.

Explore

Figure 16.2 What effect does the sun have on your life?

Materials (per team of 4)

Macintosh computer or IBM-compatible computer with Windows™
Ozone: The Game computer simulation

Process and Procedures

Part A Using a Simple System to Identify a Relationship

1. Participate in a class discussion about the effects of sunlight on organisms. Consider the following questions in your discussion:

   a. What benefits does the sun provide for organisms?

b. What negative effects can the sun have on organisms?

a.

Figure 16.3 What are the dangers of overexposure to the sun? What role does ozone play in complex interactions that occur between sunlight and the earth's living organisms? As shown in (b.), the ozone layer at the earth's South Pole has thinned significantly.

b.

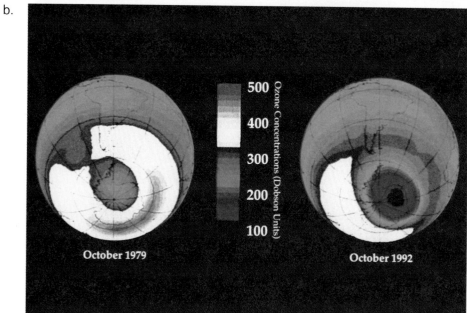

October 1979   October 1992

Ozone Concentrations (Dobson Units)

2. Yeast can be used as a model system to test the effect of sunlight on organisms. For example, suppose that you were given a strain of yeast that is sensitive to ultraviolet light and were asked to design an experiment to test the effect of sunlight on its growth. In your journal record several questions that you could investigate.

   Examine the materials that your teacher has provided.

3. Contribute your ideas to a class discussion of the questions that you could test, the variables involved, and the controls that you would need if you tested these questions.

4. With your class decide which variable each team will test.

5. To test your ideas, you will need to grow yeast cells on a solid medium that you can manipulate and observe. With your team look over the *Yeast Protocol*, Figure 16.4, which outlines the procedures for preparing yeast cultures. Take a moment to discuss what you will be testing (and how you will do it), and write a five- to six-sentence summary of your experiment in your journal.

You will know that you have written a good summary if it

- *identifies the question that you will be trying to answer,*

- *describes how you will treat each of the four yeast cultures that you will establish, and*

- *describes what you will look for as an indication of the effect of the variable that you tested.*

Figure 16.4 **Yeast Protocol**

Yeast Protocol

Materials (per team of 4)

4 pairs of safety goggles
4 sterile, disposable dropping pipets
transparent tape
glass-marking pencil
container of sterile flat toothpicks
zip-type plastic bag labeled *Waste*
4 YED medium agar plates
sterile test tube containing 8 mL of yeast suspension

SAFETY: Put on your safety goggles.

1. Use a pipet to transfer 1 mL of the yeast suspension from the test tube directly onto the surface of each of four YED medium agar plates.

2. Replace the lid of the plate to maintain the sterile environment, and tilt and rotate the plate to spread the yeast cells over the surface of the agar.

   If there are places that the liquid did not cover, use the flat end of a sterile toothpick to gently smear the suspension over those areas. Be careful not to puncture or tear the surface of the agar. Discard the toothpick in the plastic bag labeled Waste.

3. Allow the agar to absorb the liquid until the liquid disappears (about 10 minutes).

4. Secure the lid to the bottom of the petri dish by placing a small piece of tape on each side of the dish. Do not extend the tape over the top of the plate.

5. Wash your hands thoroughly.

---

6. After the teacher approves your experimental design, begin your experiment.

   Be sure to mark your cultures appropriately so you will know how each was treated.

7. After you have treated the yeast cultures according to your experimental design, incubate the cultures in the dark for three days at room temperature.

8. Examine your cultures each day to determine the effect of the variable that you tested. Record your data in a table in your journal.

9. On the third day, discuss your observations with the other members of your team. Write one or two paragraphs in your journal that interpret the results of your experiment.

10. Choose one member of your team to post your team's data table so that other students in the class can examine it.

Part B Using a Simulation to Examine Complex Relationships

Scientists often use computer programs that model the relationships that exist among organisms and between organisms and the physical components in an ecosystem. Working with such simulations can help scientists better understand these relationships and also can help us predict how such systems will behave under specific conditions.

The computer simulation that you will use in this part of the activity (and in the activity *The Ozone Layer: A Disappearing Act?*) will give you an opportunity to see how the components of a complex system can interact. Specifically, the simulation will allow you to explore a set of relationships and interactions among the earth's ozone layer, human population growth, and human behavior. You may find that some of these interactions surprise you. As you use the simulation, remember that its purpose is to allow you to explore a complex system; it will *not* enable you to make precise predictions about the earth's future. Research can provide scientists with better information about the relationships among atmospheric ozone, human population growth, and human behavior, but no simulation can describe such a complex system completely. Nevertheless, working with the software should help you see some of the challenges involved in trying to understand and describe the behavior of complex systems.

Process and Procedures

1. View the videodisc segment *Global Ozone*, take notes, and reflect on the most important points that the segment makes about ozone and the earth's atmosphere. Share your ideas with the class.

Figure 16.5 illustrates the chemical changes that occur between oxygen and ozone.

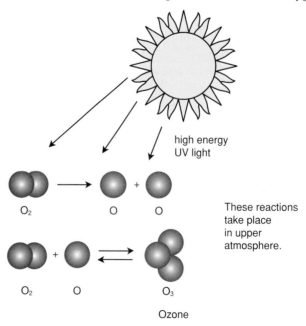

Figure 16.5 Natural ozone production Normal chemical reactions between oxygen atoms in the upper atmosphere produce ozone.

high energy UV light

O₂   O   O

O₂   +   O   O₃

These reactions take place in upper atmosphere.

Ozone

2. Form work teams around the available computers and access Level 1 of the computer simulation *Ozone: The Game.*

- Double click on the icon labeled *Ozone.*

- Click on the arrow to move from the opening screen to the Main Menu.

    *Think about the question that appears on the screen. In the future, you can press **Command-M** (Macintosh) or **Control-M** (IBM) to skip the long fade between the opening screens of the program.*

- Click on *Level 1 - A First Look at Ozone and Population.*

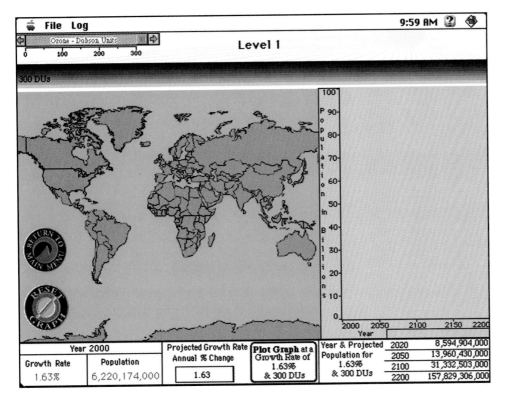

Figure 16.6 **Level 1 of** *Ozone: The Game*

3. Take a moment to study the screen (refer to Figure 16.6) and to discuss with the other members of your team what you see. Then use the simulation to answer the questions embedded in Steps a–e below. Record your answers in your journal.

a. Click on the button labeled **Plot Graph at a Growth Rate of 1.63% and 300 DUs (Dobson Units).** (The Dobson Unit (DU) is a measure of ozone in the stratosphere. The average value for the entire earth is about 300 DUs.) The graph that appears shows the increase in population that would be expected during the next 200 years, assuming that the current growth rate and thickness of the ozone layer do not change. Under these conditions, what would the world's population be in the year 2200?

Notice that the table below the graph displays the estimated populations that would exist under the specified conditions. Write the ozone thickness and population value for the year 2200 in your journal.

b. Click on the **scroll bar** in the upper left corner of the screen or drag the **indicator** to a new position on the bar to change the thickness of the ozone layer.

The current thickness setting is displayed in the shaded area just under the scroll bar, and this value changes as you move the indicator. When Level 1 opens, the value listed on the scroll bar is 300 DUs.

*Notice that the ozone thickness in the button labeled **Plot Graph** also changes as you change the position of the indicator on the ozone scroll bar.*

   c. Click on the button labeled **Plot Graph** to generate a graph of the earth's population growth at the new ozone thickness. Record the corresponding ozone thickness and population values (for the year 2200) in your journal.

   d. Repeat Steps b and c for four additional thicknesses of the ozone layer and record the corresponding ozone thickness and population value for each entry in your journal.

     *If you wish to clear all previously graphed lines from the screen display, click on the button labeled **Reset Graph**.*

   e. When you are finished working in Level 1, return to the Main Menu by clicking on the button labeled **Return to Main Menu**.

4. Use the values that you have collected to create a graph in your journal that illustrates the relationship between the thickness of the ozone layer and the world's population in the year 2200.

Make a line graph with the thickness of the ozone layer on the x-axis and the size of the world's population on the y-axis.

5. Answer the following questions in your journal:

   a. What does your graph suggest about the relationship between the thickness of the ozone layer and the earth's population?

   b. Suggest an explanation for your observations in Step 5a.

6. Access Level 2 of the computer simulation by clicking on *Level 2 - Refrigerators and Population Growth* on the Main Menu.

Level 2 allows you to manipulate two variables, growth rate and percentage of households with refrigerators. Explore the effects of each of these variables independently. That is, first consider the effect on population of changes in the average growth rate of the world's population. Then consider the effect on population of changes in the percentage of households with refrigerators.

Chlorofluorocarbons (CFCs) are a group of gases composed of carbon, chlorine, and fluorine that have been shown to participate in ozone-destroying reactions. They commonly are used in household refrigerators, freezers, and air conditioners, including those in cars, and they are widely used as solvents in the electronics industry.

7. Take a moment to study the screen and to discuss with the other members of your team what you see. For example, consider how this screen differs from the screen you saw in Level 1. Then use the simulation to answer the questions about growth rate and world population size embedded in Steps a–h. Record your answers in your journal.

Typical Chlorofluorocarbons

a.

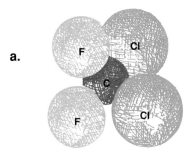

dichlorodifluoromethane
(CF_2Cl_2)

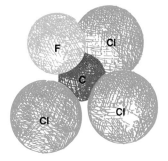

trichlorofluoromethane
($CFCl_3$)

CFC reactions that destroy ozone

b.

1. $CF_2Cl_2 + \text{Solar energy} \longrightarrow CF_2Cl^* + Cl^*$

2. $Cl^* + O_3 \longrightarrow ClO^* + O_2$

3. $ClO^* + O \longrightarrow O_2 + Cl^*$

*represents a very reactive unpaired electron called a radical electron

Figure 16.7 a. Typical chlorofluorocarbons (CFCs) b. CFC reactions that destroy ozone What do you think happens to the chlorine radical (Cl^*) after Reaction 3 is complete?

a. Click on the button labeled **Plot Graph** to generate a graph at the baseline settings of 1.63 percent growth rate and 10 %HENR.

   %HENR is an abbreviation used in this simulation; it stands for Percentage of Households in Emerging Nations with Refrigerators. This is an important variable that you will manipulate later. (The simulation uses the term "emerging" to refer to those nations that are not highly industrialized. This term is used in an economic sense and is not meant to define cultural, religious, or political characteristics of a nation.)

b. Change the growth rate of the world's population by typing in a new rate in the blue box labeled **Projected Growth Rate/Annual % Change** (located to the left of the **Plot Graph** button).

   The simulation is programmed to accept growth rates only in the range 1.0 percent to 2.0 percent.

   *Notice that the growth rate displayed in the **Plot Graph** button changes automatically as you change the growth rate displayed in the **Projected Growth Rate** box.*

c. Plot a graph that shows the increase in population that would be expected during the next 200 years if this new rate were the growth rate. Under these conditions, what would the world's population be expected to reach in the year 2200? Record the corresponding growth rate and population value in your journal.

   Leave the percentage of households with refrigerators (%HENR) at 10 percent as you investigate this value.

d. Investigate the relationship between population growth rate and population size by setting four additional growth rates and recording the population predicted in the year 2200 for each.

   Your growth rates should include values both above and below the current growth rate.

e. Use the values that you have collected to create in your journal a graph that illustrates the relationship between the growth rate of the world's population and the total number of people on earth in the year 2200.

You will have a line graph with the growth rate on the x-axis and the size of the world's population on the y-axis.

f. What does your graph suggest about the relationship between growth rate and population size? Record your answer in your journal.

g. Click on the button labeled **Reset Graph** to clear all previously graphed lines from the screen display.

h. Set the **Projected Growth Rate** back to 1.63 percent.

8. Use this simulation to answer the questions embedded in Steps a–e below about the relationship between the number of CFC-containing refrigerators and the thickness of the ozone layer. Record your answers in your journal.

a. Record in your journal the baseline settings for DUs and %HENR.

The ozone thickness projected for the year 2020 is displayed in the shaded area under the scroll bar.

The current %HENR setting is displayed to the right of the scroll bar. (All of these refrigerators are assumed to be CFC-containing refrigerators.)

b. Click on the arrows on either end of the **scroll bar** or drag the **indicator** to a new position on the bar to change the percentage of households with refrigerators. Under this condition, what would be the thickness of the ozone layer?

The projected thickness changes as you change the percentage of households that have refrigerators.

c. Investigate the relationship between households with refrigerators (%HENR) and ozone thickness by setting the simulation at four additional percentages of households with refrigerators. Record the percentages and associated thicknesses of the ozone layer in your journal.

Leave the growth rate at 1.63 percent as you investigate these values.

d. Use the values that you have collected to create in your journal a graph that illustrates the relationship between the percentage of households with CFC-containing refrigerators and the thickness of the ozone layer.

Make a line graph with the percentage of households with refrigerators on the x-axis and the thickness of the ozone layer on the y-axis.

e. What does your graph suggest about the relationship between the percentage of households with refrigerators and the thickness of the ozone layer? Record your answer in your journal.

9. Think about the relationships between ozone and population in Level 1 and between ozone and %HENR in Level 2. Based on these relationships, what prediction can you make about the relationship between the percentage of refrigerators and the population size at any constant growth rate? Use the simulation to test your prediction, and record the results in your journal.

Analysis

In order to explore the interactions between sunlight and life, you first examined a simple relationship (yeast and ultraviolet light), then a more complex relationship involving ozone, CFCs, ultraviolet light, and human population. Questions 1–4 refer back to the yeast experiments, whereas Question 6 examines ozone dynamics. Question 5 bridges the two parts of this activity.

Discuss the following questions with the other members of your team, and record your answers in your journal. Be prepared to participate in a class discussion.

Be sure to support your comments with specific evidence drawn from the data that your class has collected.

1. What do the collective results from all of the teams suggest about the effect of sunlight on this strain of yeast?

2. What do your data suggest about how certain variables, such as the length of exposure, time of day, and the angle of light, affect the yeast?

3. What do your data indicate about the effectiveness of various blocking agents or techniques?

4. There are significant differences between the blocking effects of various types of sunglasses. What information does this give you about the component of sunlight that is responsible for its effect on these yeast?

5. Record your answers to the following questions in your journal.

   Use the essay on page E252, Ultraviolet Light and the Ozone Layer, *as a resource.*

   a. The yeast strain that you used in this experiment has mutations that have inactivated three DNA repair genes. Use this information as well as the information provided in the essay to explain why the yeast responded to sunlight as it did.

   b. The relationship between ultraviolet light and the survival and growth of this strain of yeast appears to be relatively simple. Ordinary yeast cells (cells without the mutations mentioned above) repair their DNA in the same way that your skin cells do. Does this mean that the relationship between ultraviolet light and human survival is equally as straightforward as the relationship between ultraviolet light and yeast survival? Use the information from the essay and your experience to explain your answer.

   Try to list some of the important ways that humans are similar to and different from these yeast. Consider such factors as short-term and long-term survival in relation to exposure to ultraviolet light; the presence or absence of protective agents (internal as well as external); and the types of cells typically damaged by ultraviolet light (reproductive or nonreproductive).

6. Answer the following questions individually in your journal, and then discuss them with your class.

   a. According to the simulation, what relationship exists between human choices and ozone?

b. Review your answer to Question 5b, and then think about your answer to Question 6a. Do you think that this relationship is as straightforward as the simulation implies? Provide specific reasons for your answer.

c. Now that you understand the relationship between ozone and population growth, predict a course of action that will avoid the destruction of the ozone layer.

WHERE DO WE GO FROM HERE?

As you sit on the pier looking out at the ocean, you notice a trash barge hauling a load of garbage. From this distance it appears small, almost toylike. If the landfills are overflowing and the oceans are so huge, you think to yourself, why can't we just dump all our trash in the ocean? After all, what harm could a little trash in the water do? As you think more about this, you realize how difficult it would be to determine all the possible effects of different types of trash in an ocean ecosystem. Ultimately, you decide that any harm probably would be minimal.

In Part B of the activity *The Sun and Life (or Death?)*, you explored a complex interaction among refrigerators, the ozone layer, and human population. In this activity you will examine a complex ecological system that illustrates the importance and difficulty in understanding the long-term consequences of our actions in the ecosystem. Then you will reexamine parts of this ecological system with techniques that scientists use to analyze complex situations.

Process and Procedures
Part A A History of Controversy

Use the essays listed in each step to understand some of the history of the development of chemical pesticides.

1. Imagine that you are a farmer in the United States in 1910, and answer the following questions in your journal. Work individually.

   The historical summary in the essay The Farmers Are Saved! *(page E255) will give you an appreciation for the difficulties of farming in the early 1900s.*

   a. Describe your impressions of the "new" chemical method for controlling insects.

   b. Based on the information in the essay, would you use the cultural control method or calcium arsenate to rid your fields of insects? Explain your answer.

      Remember, for this question you are assuming the role of a farmer in the early 1900s.

2. Imagine that you are listening to the evening news as it is broadcast over the radio in 1946. Read or listen to the transcript of this broadcast in the essay *Breakthrough Insecticide Benefits the Boys Overseas and Farmers Alike* (page E256).

3. Now assume the perspective of a United States citizen in 1946, and work individually to answer the following questions in your journal.

   a. Why are the entomologists so excited? Explain.

   b. Based on the information in the essay, should public health officials approve the use of DDT to kill moths, mosquitoes, and houseflies in neighborhoods? Be sure to justify your answer.

4. Imagine that you are reading the local newspaper in 1958, and you come across the editorial printed in the essay *Conservationists in Hysterics over DDT* (page E257). Assume the perspective of a reader of this editorial in 1958, and work individually to complete the following tasks in your journal.

   a. In two sentences, comment on the reaction of the author of this editorial to the conservationists who opposed DDT.

   b. Based on the information you have read so far, would you support the use of DDT in the neighborhood or impose a ban on DDT? Explain.

5. Imagine that you are reading the newspaper in 1972 as you read the essay *DDT Banned!* (page E257) to yourself.

6. Answer the following questions in your journal, and be prepared to offer your answers in a class discussion.

   a. Farmers and public health officials enthusiastically greeted the introduction of DDT, but eventually people viewed this pesticide as the cause of tremendous problems. Why did the perception of DDT change?

   b. Who was responsible for the damage caused by DDT? Explain your response.

   c. Considering the information available at the time when people were making important decisions concerning the use of DDT, could the negative effects of DDT have been predicted? Could the negative effects of DDT have been prevented?

Part B A Systems Look at DDT

The history of DDT and its effects on the environment illustrate the changing nature of science as well as the complexity of interactions in ecosystems. Biology in the nineteenth and early twentieth centuries was not as strongly grounded in experimentation as it is today. Earlier biologists approached their work from a natural history perspective—that is, they tried to understand the natural world by making careful observations of living organisms in nature. By 1942 experimental biologists were just beginning to recognize the web of interdependence that exists between the biotic and abiotic world, but the idea that one new factor could dramatically change many others was not well established. Today scientists sometimes use methods known as systems analysis to try to extend their understanding of interdependence. With these methods, scientists can begin to model complex interactions, such as the effects of pesticides in the environment, and make predictions about potential consequences.

1. Make a table in your journal that lists and summarizes the steps of a systems analysis.

   Use the information in the essay Systems Analysis *(page E258) to help you create your table.*

2. Work with your teammates to conduct a retrospective ("looking backward") systems analysis of DDT in the environment.

   a. Decide with your teammates whether you wish to structure your analysis as

      • a written report to be handed in to your teacher or

      • an oral report to be presented to the class.

   b. Use the *Guidelines for a Systems Analysis of DDT* (Figure 16.8) to help you complete your report.

      You may find the information in Data about DDT *(Figure 16.9) useful, and you may use outside resources. Some suggested resources include*

      • Silent Spring *(1962), by Rachel Carson*

      • Since Silent Spring *(1970), by Frank Graham, Jr.*

      • DDT: Scientists, Citizens, and Public Policy *(1981), by Thomas R. Dunlap*

      • The Recurring Silent Spring *(1989), by H. Patricia Hynes*

3. With your class, discuss each team's explanation of the interactions that were disrupted by DDT and the effect this had on each team's limited system. When all of the individual explanations have been discussed, use a *combination* of all of the team's explanations to develop one consistent class explanation that summarizes the overall effects of DDT on the interactions in the ecosystem.

Analysis

Complete the following tasks in your journal:

1. In your systems analysis, you had an advantage over the scientists who were discussed in the essays because you had access to information that has been collected through many years. What information did your class use in this retrospective analysis of DDT in the environment that was not available at the time DDT was introduced? How did the appearance of negative effects slowly, through a long time period, affect the attitudes of scientists and the public toward DDT?

2. Briefly summarize the difference between an explanation based on a single, limited analysis and one based on a combination of related analyses. Explain how this difference should affect the way you interpret scientific studies that are reported in the news.

3. Write a general statement that assesses our ability to predict the consequences of introducing new, manmade components into the environment. Provide reasons that support your statement.

Figure 16.8 Guidelines for a Systems Analysis of DDT

Guidelines for a Systems Analysis of DDT

You will be able to make a complete report when you can do each of the following:

1. Identify at least 20 components of the DDT/environment system.

    What 20 biotic and abiotic components are important in the interactions of this complex DDT/environment system?

2. Identify a smaller but related set of your 20 components, and describe the question that the analysis would investigate.

    Clearly explain the specific ecological interactions that your simulated analysis would address. For instance, you might choose six or seven living organisms from your original 20 components and ask, Did DDT enter the food chain of any of these organisms and, if so, how did this happen?

3. Identify which components you intentionally did *not* include in Step 2, and explain why these were omitted.

    For instance, if you are analyzing only biotic components, you might choose to ignore the effect of rainfall because water is an abiotic component.

4. Describe how the components in your structured system interact under normal circumstances.

    For example, the animals in a pond ecosystem derive certain nutrients from the food chain in which they exist. If your structured system involves some of these animals, the specific feeding pattern is part of the system's initial behavior. You may need to consult some outside resources to adequately describe the behavior of the structure you chose. It is important that you have as much information as possible and that you consider all possible interactions.

5. Describe the overall effect that widespread use of DDT had on the behavior of your structured system. In other words, in what ways did DDT *change* the initial behavior of your system?

6. Explain the specific interactions in your structured system that were disrupted by DDT. In other words, *where* and *how* did DDT change the interaction of the components in your system and thus alter the system's behavior?

Data about DDT

DDT → decomposition → DDE + HCl

Figure 16.9 Data about DDT

Figure 16.9A DDT breaks down to DDE, a very stable form of the pesticide that breaks down very slowly in the environment. High levels of DDT and DDE have been found in the fatty tissues of many organisms. Despite the ban on DDT in 1972, a study in 1986 found measurable levels of DDE in every sample of human breast milk that was tested. This is an indication of how long-lasting this chemical is in the environment.

Figure 16.9B **DDT production after World War II**

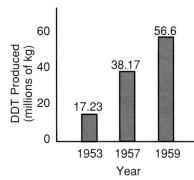

DDT Production after World War II

Figure 16.9C **Pesticide production in the United States**
After the introduction of DDT in 1942, the use of chemical pesticides became very common in American farming; pesticide production increased steadily during the next 30 years. The decline in production is due largely to an increase in the potency of chemicals produced after 1975.

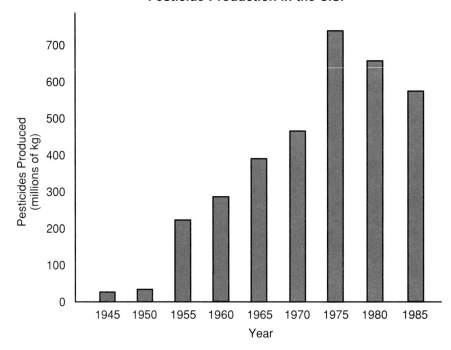

Pesticide Production in the U.S.

Source: Data from *Ecology, Economics, Ethics: The Broken Circle* by F.H. Bormann and S.R. Kellert (editors). Copyright 1991 by Yale University Press.

Figure 16.9D **Annual U.S. crop losses due to insects** Crop losses due to insects have *increased* despite the widespread use of pesticides. What do you suppose contributed to the increased crop loss between 1942–1951 and 1951–1960?

Annual U.S. Crop Losses Due to Insects

Period	Percentage of Crops Lost to Insects
1904	9.8
1910–1935	10.5
1942–1951	7.1
1951–1960	12.9
1974	13.0
1986	13.0

Source: From *Ecology, Economics, Ethics: The Broken Circle* by F.H. Bormann and S.R. Kellert (editors). Copyright 1991 by Yale University Press. Reprinted with permission.

Pesticide Resistance

>500

Number of Species Resistant to One Pesticide

500	
400	
300	
200	
100	

1970 1990
Insects

1990
Fungi

1990
Weeds

Figure 16.9E Pesticide resistance Resistant pests can lead to greater health threats: by 1986, 50 of the 61 species of malaria-carrying mosquitoes were resistant to three common insecticides, including DDT. Based on the increased resistance of insects to pesticides, what do you think happened to the number of resistant fungi and weeds between 1970 and 1990? How might the number of resistant organisms in 1970 compare with the number in 1920?

Concentration of Toxins in a Food Chain

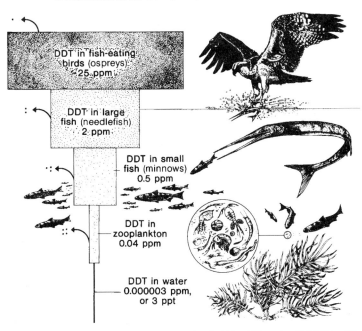

DDT in fish-eating birds (ospreys) 25 ppm

DDT in large fish (needlefish) 2 ppm

DDT in small fish (minnows) 0.5 ppm

DDT in zooplankton 0.04 ppm

DDT in water 0.000003 ppm, or 3 ppt

Figure 16.9F Concentration of toxins in a food chain The concentration of DDT becomes amplified as it moves up the food chain.

Source: From *Living in the Environment,* third edition, by G. Tyler Miller, Jr., © 1982 by Wadsworth Publishing Company. Reprinted with permission of Wadsworth Publishing Company, Belmont, CA.

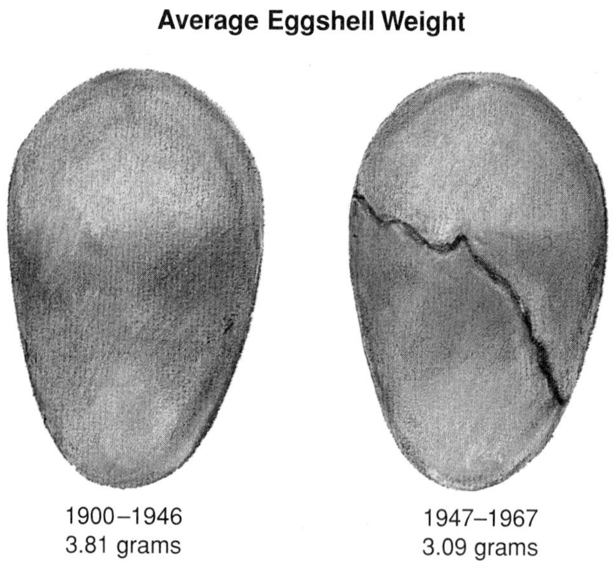

Figure 16.9G **Decrease in the mass of bird eggshells after introduction of DDT** Although the size of bird eggshells remained the same, the mass of the shells decreased by 19 percent; this weakened the eggs, and many birds died before hatching because their shells cracked. DDT inhibits carbonic anhydrase, an enzyme that controls the supply of calcium used to make eggshells.

Figure 16.9H **Pesticides that may cause cancer in humans** These estimates were made in 1987 by the National Academy of Sciences.

Average Eggshell Weight

1900–1946
3.81 grams

1947–1967
3.09 grams

Pesticides That May Cause Cancer in Humans

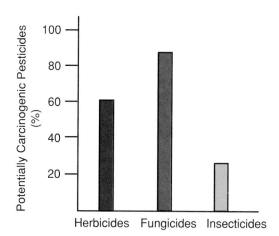

THE OZONE LAYER: A DISAPPEARING ACT?

In the activity *Where Do We Go from Here?*, you discovered how a single chemical, DDT, acted as an environmental toxin when it was introduced in large quantities into a complex ecosystem. You also conducted a systems analysis of the disrupted interactions in this ecosystem by using techniques that allow complex problems to be studied in small pieces that later are combined. Even though solutions to complex problems cannot always be found through a systems analysis, the identification of interacting components and the critical thinking necessary to conduct a systems analysis help scientists understand the level of complexity involved.

In this activity you will have the opportunity to use the ozone simulation to chart a course of world development that will not destroy the earth's ozone layer or population. This ozone simulation is quite simple compared with the ozone simulations that atmospheric scientists use, but even complex simulations are not able to predict the future with complete accuracy. Despite the drawbacks of these simulations and the fact that the data on the initial status of world ozone is incomplete, we must think about the threat to ozone and take steps to minimize the potential hazards. We may not have time to wait until we have all of the data because our window of opportunity to act effectively may be quite narrow.

Materials (per team of 4)

Macintosh computer or IBM-compatible computer with Windows™
Ozone: The Game computer simulation

Figure 16.10 *Ozone: The Game* computer simulation

Process and Procedures
Part A Setting the Limits

You and your classmates will assume the role of atmospheric scientists interested in ozone research. Your research already has suggested that a diminishing ozone layer might adversely affect life on earth. A world trade group representing the manufacturers of ozone-friendly appliances has asked the organization to which you belong, the Atmospheric Scientists and Ecologists Organization (ASEO), to use its expertise to recommend an upper limit on the number of ozone-damaging (CFC-containing) refrigerators that can be produced without causing serious harm to the ozone layer. ASEO agrees to provide a recommendation because the member scientists believe that their work will contribute meaningfully to an important public policy decision.

1. With your teammates return to Part B of the activity *The Sun and Life (or Death?)*, and make certain that you understand the following relationships, which you studied in *Ozone: The Game*, Levels 1 and 2.

   - the relationship between ozone and population

   - the relationship between growth rate and population

   - the relationship between refrigerators and ozone

2. Study the population and consumer products statistics in Figure 16.11, and copy the table into your journal. Discuss these data with your teammates.

Households with Refrigerators in the Year 2000

Figure 16.11 Households with Refrigerators in the Year 2000

	Number of Households (in millions)	Number of Refrigerators (in millions)	Percentage of Households with Refrigerators
Developed Nations	236	236	
Emerging Nations	1004	100	

3. Use the data in Figure 16.11 to calculate the percentages of households in the developed nations and in the emerging nations that will own a refrigerator in the year 2000. Record these percentages in the last column of Figure 16.11 in your journal.

   In this portion of the simulation, assume that the refrigerators are chlorofluorocarbon-containing (CFC) refrigerators, which damage the ozone layer. Because ozone-friendly refrigerators are currently quite expensive, very few refrigerators in households are ozone-friendly.

4. Discuss the following questions with your teammates, and record your answers in your journal.

   a. Do you think the emerging nations would be satisfied with this distribution of refrigerators? Explain why or why not.

   b. Should the percentage of households in emerging nations with refrigerators change? If so, suggest a new percentage to which it should increase or decrease, and record this suggested percentage.

5. Open the computer simulation *Ozone: The Game* to Level 2.

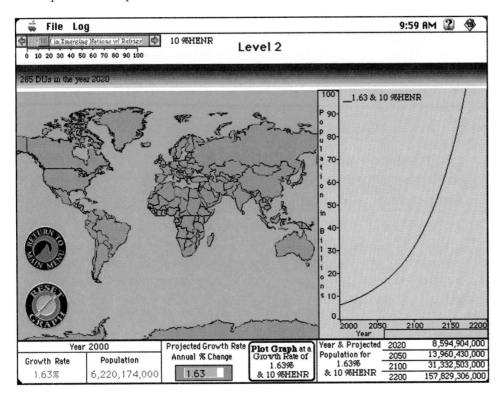

Figure 16.12 **A population growth graph from level 2 of** *Ozone: The Game* **simulation**

6. Plot a population graph using the following default values:

   Growth Rate = 1.63 percent

   Percentage of Households in Emerging Nations with Refrigerators (%HENR) = 10 percent

   Do not change any numbers in this step. The default values will be present when you open Level 2.

7. Copy the data table in Figure 16.13 into your journal, and fill in the values for population and ozone levels. Leave room for six to eight new rows.

Figure 16.13 **Tracking the relationship between ozone and population**

8. Change the %HENR on the scroll bar to match the new %HENR that you suggested in Step 4b. Let the computer plot the graph, then record your new data in the table.

*Use the **Plot Graph** button after you change the default values.*

9. With your teammates analyze the effect of this new %HENR, and answer the following question in your journal:

What effect does the new %HENR that you entered have on the ozone layer? on population?

10. Now consider the effect that changing the growth rate will have on the number of refrigerators that will be needed in the future.

 • The current world growth rate of 1.63 percent may increase or decrease in the future. Enter a growth rate other than 1.63 percent, and have the computer plot the graph to see if the new growth rate, combined with your new %HENR, will have an effect on ozone.

 You may want to try some values of growth rate that are higher than 1.63 percent and some values that are lower.

 • Record in your table the new values for growth rate, ozone, and population in the year 2020.

11. Run the simulation several times using different combinations of growth rate and %HENR until you find a combination that maximizes the %HENR but still predicts an acceptable level of ozone and population in the year 2020.

Each team's definition of acceptable may be different. For instance, your team might decide that an acceptable level of ozone and population allows for only a slight increase in %HENR. Another team might decide that a large increase in %HENR is reasonable, even if that means there will be a greater effect on population growth and ozone.

12. Participate in a class discussion to arrive at consensus levels of growth rate and %HENR. Record the consensus percentages in your journal.

These numbers will be ASEO's recommendation to the trade group. They will form the basis of a trade agreement that sets limits on the maximum number of CFC-containing refrigerators that households in developed and emerging nations can own in the year 2020.

Part B There's No Time to Waste!

From your experience with the simulation *Ozone: The Game* in Part A, you realize that your recommendations to the trade group are a compromise solution. These recommendations do not offer a perfect solution to the problems of ozone depletion caused by CFC-containing refrigerators and the desire for more refrigerators by those living in the emerging nations of the world. Even though your recommendations represent a compromise, you can work within their constraints to maximize the common good by controlling a number of variables that influence the complex system of ozone, population, and ownership of refrigerators.

In this part of the activity, you will try to maximize the common good by using the ozone simulation to manipulate the number of CFC-containing and ozone-friendly refrigerators purchased by developed and emerging nations. In the process, you will score points (a *World Planning Merit Score*) that measure your success in providing as many refrigerators to emerging nations as possible and in preserving ozone thickness (and hence, population).

Process and Procedures

1. Read this short definition of the common good as it is applied to this simulation. Discuss with your teammates the reasons why developed and emerging nations might agree to the two parts of this definition.

Common Good

Developed nations and emerging nations each have an interest in: (1) improving, or at least maintaining, the world's standard of living and (2) protecting the ozone layer. In this simulation the greatest common good (and therefore the highest scores) will result when a strategy of both refrigerator production and purchase (both CFC-containing and ozone-friendly) results in high ozone levels and a high percentage of households in emerging nations with refrigerators.

2. Participate in a class discussion to identify factors that might influence the common good.

3. Open the simulation *Ozone: The Game* to Level 3, and follow the directions in the red box on your screen.

   You must enter the growth rate and %HENR recommendations of the class from Part A, Step 12 before you can enter Level 3.

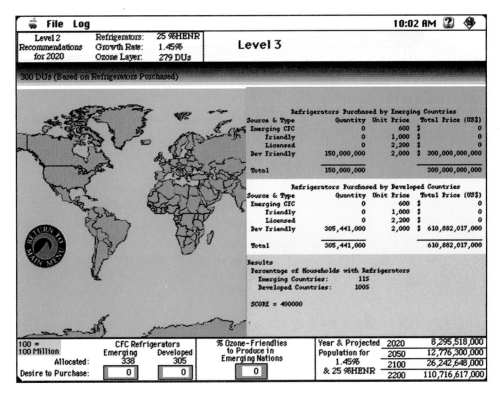

Figure 16.14 **Level 3 screen**

4. Read the letter from the Atmospheric Scientists and Ecologists Organization (ASEO).

ATMOSPHERIC SCIENTISTS AND
ECOLOGISTS ORGANIZATION

An Open Letter to World Leaders:

Thank you for accepting the recommendations in the International Agreement for Preserving the Ozone Layer. By signing this agreement, you agree to voluntarily limit the number of chlorofluorocarbon-containing (CFC) refrigerators that you will allow in your nation. Because the Atmospheric Scientists and Ecologists Organization (ASEO) is solely a scientific organization, we have no method to police this agreement. We are counting on your good faith to purchase only the number of CFC-containing refrigerators that we have allocated.

Your citizens may purchase as many ozone-friendly refrigerators as they can afford. We understand that this may be a burden to those of you in emerging nations because ozone-friendly refrigerators cost nearly twice as much as CFC-containing refrigerators. Currently, a considerable amount of research is being conducted throughout the world to try to lower the cost of ozone-friendly refrigerators, and we project that manufacturers will offer a low-cost alternative in several years. Until then, we appreciate your cooperation.

Notice that the price of refrigerators varies by type of unit and place of manufacture. Generally, refrigerators produced in emerging nations are less expensive than refrigerators produced in developed nations because labor costs are lower. Due to international tariffs and licensing agreements, however, emerging nations can produce only a limited number of low-cost ozone-friendly refrigerators. If these nations want to produce more, these additional refrigerators (known as licensed ozone-friendly refrigerators) will be significantly more expensive.

Because the agreements and calculations are very complex, we have provided a computer program to help you determine the best method to produce and purchase CFC-containing and ozone-friendly refrigerators. The program also monitors your success at maximizing the common good by assigning points based on the level of ozone protection and the percentage of households that have refrigerators.

The prices of refrigerators (in U.S. dollars) are as follows:

	Cost and Place of Production	
Type of Refrigerator	**Emerging Nations**	**Developed Nations**
Ozone-Friendly Refrigerators	$1000.00	$2000.00
CFC-Containing Refrigerators	$600.00	not applicable*
Licensed Ozone-Friendly Refrigerators	$2200.00	

*The agreement forbids developed nations from producing CFC-containing refrigerators.

We encourage cooperation among developed and emerging nations to reduce the total number of CFC-containing refrigerators in the world. Thank you again for your cooperation.

Sincerely,

Clarissa F. Carbonetti

Clarissa F. Carbonetti
President, Atmospheric Scientists
and Ecologists Organization

Oliver O. Radicalitz

Oliver O. Radicalitz
President, World Manufacturers of
Ozone-Friendly Appliances

5. Try to maximize the common good by entering

   a. the number of CFC-containing refrigerators that you want to purchase for both developed and emerging nations.

   b. the percentage of ozone-friendly refrigerators that you would like to produce in emerging nations.

   Enter these numbers in the blue boxes at the bottom of the screen. The worksheet on the right side of the screen will temporarily record the results of your entries.

6. Use the pull down menu labeled **Log** to save the results of your entries. Your changes will not be saved unless you use this feature.

   a. Select **Log a Data Set** to save the results of your entry. The data will not be visible at this time, but the computer will save the data with a label that identifies the level of the program, a data set number, and the recommendations that you entered.

   b. To see your data set, select **Show Log** from the pull down menu. To save time, you may choose **Log a Data Set & Show Log** to accomplish both steps at once. To close the log, select **Hide Log** from the pull down menu.

   c. To erase the entire log, select **Reset Log**.

   d. To print your results (if your computer is connected to a printer), select **Print Log** from the **File** menu.

7. Repeat Step 5 as often as necessary to maximize the common good and improve your score.

   A score above 700,000 is considered very good.

   Use the on-screen log and the permanent log to track the effects of each of your changes. Use these data to help you make efficient choices as you try to maximize the common good.

8. When you are satisfied with your strategy to maximize the common good, participate in a class discussion about effective strategies.

Analysis

Complete the following tasks in your journal:

1. Identify parameters that might influence ozone and population that were not included in the simulation, and explain how these parameters might affect the system.

2. The simulation used in this activity is based on certain assumptions for the purpose of simplicity; it cannot be expected to predict the future with any reliability. Advanced scientific computer simulations also contain simplifying assumptions, but scientists and policymakers still use them to guide their thinking about the future. Explain the advantages and disadvantages of using a well-reasoned but incomplete understanding of a complex situation to guide policymaking.

3. Use the guidelines from the essay *Ethical Analysis* in Chapter 6 (page E87), and write a one- to two-page essay that addresses one of the following questions:

- Are developed nations morally obligated to help emerging nations deal with the unequal distribution of wealth that exists in the world? Why or why not?

- Do all humans have an ethical responsibility to follow ozone protection guidelines? Why or why not?

- Are there ethical reasons to limit human consumption of energy and material goods? Why or why not?

- Are mandatory restrictions on human population growth ethical? Why or why not?

- Do all humans have the ethical responsibility to protect the natural habitats of other species from the effect of human population growth? Why or why not?

Further Challenges

This activity uses a number of assumptions in order to provide a workable model for the computer simulation. These assumptions are open to discussion and challenge. Your teacher may decide to use the copymaster *Making Assumptions* as a basis for discussing these assumptions.

THE LIMITS OF ABUNDANCE

Evaluate

In September 1994 representatives from more than 160 nations met in Cairo, Egypt, for an international conference on population and development. The agreement that these representatives reached—an agreement that is described in the Worldwatch Institute's *State of the World, 1995*[1] as one of the boldest international efforts that the United Nations has ever undertaken—set policies and initiated programs aimed at stabilizing the world's population by the year 2050.

In meetings before the conference, the delegates overwhelmingly rejected the notion that the earth's population should be allowed to rise at its current high-growth rate. This strong starting position set the stage for the ambitious plan that emerged from the conference talks, talks that were fueled by a rising sense of urgency. The concern the delegates felt was clear: unless human population growth can be slowed quickly, the delegates fear that it will push the collective level of human needs and demands beyond the earth's capacity to meet them.

In this activity you will examine some of the data that have led some world leaders to call for population control. As you examine the data, ask yourself what this information says about the earth's capacity to support human life and about the growing demands that are moving us toward (or beyond?) the limits of this capacity. Suppose *you* had to estimate the earth's carrying capacity for humans. What are the factors that you would take into consideration and how would you determine what that number of people might be?

---

[1]The Worldwatch Institute is an organization that monitors economic, demographic, and environmental indicators.

Process and Procedures

Your goal in this activity is to develop your own, individual answer to the *Analysis* question, What is the earth's carrying capacity for human life? The following procedural steps will help you develop your answer.

1. To help you establish a foundation on which you can construct an answer, briefly review the information provided in the essay *Interdependence Involves Limiting Factors and Carrying Capacity* in Chapter 15 (page E232) about carrying capacity and the information provided in the essay *Systems Analysis* in Chapter 16 (page E258) about complex systems and how they can be analyzed.

2. As you complete your review, record in your journal any definitions or ideas that might help you think about the earth's carrying capacity. Record as well any questions about your task that occur to you as you read.

3. Share and discuss the results of your review with the other members of your team. Some of the related issues that you also may wish to discuss include the following:

    a. You are not likely to be successful in your effort to determine the earth's carrying capacity unless you can agree on what carrying capacity for human life means. For example, how will you define carrying capacity? And, what does carrying capacity mean when the term is applied to *human* life?

    b. You also need to be able to describe the conditions you might expect to see if the earth's population exceeded its carrying capacity. What types of evidence would indicate to you that this had occurred? (That is, how would you recognize the fact that the earth's population had grown too large?)

    c. It is one thing to recognize that a population has exceeded the carrying capacity of its environment *after* it has done so. In that case the population size just before the population crash gives you an upper limit for the environment. It is quite another thing, however, to determine the maximum population that a given environment can support *before* the population reaches this size. How could you use data about past and present conditions on the earth to determine current trends and to estimate at what point those trends might reach a danger point? What might this say about the maximum population size the earth is likely to be able to support?

4. Work with your team to create a short position paper on the earth's carrying capacity that you will present to the rest of the class.

You will know that you have developed a good paper when it

- *lists all of the factors that you think would have to be considered to determine the earth's carrying capacity (that is, all of the components that you think make up this complex system),*

- *provides an answer to the question, What is the earth's carrying capacity for human life?,*

- *supports that answer with relevant data and logical arguments, and*

- *identifies additional data that you would need to improve your answer and identifies the areas of uncertainty that may prevent you from answering the question more precisely.*

The data provided in Environmental Indicators, *Figure 16.15, may help you develop your position paper.*

5. Participate in a general class discussion about the earth's carrying capacity. As part of this process, you will need to

- present your position paper to the rest of your class,

- listen carefully as other teams present their papers,

- take notes in your journal about relevant data or arguments that (a) provide additional support for your position, (b) could be challenged or questioned, or (c) might cause you to change your position, and

- after all of the groups have presented their papers, join the class in a discussion of the various positions represented.

Active participation and careful listening during this discussion will help you to prepare for the Analysis. You may wish, for example, to add to the data or the arguments presented, to challenge the interpretation or application of the data or the arguments presented, and/or to raise questions that you think the group has not considered. Use this opportunity to consider the question from every possible perspective and to clarify your thinking about each aspect of your answer.

Analysis

Write an essay in your journal that presents your individual position on the question, What is the earth's carrying capacity for human life? If your position has not changed from the one presented by your team, explain why data and arguments proposed by other teams were not convincing to you. If your position has changed, explain what data and arguments caused you to reevaluate your position and why. Your teacher will evaluate this journal entry as an indication of your understanding of the major issues in this unit.

Environmental Indicators

Population

The following figures contain data that describe various aspects of the world's population (both past and present). What trends do you see in the data? What questions do the data raise?

Population of the World, 1950–1993

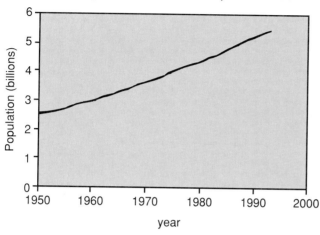

Source: U.S. Bureau of the Census, Center for International Research. November 1993.

Average World Population Growth Rate, 1950–1993

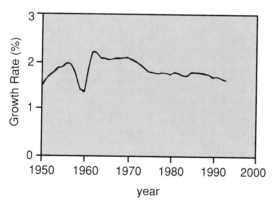

Source: U.S. Bureau of the Census, Center for International Research. November 1993.

Figure 16.15 **Environmental Indicators**

Figure 16.15A **Population of the World, 1950–1993** What does this graph indicate about the world's population?

Figure 16.15B **Average World Population Growth Rate, 1950–1993** In approximately what year was the average annual growth rate the highest? Does this mean that the total number of people added to the world's population that year was more than were added in 1993? Why or why not?

Figure 16.15C **Population Growth for Selected Nations, 1950–1990, with Projections to 2030**

Population Growth for Selected Nations, 1950–1990, with Projections to 2030*
(in millions of people)

Nation	Year			Increase (%)	
	1950	1990	2030	1950–1990	1990–2030
Slowly Growing Nations					
Japan	84	124	123	48	−1
Russia/USSR	114	148	162	30	9
United Kingdom	50	58	60	16	3
United States	152	250	345	64	38
Rapidly Growing Nations					
Bangladesh	46	114	243	148	113
Brazil	53	153	252	189	65
China	563	1134	1624	101	43
India	369	853	1443	131	69
Indonesia	83	189	307	128	62
Iran	16	57	183	256	221
Mexico	28	85	150	204	76
Nigeria	32	87	278	172	220

*The projection to 2030 assumes that the populations involved will continue to grow at their current rate.

Source: From United Nations (1993) *World Population Prospects, The 1992 Revision,* New York; Population Reference Bureau (1993) *1993 World Population Data Sheet,* Washington, DC.

Figure 16.15D **Life Expectancies at Birth in Selected Nations, 1950 and 1993**

Life Expectancies at Birth in Selected Nations, 1950 and 1993
(in millions of people)

Nation	1950	1993	Increase (%)
	(age in years)		
Slowly Growing Nations			
Japan	64	79	23
Russia/USSR	64	69	8
United Kingdom	69	76	10
United States	69	75	9
Rapidly Growing Nations			
Bangladesh	37	53	43
Brazil	51	67	31
China	41	70	71
India	39	59	51
Iran	46	62	35
Nigeria	37	53	43
Regions			
World	46	65	41
Developed Nations	66	74	12
Emerging Nations	41	63	54

Source: From United Nations (1993) *World Population Prospects, The 1992 Revision,* New York; Population Reference Bureau (1993) *1993 World Population Data Sheet,* Washington, DC.

Population in Developed and Emerging Geographic Regions, 1950–1990
(in millions of people)

Region	1950	1990	Increase 1950–1990 (actual)	Increase 1950–1990 (%)
Developed Regions				
Australasia	11	20	9	82
Europe	393	550	157	40
North America	166	275	109	66
Russia/USSR	180	288	108	60
Total	**750**	**1133**	**383**	**51**
Emerging Regions				
Africa	224	642	418	187
Asia	1375	3112	1737	126
Latin America	165	448	283	172
Total	**1764**	**4202**	**2438**	**138**

Source: United Nations (1990) *Demographic Yearbook 1988*, New York.

Figure 16.15E Population in Developed and Emerging Geographic Regions, 1950–1990

Food

The following figures contain data that describe various aspects of the world's food supply (both past and present). What trends do you see in the data? What questions do the data raise?

Worldwide Grain Production, 1950–1993

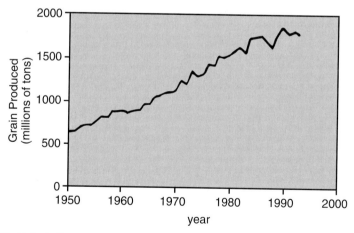

Source: World Grain Database, USDA. 1992.
World Grain Situation and Outlook, USDA. 1993.

Figure 16.15F Worldwide Grain Production, 1950–1993 How has the total amount of grain produced in the world per year changed since 1950?

Figure 16.15G Worldwide Grain Production Per Person, 1950–1993 How has the total amount of grain produced per year *per person* changed since 1950?

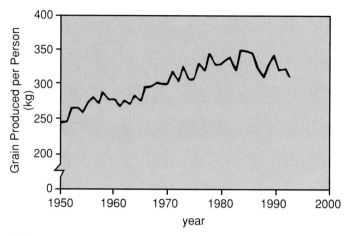

Worldwide Grain Production Per Person, 1950–1993

Source: *World Grain Database*, USDA. 1992.
World Grain Situation and Outlook, USDA. 1993.

Figure 16.15H Grain Carryover Stocks, Worldwide, 1963–1994 Carryover grain stocks are the amounts of grain that are in storage each year when the new harvest begins. What factors do you think influence this number?

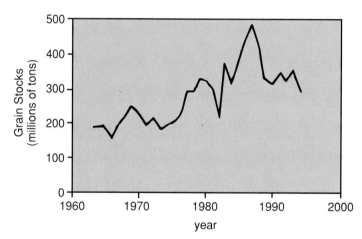

Grain Carryover Stocks, Worldwide, 1963–1994

Source: *World Grain Situation and Outlook*, USDA. November 1993.

Figure 16.15I Grainland Per Person, Worldwide, 1950–1990

Grainland Per Person, Worldwide, 1950–1990

Year	Grainland Per Person in Hectares*
1950	0.23
1960	0.21
1970	0.18
1980	0.16
1990	0.13

*One hectare is equal to an area of 10,000 square meters or approximately 2.47 acres.

Source: From U.S. Department of Agriculture.

Worldwide Use of Fertilizers and Irrigation, 1950–1990

Year	Total Fertilizer Used (millions of tons)	Fertilizer Used Per Person (kilograms)	Total Irrigated Area (millions of hectares*)	Irrigated Area Per Person (hectares per per 1000 people)
1950	14	5.5	data not available	data not available
1961	28	9.1	139	45.3
1970	66	17.8	169	45.5
1980	112	25.1	211	47.5
1990	143	27.0	240	45.2

*One hectare is equal to an area of 10,000 square meters or approximately 2.47 acres.

Source: From Worldwatch Institute (1994) *Vital Signs.* New York: W.W. Norton & Company, Inc.

Grain Yield Per Hectare, Worldwide, 1950–1992*

Year	Yield (tons)
1950	1.06
1960	1.28
1970	1.65
1980	2.00
1990	2.54
1991	2.46
1992	2.58

*One hectare is equal to an area of 10,000 square meters or approximately 2.47 acres.

Source: From USDA (November, 1993) *Production, Supply, and Demand View,* electronic database, Washington, DC; USDA (1992) *World Grain Database,* unpublished printout, Washington, DC.

Fish Harvest, Worldwide, 1950–1992

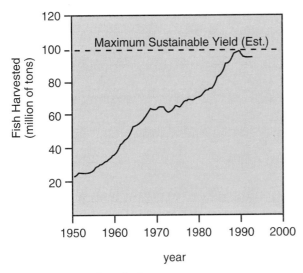

Source: U.N. Food and Agriculture Organization

Figure 16.15J Worldwide Use of Fertilizers and Irrigation, 1950–1990

Figure 16.15K Grain Yield Per Hectare, Worldwide, 1950–1992

Figure 16.15L Fish Harvest, Worldwide, 1950–1992
Between 1950 and 1992, the world fish catch increased steadily at a rate approximately three times the rate of human population growth during those years. The dotted line shows the maximum yield of fish that biologists at the United Nations Food and Agriculture Organization (FAO) estimate can be sustained through time, unless marine specialists find better ways to manage these resources. What does this limit suggest about further increases in the size of the world's annual fish catch?

Figure 16.15M **Change in
Selected Major Marine Fishing
Regions**

Change in Selected Major Marine Fishing Regions

Region	Peak Year	Peak Catch (millions of tons)	1992 Catch (millions of tons)	Change from Peak Year to 1992 (%)
Atlantic Ocean				
Northwest	1973	4.4	2.6	−41
West Central	1984	2.6	1.7	−35
East Central	1990	4.1	3.3	−20
Southwest	1987	2.4	2.1	−13
Pacific Ocean				
Northwest	1988	26.4	23.8	−10
West Central	1991	7.8	7.6	−3
East Central	1981	1.9	1.3	−32
Southwest	1991	1.1	1.1	0
Indian Ocean				
Western	still rising		3.7	not applicable
Eastern	still rising		3.3	not applicable

Source: From Fisheries Statistics Division, U.N. Food and Agriculture Organization (1994) (*FISHSTAT–P*, fisheries database), Rome.

Figure 16.15N **Grain Use and
Consumption of Livestock
Products in Selected Nations,
Per Person, 1990**

Grain Use and Consumption of Livestock Products in Selected Nations, Per Person, 1990

Nation	Grain Use (in millions of tons)	Consumption (in kilograms per person)						
		Beef	Pork	Poultry	Mutton	Milk	Cheese	Eggs
China	300	1	21	3	1	4	–	7
India	200	–	0.4	0.4	0.2	31	–	13
Italy	400	16	20	19	1	182	12	12
United States	800	42	28	44	1	271	12	16

Source: From U.N. Food and Agriculture Organization (1991) *Production Yearbook 1990*, Rome.

Land Use

The following figures contain data that describe various aspects of the world's use of its land resources. What trends do you see in the data? What questions do the data raise?

Population Size and Land Use, 1990, with Projections to 2010*

Category	1990 (in millions)	2010 (in millions)	Total Change (%)	Per Person Change (%)
Population	5290	7030	+33	—
Cropland (hectares)	1444	1516	+5	−19
Rangeland and Pasture (hectares)	3402	3540	+4	−22
Forests (hectares)	3413	3165	−7	−31

*The projection to 2010 assumes that current trends continue.

Source: From U.S. Bureau of the Census, Department of Commerce (November 2, 1993) *International Data Base,* unpublished printout; U.N. Food and Agriculture Organization (1992) *Production Yearbook 1991,* Rome; M. Perotti, Statistics Branch, Fisheries Department, U.N. Food and Agriculture Organization (November 3, 1993) private communication; U.N. Food and Agriculture Organization (1992 and 1993) *Forest Resources Assessment 1990,* Rome.

Figure 16.15O **Population Size and Land Use, 1990, with Projections to 2010**

Human-Induced Land Degradation, 1945–1991

Region	Total Amount of Land Damaged* (millions of hectares)	Degraded Area as Share of Total Vegetated Land (%)
Africa	494	22
Asia	746	20
Europe	220	23
North and Central America	158	8
South America	244	14
World	1965	17

*Numbers represent the total amount of land damaged by activities such as overgrazing, deforestation, and agricultural mismanagement.

Source: From Worldwatch Institute (1994) *Vital Signs.* New York: W.W. Norton & Company, Inc.

Figure 16.15P **Human-Induced Land Degradation, 1945–1991**

Energy

The following figures contain data that describe various aspects of the world's use of energy resources (both past and present). What trends do you see in the data? What questions do the data raise?

World Production of Nonrenewable Energy Sources, 1950–1990

Figure 16.15Q World
Production of Nonrenewable
Energy Sources, 1950-1990

Year	Total Crude Oil Production (millions of tons)	Total Natural Gas Production (millions of tons*)	Total Coal Production (millions of tons*)
1950	518	168	884
1960	1049	411	1271
1970	2281	920	1359
1980	2976	1357	1708
1990	2963	1865	2115

*Numbers for natural gas and coal production represented as millions of tons of equivalent oil.

Source: From API (1993) *Basic Petroleum Data Book,* Washington, DC; U.S. Department of Energy, *Annual Energy Review, 1992,* electronic database; and *United Nations World Energy Supplies and Energy Statistics Yearbook.*

World Production of Renewable Energy Sources, 1960–1990

Figure 16.15R World
Production of Renewable
Energy Sources, 1960-1990

Year	Electrical Generating Capacity of Nuclear Power Plants (billions of watts)	Electrical Generating Capacity of Photovoltaic Cells (millions of watts)	Electrical Generating Capacity of Wind Systems (millions of watts)
1960	0.8	data not available	data not available
1970	24.0	0.1	data not available
1980	135.0	6.5	4.0
1990	329.0	46.5	1789.0

Source: From Worldwatch Institute (1994) *Vital Signs.* New York: W.W. Norton & Company, Inc.

Miscellaneous Indicators

The following figures contain data that describe a variety of other environmental indicators (both past and present). What trends do you see in the data? What questions do the data raise?

Average Temperature and Atmospheric Concentration of Carbon Dioxide, Worldwide, 1950–1990

Year	Temperature (degrees Celsius)	Carbon Dioxide (parts per million)
1950	14.86	data not available
1960	14.98	316.8
1970	15.04	325.3
1980	15.28	338.5
1990	15.47	354.0

Source: From Worldwatch Institute (1994) *Vital Signs.* New York: W.W. Norton & Company, Inc.

Figure 16.15S Average Temperature and Atmospheric Concentration of Carbon Dioxide, Worldwide, 1950–1990

Carbon, Sulfur, and Nitrogen Emissions, Worldwide, 1950–1990
(in millions of tons)

Year	Carbon Emissions	Sulfur Emissions	Nitrogen Emissions
1950	1620	30.1	6.8
1960	2543	46.2	11.8
1970	4006	57.0	18.1
1980	5172	62.9	22.3
1990	5941	68.4	26.5

Source: From Worldwatch Institute (1994) *Vital Signs.* New York: W. W. Norton & Company, Inc.

Figure 16.15T Carbon, Sulfur, and Nitrogen Emissions, Worldwide, 1950–1990

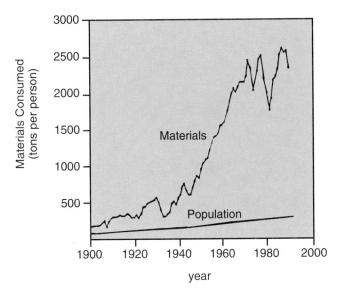

Materials Consumption and Population in the United States, 1900–1991

Source: U.S. Bureau of Mines

Figure 16.15U Materials Consumption and Population in the United States, 1900–1991 How has U.S. consumption of raw materials changed in relation to the growth in population since 1900? What implications does this have for emerging nations (that is, what do you think will happen to their demand for raw materials as these nations undergo industrial and economic development)?

Figure 16.15V **Number of Refugees, 1960–1993** How has the number of refugees changed in recent years? What explanations can you offer for this change?

Figure 16.15W **Demographic Indicators of Hunger, 1991**

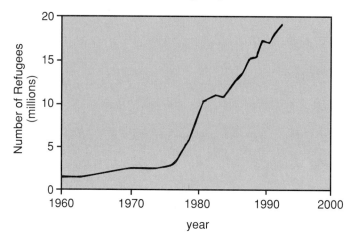

Number of Refugees, 1960–1993

Source: U.N. High Commissioner for Refugees

Demographic Indicators of Hunger, 1991

Region	Infant Mortality Per 1000 Live Births (actual)	Mortality for Children under 5 Per 1000 Children (actual)	Children under 5 Suffering from	
			Under-weight (%)	Wasting and/or Stunting (%)
Africa	102	180	26	47
Developed Nations (including the United States)	12	17	data not available	data not available
East Asia and the Pacific	35	42	26	39
Latin America	47	57	12	24
Middle East	55	83	14	40
South Asia	97	131	62	75

Source: From Bread for the World Institute, *Hunger 1994 Transforming the Politics of Hunger, Fourth Annual Report on the State of World Hunger,* Silver Spring, MD: The John D. Lucas Printing Company.

Selected Indicators of Health, Nutrition, and Welfare, 1988–1990

Region	Infants with Low Birthweight (%)	1-year-olds Fully Immunized (%)	Daily Calorie Supply (% of requirements)	Individuals without Access to Health Facilities (%)	Individuals without Access to Safe Water (%)
Africa	16	61	93	40	59
Developed Nations (including the United States)	6	80	133	data not available	data not available
East Asia and the Pacific	11	94	112	50	33
Latin America	11	79	114	14	19
Middle East	10	87	126	11	11
South Asia	34	86	99	22	29

Source: From Bread for the World Institute, *Hunger 1994 Transforming the Politics of Hunger, Fourth Annual Report on the State of World Hunger,* Silver Spring, MD: The John D. Lucas Printing Company.

Figure 16.15X **Selected Indicators of Health, Nutrition, and Welfare, 1988–1990**

> **Not to know is bad, but not to wish to know is worse.**
>
> African proverb

Evaluate
THINKING LIKE A BIOLOGIST

This final section in *BSCS Biology: A Human Approach* helps you evaluate what you have learned about biology and about the process of scientific thinking. This section focuses on the six unifying principles of biology that we used to organize this study of the biological sciences. There are several opportunities to express your understanding of these principles, including two scenarios based on actual events, a series of questions under the *Chapter Challenges* heading, and a portfolio that your teacher may assign as an additional activity or an alternate activity.

ACTIVITIES

Part A Recognizing Biology in Medicine

Part B Chapter Challenges

Alternate Building a Portfolio of Scientific Literacy

RECOGNIZING BIOLOGY IN MEDICINE

Part A

One of the organizing themes for this program has been that all of biology is united by six unifying principles; each unit in the program focused on one of these principles. Each principle illustrates a different aspect of biology, but remember that all of these principles act together in living systems. If you have been successful in learning the concepts associated with each principle, you now should be able to think about any biological topic in the light of these principles. Doing so will help you understand the world from a biological point of view.

Process and Procedures

1. Review the scoring rubric in Figure Ev.1 to help you understand what you can do to complete this activity successfully.

2. Identify each of the six unifying principles, and write a short statement in your journal describing what each one means.

    The unifying principles (Figure Ev.2) were the foundation for this course, and they are in the title of each unit of the course.

3. Review both scenarios that follow, and choose one that you will use to complete the Analysis.

4. In a few sentences, describe two or more interesting things that you learned from reading the scenario.

Analysis

Write an essay of at least five pages that describes how each of the six unifying principles is evident in the scenario, and describe how the principles overlap and act together in this situation.

Remember, all six unifying principles are represented either directly or indirectly whenever you study living systems.

Scenario 1 Iguanas and Aspirin, Shots and Antibiotics

"I hate going to the doctor. It's probably just a cold anyway," Miguel complained as he sat in the waiting room with his father.

"Oh come on, it's not a tragedy. Besides, there's been a lot of flu going around, so I don't want to take any chances," his father replied. Just then the nurse called, "Miguel Hernandez, Dr. Beuf can see you now."

Before seeing the doctor, Miguel was examined by a nurse who took some measurements and asked him a few questions.

Scoring Rubric for Recognizing Biology in Medicine

Level of Achievement	General Presentation	Conceptual Understanding	Critical Thinking: Ability to Support Ideas
Excellent-Good (full–3/4 credit)	• Shows large concepts • Uses specifics to justify ideas • Applies learning to novel situations • Synthesizes information • Makes connections	• Identifies and describes all six unifying principles • Uses specific evidence from scenario to illustrate unifying principles • Uses more than one example for each unifying principle • Makes strong connections among the unifying principles • Makes connections to other examples not in the scenario	• Uses specific data to support ideas in essay • Connects data in scenario to preexisting scientific knowledge • Uses inference to make strong connections between scenario and unifying principles
Adequate-Needs Improvement (3/4–1/2 credit)	• Grasps major concepts • Uses generalities but lacks specifics • Cannot effectively apply knowledge to new situations	• Identifies all and describes most of the unifying principles correctly • Uses evidence from scenario inconsistently • Makes connections to only 3 or 4 unifying principles • Makes vague connections among unifying principles • Cannot state significance of scenario	• Lacks consistent use of specific data from scenario to support ideas in essay • Rarely connects data from scenario to preexisting knowledge • Makes incomplete or illogical connections between scenario and unifying principles
Inadequate (<1/2 credit)	• Presents fragments of concepts • Cannot synthesize ideas • Makes erroneous assumptions	• Cannot identify all unifying principles • Describes unifying principles in confusing way • Makes fragmented or no connections between scenario evidence and unifying principles	• Fails to use logical reasoning in stating significance of scenario • Fails to use specific data or examples to support statements • Uses illogical connections between evidence and unifying principles
Relative Percentage	25 percent	50 percent	25 percent

BSCS

Evolution: Patterns and Products of Change in Living Systems

Homeostasis: Maintaining Dynamic Equilibrium in Living Systems

Energy, Matter, and Organization: Relationships in Living Systems

Continuity: Reproduction and Inheritance in Living Systems

Development: Growth and Differentiation in Living Systems

Ecology: Interaction and Interdependence in Living Systems

Figure Ev.1 Scoring Rubric for Recognizing Biology in Medicine The criteria described in this table will help you determine how to complete this assignment successfully.

Figure Ev.2 Unifying Principles of Biology These principles should look very familiar by now. They have been the basis for organizing the ideas of this program.

385

Figure Ev.3 The nurse recorded these data for Miguel.

Patient	Miguel Hernandez
Temperature	38.3°C (101°F)
Blood Pressure	122/80
Pulse	77 beats/min
Mass/Weight	52.3 kg (115 lbs.)
Comments	Complains of sore throat, headaches, lack of appetite

The doctor then examined Miguel, looking in his throat and ears and listening to his heart and lungs as he breathed slowly. When she had finished, she began discussing the treatment she was recommending.

"I suspect you have a *Streptococcus* bacterial infection. I'll give you a prescription for penicillin, but first we need to take a throat swab and blood sample for the lab. They will grow bacteria from the swab on an agar plate and use antibiotic disks to test for resistance; they'll also check the blood for antibodies."

"How do you know it isn't just a cold?" Miguel asked.

"The rhinovirus that causes a common cold usually doesn't cause a fever. And you don't have congestion. Also, the spots on your throat look like a strep infection to me."

"Are you sure it's not the flu?" Miguel's dad asked. "Maybe he should have had a flu shot. He wasn't vaccinated for German measles either."

"Flu is caused by an infection of body cells by the influenza virus. It infects humans and some other animals such as ducks. Influenza does produce fever, but it also causes extreme body aches and many other symptoms that Miguel doesn't have. I don't think Miguel necessarily needs a flu shot; he is young and generally healthy. Vaccination against the rubella virus, which causes German measles, is probably a good idea for all youngsters, though the danger of infection is much lower than it used to be. Let me see if I can find those data . . . Here we are; look at these data showing a recent history of rubella cases."

Figure Ev.4 Rubella occurrence in the United States Data from the past several years show an increase in the incidence of rubella. Why do you think rubella is becoming more common again?

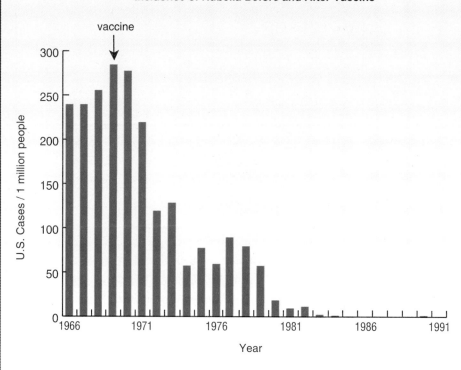

Incidence of Rubella Before and After Vaccine

"Rubella vaccination is most important, however, for girls because of the risk of infection later in life. If a pregnant woman gets rubella, it can seriously damage her developing fetus."

"Anyway, if it were the flu, the penicillin would wipe it out, right, Dr. Beuf?" Miguel asked.

"No, penicillin works by interfering with the production of the bacterial cell wall as the bacterial cells divide. Because viruses such as the influenza virus live inside the host's cells, antibiotics do not affect them. And it is not a good idea to use antibiotics when they aren't needed because some bacteria carry genes whose products make the bacteria resistant to the action of a particular antibiotic. These resistant bacteria are usually only a tiny percentage of the population infecting you. In addition, you should always finish a prescription of antibiotics. If you stop short, you could give resistant bacteria a chance to reproduce and repopulate your tissues."

"Should Miguel take aspirin or ibuprofen to reduce his fever? Is the fever bad for him?" Mr. Hernandez asked.

"Well, it certainly makes him feel bad. But his fever is not at a dangerous level." Dr. Beuf points to a wall chart showing grades of temperature for human beings.

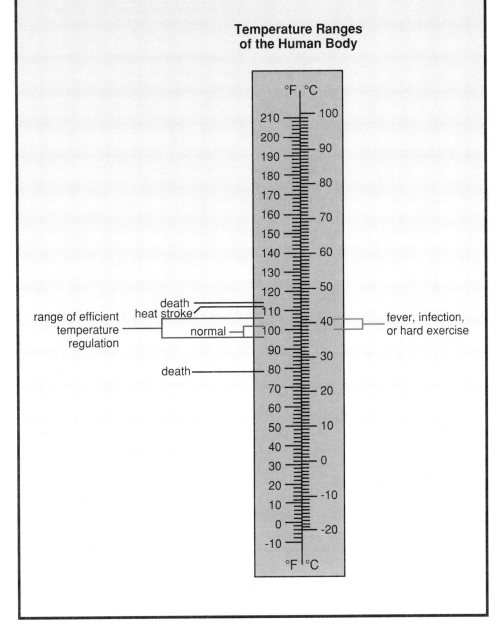

Temperature Ranges of the Human Body

"As long as his fever is at a reasonable level and appears to be caused by an infection instead of a head injury or overheating, it won't hurt him. Aspirin and ibuprofen can be effective in reducing fevers, and they also reduce inflammation, which can speed the healing of certain types of injuries. Of course, there are drawbacks as well. Anti-inflammatory drugs can suppress some immune system responses. In addition, taking aspirin and maybe ibuprofen can be dangerous for children or teenagers who have certain viral infections such as chicken pox or influenza. Treating these viral diseases with these drugs can trigger a life-threatening pressure on the brain in what is known as Reye syndrome.

"But in Miguel's case today, it is difficult to say whether it is helpful or harmful to lower his fever. Fever may be helping to fight his infection. Read this interesting research report." Dr. Beuf shows this account to Miguel and his father:

Fever in Animals

Does fever act as a defense mechanism against disease? Several studies have shown that infections may last longer and be more severe when people are treated with drugs such as aspirin. Aspirin not only reduces fever, it also blocks pain and reduces inflammation. Scientists have conducted experiments with animals to try to isolate the effects of fever. Rabbits treated with drugs that reduce fever are more likely to survive bacterial infections if an external heat source is applied to the animals.

In humans and other mammals, fever is produced when the chemical signals of infection cause the brain to increase the set point for body temperature. The rate of the body's metabolism increases. The organism shivers to warm its body, and blood vessels near the skin constrict, thus reducing heat loss. Fever may enhance the immune system's response. It also induces sleepiness and pain, so the victim is less active and saves energy for fighting the infection. When the infection is over, sweating occurs, and the body temperature returns to normal.

Some animals, such as reptiles, do not rely on metabolic warming to produce fever. They rely on behavioral changes to help adjust their body temperature. For example, they can move to warm or cool locations. In one study scientists infected desert iguanas (*Dipsosaurus dorsalis*) with the bacteria *Aeromonas hydrophilia*. The iguanas moved to locations that increased their temperature above normal. When the infected iguanas were prevented from moving to the warmer, fever-inducing locations, the infections worsened.

"What about the old saying, starve a fever, feed a cold?" Mr. Hernandez asks.

"Well, I wouldn't recommend starving, but don't force too much food. Do drink lots of juices and water to keep hydrated. You use energy to produce the fever, but remember that you have some body stores of glycogen and fat to use for energy. By the way, you may suffer from diarrhea as the antibiotic kills off the bacteria that normally live in your intestines. If so, you can eat yogurt, which contains lactobacillus cultures, or take tablets containing normal intestinal bacteria to stop the diarrhea. These bacteria also help keep the growth rate of pathogens low by competing for resources."

Scenario 2 Cystic Fibrosis and Cholera

Most harmful genetic disorders are quite rare among human populations. In many cases, the most serious mutations never show up because the changes are so harmful that the embryo does not survive. So it may seem surprising that cystic fibrosis (CF), an inherited disease that causes severe problems with the gas exchange and digestive systems, is much more common than many other serious genetic disorders. Cystic fibrosis is more common among Caucasian children (affecting about 1 in every 2000 infants) than among African-American, Asian, or Jewish children, in whom it is very rare. The victims of CF suffer from chronic coughs, lung infections, pneumonia, and digestion difficulties. These problems generally become worse as patients become older, and only 25 percent of all CF patients survive into their thirties, although new treatments may help extend their lifetimes.

Why has the mutant gene for CF remained in certain populations at a relatively high frequency? One possibility is that it provides some advantage in addition to the problems it causes. Scientists examined the mechanism by which the CF mutation causes disease and looked for the possibility that it also could confer an advantage. One way to investigate this was to consider what happens in heterozygotes, who carry one copy of the CF allele but who do not suffer the disease symptoms. (The pedigree, or family tree, in Figure Ev.6 shows a possible inheritance pattern of CF.)

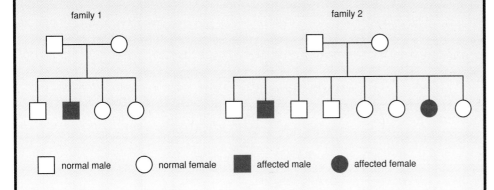

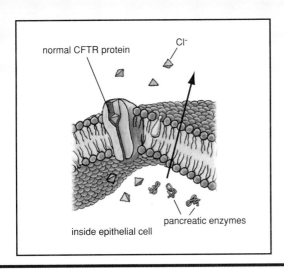

Figure Ev.6 Pedigree for a family with cystic fibrosis

This is what the scientists found. Cystic fibrosis is caused by mutations in the gene for a protein known as the cystic fibrosis transmembrane conductance regulator protein, or CFTR protein. This protein normally has 1480 amino acids and is found in the membranes of lung and intestinal cells. The normal form of the CFTR protein is a transport protein; it acts as a gate in certain membranes, much like a gate that allows people to go in and out of a stadium. This protein controls the exchange of chloride ions across the membranes of cells in the gas exchange and digestive systems (Figure Ev.7).

Figure Ev.7 Normal CFTR protein regulates chloride ion flow across the membrane, which indirectly affects the secretion of pancreatic enzymes.

In cystic fibrosis patients, the protein is altered and does not function properly. For example, one particular mutation causes the deletion of just one amino acid, the phenylalanine at position 508 in the amino acid sequence of the CFTR protein. This mutation, called ΔF508, accounts for 75 percent of all CF cases.

Although the CF allele results in only a small change in the amino acid sequence of the CFTR protein, it has a large effect on the protein's ability to function. Mutant versions of CFTR protein are not able to regulate the exchange of chloride ions. The loss of chloride ion regulation means that the regulation of water balance is lost as well because the concentration of chloride ions outside of the cell affects the amount of water moving into or out of the cell. (This movement occurs by the process of osmosis.) As a result, thick mucus secretions build up outside the cells of the gas exchange and digestive systems in people with CF, that is, those persons who are homozygous for the CF allele.

Mucus in the digestive ducts of the pancreas interferes with the release of certain digestive enzymes, as shown in Figure Ev.8. Mucus in the lungs interferes with breathing and makes CF patients more vulnerable to lung infection. This weakness occurs because the immune system cells, which normally destroy harmful invaders, are unable to pass through the mucus surrounding lung cells of CF victims. Without the help of the immune system, even relatively minor infections can become quite serious.

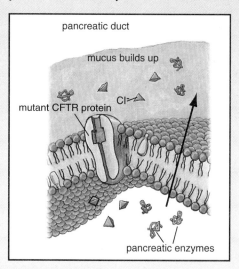

Figure Ev.8 The effect of cystic fibrosis on the pancreas
Mutant CFTR protein interferes with Cl⁻ movement across the membrane, which, in turn, interferes with water movement across the membrane. Pancreatic enzymes get stuck in mucus that builds up outside the cell.

You may wonder, along with many scientists, how such a harmful allele could continue in the Caucasian population at the rate of roughly 1 allele in every 25 individuals. Evidence that suggests a reason for this widespread occurrence of the CF allele comes from studies of the disease cholera. Cholera is an infectious disease that is spread by bacteria that infect the intestines. The disease causes severe diarrhea that, in humans, often leads to death if medical treatment is not available. Individuals who are heterozygous for the CF allele may be more likely to survive cholera than those individuals who are homozygous normal for the CFTR gene.

The first step in understanding the possible relationship between CF and cholera was to consider what causes the symptoms of cholera. The invading pathogen is the bacterium *Vibrio cholerae*. It produces a toxin called cholera toxin, CT, that interferes with the *normal* CFTR protein, causing it to alter its regulatory activity and secrete too much chloride ion and water from intestinal cells. This alteration causes diarrhea and dehydration in otherwise healthy mice (and humans).

Once the cause of cholera's symptoms was understood, scientists set up a model using mice that had mutant CFTR genes and that could be infected with cholera. With this model, the scientists tested how mice that are heterozygous for the CF allele would react to the disease cholera. Such heterozygous mice have only half as much normal CFTR protein as homozygous normal mice. This reduction in functional regulatory protein occurs because there is only one allele for normal CFTR; the other allele produces nonfunctional (mutant) CFTR. This difference is illustrated by the data shown in Figure Ev.9.

The Effect of Cholera in a CF Mouse Model

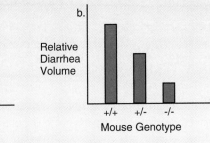

a. Amount of CFTR Protein in Mouse Intestine

Mouse Genotype
+/+   +/-   -/-

b. Relative Diarrhea Volume

Mouse Genotype
+/+   +/-   -/-

+/+ homozygous normal CFTR
+/- heterozygous CFTR
-/- homozygous mutant CFTR

Source: From S.E. Gabriel, K.N. Brigman, B.H. Koller, R.C. Boucher, and M.J. Stutts (7 October 1994) Cystic Fibrosis Heterozygote Resistance to Cholera Toxin in the Cystic Fibrosis Mouse Model, *Science*, 266, pp. 107-109.

When infected with the cholera pathogen, the heterozygous mice secrete only half as much fluid (diarrhea) as normal (non-CF) mice. Because their diarrhea is less severe than normal mice, as shown in Figure Ev.9b, the mice that are heterozygous for CF do not become as sick from cholera as do mice lacking the mutant CF allele.

The scientists who conducted this experiment concluded that their data supported the hypothesis that the presence of the CF allele may increase the ability to survive cholera. They also concluded that these data suggest that the high incidence of CF in certain human populations may be related to an increased resistance to cholera. They submitted a report of their experiments to the journal *Science*. Their report was reviewed by a panel of scientists who decided that their results and conclusions were worthy of publication. As a result the science and medical community had access to this new information.

Figure Ev.9 The effects of cholera in relation to the presence of the CF allele Figure Ev.9a shows the amount of CFTR protein in normal mice, mice with one copy of the CF allele (heterozygotes), and mice with two copies of the CF allele (homozygotes). Figure Ev.9b shows how these groups of mice react to having cholera.

CHAPTER CHALLENGES

In this section you will find at least one challenging question or problem for each chapter in the program. You can use these questions to test your understanding of the specific ideas in the chapters you studied. Keep in mind that although a question may be listed under a particular chapter heading, you can respond by including material from other chapters.

Unit 1 Evolution: Patterns and Products of Change in Living Systems
Chapter 1 The Human Animal

1. Long before humans developed ways to write their languages, they passed on information about their lives and the world around them by telling stories, drawing pictures, acting out plays, dancing, or singing. As the collection of information grew, people needed very good memories to pass along all the information.

   How have written language and modern electronic communication altered the way information is stored and transmitted? Explain how you think these changes may have affected human cultures.

Figure Ev.10 There are many ways to transmit information.

Chapter 2 Evolution: Change across Time

1. Describe two examples of how evolution is happening today.

2. How does technology play a role in helping scientists collect evidence of biological change across long periods of time?

3. Scientists have hypothesized that modern whales are descended from land mammals that moved into the water environment between 50 to 60 million years ago. In 1994 scientists found two exciting fossil discoveries, both whalelike creatures with legs.

Explain how the data presented in Figure Ev.11 support an evolutionary explanation for the origin of modern whales from an ancestor that lived on land and had legs.

Occurrence (in millions of years ago)	Appearance*	Name	Year of Discovery	Description*
55		*Mesonychid*	prior to 1989	• Hyena-like land mammal • 4 long legs • Slender tail
50		*Ambulocetus*	1989	• Land and sea mammal • 4 short legs with feet • No tail fluke but probably swam like an otter
46		*Rodhocetus*	1994	• Whale-like sea mammal • Legs shorter than *Ambulocetus* • May have been able to move awkwardly on land • Strong tail for swimming
40		*Prozeuglodon*	1994	• 15-foot long aquatic mammal • Tiny 6-inch hind legs that could not support weight on land • Tail fluke for swimming

* Based on fossils of skeleton

Source: From C. Zimmer, Back to the Sea, January 1995, *Discover* magazine, pp. 82–84.

Figure Ev.11 New fossil evidence for whale ancestors reveals a link to land-based mammals.

Chapter 3 Products of Evolution: Unity and Diversity

1. Develop a reasonable scientific explanation for this statement: Birds are related to dinosaurs.

2. The Nature Conservancy is an example of a conservation organization that seeks to protect land from human development. How could the activities of an organization such as this have an effect on biological diversity? Give specific examples using this organization or a similar one.

EVALUATE
Thinking Like a Biologist

Unit 2 Homeostasis: Maintaining Dynamic Equilibrium in Living Systems

Chapter 4 The Internal Environment of Organisms

1. An astronaut in space depends on a space vehicle or a space suit to create an environment that can support his or her life. The data in Figure Ev.12 show the average daily dietary and metabolic needs of an astronaut in space.

An Astronaut's Daily Dietary and Metabolic Needs

Input	Amount Needed per Day
oxygen	0.84 kg/day
food solids	0.62 kg/day
dietary water (includes drinking water and water in food)	2.77 kg/day
water for washing and food preparation	25.26 kg/day

Source: From P.O. Wieland (1994) "Designing for Human Presence in Space: An Introduction to Environmental Control and Life Support Systems," *NASA Reference Publication 1324*. Alabama: George C. Marshall Space Flight Center.

Figure Ev.12 An astronaut's average daily dietary and metabolic needs What do these data tell you about an astronaut's needs?

    a. Make a table that shows the types of output that each astronaut would produce given the input shown. (You do not need to use numerical values, but you can identify what types of waste will be produced.)

    b. Considering the output you listed in your chart, what technological adaptations would be necessary to maintain a healthy and clean living environment in the space vehicle? How does your answer relate to the concepts of this chapter?

Chapter 5 Maintaining Balance in Organisms

1. What role does the brain have in maintaining the body's balance of temperature, water, gas exchange, and blood pressure?

Chapter 6 Human Homeostasis: Health and Disease

1. A human body is continuously subjected to changes in its external environment. Use one or two examples to explain why these changes normally do not cause problems for the body and under what circumstances homeostasis can become disrupted.

2. How do AIDS and autoimmune diseases keep the body from maintaining a healthy condition?

1. Have you ever watched someone race-walk? The unusual twisting motion
    of the hips makes it look as if it would be so much easier for the racers to
    pick up their feet and run. What is the difference between running and
    walking? Humans, like many other vertebrates including horses, dogs,
    cats, and deer, have different ways of moving their legs for locomotion. If
    you ever have ridden a horse you know that the jarring bounce you feel
    when the horse trots is very different from a gallop or a canter. Use the
    data displayed in Figure Ev.13 to answer the questions below.

    a. At a speed of 5 km/hour (about 3 miles per hour), is running or
        walking more energy efficient? Is the same true at 8 km/hour (about
        5 miles per hour)? Explain how the data in Figure Ev.13 support
        your answers.

    b. During very fast running, above 16 km/hour (10 miles per hour),
        most people are exercising anaerobically. If this is the case, would the
        type of data measured in Figure Ev.13 be an appropriate way to
        determine their energy efficiency? Explain your answer.

Energy Use and Locomotion

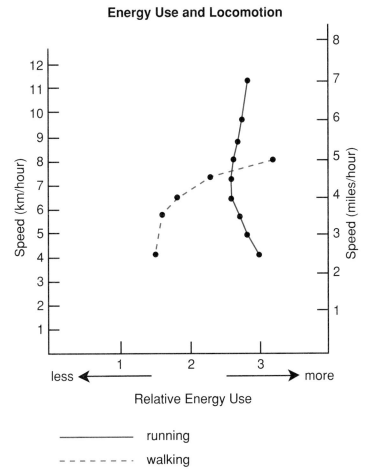

Figure Ev.13 **Energy use in
different modes of locomotion
in humans** These data are
based on measurements of
oxygen consumption of the
person running or walking.

Source: From A.R. McNeill (1992) *Exploring Biomechanics: Animals in Motion.* New York: Scientific
American Library.

EVALUATE
Thinking Like a Biologist

395

Chapter 8 The Cellular Basis of Activity

1. Metabolism includes the series of chemical reactions that break down macromolecules, such as glycolysis, fermentation, and aerobic respiration, and the reactions that build up macromolecules, such as protein synthesis or (in plants and certain prokaryotes) photosynthesis.

    Can both types of reactions happen in the same cells of the same organism? Explain your answer, and show how it relates to activity in a living system.

2. In order to grow and develop, a seed must germinate (sprout) as the embryonic plant begins to grow. Germination requires water. For example, lettuce seeds germinate in about 48 hours after being soaked in water. Germination also may be sensitive to light. Some seeds require light to germinate, whereas light inhibits germination in other species.

    Design and carry out an experiment to test whether the germination of common garden seeds is affected by light. In order to have time to do this test, you need to use seeds that germinate quickly, such as Grand Rapids lettuce seeds.

    a. Report your results and conclusions, and include a possible explanation of why the seeds you selected behaved as they did.

    b. Explain the sources of energy and matter for a germinating seed (1) *before* the sprout emerges and leaves grow and (2) *after* the sprout is above ground and the leaves are open.

Chapter 9 The Cycling of Matter and the Flow of Energy in Communities

1. Explain the flow of energy in a compost pile. What happens to the matter? What is the connection between the flow of energy and the changes in the matter?

2. Identify a population of organisms that lives in your community. Draw or construct a food web that shows the connections among the organisms in that population. Include a discussion of what happens to the energy as you move to higher trophic levels, and illustrate these relationships.

Unit 4 Continuity: Reproduction and Inheritance in Living Systems
Chapter 10 Reproduction in Humans and Other Organisms

1. Use the population data shown in Figure Ev.14 to answer the following questions:

   a. Explain how birth rate and death rate work together to determine both continuity and size of a population of organisms.

   b. What other factors contribute to the size of a population of organisms? How?

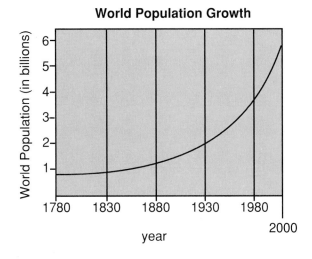

World Population Growth

Figure Ev.14 How does reproductive success affect population size?

- Starting in the year A.D. 1, it took over 1500 years for the population size to double.
- Starting in the year 1900, it took about 75 years for the population size to double.
- In 1980, the number of babies dying in their first year of life was 1/4 the number that died in 1940.

2. The behavior of animals such as crickets plays an important role in reproductive success. Read this short description from *Science* magazine of the results of a research project that used a robot model of a cricket.

A psychologist in the United Kingdom, Barbara Webb, used a robot cricket to model the way crickets use call songs to find a mate. Webb had hypothesized that the

Figure Ev.15. Modeling behavior of even relatively simple organisms such as crickets has proven difficult. Why do you think this is so?

EVALUATE
Thinking Like a Biologist

397

> crickets' complex behavior of moving toward a calling mate may result from simple reflexes that could be mimicked in a robot cricket that had sound sensors.
>
> The robot was programmed to respond to a specific set of syllables from a recording of a male cricket's song. When this recording was played, the robot moved toward the speakers. When the syllables of the recorded cricket song were altered, the robot became "confused." And when key syllables were separated between two speakers, the robot went to a spot halfway between the speakers before "choosing" one and heading toward it.

Explain how these results relate to what you observed with live crickets and what you know of mating behavior in other animals, including humans.

3. If you were to invent two new types of contraceptives, one hormonal and the other physical, what would they be and how would they work?

Chapter 11 Continuity of Information through Inheritance

1. In the laboratory, scientists can use enzymes to remove the cell walls from plant cells in growing plant tissue; this produces protoplasts. Protoplasts then can be placed on a growth medium in a petri dish. If the medium contains the proper mixture of plant hormones, a single isolated protoplast can divide and eventually produce a whole plant.

   What does this observation tell you about the genetic material of a plant cell? How does genetic material affect continuity in sexually reproducing and asexually reproducing organisms?

Chapter 12 Gene Action

1. How is the molecular structure of genetic information important to the replication and expression of genes? In your response, describe the similarities and differences among the following molecular processes: replication, transcription, and translation.

2. How is it possible that a change in a single nucleotide of a gene can produce a mutant protein? Will such a change in a nucleotide always produce a mutant protein? Explain.

Unit 5 Development: Growth and Differentiation in Living Systems
Chapter 13 Processes and Patterns of Development

1. Plants and animals adjust to environmental changes in many ways. Read the following brief description of several adaptations to change, and complete the task below.

> Have you ever noticed that you are bothered by fewer flying insects, such as flies and mosquitoes, in the winter than in the summer? In regions where winters are very cold and the land often is covered by snow or lakes are frozen over, organisms display a wide variety of adaptations that help them survive the harsh conditions. For example, the sap in broad-leafed deciduous trees, such as maple, apple, and oak, withdraws into

the roots and trunk. Leaves fall off and new ones grow in the spring. Animals may migrate, as do many birds; they may slow their activity and enter a prolonged sleep (as do bears and skunks); or they may enter a dormant state of hiberation (as do ground squirrels), in which metabolic activity and body temperature drop significantly.

Adult insects have hard exoskeletons and their bodies are filled with bloodlike liquid in which their organs are suspended. Insects cannot easily regulate body temperature and can be damaged by freezing. With their small size, migration is difficult, though monarch butterflies manage to use this strategy successfully. Some insects bury themselves in a protected place, but most insects in harsh climates go through winter in an immature developmental stage. These species lay eggs prior to cold seasons, and then the adults die. The eggs, larvae, or pupae spend the winter in a protected spot, becoming active and continuing their development in the spring.

Using either deciduous trees or insects as an example, describe how their pattern of development represents an adaptation for extreme seasonal differences in temperature.

2. A tumorous disease of tomato and tobacco plants known as crown gall can result from the action of a bacterium, called *Agrobacterium tumefaciens*, that infects a wound on the plant stem. These bacteria contain copies of a DNA plasmid with special genes that can transform the growth regulation of plant cells. The bacterial genes incorporate into the plant's genetic material, where these genes direct plant cells to make unusual amino acids that are useless to the plant but that serve as food for the bacteria. A mass of undifferentiated tissue at the infected site soon grows into a tumor.

Describe how this bacteria-induced tumor compares with and is different from cancer in humans.

a.

b.

Figure Ev.16. Crown gall on (a.) tomato plant and (b.) laurel tree

1. Although people go through the same life stages, each generation does so at a different time in history. A high school student named Rachel was 15 years old in 1996, and her parents were 15 in 1970. They listened to music by the Beatles and the Rolling Stones recorded on large vinyl records. Rachel's grandparents were 15 in 1946, just after World War II. They liked to dance to a big band sound and listen to shows on the radio.

Make a table that compares the physical biology and the cultural setting for a single life stage of three generations. For example, your chart could include you, your parents at your age, and your grandparents at your age. (If you prefer not to use your own family, use a family that you know about or have read about.) You can use any life stage as long as you have information that corresponds to each generation.

a. Make two columns, one labeled *Physical Biology* and one labeled *Cultural Setting*. Under each column describe the characteristics that apply to each of your individuals.

b. Write a few sentences that describe the differences in the life stage of the various individuals according to the historical setting in which they take place.

Unit 6 Ecology: Interaction and Interdependence in Living Systems

Chapter 15 Interdependence among Organisms in the Biosphere

1. An instrument aboard NASA's *Nimbus-7* satellite records data from the surface water of the Atlantic Ocean off the U.S. coast. This instrument, called a Coastal Zone Color Scanner or CZCS, measures infrared radiation and concentrations of chlorophyll pigments. The pattern of radiation corresponds to water temperature and the patterns of both radiation and chlorophyll correspond to currents and tidal mixing. The highest concentrations of chlorophyll pigments are found near the shore, while the lower concentrations are found in the relatively unmixed waters of the warm Gulf Stream current.

   To answer the following questions about interdependence, keep in mind that some places in the ocean have high concentrations of phytoplankton, photosynthetic microorganisms that drift with ocean currents. (You also may find it useful to look back at the discussion of photosynthesis in Chapter 8.)

   a. The CZCS instrument on board the *Nimbus-7* can take measurements only from the surface water of the ocean. Is this where you would expect to find phytoplankton? Explain the basis for your answer.

   b. How could the *Nimbus-7* data be useful for management of commercial fisheries?

2. Population growth rates are easy to see on graphs, and they typically have a characteristic shape that reflects the involvement of limiting factors. Use your knowledge of these types of graphs to complete the following task:

   a. Draw a graph that illustrates the growth of any population of organisms over a time period of at least 10 generations.

      You can pick a general type of organism such as "insects" for your graph.

   b. Label your graph so that the reader can tell how many generations have passed and what type of organism is represented.

   c. Pick three distinct points on different parts of your graphed line and label them *A, B, C*. Describe what is happening to the population at points *A, B,* and *C*.

      Include a discussion of limiting factors to help explain your answer.

   d. What point on the graph represents the carrying capacity for your population? Explain.

Chapter 16 Decision Making in a Complex World

1. A panel of lawmakers has met to consider zoning for 200 acres of land near a large forest that currently includes wetlands and a meadow. One development company wants to build a shopping center and apartments on most of this land. Another company proposes building a manufacturing plant there. Conservation groups want to protect the

wetlands. Various specialists have provided reports that mention the following observations about natural land and human development:

- Nitrogen acts as a fertilizer for leafy parts of plants and for some microorganisms. Too much of this fertilizer causes plants to grow large tops and insufficient roots.

- Forests may help protect the world from global warming by converting carbon dioxide to plant mass by way of photosynthesis.

- Acid rain results from certain air pollutants produced by industrial processes. Its effects on living systems are complex.

- Nitrogen cycling is the conversion of atmospheric nitrogen into nitrogen that living organisms can use. In this process, nitrogen-fixing bacteria convert nitrogen gas, N_2 into chemical forms that biological systems can use (such as ammonium, nitrates, and nitrites). When an organism dies, different microorganisms (called denitrifying bacteria) decompose the organism's body and release some of its nitrogen back into nonbiological systems. Many denitrifying bacteria live in boggy wetland environments.

The panel is ready to make a decision, but now a new finding is presented.

In a study in Germany, where acid rain is a serious problem, scientists used radioisotope labels to trace nitrates leaving a forest through runoff water. Figure Ev.17 shows what happened to the nitrates in a stand of trees that appear healthy, a slightly declining stand, and a dying forest. For a control, the scientists compared their results with a tropical forest undisturbed by human habitation.

If nitrates move through soil without being taken up by living systems, then plants are deprived of the benefits of nitrogen. In addition, the free nitrates can remove other important nutrients that plants need, such as calcium and magnesium. The soil left behind is acidic enough to harm tree roots and kill microorganisms. In these conditions old trees may die, and young trees grow very slowly and may be stunted.

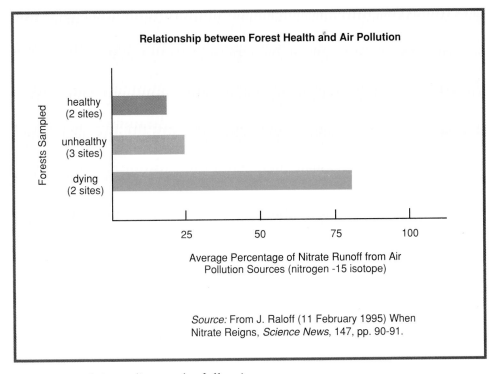

Relationship between Forest Health and Air Pollution

Forests Sampled:
- healthy (2 sites)
- unhealthy (3 sites)
- dying (2 sites)

Average Percentage of Nitrate Runoff from Air Pollution Sources (nitrogen -15 isotope)

Source: From J. Raloff (11 February 1995) When Nitrate Reigns, *Science News*, 147, pp. 90-91.

Figure Ev.17 **Tracking nitrate pollution**

Your task is to discuss the following:

a. How do the data in Figure Ev.17 support the concerns of scientists who claim that acid rain and excessive nitrates from pollution cause damage to trees?

b. List all of the possible biological and nonbiological consequences of (1) developing the land and (2) not developing the land.

c. How might the findings reported in Figure Ev.17 and the findings from earlier reports influence the panel's decision about how or whether the land in question should be developed?

d. What would you decide to do, and why?

BUILDING A PORTFOLIO OF SCIENTIFIC LITERACY

In this activity you will prepare a portfolio that shows your qualifications to be a scientifically literate citizen. To fill the job, you need to be able to reason, use evidence to support your ideas, and think conceptually about biology. This means you need more than just facts about biology; you must demonstrate that you understand those facts and how to use them. You will choose samples of your work from throughout the year, some of which you may want to improve to accurately show your progress. In addition, you may create new work to make your presentation the best show of your success.

Process and Procedures

1. Review the chapters, and record in your journal what you think are the main concepts from each chapter. To do so, make a table with one

EVALUATE
Thinking Like a Biologist

column for the titles of the chapters in each unit and a second column to record the main concepts or ideas from each chapter.

2. Let your teacher review your table of concepts before you prepare your portfolio.

3. Look over your work for each chapter, and select the examples you think will make up the best portfolio for your interview as a scientifically literate citizen. Follow these instructions:

    a. Decide how to display your samples of work to make a portfolio for your interview.

    b. Find an example of work to illustrate the key concept(s) of each chapter. Use examples of your work just as they are, or make corrections or additions if you think you understand the concepts and the particular activity better than when you did it.

        Remember that you can use resources, such as the essays, your journal, your teammates, or outside references, to help you improve an existing activity.

    c. Add new work to your portfolio by choosing at least one question from each unit from the collection of questions in the *Chapter Challenges*.

        You can put this new work in the portfolio along with the samples of your previous work.

4. Prepare a caption for each example in the portfolio. Each caption should include the following:

    a. A label identifying which chapter the entry represents and what the entry is (for example, a lab report, a graph, or a summary of data).

    b. A sentence or two for each example of work to show how it represents concepts from the chapter.

Include this explanation for all work in the portfolio, both existing work and new work.

c. A sentence or two that explains *why* you chose the work, why you improved it, or why you added new work.

Analysis

1. When you think your portfolio is complete, take time to look through it and decide if you would get the job as a scientifically literate citizen. Exchange portfolios with a partner and determine the level of literacy evident in the portfolio you examine.

2. Revise your portfolio based on your self-evaluation and your partner's feedback.

APPENDIX A

LABORATORY SAFETY

The laboratory has the potential to be either a safe place or a dangerous place. The difference depends on how well you know and follow safe laboratory practices. It is important that you read the information here and learn how to recognize and avoid potentially hazardous situations. Basic rules for working safely in the laboratory include the following:

1. Be prepared. Study the assigned activity before you come to class. Resolve any questions about the procedures before you begin to work.

2. Be organized. Arrange the materials you need for the activity in an orderly way.

3. Maintain a clean, open work area, free of anything except those materials you need for the assigned activity. Store books, backpacks, and purses out of the way. Keep laboratory materials away from the edge of the work surface.

4. Tie back long hair and remove dangling jewelry. Roll up long sleeves and tuck long neckties into shirts. Do not wear loose-fitting sleeves or open-toed shoes in the laboratory.

5. Wear safety goggles and a lab apron whenever you work with chemicals, hot liquids, lab burners, hot plates, or apparatus that could break or shatter. Wear protective gloves when working with preserved specimens, toxic and corrosive chemicals, or when otherwise directed to do so.

6. Never wear contact lenses while conducting any experiment involving chemicals. If you must wear them (by a physician's order), inform your teacher *before* conducting any experiment involving chemicals.

7. Never use direct or reflected sunlight to illuminate your microscope or any other optical device. Direct or reflected sunlight can cause serious damage to your retinas.

8. Keep your hands away from the sharp or pointed ends of equipment, such as scalpels, dissecting needles, or scissors.

9. Observe all cautions in the procedural steps of the activities. **CAUTION** and **WARNING** are signal words used in the text and on labeled chemicals or reagents that tell you about the potential for harm and/or injury. They remind you to observe specific safety practices. *Always read and follow these statements.* They are meant to help keep you and your fellow students safe.

CAUTION statements advise you that the material or procedure has *some potential risk* of harm or injury if directions are not followed.

WARNING statements advise you that the material or procedure has a *moderate risk* of harm or injury if directions are not followed.

10. Become familiar with the caution symbols identified in Figure A.1.

caution

The caution symbol alerts you to procedures or materials that may be harmful if directions are not followed properly. Following are the common hazards that you may encounter during this course.

sharp object

Sharp objects can cause injury, either a cut or a puncture. Handle all sharp objects with caution and use them only as your teacher instructs you. Do not use them for any purpose other than the intended one. If you do get a cut or puncture wound, call your teacher and get first aid.

irritant

An irritant is any substance that, on contact, can cause reddening of living tissue. Wear safety goggles, lab apron, and protective gloves when handling any irritating chemical. In case of contact, flush the affected area with soap and water for at least 15 minutes and call your teacher. Remove contaminated clothing.

reactive

These chemicals are capable of reacting with any other substance, including water, and can cause a violent reaction. **Do not** mix a reactive chemical with any other substance, including water, unless directed to do so by your teacher. Wear your safety goggles, lab apron, and protective gloves.

corrosive

A corrosive substance injures or destroys body tissue on contact by direct chemical action. When handling any corrosive substance, wear safety goggles, lab apron, and protective gloves. In case of contact with a corrosive material, *immediately* flush the affected area with water and call your teacher.

biohazard

Any biological substance that can cause infection through exposure is a biohazard. Before handling any material so labeled, review your teacher's specific instructions. **Do not** handle in any manner other than as instructed. Wear safety goggles, lab apron, and protective gloves. Any contact with a biohazard should be reported to your teacher immediately.

Figure A.1 Safety symbols used in this program

safety goggles

Safety goggles are for eye protection. Wear goggles whenever you see this symbol. If you wear glasses, be sure the goggles fit comfortably over them. In case of splashes into your eyes, flush your eyes (including under the lid) at an eyewash station for 15 to 20 minutes. If you wear contact lenses, remove them *immediately* and flush your eyes as directed. Call your teacher.

lab apron

A lab apron is intended to protect your clothing. Whenever you see this symbol, put on your apron and tie it securely behind you. If you spill any substance on your clothing, call your teacher.

gloves

Wear gloves when you see this symbol or whenever your teacher directs you to do so. Wear them when using **any** chemical or reagent solution. Do not wear your gloves for an extended period of time.

flammable

A flammable substance is any material capable of igniting under certain conditions. Do not bring flammable materials into contact with open flames or near heat sources unless instructed to do so by your teacher. Remember that flammable liquids give off vapors that can be ignited by a nearby heat source. Should a fire occur, **do not** attempt to extinguish it yourself. Call your teacher. Wear safety goggles, lab apron, and protective gloves whenever you handle a flammable substance.

poison

Poisons can cause injury by direct action within a body system through direct contact with skin, inhalation, ingestion, or penetration. Always wear safety goggles, lab apron, and protective gloves when handling any material with this label. If you have any preexisting injuries to your skin, inform your teacher before you handle any poison. In case of contact, call your teacher immediately.

11. Never put anything in your mouth and never touch or taste substances in the laboratory unless your teacher specifically instructs you to do so.

12. Never smell substances in the laboratory without specific instructions from your teacher. Even then, do not inhale fumes directly; wave the air above the substance toward your nose and sniff carefully.

13. Never eat, drink, chew gum, or apply cosmetics in the laboratory. Do not store food or beverages in the lab area.

14. Know the location of all safety equipment, and learn how to use each piece of equipment.

15. If you witness an unsafe incident, an accident, or a chemical spill, report it to your teacher immediately.

16. Use materials only from containers labeled with the name of the chemical and the precautions to be used. Become familiar with the safety precautions for each chemical by reading the label before use.

17. To dilute acid with water, *always add the acid to the water.*

18. Never return unused chemicals to the stock bottles. Do not put any object into a chemical bottle, except the dropper with which it may be equipped.

19. Clean up thoroughly. Dispose of chemicals, and wash used glassware and instruments according to the teacher's instructions. Clean tables and sinks. Put away all equipment and supplies. Make sure all water, gas jets, burners, and electrical appliances are turned off. Return all laboratory materials and equipment to their proper places.

20. Wash your hands thoroughly after handling any living organisms or hazardous material and before leaving the laboratory.

21. Never perform unauthorized experiments. Do only those experiments your teacher approves.

22. Never work alone in the laboratory, and never work without your teacher's supervision.

23. Approach laboratory work with maturity. Never run, push, or engage in horseplay or practical jokes of any type in the laboratory. Use laboratory materials and equipment only as directed.

In addition to observing these general safety precautions, you need to know about some specific categories of safety. Before you do any laboratory work, familiarize yourself with the following precautions.

Heat

1. Use only the heat source specified in the activity.

2. Never allow flammable materials such as alcohol near a flame or any other source of ignition.

3. When heating a substance in a test tube, point the mouth of the tube away from other students and you.

4. Never leave a lighted lab burner, hot plate, or any other hot objects unattended.

5. Never reach over an exposed flame or other heat source.

6. Use tongs, test-tube clamps, insulated gloves, or potholders to handle hot equipment.

Glassware

1. Never use cracked or chipped glassware.

2. Use caution and proper equipment when handling hot glassware; remember that hot glass looks the same as cool glass.

3. Make sure glassware is clean before you use it and when you store it.

4. To put glass tubing into a rubber stopper, moisten the tubing and the stopper. Protect your hands with a heavy cloth when you insert or remove glass tubing from a rubber stopper. Never force or twist the tubing.

5. Immediately sweep up broken glassware and discard it in a special, labeled container for broken glass. *Never pick up broken glass with your fingers.*

Electrical Equipment and Other Apparatus

1. Before you begin any work, learn how to use each piece of apparatus safely and correctly in order to obtain accurate scientific information.

2. Never use equipment with frayed insulation or loose or broken wires.

3. Make sure the area in and around the electrical equipment is dry and free of flammable materials. Never touch electrical equipment with wet hands.

4. Turn off all power switches before plugging an appliance into an outlet. Never jerk wires from outlets or pull appliance plugs out by the wire.

Living and Preserved Specimens

1. Be sure that specimens for dissection are properly mounted and supported. Do not cut a specimen while holding it in your hand.

2. Wash your work surface with a disinfectant solution both before and after using live microorganisms.

3. Always wash your hands with soap and water after working with live or preserved specimens.

4. Care for animals humanely. General rules for their care are listed below.

   a. Always follow carefully your teacher's instructions about the care of laboratory animals.

   b. Keep the animals in a suitable, escape-proof container in a location where they will not be disturbed constantly.

   c. Keep the containers clean. Clean cages of small birds and mammals daily. Provide proper ventilation, light, and temperature.

   d. Provide water at all times.

   e. Feed regularly, depending on the animals' needs.

   f. Treat laboratory animals gently and with kindness in all situations.

   g. If you are responsible for the regular care of any animals, be sure to make arrangements for their care during weekends, holidays, and vacations.

   h. Your teacher will provide a suitable method to dispose of or release animals, if it becomes necessary.

5. Many plants or plant parts are poisonous. Work only with the plants your teacher specifies. Never put any plant or plant parts in your mouth.

6. Handle plants carefully and gently. Most plants must have light, soil, and water, although the specific requirements differ.

7. Wear the following personal protective equipment when handling or dissecting preserved specimens: safety goggles, lab apron, and plastic gloves.

Accident Procedures

1. Report *all* accidents, incidents, and injuries, and all breakage and spills, no matter how minor, to your teacher.

2. If a chemical spills on your skin or clothing, wash it off immediately with plenty of water and have a classmate notify your teacher immediately.

3. If a chemical gets in your eyes or on your face, wash immediately at the eyewash fountain with plenty of water. Flush your eyes for at least 15 minutes, including under each eyelid. Have a classmate notify your teacher immediately.

4. If a chemical spills on the floor or work surface, do not clean it up yourself. Notify your teacher immediately.

5. If a thermometer breaks, do not touch the broken pieces with your bare hands. Notify your teacher immediately.

6. In case of a lab table fire, notify your teacher immediately. In case of a clothing fire, drop to the floor and roll. Use a fire blanket if one is available. Have a classmate notify your teacher immediately.

7. Report to your teacher all cuts and abrasions received in the laboratory, no matter how small.

Chemical Safety

All chemicals are hazardous in some way. A hazardous chemical is defined as a substance that is likely to cause injury. Chemicals can be placed in four hazard categories: flammable, toxic, corrosive, and reactive.

In the laboratory investigations for this course, every effort is made to minimize the use of dangerous materials. However, many "less hazardous" chemicals can cause injury if not handled properly. The following information will help you become aware of the types of chemical hazards that exist and of how you can reduce the risk of injury when using chemicals. Before you work with any chemical, be sure to review safety rules 1 through 23 described at the beginning of the appendix.

Flammable substances. Flammable substances are solids, liquids, or gases that will burn. The process of burning involves three interrelated components—fuel (any substance capable of burning), oxidizer (often air or a specific chemical), and ignition source (a spark, flame, or heat). The three components are represented in the diagram of a fire triangle in Figure A.2. For burning to occur, all three components (sides) of the fire triangle must be present. To control fire hazard, one must remove, or otherwise make inaccessible, at least one side of the fire triangle.

Flammable chemicals should not be used in the presence of ignition sources, such as lab burners, hot plates, and sparks from electrical equipment or static electricity. Containers of flammables should be closed when not in use. Sufficient ventilation in the laboratory will help to keep the concentration of flammable vapors to a minimum.

Figure A.2 **The fire triangle** To control a fire, one must remove or make inaccessible at least one side of the fire triangle.

ignition source

412

Toxic substances. Most of the chemicals you encounter in a laboratory are toxic, or poisonous to life. The degree of toxicity depends on the properties of the specific substance, its concentration, the type of exposure, and other variables. The effects of a toxic substance can range from minor discomfort to serious illness or death. Exposure to toxic substances can occur through ingestion, skin contact, or inhalation of toxic vapors. Wearing a lab apron, safety goggles, and plastic gloves is an important precautionary measure when using toxic chemicals. A clean work area, prompt spill cleanup, and good ventilation also are important.

Corrosive substances. Corrosive chemicals are solids, liquids, or gases that by direct chemical action either destroy living tissue or cause permanent changes in the tissue. Corrosive substances can destroy eye and respiratory-tract tissues. The consequences of mishandling a corrosive substance can be impaired sight or permanent blindness, severe disfigurement, permanent severe breathing difficulties, and even death. As with toxic substances, wear a lab apron, safety goggles, and plastic gloves when handling corrosive chemicals to help prevent contact with your skin or eyes. Immediately wash off splashes on your skin or eyes while a classmate notifies the teacher.

Reactive substances. Under certain conditions, reactive chemicals promote violent reactions. A chemical may explode spontaneously or when it is mechanically disturbed. Reactive chemicals also include those that react rapidly when mixed with another chemical, releasing a large amount of energy. Keep chemicals separate from each other unless they are being combined according to specific instructions in an activity. Heed any other cautions your teacher may give you.

TECHNIQUES

Technique 1   Journals 414
Technique 2   Graphing 416
Technique 3   Measurement 422
Technique 4   The Compound Microscope 424

TECHNIQUE 1 JOURNALS

In this program you will use a journal on a regular basis. Scientific journals have many purposes; they provide a place to record data, take notes, reflect on your progress, or respond to questions. This journal will become your permanent record of your work, and you will refer to it often during discussions and assessments. The more complete your journal is, the more valuable it will be for you.

Your journal should be a spiral notebook or a hardcover book that is permanently bound (do not use a loose-leaf notebook). A notebook with square-grid (graph-paper) pages will make any graphing that you do much easier.

Following is a description of the major ways in which you will use your journal in this program.

Recording Data

Science depends on accurate data. No one—not even the original observer—can trust the accuracy of confusing, vague, or incomplete data. Scientific record keeping is the process by which you maintain neat, organized, and accurate records of your observations and data. Use a pen to record data. Although your interpretation of data may change, *the original data are a permanent record.*

Keep records in a diary form, and record the date at the beginning of each entry. Keep the records of each activity separate. Be brief but to the point when recording data in words. It may not be necessary to use complete sentences, but single words seldom are descriptive enough to represent accurately what you have observed or done.

Sometimes the easiest way to record data is by means of a drawing. Such drawings need not be works of art, but they should be accurate representations of what you have observed. Keep the drawings simple, use a hard pencil, and include clearly written labels. Often the easiest way to record numerical data is in the form of a table. When you record data numerically for counts or measurements, include the units of measurements you used, for example, degrees Celsius or centimeters.

Do not record your data on other papers and then copy it into your journal. Doing so may increase neatness, but it will decrease accuracy.

Your journal is *your* book, and blots and stains are a normal circumstance of field and laboratory work.

You will do much of your laboratory work as a member of a team. Your journal, therefore, will contain data that other team members have contributed. Keep track of the source of observations by circling (or recording in a different color) the data that others reported.

Responding to Questions

When you answer discussion or *Analysis* questions, record the date and activity title. Then number each response. You also may find it useful to record the questions. Sometimes you will respond to questions individually and sometimes with your team; indicate whether your responses are your own or your team's. As you are writing your responses, practice writing in complete sentences; this will help you when you synthesize and present ideas.

Taking Notes

Always begin with the date. Then record the source of information. Usually this is a person or a book, but it could be a videodisc, a video, or a computer program. When recording notes, change lines for each new idea. Try to group related ideas under broad headings that will help you remember the important ideas and how they are connected. Write down more than you think you will need; it is hard to make sense of a few words when you look back at them later.

You may find it valuable to take notes during team discussions and class discussions as well as when your teacher is presenting ideas or instructions. You also may find that taking notes in your journal as you read helps you better absorb the written information.

You can use the information in your journal to prepare for discussions or to review what you have learned. At times you also will use the information that you have recorded in your journal to complete assessment activities.

Keeping Track of Your Questions

Often, as you read or work through an activity, a question will come to mind or you will find that you are confused about something. If you are not able to talk with your teammates or your teacher right away, jot down your question or confusion in your journal so that you will remember to ask about it when you have the opportunity. You also may use this technique to record questions that you want to answer yourself.

Keeping Track of Your Responsibilities

Because you will use your journal every day in science class, it is a good place to record your class assignments and responsibilities. Each day, you may want to record these in red in the upper corner of your journal page.

Using Your Journal during Assessment

At times throughout this program, you will use your journal during assessments—both ongoing assessments, such as class discussions and team presentations, and more formal, end-of-unit assessments. Your teacher may collect your journal from time to time to assess your progress. Using a journal for assessment will be a rewarding experience if your entries are complete and detailed. Keep this in mind as you make entries in your journal.

TECHNIQUE 2 GRAPHING

Graphs are visual representations of numerical data. They help us see patterns that we might not see if we looked at numbers alone. Graphs make it easy to see at a glance what happened in an experiment. We also can use the patterns we observe in graphs to predict future changes or events.

Different graphs serve different purposes. For example, a line graph is a good way to show the relationship between two sets of numbers. Figure B.1 shows a line graph that relates the number of days and the number of individuals in a laboratory population of microorganisms. A bar graph is a good way to show the distribution of measurements among a particular group of objects. Figure B.2 shows a bar graph of shoe sizes among the girls in a high school biology class.

A graph has two major lines, one that runs horizontally and one that runs vertically on the page. These lines are called the *axes* (singular, *axis*). The horizontal line is the *x* axis and the vertical line is the *y* axis. The point at which these two lines meet is the place where the graph begins.

Both axes include a sequence of numbers called a *number scale*. The numbers on the *x* axis read from left to right and those on the *y* axis from bottom to top. The number scale is not necessarily the same on both axes. Notice in Figure B.1 that the number scale on the *x* axis reads from 0 to 30; in Figure B.2 the scale reads from 5 to 9. What is the number scale on the *y* axis of each graph?

Graphs should be labeled to help the reader understand them. Look again at Figures B.1 and B.2. Each axis is labeled to explain to the reader what the numbers represent. And each graph has a title that describes the relationship that the graph displays.

Figure B.1 Line graph

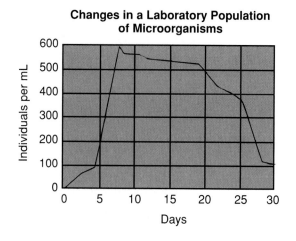

416

Shoe Sizes among Girls in a Biology Class

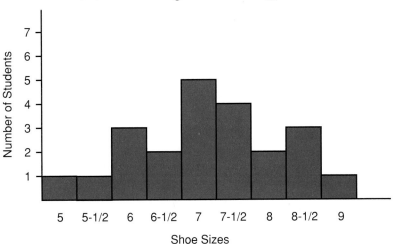

Figure B.2 **Bar graph**

Making a graph involves several important steps. The steps in Part A will help you draw any line graph; the steps in Part B will help you draw any bar graph.

Materials (per person)

2 sheets of graph paper
ruler

Process and Procedures
Part A Line Graphs

1. Review the data in Figure B.3.

   Before you can make a graph using data that you have collected, you need to organize your data into a data table. For this practice in graphing, you will use the data in Figure B.3.

2. Draw the *x* axis and the *y* axis for the graph.

   Use a ruler and graph paper so that your lines will be straight and perpendicular (see Figure B.4a).

3. Identify what information will go on the *x* axis and what information will go on the *y* axis.

   Information on the x axis is usually constant in nature—for instance, items such as dates, numbers, miles, and sizes. Information on the y axis is variable in nature—for instance, the number of mice caught, the number of people of a certain height, or the number of butterflies captured.

4. Label each axis of the graph using the headings in the data table.

Number of Mice Caught in a Field

Day	Number of Mice Caught per 100 Traps per Night
0	25
30	45
60	38
90	30
120	20
150	14
180	13
210	8
240	7
270	11
300	4
330	13

Figure B.3 **Table showing the number of mice caught**

5. Set up the number scales on each axis.

   Allow space on each axis for all the numbers that are included in the data table (see Figure B.4b). (A number scale does not have to start with the number 1, but the numbers do need to be spaced in equal increments.)

6. Give your graph a descriptive title.

7. Plot the data on your graph by doing the following:

   a. Read one row of data from the data table, for example, day 0 and 25 mice caught.

   b. Find the number on the x axis where the piece of corresponding data fits (for example, 0).

   c. Move up from the number on the x axis to the place on the y axis where the corresponding piece of data fits (for example, 25).

   d. Draw a dot, called a **data point**, at that place.

   e. Repeat Steps a–d for all the pieces of data in the data table.

      Figure B.4c illustrates five data points. Can you determine where the remaining data points go?

8. Draw a smooth line from left to right that connects the data points. This should help you see the relationship between the data points.

   In this case, the line illustrates how the number of mice caught changed throughout the year. See Figure B.4d.

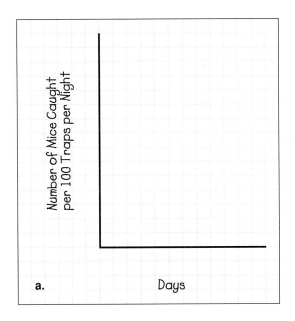

a. Days

Number of Mice Caught per 100 Traps per Night

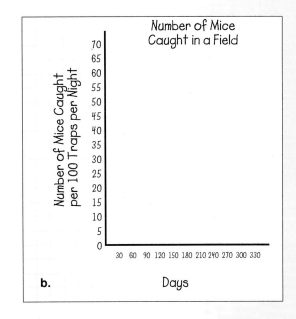

b.

Number of Mice Caught in a Field

Days

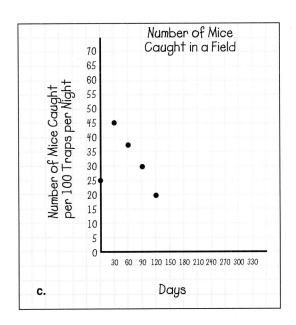

c.

Number of Mice Caught in a Field

Days

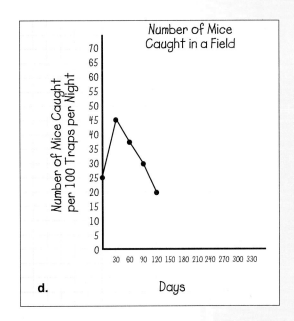

d.

Number of Mice Caught in a Field

Days

419

Part B Bar Graphs

1. Review the data in Figure B.5.

Population of Heath Hens, Martha's Vineyard, MA

Year	Population Size
1900	90
1905	45
1910	280
1915	2010
1920	550
1925	40
1930	10

Figure B.5 Table showing the population of heath hens

2. Draw the *x* and *y* axes for the graph.

3. Decide which information goes on each axis (refer to Figure B.6a).

4. Label each axis, using the headings in the data table.

5. Decide on the number scales or labels for each axis. Position the numbers on the *x* axis so that you can draw bars in the spaces between the lines. Place the numbers on the vertical axis next to the lines so that you can end a bar between two numbers, if necessary.

   Figure B.6b will help you with this step.

6. Add a title to your graph.

7. Plot the data by following these steps:

   a. Read one row of data from the data table, for example, year 1900 and population size 90.

   b. Find the label for the corresponding piece of data on the x axis of the graph (for example, 1900).

   c. Move up the column above the label to the appropriate number for that piece of data on the y axis (for example, 90).

   d. Draw a horizontal line at that number to make the top of the bar.

   e. Color in the bar from that line down to the x axis.

   f. Repeat Steps a-e for all the pieces of data in the data table (see Figure B.6c).

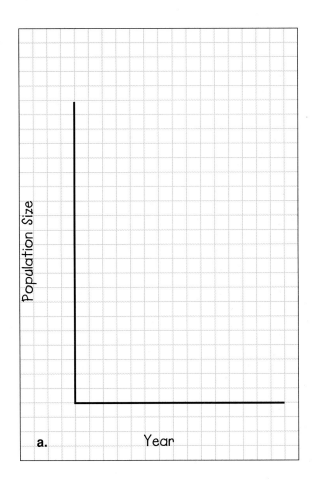

a. Year

Population Size

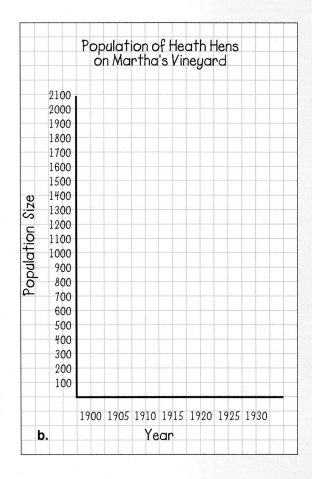

Population of Heath Hens
on Martha's Vineyard

b. Year

Population Size

2100
2000
1900
1800
1700
1600
1500
1400
1300
1200
1100
1000
900
800
700
600
500
400
300
200
100

1900 1905 1910 1915 1920 1925 1930

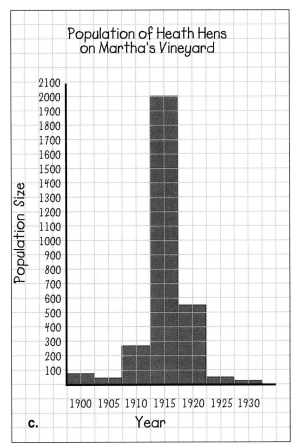

Population of Heath Hens
on Martha's Vineyard

c. Year

Population Size

2100
2000
1900
1800
1700
1600
1500
1400
1300
1200
1100
1000
900
800
700
600
500
400
300
200
100

1900 1905 1910 1915 1920 1925 1930

Figure B.6 How to draw a bar graph

TECHNIQUE 3 MEASUREMENT

Scientists measure things according to the *Système Internationale d'Unités* (International System of Units), more commonly referred to as "SI." A modification of the older metric system, SI was used first in France and now is the common system of measurement throughout the world.

Among the basic units of SI measurement are the meter (length), the kilogram (mass), the kelvin (temperature), and the second (time). All other SI units are derived from these four. Some of these units are described in following sections. Note that you will use some of these units in your laboratory work.

Length

1 kilometer (km) = 1000 meters
1 hectometer (hm) = 100 meters
1 dekameter (dkm) = 10 meters
1 meter (m)
1 decimeter (dm) = 0.1 meter
1 centimeter (cm) = 0.01 meter
1 millimeter (mm) = 0.001 meter
1 micrometer (μm) = 0.000001 meter
1 nanometer (nm) = 0.000000001 meter

The meter is the basic unit of length, with increments increasing or decreasing by the power of 10. For example, measurements under microscopes often are made in micrometers, which is one-millionth of a meter. Still smaller measurements, such as those used to measure wavelengths of light that plants use in photosynthesis, are made in nanometers. The units of length you will use most frequently in the laboratory are centimeters (cm).

Units of area are derived from units of length by multiplying two lengths. One hectometer squared is one measure that often is used in ecological studies; it commonly is called a hectare and equals 10,000 m^2. Measurements of area made in the laboratory most frequently are in centimeters squared (cm^2).

Mass

1 kilogram (kg) = 1000 grams
1 hectogram (hg) = 100 grams
1 dekagram (dkg) = 10 grams
1 gram (g)
1 decigram (dg) = 0.1 gram
1 centigram (cg) = 0.01 gram
1 milligram (mg) = 0.001 gram
1 microgram (μg) = 0.000001 gram
1 nanogram (ng) = 0.000000001 gram

Like the meter, measurements of mass are based on the gram, and basic units of mass increase or decrease by the power of 10. In the biology laboratory, measurements usually are made in kilograms, grams, centigrams, and milligrams.

Volume

1 kiloliter (kL) = 1000 liters
1 hectoliter (hL) = 100 liters
1 dekaliter (dkL) = 10 liters
1 liter (L)
1 deciliter (dL) = 0.1 liter
1 centiliter (cL) = 0.01 liter
1 milliliter (mL) = 0.001 liter

SI units of volume are derived from units of length by multiplying length by width by height; one meter cubed (m^3) is the standard unit. Although not officially part of SI, liters are often used to measure the volume of liquids. There are 1000 liters in one meter cubed (m^3), that is, 1 L = 0.001 m^3. One milliliter equals one centimeter cubed, that is 1 mL = 1 cm^3.

Because one meter cubed (m^3) is too large for practical use in the laboratory, centimeters cubed (cm^3) are used. Volume measurements in the laboratory usually are made in glassware marked for milliliters and liters.

Temperature

Units of temperature that you will use in this course are degrees Celsius, which are equal to kelvins. On the Celsius scale, 0°C is the freezing point of water, and 100°C is the boiling point of water. Figure B.7 illustrates the Celsius scale alongside the Fahrenheit scale, which still is used in the United States. On the Fahrenheit scale, 32°F is the freezing point of water and 212°F is the boiling point of water. Figure B.7 is useful for converting from one scale to the other.

Another type of measurement you will encounter is molarity (labeled with the letter M). Molarity measures the concentration of a dissolved substance in a solution. A high molarity indicates a high concentration. Some of the solutions that you will use in the activities are identified by their molarity.

If you wish to learn more about SI measurement, write to the U.S. Department of Commerce, National Institute of Standards and Technology (NIST), Washington, DC 20234.

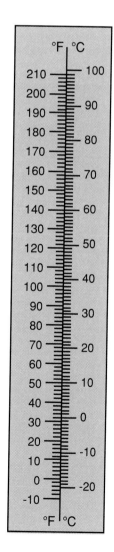

Figure B.7 **A comparison of Fahrenheit and Celsius temperature scales**

TECHNIQUE 4 THE COMPOUND MICROSCOPE

The human eye cannot distinguish objects much smaller than 0.1 mm in diameter. The compound microscope is a technology often used in biology to extend vision and allow observation of much smaller objects. The most commonly used compound microscope (pictured in Figure B.8) is monocular (that is, it has one eyepiece). Light reaches the eye after it has passed through the objects being examined. In this activity you will learn how to use and care for a microscope.

Materials (per person or team of 2)

3 cover slips
3 microscope slides
100-mL beaker or small jar
dropping pipet
compound microscope
scissors
transparent metric ruler
lens paper
newspaper
water

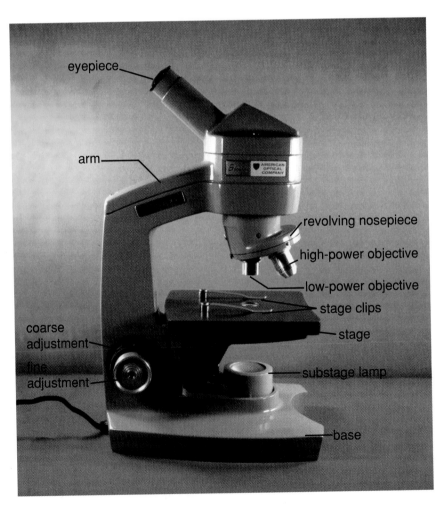

Figure B.8 **Parts of a compound microscope**

424

Process and Procedures

> ## Care of the Microscope
>
> 1. The microscope is a precision instrument that requires proper care. Always carry the microscope with both hands, one hand under its base, the other on its arm.
>
> 2. Keep the microscope away from the edge of the table. If a lamp is attached to the microscope, keep its wire out of the way. Move everything not needed for microscope studies off your lab table.
>
> 3. Avoid tilting the microscope when using temporary slides made with water.
>
> 4. The lenses of the microscope cost almost as much as all the other parts put together. Never clean lenses with anything other than the lens paper designed for this task.
>
> 5. *Always* return the microscope to the low-power setting before putting it away. The high-power objective extends too close to the stage to be left in place safely.

Part A Setting Up the Microscope

1. Rotate the low-power objective into place if it is not already there. When you change from one objective to another you will hear the objective click into position.

2. Move the mirror so that you obtain even illumination through the opening in the stage, or turn on the substage lamp. Most microscopes are equipped with a diaphragm for regulating light intensity. Some materials are best viewed in dim light, others in bright light.

   WARNING: **Never use a microscope mirror to capture direct sunlight when illuminating objects under a microscope. The mirror concentrates light rays, which can permanently damage the retina of the eye. Always use indirect light.**

3. Make sure the lenses are dry and free of fingerprints and debris. Wipe lenses with lens paper only.

Part B Using the Microscope

1. In your journal, prepare a data table similar to the one in Figure B.9.

Microscopic Observations

Object Being Viewed	Observations and Comments
Letter *o*	
Letter *c*	
Letter *e* or *r*	
mm ruler	

Figure B.9 Microscopic observations

2. Cut a lowercase letter *o* from a piece of newspaper. Place it right side up on a clean slide. With a dropping pipet, place one drop of water on the letter. This type of slide is called a wet mount.

425

3. Wait until the paper is soaked before adding a cover slip. Hold the cover slip at about a 45° angle to the slide and then slowly lower it. Figure B.10 shows these first steps.

4. Place the slide on the microscope stage and clamp it down with the stage clips. Move the slide so that the letter is in the middle of the hole in the stage. Use the coarse-adjustment knob to lower the low-power objective to the lowest position.

5. Look through the eyepiece and use the coarse-adjustment knob to raise the objective slowly, until the letter *o* is in view. Use the fine-adjustment knob to sharpen the focus. Position the diaphragm for the best light. Compare the way the letter looks through the microscope with the way it looks to the naked eye.

6. To determine how magnified the view is, multiply the number inscribed on the eyepiece by the number on the objective being used. For example, eyepiece (10X) × objective (10X) = total (100X).

7. Follow the same procedure with a lowercase *c*. Describe in your journal how the letter appears when viewed through a microscope.

8. Make a wet mount of the letter *e* or the letter *r*. Describe how the letter appears when viewed through the microscope. What new information (not revealed by the letter *c*) is revealed by the *e* or *r*?

9. Look through the eyepiece at the letter as you use your thumbs and forefingers to move the slide slowly *away* from you. Which way does your view of the letter move? Move the slide to the right. Which way does the image move?

10. Make a pencil sketch of the letter as you see it under the microscope. Label the changes in image and movement that occur under the microscope.

11. Make a wet mount of two different-colored hairs, one light and one dark. Cross one hair over the other. Position the slide so that the hairs cross in the center of the field. Sketch the hairs as they appear under low power; then go to Part C.

Part C Using High Power

1. With the crossed hairs centered under low power, adjust the diaphragm for the best light.

2. Turn the high-power objective into viewing position. Do *not* change the focus.

3. Sharpen the focus with the *fine-adjustment knob only. Do not focus under high power with the coarse-adjustment knob. This can damage the objective and the slide.*

Figure B.10 Preparing a wet mount with a microscope slide and cover slip

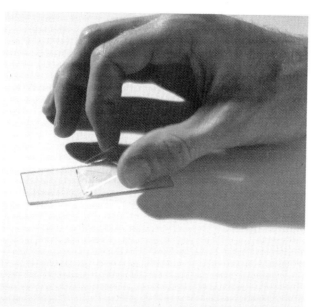

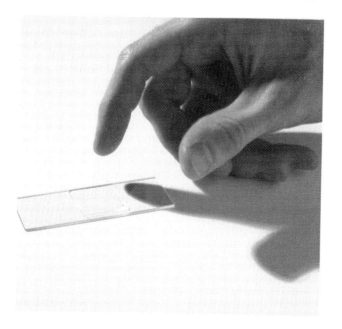

4. Readjust the diaphragm to get the best light. If you are not successful in finding the object under high power the first time, return to Part C, Step 1, and repeat the entire procedure carefully.

5. Using the fine-adjustment knob, focus on the hairs at the point where they cross. Can you see both hairs sharply at the same focus level? How can you use the fine adjustment knob to determine which hair is crossed over the other? Sketch the hairs as they appear under high power.

Part D Measuring with a Microscope

1. Because objects examined with a microscope usually are small, biologists use units of length smaller than centimeters or millimeters to make microscopic measurements. One such unit is the micrometer, which is one thousandth of a millimeter. The symbol for micrometer is µm, the Greek letter µ (called mu) followed by the letter *m*.

2. You can estimate the size of a microscopic object by comparing it with the size of the circular field of view. To determine the size of the field, place a transparent metric ruler on the stage. Use the low-power objective to obtain a clear image of the divisions on the ruler. Carefully move the ruler until its marked edge passes through the exact center of the field of view. Now, count the number of divisions that you can see in the field of view. The marks on the ruler will appear quite wide; 1 mm is the distance from the center of one mark to the center of the next. Record the diameter, in millimeters, of the low-power field of your microscope.

3. Remove the ruler and replace it with the wet mount of the letter *e*. (If the mount has dried, lift the cover slip and add water.) Using low power, compare the height of the letter with the diameter of the field of view. Estimate as accurately as possible the actual height of the letter in millimeters.

Analysis

1. Summarize the differences between an image viewed through a microscope and the same image viewed with the naked eye.

2. When you view an object through the high-power objective, not all of the object may be in focus. Explain.

3. What is the relationship between magnification and the diameter of the field of view?

4. What is the diameter in micrometers of the low-power field of view of your microscope?

5. Calculate the diameter in micrometers of the high-power field. Use the following equations:

$$\frac{\text{magnification number of high-power objective}}{\text{magnification number of low-power objective}} = A$$

$$\frac{\text{diameter of low-power field of view}}{A} = \text{diameter of high-power field of view}$$

For example, if the magnification of your low-power objective is 12X and that of your high-power is 48X, A = 4. If the diameter of the low-power field of view is 1600 µm, the diameter of the high-power field of view is 1600 ÷ 4, or 400 µm.

EXPLORING BIOLOGY

ESSAY SECTION

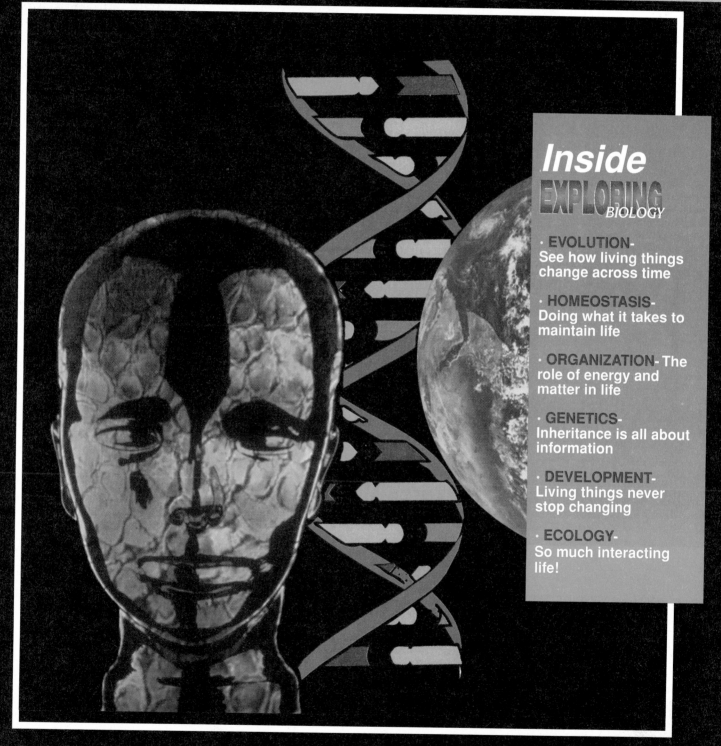

Inside EXPLORING BIOLOGY

- **EVOLUTION-** See how living things change across time

- **HOMEOSTASIS-** Doing what it takes to maintain life

- **ORGANIZATION-** The role of energy and matter in life

- **GENETICS-** Inheritance is all about information

- **DEVELOPMENT-** Living things never stop changing

- **ECOLOGY-** So much interacting life!

A collection of essays to help explain the living world around you

The Chimp Scientist

. . . I have often wondered exactly what it was I felt as I stared at the wild country that so soon I should be roaming. Vanne admitted afterward to have been secretly horrified by the steepness of the slopes and the impenetrable appearance of the valley forests. And David Anstey told me several months later that he had guessed I would be packed up and gone within six weeks. I remember feeling neither excitement nor trepidation but only a curious sense of detachment. What had I, the girl standing on the government launch in her jeans, to do with the girl who in a few days would be searching those very mountains for wild chimpanzees?

Jane Goodall, *In the Shadow of Man*

When Dr. Jane Goodall began to observe and study the chimpanzees of Gombe Stream in Tanzania, Africa, (see Figure E1.1) more than 30 years ago, some people thought that it was not scientific to talk about an animal's mind or personality or to give funny names to their subjects. Today, however, we recognize Goodall's work as one of the great contributions to science made during the twentieth century.

Thirty years ago, some people thought of science as experiments that men in white coats conducted in laboratories. Indeed, some people still have that mistaken image today. As you have been discovering, however, science is a particular way of knowing and learning about the world, and we each can learn the methods of science and put them into practice. It likely is becoming clear to you that observation and reflection are key aspects of science. Dr. Goodall's work is an exemplary model of both of these aspects.

In order to learn about the true nature of chimpanzees in the wild, you have to go into the wild—into their natural habitat—and observe them across long periods of time. It took Goodall close to four years to collect a significant amount of information. The chimpanzees were elusive and resisted her presence for the first several years, but eventually they came to trust her—this itself was a key observation. Goodall recorded meticulously their every move, interaction, gesture, and grunt. She discovered the intricacies of their way of living and of their social nature. She discovered through the years that not only are chimpanzees caring, clever, and capable of lasting attachments, but some individuals are capable of extreme aggression, and one practiced cannibalism.

Thirty years ago, scientists thought that humans were the only tool-making animals and generally assumed that humans were the only organisms that could think or have emotions. Today, due to the work of Dr. Goodall and other scientists in related fields, we recognize that many intellectual abilities once thought to be unique to humans are present in other animals. For instance, Goodall observed over and over the ability of chimpanzees not only to use tools but to recognize a

Mapping the Brain

In the mid-1800s, an engineer named Phineas Gage made a spark while tapping a metal rod to set a dynamite charge, and in that instant his whole life changed. The charge exploded, and the metal rod shot up with enormous force and passed through the young man's cheek, eye, and the front part of his brain. He fell backward. The other workers thought he was dead. Remarkably, he survived and it seemed that his only lasting injury would be the loss of sight in one eye. Unfortunately, his loss of sight was only a small part of the lasting effect. As time passed, people who worked with Gage found him to be an entirely different person. His good nature was a thing of the past; he had survived, but not as himself. Instead, he was a foul-mouthed, rude, and untrustworthy person.

For more than 100 years, the strange effects of Gage's remarkable injury and recovery remained a

Figure E1.2 Mike bangs kerosene cans together. Although Mike was not the largest or strongest male, he made his way to the top by creating frightful noise with these empty cans.

need for them and make one ahead of time. One such observation involved a chimp named Mike. He finally solved the problem of how to get a banana that Dr. Goodall held out to him. Apparently too nervous to take the banana from her hand, he finally used a stick to knock it to the ground. Only then did he grab it and run. And even though Mike was not the biggest or strongest male, he was clever. He was able to bluff his way to become the top male by banging together empty kerosene cans that he found in Goodall's camp (see Figure E1.2).

Without careful, continuous, and unobtrusive observations such as Dr. Goodall's, we would not be able to construct our understanding of the true nature of the chimpanzee. And because the chimpanzee is the closest living relative of the human species, we have the possibility to learn more about our common evolutionary heritage when we study the chimp.

Figure E1.1 The Gombe Stream Chimpanzee Reserve The Gombe Stream Chimpanzee Reserve is located near Kigoma, on Lake Tanganyika in Tanzania.

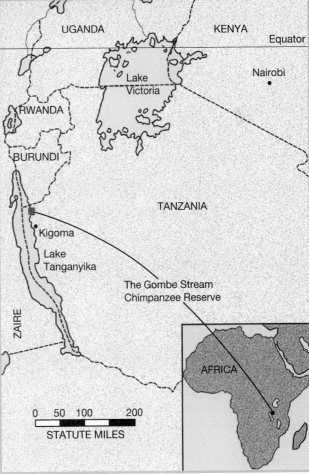

mystery, but in the 1990s, scientists combined their improved understanding of brain function, computer modeling techniques, and new data from Gage's skull to find the answer. The scientists found that the accident damaged both hemispheres of Gage's frontal lobes, which is the part of the brain that influences social behavior (see Figure E1.3).

This dramatic and unusual story is just one example of the way the human brain and the brains of other animals have specialized regions that control different activities. For instance, the victim of a motorcycle accident who sustains a severe

brain injury may continue to breathe and to have a regular heartbeat, yet he or she may have no awareness of the surroundings, no conscious thoughts, and no voluntary control over the movement of arms and legs. In a situation like this, the family members, the physicians, and the legal experts are faced with a difficult question: Is this patient a living person, even though there is no thought and the body can stay alive only through artificial feeding? This dilemma raises many

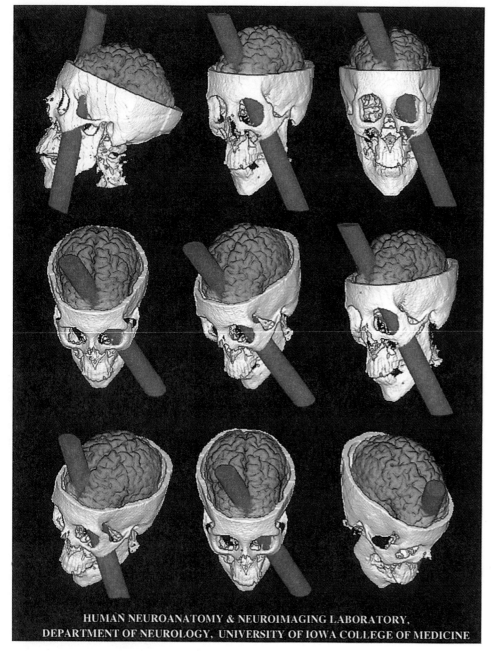

HUMAN NEUROANATOMY & NEUROIMAGING LABORATORY,
DEPARTMENT OF NEUROLOGY, UNIVERSITY OF IOWA COLLEGE OF MEDICINE

Figure E1.3 In 1994 scientists at the University of Iowa used measurements from Gage's skull along with modern neuroimagery technology to revisit the accident and determine the likely location of damage to the brain. Dr. Hanna Damasio created this series of digital images showing the results of their work from different angles.

Reprinted with permission from *Science 264*, (5162) "The Return of Phineas Gage: Clues About the Brain from the Skull of a Famous Patient." © 1994 American Association for the Advancement of Science

cerebrum

brainstem   cerebellum

complex, ethical questions, but it also shows an important biological fact: the brain has specialized functions, and these functions often can be mapped to specific regions of the brain.

We have learned quite a bit about the role of different parts of animal brains. The drawing in Figure E1.4 shows only some of what we know about the different functions of various regions of the brain. The large upper part of the brain, the **cerebrum**, is responsible for complex reasoning, thought, language, voluntary movement, and the initial processing of sensations.

The structure at the back of the brain, which in certain views looks like a clamshell, is called the **cerebellum**; it controls posture and balance. The **brainstem** directs critical, life-sustaining activities such as breathing and heartbeat. This part of the brain evolved long ago, and it occurs in one form or another in many animals. In the example of the victim of the motorcycle accident mentioned previously, the brainstem may have been uninjured. If this were so, the victim would continue to breathe, the heart would continue to beat, and other body functions would continue, but the cerebrum may have been damaged so seriously that the victim would be unable to think.

Figure E1.4 Regions of the brain The cerebrum is responsible for conscious thought, language, and voluntary movement. The cerebellum is responsible for balance and posture.

E4

Do You Have a Grip on That?

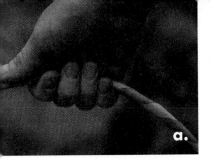

a. **Hook grasp:** the fingers are used as a hook; the thumb is not used.

b. **Cylindric grasp:** the palm is used to grasp a cylindrical object; the thumb opposes all of the fingers.

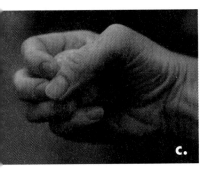

c. **Fist grasp:** the fingers are wrapped around a narrow or small object, the thumb opposes the fingers, lying on top of them and securing the grip.

d. **Spheric grasp:** the palm is used to grasp a spherical object; the thumb opposes the fingers.

Tying a shoe, threading a needle, throwing a baseball, taking notes, or carrying a bucket. Think of all the tasks—large and small—that you do with your hands every day. Have you ever thought about the amazing number of ways in which you can use your hands to grip something?

Scientists have studied the different grips of primates for some time and have more than one way of classifying what they have observed. One school of thought divides grips into two major categories: power grips and precision grips. You use the power grip when you grasp an object with your palm and then curve your fingers around it. The power from the power grip comes as you apply pressure with your whole hand. This power is especially evident in the muscles of your palm that are located at the base of your thumb and that are responsible for moving your thumb. The power grip requires the full grip of the hand—when you use a hammer or open a tightly closed jar, for example.

The precision grip is a more intricate grip that requires a specific alignment of the thumb with one or more fingers. The thumb applies pressure against another finger or fingers in order to accomplish a precise motion, such as picking up a coin, or to accomplish a precise motion that also requires strength, such as placing a key in a lock and turning it.

Another school of thought subdivides the power and precision grips into as many as twelve categories by defining the grip according to the exact placement of the palm, fingers, and thumb. Figures E1.5 a-g show 7 of the 12 grips.

Figure E1.5 a-g Types of grips These photographs show 7 of the 12 ways humans can grip objects. Which grips did you see the primates use in the video *Observing Primates*? Which grip did you test in *Primates Exploring Primates*?

e. **Tip prehension:** the tip of the thumb and one or more fingers are used like tweezers to pick up small objects; the thumb opposes one or more fingers.

f. **Palmar prehension:** the flat surface of the thumb is pressed against the flat surface of the finger or fingers; this grip is similar to the way the tip of a pair of pliers holds an object.

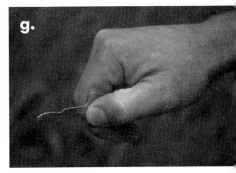

g. **Lateral prehension:** the thumb presses against the side of the index finger with other fingers providing additional support. Another type of lateral grip involves the sides of two fingers pressing against each other.

On Being Human

Have you ever seen a cow laugh or a cat tap its foot in time to music? Have you ever heard a human purr? Humans are animals, yet humans do things that you would never expect to see in other animals, and other animals do things that you would never expect to see a human do. You are quite capable of *being* human, but you may not have taken time to really *think* about what it means. (The cow and cat haven't thought about themselves either, but as far as we know, they cannot do so. The fact that you can is just one of many ways in which humans can do unusual things.)

A surgeon reaches carefully past his assistant to suture in a small bit of vein that will bypass a block in the coronary artery on the patient's heart. The assistant firmly places the correct steel tool in the surgeon's palm and watches as the delicate procedure is completed.

Across town, in a concert hall, a guitarist waits for the applause to subside before she begins the next song. The fingers of her left hand push against the frets to produce a chord, while the fingers of her right hand rapidly pluck the strings in a set pattern. The audience recognizes the song and applauds.

These examples show some of the characteristics that distinguish humans from other animals—even from other closely related animals, the primates. Humans, like many other primates, can grasp things with their hands and fingers. Primates such as gibbons, chimps, and spider monkeys also can grasp with their feet, a trait most humans do not share. However, the fine dexterity required to suture a vein or to

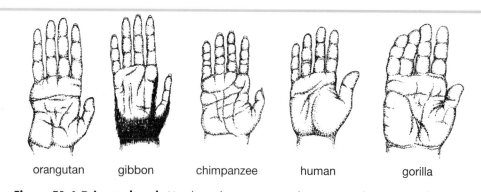

orangutan   gibbon   chimpanzee   human   gorilla

Figure E1.6 Primate hands Use these drawings to supplement your observations of primate hands. Which hands look capable of a power grip? Do you notice anything different about muscle development in the hands? Which hands do you think could use the precision grip? (See the essay *Do You Have a Grip on That?*)

play a complicated pattern of notes on the guitar appears to be unique to human hands (the drawings in Figure E1.6 allow you to compare a series of primate hands). This precision is possible because humans have fully opposable thumbs that can move to touch the pad of any of the other fingers with precision. Not only is our grasp precise, it is strong—strong enough to grasp a baseball bat and hold on to it as we swing it with force or to grasp a hammer and use it to pound a nail into a solid piece of wood.

Other primates share some degree of these various grips. A baboon that had been stung by a scorpion was able to grip the stinger and remove it. This action, however, does not require that the grip be strong. Human hands are shorter and broader than those of the baboon; our thumbs are longer, stronger, and somewhat more flexible. The structure of the human hand

allows room for additional muscle attachment, and this results in a more flexible, more precise grip than that of the baboon.

Opposable thumbs are not the only characteristic that distinguishes humans from other primates and all other animals. Humans stand on two feet and walk fully upright with the body balanced directly over the feet. This **bipedal** method of moving is very different from other animals such as horses, spiders, or animals that fly. Although other primates sometimes walk upright, humans are the only ones that are *consistently* bipedal.

For any animal with a skeleton, the skeleton has features particularly suited to the animal's way of moving. Bird bones, for example, are mostly hollow and consequently are light—a property important for flying. Humans have large leg bones that support their weight, and the bones of the spine form an S curve that is unlike the rounded arch of the gorilla's back (refer to Figure E1.7). Standing erect puts a lot of force and downward stress on the body. The S curve helps the human body tolerate that stress. In contrast, the gorilla usually uses all four limbs and walks on its feet and on the knuckles of its hands. For brief periods of time, the gorilla can walk upright, but

the arch in its back is better suited to walking on all fours (see Figure E1.8).

Humans have other adaptations suited to bipedal walking. Look at your foot. The average human foot has rounded heels, an arch at mid-foot, and shorter toes than those of other primates (see Figure E1.9). These features enable the foot to balance well and support the body weight even while walking on flat surfaces. Our legs are close together, allowing our weight to be balanced over each foot as we walk. A gorilla's foot certainly is more similar to a human foot than that of a cow or a lizard, but the gorilla's foot has longer toes and a big toe that looks and functions more like a thumb.

The upright posture of humans has an important advantage: it frees our hands and arms to do something other than to carry our bodies around. This way of moving about enables humans to gather and carry food or to carry and use tools.

Think back to the example of a surgeon (a tool user) repairing a damaged blood vessel. Even if a chimpanzee or gorilla could hold the surgical tools, would you want it to perform surgery on you? The human surgeon has a characteristic that makes such activities possible: humans have large and complex brains. It is important to remember, however, that the human brain is not the largest on earth—an elephant's brain and a whale's brain are larger. The brains of all animals control similar functions such as breathing, movement, and other body activities, but the part of the brain that controls abstract thinking and reasoning is very

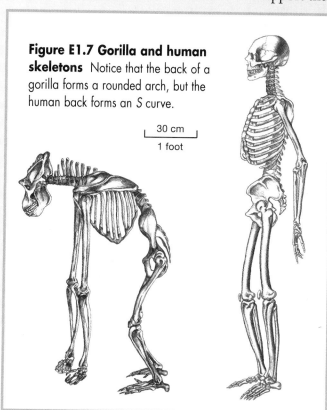

Figure E1.7 Gorilla and human skeletons Notice that the back of a gorilla forms a rounded arch, but the human back forms an S curve.

30 cm
1 foot

large and well developed in humans. In this function the human brain appears to surpass all other animal brains—even those of relatively intelligent animals such as chimpanzees or dolphins. As a human, you have special abilities to find solutions to puzzles, to use complex language that is based on symbols, to use memories of the past to plan the future, and to reflect on all of these many abilities.

Although other animals may have some small degree of these characteristics, these abilities are very well developed in humans, and as a collection, they distinguish humans from other organisms. Every animal and, indeed, every organism has a combination of characteristics that makes it distinctive; it also shares many characteristics with other living things. For example, human vision is better than that of a dog, but not nearly as acute as the vision of an eagle, which must spot prey on the ground from high up in the air. The dog, however, has an acute sense of smell to help it hunt. Humans are not nearly as fast as a cheetah, but humans are much faster than an armadillo or a slug. Each organism shares many basic life properties but each can be recognized by a collection of special traits.

In the huge and diverse group of animals, humans are most similar to mammals, which are animals that have hair, give birth to live young, and nurse their offspring. Within the group known as mammals, humans are most closely related to the primates. And now, having read this essay, you have accomplished something that as far as we know, no organism other than a human can do: you can read.

Figure E1.8 Methods of movement The gorilla is quadrupedal and walks on its hind feet and knuckles. Humans are completely bipedal, except as infants.

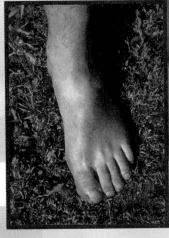

Figure E1.9 A gorilla foot and a human foot Notice how the gorilla foot is similar to the gorilla hand (Figure E1.6), but the human foot, adapted for upright bipedal movement, is significantly different from the human hand.

Brains and a Lot of Nerve

Think hard! Now, which part of the cerebrum did you use? It depends in part on what you were thinking about. If you were trying to calculate your expenses this week, the left half of your cerebrum may have been the most active. On the other hand, if you were imagining how you would draw a picture, you probably used more of the right half of your brain.

Not only are the physical structures of the brain specialized, but within each of the structures there is further specialization. The cerebrum has two distinct parts, or sides, called the left and right hemispheres (see Figure E1.10). The hemispheres individually control certain behaviors: sensory and movement functions on the left side of the body are controlled by the right hemisphere of the cerebrum and vice versa. The specific area of the cerebrum that is responsible for controlling the movements of your body is called

motor cortex
sensory cortex

Figure E1.11 In humans, the motor cortex and the sensory cortex are located alongside each other, across both hemispheres of the cerebrum. The cortex is the top layer of the cerebrum.

the motor cortex, and the specific area of the cerebrum that is responsible for letting you feel sensations from your body is called the sensory cortex (see Figure E1.11).

Scientists have been able to map many sites on the motor cortex to a particular part of the human body, and they have been able to do the same for the sensory cortex. That is, when you move a certain part of your body, scientists know what part of your motor cortex is making that movement possible. And when you smell a rose, scientists know what part of your sensory cortex is allowing you to experience that fragrance. Figure E1.12 shows the results of this mapping process. The figure of the person looks distorted because the size of the area that is mapped to a particular part of the body is not related to the actual size, but rather to the amount of precision required for that part of the body to do its job. In humans, a good deal of space is dedicated to the face, tongue, and hands. Why do you think this is so?

In most individuals, the left hemisphere of the human brain controls language and speech as well as mathematical and other analytical abilities. The right hemisphere is responsible for visual and spatial processing. Many functions, including various aspects of those mentioned above, are controlled jointly by both hemispheres. Communication between the two hemispheres is possible because a bundle of nerve fibers, called the corpus callosum, connects them.

left hemisphere   **right hemisphere**

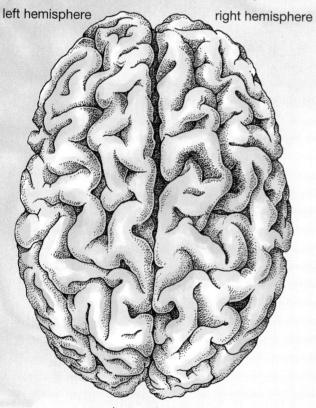

human brain

Figure E1.10 Top view of human brain (approximately 65 percent of life-size) The human cerebrum is split into halves or hemispheres.

An unusual surgical procedure illustrates an interesting exception to this rule of joint hemisphere control and reveals the complexity of the brain. In extreme cases of epilepsy—a human condition that causes seizures—the fibers of the corpus callosum are surgically severed, separating the two hemispheres. This procedure greatly reduces the number and severity of seizures, and most patients notice no impairment in perception or movement. Their perception, however, *is* impaired. Messages from their right hand are received only by the left hemisphere, messages from their left hand are received only by the right hemisphere, and because of the surgery, the hemispheres no longer can share information.

The patients generally do not notice this impairment because they compensate by using their eyes. Unlike messages from the hands, messages from *each* eye are received in *both* hemispheres. Some interesting studies have been done on these patients that reveal the nature of their impairment. When a researcher places a blindfold on the patient and then places a common object, perhaps a toothbrush, in his or her left hand, the patient can describe what the object feels like ("it has a long, hard handle and soft bristles at the tip"). Sometimes the patient can describe how he or she would use the object ("you use it every morning"), but cannot name the object. When the researcher places an object in the patient's right hand he or she can name it ("it's a toothbrush") but cannot describe to the researcher how to use it.

Again, under normal circumstances in which the patients are able to use both of their hands and their eyes, their brains are able to compensate for their impairments and the patients experience little difficulty.

This example, and observations obtained from many other surgical procedures and experiments, suggests that the brain can process information in more than one way. In addition, particularly when the individual is young, certain functions normally performed by one region can be taken over by other regions. For instance, when young children suffer strokes that damage their left hemisphere, instead of losing some or all of their language ability, the right hemisphere often takes over this function. Also, if input from one of the senses is impaired, input from another sense will compensate. For these reasons, the brain often is said to be *plastic*—that is, it can be molded by certain events and experiences to accommodate particular circumstances.

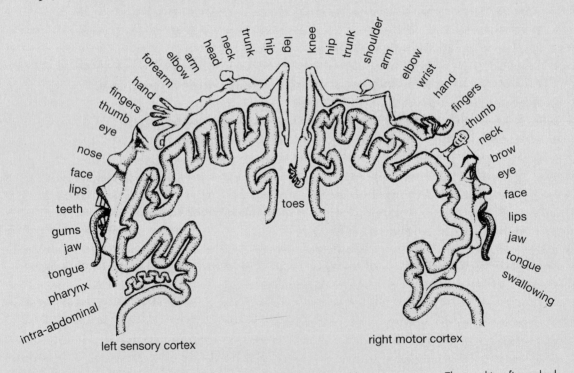

left sensory cortex   right motor cortex

Figure E1.12 Most parts of the human body can be mapped to the motor and sensory cortices. The resulting figure looks distorted because the area of the cortex that is devoted to a particular body part is not related to the actual size of that part but to the amount of precision required to control it. In this drawing only the left sensory cortex and the right motor cortex are shown.

A Complexity of Interactions. It may surprise you to know that in many ways you are capable of thinking better than a computer can. In fact, it has been very difficult to get computers to do things that are easy for humans to do, such as to recognize the faces of different people. Computer programs only very recently have been able to beat master chess players. Some scientists have used computer programs to try to design artificially intelligent robots that model the abilities and processes of the human brain. So far these scientists only have been able to model the way a cockroach processes information, but even this task was very challenging. Imagine how hard it would be to build a computer and develop a program that models all of the abilities of the human brain.

One reason it is so difficult for scientists to build a computer model of the human brain is that the human brain contains many nerve cells or neurons, and the number of interconnections they form in the brain is astounding. Your brain contains about 10^{12} (1 trillion) **neurons**, and the many ways they can interconnect result in trillions of potential associations. Until someone develops computer technology that can form many more interconnections, it will be difficult to model thinking that is similar to that of a human brain.

In order to understand how these neuronal interconnections work, it is useful to look closely at the structure of a neuron. Figure E1.13 shows how unusual this cell is. At one end of a neuron are branched extensions, or dendrites, that act like antennae to receive incoming messages. At the other end of the neuron is an extension called an axon, which sends outgoing messages. Each neuron connects to hundreds or thousands of other neurons. This arrangement means that a single neuron can send or receive messages in many combinations with other neurons. These neurons form an organized network of interconnections that extends throughout the brain and body. All together, this **nervous system** provides a link between the outside world and the inner world of the body.

Incoming information that causes the body to respond is called a **stimulus**. The human nervous system, like those of all animals, collects this information from stimuli and signals the brain. The brain then processes the information and responds by sending back a signal to the appropriate part of the body. For example, the stimulus of seeing the face of a friend allows the brain to recognize the person and signal the face to smile.

How is a message sent from one neuron to another? The message is a signal that is caused by changes in the electrical charge of the neuron. An electrical impulse moves along an axon to the end of the neuron. There the electrical impulse triggers the release of chemicals called **neurotransmitters**. Neurotransmitters act as messengers that transmit the signal across the gap between neurons. This gap is called the **synapse**. When the neurotransmitters encounter the dendrites of the next neuron, the chemical signal is converted into an electrical signal. The electrical signal moves along the length of the axon and is ready to release neurotransmitters into the next synapse. This process is repeated to form paths of signals

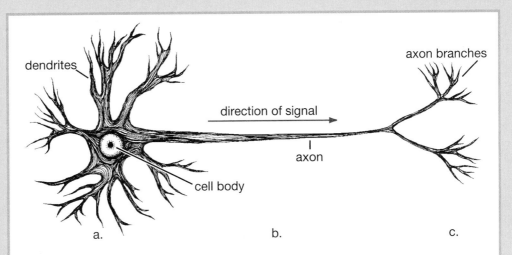

Figure E1.13 A neuron is a nerve cell. It has (a.) extensions called dendrites, which receive nerve signals, (b.) a long axon, which carries the nerve signals, and (c.) axon branches, which transmit the nerve signals to the next neuron.

along the routes of various neuronal networks. These events happen very quickly, so you are able to have a thought and signal your mouth to speak or your legs to run in an amazingly short time. This system of interconnected neurons and their ability to send and receive signals is found generally throughout different types of animals, not just humans.

Not all responses require processing through the thinking centers of the brain. If you touch a hot object, you immediately will withdraw your finger without having to think at all. Such a simple and quick response to a stimulus is called a **reflex**. The messages involved in this response travel from sensory detectors in your finger, through neurons in your arm, to the spinal cord, then back to the muscles in your arm through other neurons, along a pathway called a reflex arc (see Figure E1.14). At the same time, messages are sent from the spinal cord to the brain. When your brain receives these messages, you interpret them as "that thing is HOT!" By this time, however, you already have removed your finger from the hot object because of your reflex response. The advantage of a reflex over a signal the brain processes is speed. In such a situation, you would need to move fast to prevent damage to your body. Reflex responses can save your life.

Newborn human babies turn their heads toward a touch on their cheek. This reflex encourages newborns to nurse in the first few hours of life, but they lose this reflex as time goes by. Most human activity, however, is processed through the brain and is not carried out as reflexes. The human brain, with its trillions of interconnections between neurons, is well-equipped to handle the everyday business of directing the activities of human life. Even simpler animals such as earthworms respond reflexively to external stimuli. The ability to respond to heat, light, and moisture helps the worm to find food and to keep from being dried out on a hot, sunny day.

Because the brain is responsible for so much complex human behavior, it not only can respond to a signal from an outside stimulus, but it also must be able to store and retrieve certain patterns of information. For example, when you pick up a book and begin to read, you do not have to consciously think about how to read, you do it naturally. When you first were learning how to read, however, you needed to think about what sounds the letters represented and how they went together to form a word, and how the words went together to form a meaningful sentence. These processes only seem natural to you now because your brain has learned and stored

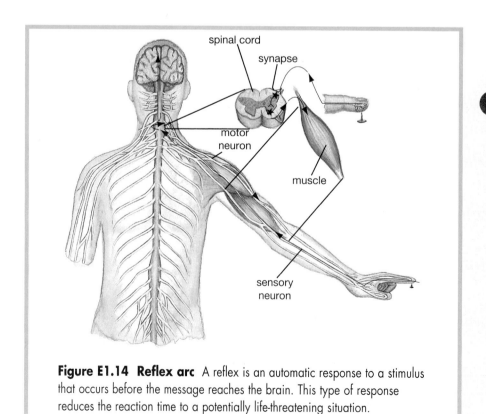

Figure E1.14 Reflex arc A reflex is an automatic response to a stimulus that occurs before the message reaches the brain. This type of response reduces the reaction time to a potentially life-threatening situation.

the patterns for thousands of words and the patterns for a variety of sentence structures.

Now that you are at the end of this essay, what do you think your eyes and brain have been doing while you were reading?

Figure E1.15 During our long childhoods, we learn about ourselves and our societies.

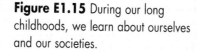

The Importance of Being Children

"Will I be a grown-up tomorrow?"

This is a question that 4-year-olds all over the world ask in hundreds of different languages. The words may be different, but the question is the same—When will I know all that I need to know to be me?

Growing up and learning are intriguing and often puzzling processes, but even as adults, humans never stop learning. The complexity of interconnections in the human brain is one reason that humans have such a tremendous capability for learning. In fact, scientists who study the brain and the nervous system think that certain connections between neurons form *in response to* learning. That is, the physical structure of the brain may be influenced, in part, by *what* and *how much* we learn. This plasticity, or ability to be molded, is particularly true of infants and young children because their brains (and bodies) still are growing at a rapid rate. Indeed, our longer period of childhood dependency may contribute significantly to making our capacity for learning greater than that of other animals.

Our long period of childhood dependency and our great capacity for learning also give us time to acquire language, which is perhaps one of the most important things that humans ever learn. Most languages have numerous intricate rules of grammar and thousands or tens of thousands of vocabulary words. As we learn language, we can communicate an infinite number of thoughts and ideas, all of which are understood by others who speak the same language. Despite this apparent complexity, learning language seems to occur quite naturally. Through the use of language and in the context of a long childhood, adults are able to teach children about their **culture**—the set of shared, learned behaviors and beliefs that are passed from one generation to the next (see Figure E1.15).

Casual observations make it quite clear that it does not matter what culture or society a child is born into; a child naturally will learn to speak the language that is spoken in the household. Children's capacity for learning languages is so great that children who grow up hearing one language exclusively at home and a different language exclusively at a relative's house can become fluent in both languages with apparently little effort on their part. Language, like many cultural behaviors, is

Figure E1.16
Communication between wolves
Animals communicate in a variety of ways with different levels of complexity. Wolves communicate with each other through vocalizations such as howling, barking, and growling as well as behaviors such as licking and submissive postures.

learned, or *acquired*, naturally as part of normal development. However, the opportunity for learning language in this most efficient and natural way is quite limited. After age 8 or 9, the capacity for learning a language diminishes. This limitation explains why it is more difficult to learn a foreign language in high school or college.

Humans, of course, are not the only animals that communicate with each other (see Figure E1.16). Many organisms communicate with each other through grunts, whistles, songs, and movements, but do we consider these activities language? Experiments show that trained dolphins can understand a simple series of signals, and chimpanzees can learn some elements of symbolic language. In fact, one chimp has been trained to understand more than 100 *spoken* English words. And other chimpanzees and gorillas have learned to communicate using American Sign Language and are thought to communicate about as well as a 4-year-old child would. Some of these chimps also have taken certain signs and created their own words; one chimpanzee put together the sign for *cold* and the sign for *food* to create a sign for *refrigerator*. Such studies indicate that these animals have the capacity for languagelike behavior.

Similarly, a number of different animals appear to think and solve problems creatively. Simple tool use seems to be fairly common in the animal world. Sea otters carry rocks that they use as anvils to break shellfish open, and vultures drop stones on ostrich eggs to break them. Some ants even use bits of wood as sponges to soak up liquid so that they can transport it back to the nest. It is not clear whether these activities really are learned, or if they represent behaviors that individuals can modify quickly to fit changing circumstances.

Consider the activities of a honeybee society. This society is composed of many individuals with different roles, and all members of the society must coordinate their activities to accomplish the tasks of laying eggs, providing food for the developing larvae, and collecting food from the environment. These relationships are maintained through a complex system of communication that is heavily dependent on the release of chemicals by the queen bee and on a series of physical movements called dances. Although these behaviors are very complex, they are not learned. Instead, they are relatively inflexible, biologically determined responses that appear almost as soon as the adult bee emerges from the hive. Societies based on this type of behavior often are very successful as long as conditions are constant or change slowly, but these societies do not lend themselves to individual action or promote much creative response to novel situations.

In contrast to the social insects, such as bees, ants, and termites, the group interactions of other animals generally are more variable and less dependent on rigidly determined roles and behaviors. In primate societies, learning, flexibility, and individuality are important characteristics. For example, when a macaque monkey in Japan began washing her sweet potatoes in a stream before eating them (see Figure E1.17), the practice gradually spread to almost every member of the troop and to the next generation of macaques. Bipedal walking also became more common as the monkeys learned to carry the potatoes in their hands to the water. Later, the monkeys learned to separate sand from wheat that they had collected on the beach. They carried it to the sea and plunged it into the water; the sand sank and the wheat floated. Then they quickly retrieved the wheat and ate it. The macaques adopted these new habits after people set up a feeding station. These behaviors demonstrate creative responses to novel situations.

This type of behavioral flexibility—the capacity not only to learn how to perform established behaviors but to invent new behaviors—reaches its greatest expression in human societies. Although humans and many other animals have incompletely

Figure E1.17 Behavioral flexibility When a feeding station was established, one macaque learned to wash sweet potatoes in the water before eating them. This behavior was adopted by most of the macaques in the community and even was passed on to the next generation of macaques.

development, which gives us the opportunity to learn about our culture.

Language, in turn, helps us to record and communicate our culture more efficiently than if we had to rely exclusively on experience, imitation, and other nonverbal forms of communication. Imagine what our world would be like if we eliminated just one of our cultural uses of language, the preservation of information in libraries. Each new generation would have to rediscover or reinvent all of the complex technologies of the previous generation because oral transmission of so much detailed information would be impossible. The use of libraries, on the other hand, means that if technological or social problems arise, we have a vast resource of information that we can use. This interplay between language and culture helps to further distinguish humans from all other animals.

formed nervous systems at birth, a human's nervous system continues to develop for a much longer time. Consequently, we are influenced by our experience to a much greater degree than are other animals. Much of our tremendous behavioral flexibility is put to use during this long period of childhood

FOSSILS: TRACES OF LIFE GONE BY

Glancing around nervously, a whalelike creature waddles awkwardly on its very short legs down the sandy slope. Once it enters the water, however, it moves gracefully, swimming with its powerful tail.

Although this particular mammal is a *Rodhocetus*, one of the ancestors of modern whales, it has something very special in common with dinosaurs, bacteria, ferns, horses, and even Lucy. They all left fossil evidence behind when they died. In each case, a series of unusual events occurred that preserved their bodies in the form of fossils, which scientists sometimes find and use in their studies of biological change.

Fossils do not form easily. From your own experience, you know that dead animals often are eaten by scavengers or quickly consumed by microorganisms. A

recently deceased organism can become a fossil only under rare and unusual conditions: layers of mud or silt quickly must cover the dead organism to protect it temporarily from decomposition (see Figure E2.1). Although the tissues of the organism are now protected, they are not yet fossilized. A fossil is created when minerals in the water replace the minerals in the organism's tissues. Bones are the most common tissue for mineral replacement.

Fossils usually are found in sedimentary rocks, which form as particles of mud, sand, or small pebbles slowly filter through water and settle into layers. The particles in these layers harden into rock after they are cemented together by a chemical reaction or by evaporation. Igneous and metamorphic rocks rarely contain fossils, because these rocks form in the presence of great heat and pressure. If an organism were present during these processes, it

probably would be destroyed. Occasionally volcanic ash or molten lava that has begun to cool will encase an organism and form a fossil.

A preserved fossil might be found if it escapes destruction by geological forces and is exposed by the forces of erosion. For these reasons, fossils usually are discovered near the earth's surface, and generally only the hard parts of organisms, such as shells or bones, are fossilized. (In rare instances scientists have found soft parts of organisms, such as the extinct giant mammoths frozen in arctic ice and the prehistoric insects fossilized in hardened tree sap.) Even if a fossil survives geological change and becomes exposed at the earth's surface, it cannot serve as evidence of biological change unless someone discovers it and realizes its importance.

In most cases fossils are evidence of an organism as it appeared at the time of its death. Other types of fossils allow scientists to make inferences

b. Dodo bird These large, flightless birds became extinct in the seventeenth and eighteenth centuries.

a.

Figure E2.2 Extinct animals a. **Ornitholestes** This dinosaur is now extinct. The mass extinction of dinosaurs was 65 million years ago.

b.

a.

b.

Figure E2.1 Fossil formation a. Dead organisms or their skeletons may remain intact in underwater sediments through long periods of time. Eventually, minerals circulating in underground water replace the bone of the skeletons, forming fossils. b. Buried fossils may be brought to the surface by any of the forces that uplift segments of the earth's crust. Once near the surface, the fossil-containing rock layers are exposed to erosion, which can free the fossil from the rock.

about an organism's life and behavior. For example, footprints of dinosaurs that walked in soft sand or mud sometimes survive as recognizable tracks in solid rock. These trace fossils suggest how the animal walked, including the length of its stride and where it placed its feet.

The fossil record not only provides information about once-living organisms, it also provides an account of how the earth itself changed across very long periods of time. For instance, if we find fossils of marine animals on what now are mountaintops, we may infer that the rock in that area once was under a sea. This may be because an ancient seabed was

lifted up by mountain-building geological forces. In this way the earth in which fossils are found gives clues to physical changes in climate and land conditions, and a combination of fossils and the physical structure of the earth can serve as a geological record of change.

The geological record also reveals that large-scale geological

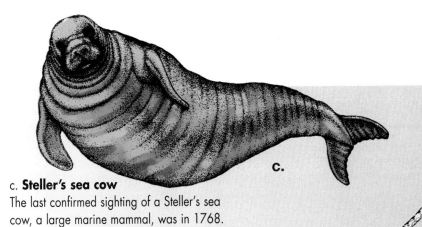

c. **Steller's sea cow**
The last confirmed sighting of a Steller's sea cow, a large marine mammal, was in 1768.

c.

d. **Tasmanian wolf** The last Tasmanian wolf, a doglike marsupial, died in 1936.

d.

changes ultimately are driven by the energy of the earth itself. Deep within the earth, the temperature is very high, and the rock is molten liquid rather than solid. As molten rock, or magma, rises, it can form new land masses or move existing continents, which results in an ever-changing organization of land on earth. Of course, such changes occur only across the vast periods of time that typify geological change. For example, about 230 million years ago, all of the land on the planet was grouped together in one gigantic continent called Pangea. Since that time, the land has broken apart and reformed into the current arrangement of continents, a process called **continental drift**. Continental drift is a geological phenomenon that affects biological change directly because the changes in land connections provide opportunities for living organisms to migrate to new areas. Many such migrations are documented in the fossil record.

A wide-scale analysis of fossil evidence reveals a fundamental feature of living organisms: throughout earth's history, the different types, or species, of organisms living on earth have changed. New species emerge and branch off from existing species, while many other species die off. The most dramatic change possible for a species—its complete disappearance—is called **extinction**. The fossil record chronicles the remains of many now extinct species, such as those illustrated in Figure E2.2. Of all the species that ever existed on earth, 99.9 percent are now extinct. Species continue to become extinct due to natural and manmade changes in the environment, but new species continue to emerge as well. These new species arise from earlier ones through gradual biological change, the change that is shown in the fossil record.

Technologies That Strengthen Fossil Evidence

From sundials to atomic clocks, humans have invented many tools to measure time as it flows continuously from seconds to minutes to hours. Certain natural processes also measure time. For example, a cut through the trunk of a fallen tree reveals growth rings; counting those rings is a way to measure the number of years the tree was alive and growing. The rings, however, do not tell how long it has been *since* the tree fell. Scientists often want to know when a rock was formed or how long ago an organism lived. This is difficult to measure, but sometimes there are clues that tell how long ago an organism lived.

Scientists who study the long-term history of life on earth are more concerned with measuring *when* an organism was alive than *how long* it lived. This task is difficult because these scientists must calculate the age of fossils that are thousands or millions of years old. The age of a fossil may be calculated two ways: indirectly, from the age of the surrounding rock layer in which it is found, or directly, from the age of the fossil itself. Indirect measurements are known as

Figure E2.3 Stratigraphy is a relative dating method. The stratification of rock in this cliff was exposed when a highway was built. Fossils of older organisms are found in the deepest layers; organisms that lived more recently are found in the upper layers.

the dead organism. Thus, scientists infer that any fossil that they find in a particular rock layer is older than the fossils above that layer and younger than the fossils below that layer. Using this knowledge and other techniques, geologists can develop estimates of the time period for each layer of rock, and therefore determine a rough idea of the age of a fossil. Direct dating methods, however, can determine the age of fossils with much greater accuracy.

Direct dating methods usually depend on naturally occurring radioactive elements, which release energy. These elements are unstable forms of matter that break down, or decay, at a steady rate into other chemical elements. The rate of decay varies depending on the element, but it is very precise for a particular element. Elements that undergo decay are called **radioactive isotopes**. For example, an isotope of uranium known as uranium-238 (^{238}U) decays into lead-206 (^{206}Pb). For a given amount of uranium, half of it is converted to lead in about 4.5 million years. Another radioactive isotope, potassium-40 (^{40}K) decays into argon-40 (^{40}A), but at a much slower decay rate. It takes 1.3 billion years for half of a sample of potassium-40 to convert to argon-40. Because the rate of potassium decay is very slow, this method only can be used to date material that is more than 0.5 million years old.

If scientists know the constant rate of decay for a radioactive element, they can measure how much of the radioactive element and its decay product are present in a material such as rock. The ratio of isotope to decay product shows how long the radioactive element has been decaying and thus how old the rock is. Likewise, if the scientist finds a fossil in this rock layer, he or she can infer that the fossil is approximately the same age as the rock that surrounds it. Potassium-argon dating is very useful for measuring the age of fossils because potassium is present in volcanic ash and lava. If fossils are found in between volcanic layers, scientists can determine the likely age of the fossils.

Another similar method is useful for less ancient material, material that is between 500 and 40,000 years old. This method is carbon-14 (^{14}C) dating. Carbon-14 occurs in the atmosphere and becomes a part of plant material as plants use atmospheric carbon dioxide

relative methods because these techniques allow a scientist to say only that a certain fossil is older or younger than another fossil.

Geologists use an indirect method called **stratigraphy** to help them determine the relative ages of fossils. In this method geologists study the strata, or layers, of the earth (see Figure E2.3) to create a rough outline of the earth's history. The relationship between layers of rock and the occurrence of certain fossils was noted nearly 200 years ago. Around 1800 William Smith, an English surveyor and civil engineer, became interested in rock strata because of its relationship to the structural success of the canals he was building. He noticed that certain layers contained fossils, and that throughout England there was a match between the type of rock layer, its placement between other layers, and the fossils it contained.

About the same time, two geologists, Georges Cuvier and Alexander Brongniart, were studying fossils in rock strata in France. When they compared the fossils they found with modern life forms, they discovered that the modern forms were more similar to the fossils from the higher rock layers than those from the lower layers. In fact, no modern species were found in the oldest and deepest layers. The reason for this is related to how fossils are formed. Sediment that settles on top of a dead organism is more recent than the sediment under

during photosynthesis. Carbon-14 also can enter animals or other organisms if those organisms feed on plant tissue. Scientists can measure the amount of carbon-14 and its decay product, carbon-12, in fossil tissue to determine the age of the fossil. Figure E2.4 helps illustrate this process.

Scientists often combine radioactive-isotope dating techniques with other methods for dating material, such as measuring magnetic properties. This improves the reliability of their estimates of fossil ages. The increased reliability occurs because a combination of techniques allows them to check for the consistency of the estimates. These multiple lines of evidence, which support scientific reasoning, strengthen scientists' confidence in the conclusions. For example, a variety of dating techniques have strengthened our estimates of

geological time and have supported the theory of continental drift.

Using these techniques, scientists have determined that the oldest fossils ever found, primitive bacteria, lived 3.5 billion years ago. Newer forms of bacteria are dated at about 2.5 billion years ago. Other microorganisms, ancestral plants, and animals without backbones (invertebrates) appeared later in the fossil record. They were followed by fish, amphibians, and reptiles (300-400 million years ago). Birds, mammals, and modern flowering plants appeared relatively late, about 225-140 million years ago, and Lucy, one of the ancestors of modern humans, appeared relatively recently, at 3.5 million years ago.

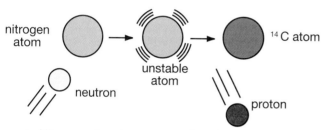

a. Nitrogen atom becomes carbon-14 atom in the atmosphere.

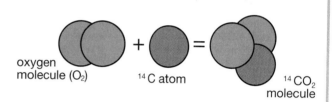

b. An O_2 molecule combines with a carbon-14 atom to form carbon-14 dioxide.

c. Living organisms absorb ^{14}C.

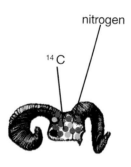

d. When the organism dies, the ^{14}C atoms begin to disintegrate.

Figure E2.4 Carbon-14 becomes part of plants and animals. Small amounts of carbon-14 exist in the atmosphere as carbon dioxide, which plants incorporate into their tissues. This carbon-14 enters animals when they eat plants. Scientists can measure the amount of carbon-14 in fossils and, because they know that the half-life of carbon-14 is 5730 years, they can calculate the age of the organism.

Modern Life: Evidence for Evolutionary Change

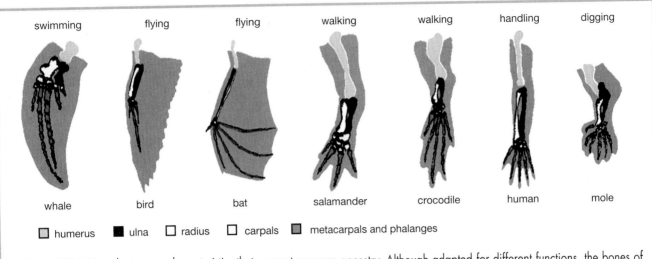

Figure E2.5 Homologies are characteristics that suggest common ancestry. Although adapted for different functions, the bones of these forelimbs (not drawn to scale) show comparable structures. What similarities can you find?

The fossil skeleton of Lucy has many of the same features as skeletons of modern humans, but modern human skeletons display some significant differences as well. Comparisons such as this, where fossils from long ago are compared with modern organisms, provide one line of evidence supporting the scientific view that living systems change. It is important to remember that the process of change is not only in the past, such as when Lucy lived; it is going on in all of the living systems alive on earth today.

On an isolated island in the Galápagos Archipelago, biologists observed two species of birds, the medium ground finch and the cactus finch. During a 12-year period, the scientists recorded measurements of the size of the birds' bodies and the thickness of their beaks. During this time there were two extreme dry spells, and many birds of each species died. At the end of the observation period, the scientists noticed that the average characteristics in the surviving populations of both species had changed slightly: the birds that survived, and their offspring, were a little larger and had beaks that were a little thicker than before. The scientists had

witnessed a small but ongoing biological change that was influenced by natural circumstances and could be inherited. These results support the conclusion that species change gradually, and they suggest that across very long periods of time, small changes might add up to significant differences.

Additional evidence for biological change and the relatedness of different organisms comes from the study of comparative anatomy, which is the branch of biology that concentrates on the similarities and differences in anatomical features of different species. By comparing the anatomy of different organisms, scientists have noticed many structural similarities among organisms. A pattern of similar characteristics may suggest evolutionary relatedness. For example, compare the structure of the animal forelimbs shown in Figure E2.5. Although they have different functions, these limbs show consistent similarities among the organisms. In addition, the limbs have the same relationship to the body (they all are forelimbs), and they develop in the same way in the young. Because of these consistent similarities, biologists infer that the

forelimb structure is a **homology**, a characteristic that is similar among different organisms because they evolved from a common ancestor.

In making anatomical comparisons, scientists also discover structures that are functional in some species, but seemingly useless in others. We refer to these dwarfed or apparently useless structures as vestigial, which comes from the word *vestige*, meaning "a remnant of." The "goose bumps" that you get when you are cold are an example of vestigial structures. Goose bumps are produced by small erector muscles in the skin. In other mammals and birds, the contraction of erector muscles causes the fur or feathers to fluff up. This mechanism helps the organism warm up. Modern humans do not have very much hair, but we have retained the erector muscles. Figure E2.6 shows three more examples of vestigial structures.

Homologies and vestigial structures are examples of similar features shared by seemingly unrelated organisms. Is there additional evidence to suggest that the tremendous diversity of life arose from the gradual accumulation of small changes in

Figure E2.6 Examples of vestigial structures a. Human tail structures

Most mammals have a well-developed tail, but this is lacking in apes and humans. Still, the tail is represented by the last three to five bones in the backbone of humans. Even though an external tail usually is not present, the muscles that move the tail in other mammals also are present in humans, and on rare occasions, a fleshy tail in a human extends a few inches beyond the caudal vertebrae.

caudal vertebrae

a.

b.

b. **Blind salamanders** These salamanders from Arkansas and Missouri have become adapted to life in deep caves where sunlight never reaches. They possess eyes, but their eyes do not function; they are blind.

c. **Snakes with legs**
Notice the bony parts of the python's pelvic girdle to which tiny limb bones attach. The legs serve no function.

c.

related organisms? The answer is yes, and this evidence is found by comparing modern species. Whether you study single-celled bacteria such as *Staphylococcus*, protists such as *Amoeba*, fungi such as mushrooms, or the more familiar plants and animals, you will find that each stores its genetic information in the same complex molecule, **DNA** (deoxyribonucleic acid). Genetic material in the form of DNA is the master plan for the diverse characteristics of each species, yet the method of storing that information, including the code that is used to interpret the information, is essentially the same in any living organism.

Scientists have used this common basis for storing genetic information to determine just how closely two organisms are related. Molecular techniques now make it possible to directly compare the information encoded in DNA. Biologists have compared the DNA among modern species and even among modern and extinct species in the rare cases when the soft tissue of an extinct organism has been preserved. In fact, new laboratory techniques enable molecular biologists to transfer DNA from one species to another, where the DNA functions successfully. Thus the genetic information from one organism, such as a bacterium, can be interpreted by a completely different organism, even an organism as different as a human. Figure E2.7 shows a comparison among primates.

The probability that the primates listed in Figure E2.7 developed an identical DNA code without being related to one another is extremely small. A much more likely explanation for this close similarity is that they inherited the genetic code from the same ancestors. The similarity in genetic code across living organisms strongly suggests a common origin for all modern life.

Animal	Percentage of DNA That Is the Same as Human DNA
Monkey (an old world primate)	93%
Gibbon (a lesser ape)	95%
Chimpanzee (a great ape)	98+%

Figure E2.7 Comparing DNA across primates

Primates Show Change across Time

A carpenter walks up to a cabinet he is building and places the metal handle in the spot he has marked. Holding the handle with his left hand, he turns the screwdriver quickly with his right hand to attach the handle to the wood. Five thousand years ago, a hunter steps quietly through a thin forest, keeping his sight fixed on a small deer a short distance away. As he walks, the hunter pulls an arrow from his quiver and sets it to the bow, ready for a shot that may win his family several good meals and a hide for clothing.

Walking upright has been advantageous for humans because it frees our fingers and hands to grasp and manipulate objects. This combination of

Fossils of the earliest known primates indicate they were very small animals, much like squirrels. Millions of years passed before monkeys, apes, and eventually humans appeared in the fossil record, but certain physical features remained quite similar between the old primates and the new.

Because of skull size and shape, scientists assigned Lucy to a group of hominids that walked upright and had brain cases more similar in size to apes than to humans. Still, hominid skulls indicate that their brains were larger for their body size than those of modern apes. The group of hominids to which Lucy belongs also had more humanlike front teeth, which are considerably smaller than those in apes. In addition, the position of the opening through which the spinal cord passes into the skull indicated an erect, bipedal posture.

Likewise, Lucy's pelvis clearly indicates that she walked fully upright. By contrast, the pelvis of a chimpanzee indicates that it is a knuckle-walker. Taken together, Lucy's skeletal remains suggest that her species was somewhere between apelike ancestors and modern humans. Her genus, named

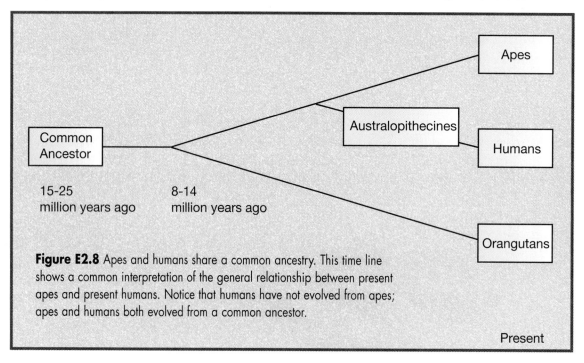

Figure E2.8 Apes and humans share a common ancestry. This time line shows a common interpretation of the general relationship between present apes and present humans. Notice that humans have not evolved from apes; apes and humans both evolved from a common ancestor.

abilities allows us to use tools even while we move about. Most primates lack these abilities, and you probably take them for granted, but the biological changes that made these skills possible took place through a very long period of time in human evolution.

The fossil record of primates supports the idea of biological changes at various rates in primate history. Today the primate group includes such mammals as monkeys, apes, and humans.

australopithecines, is considered to be hominid because of its many humanlike features. Other more recent hominids, such as *Homo habilis* and *Homo erectus*, show a more human appearance, with a more vertical face, smaller front teeth, a smaller lower jaw, and a significantly larger brain case. These specimens show a gradual change in form that is consistent with the scientific explanation that hominids evolved through time and that apes and humans share a common ancestry (see Figure E2.8).

Darwin Proposes Descent with Modification

Figure E2.9 Charles Darwin Born in 1809, Darwin studied medicine and theology to please his father, but his real interest was in natural history. Eventually he left medical school to enroll at Cambridge University to study natural history, the term then used for biology. His mentor, the Reverend John Henslow, was a famous botanist who later recommended Darwin to the captain of the *HMS Beagle*.

Charles Darwin, a 22-year-old English naturalist, was nervous and excited as he climbed up the gangplank to the scientific research ship, *HMS Beagle,* just after Christmas in 1831. Just imagine how much more excited he might have been if he had known what his voyage would mean to modern science. Although young Charles' journey would last only five years, the ideas born of that journey would change forever the way biologists understand the world.

The central mission of the *Beagle* was to chart sections of the coast of South America. While the ship's crew completed that work, Darwin spent his time on shore, collecting thousands of specimens and recording detailed observations of the interesting and exotic organisms he encountered. Darwin observed organisms in such diverse environments as the jungles of

Brazil, the harsh plains of Tierra del Fuego (near Antarctica), the grasslands of Argentina, and the heights of the Andes mountains. Before it was over, the *Beagle's* voyage took Darwin to the coasts of Australia, New Zealand, Tasmania, and to several islands in the South Pacific and Atlantic oceans (see Figure E2.10).

One of the most important questions that Darwin pondered during his travels related to the geographical distribution of the organisms that he observed. Although similar to organisms in Europe, the plants and animals that lived in South America and the South Pacific were clearly distinct. That, perhaps, was not surprising, but what perplexed Darwin was the fact that organisms that lived in temperate (mild) areas of South America were more similar to organisms living in tropical areas of South America than to organisms living in temperate regions of Europe. Orchids, army ants, marine iguanas,

Figure E2.10 Voyage of the *HMS Beagle* Darwin sailed around the world on the *Beagle*. The route included stops in South America and a stay in the Galápagos Islands. During the trip Darwin collected evidence that he later used to support the theory of evolution.

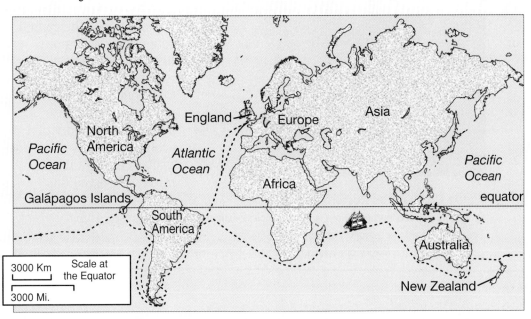

England • Europe • Asia • North America • Pacific Ocean • Atlantic Ocean • Pacific Ocean • Africa • equator • Galápagos Islands • South America • Australia • New Zealand

3000 Km
Scale at the Equator
3000 Mi.

penguins . . . each type of organism was well suited for the environment in which it lived, yet seemed to be related to organisms living in other parts of that huge continent. How could one account for both the similarities and the differences among species, and how could one account for their specific patterns of geographic distribution?

After he returned to England, Darwin spent nearly 20 years analyzing his observations and thinking about their implications before he published *On the Origin of Species*. Although this landmark book established Darwin as the author of the theory of evolution, Darwin did not use the word *evolution* in the first edition. Instead, he proposed the concept of *descent with modification*, a phrase that expressed his view that all organisms on earth are related through descent from some unknown ancestral type that lived long ago. This idea helped Darwin develop an explanation for the diversity of organisms that he had encountered on his travels and also for the patterns in geographical distributions that he had observed. The concept of descent with modification is described in more detail in the next essay, *Evolution by Natural Selection*.

The publication of Darwin's book represents an interesting twist in the history of science. If Alfred Russel Wallace, a young scientist working in the East Indies, had not written to Darwin, he might not have published the book when he did. In his letter to Darwin, Wallace enclosed a draft of a scientific paper that described a theory of evolution; his theory was less detailed than Darwin's theory, but almost identical in basic outline. Darwin was shocked to learn that other scientists not only were thinking about these ideas but might be able to

Figure E2.11 Alfred Russel Wallace Although Wallace independently reached many of the same conclusions that Darwin reached, he seldom is acknowledged for his contributions to the theory of evolution.

publish them before he did. Nevertheless, he behaved with integrity, acknowledging the excellence and importance of Wallace's work and forwarding the paper to another noted scientist for public presentation. Knowing of Darwin's nearly 20 years of study, the scientist decided to present Wallace's paper with an excerpt from an essay about evolution that Darwin wrote in 1844 but never published. This decision preserved Darwin's claim to these ideas, but Darwin still had to publish his full results. Motivated now by the concern that other scientists were about to reach the same conclusions that he had reached, Darwin worked feverishly for almost a year, finally completing *On the Origin of Species* in 1859.

Although Wallace published his paper first, Darwin's explanation of evolution was more detailed and contained more extensive supporting evidence than Wallace's paper. Darwin's journals also demonstrated that he had developed the central core of his ideas more than 15 years before reading Wallace's paper. Consequently, Darwin is known as the main author of the theory of evolution, and even Wallace thought that Darwin deserved the credit. (Wallace is pictured in Figure E2.11.)

EVOLUTION BY NATURAL SELECTION

Even before Darwin's work, several individuals already had proposed the basic idea of evolution, the concept that differences among species are the result of changes across time. Unlike these earlier scientists, however, Darwin was able to provide not only a logical argument to support evolution, but also clear evidence for his key points. Even more important, Darwin proposed a mechanism, that is, a way by which a new species could eventually appear from ancestral forms.

Although he was unable to explain the source of variation in organisms, Darwin observed that the individual members of any species show a great deal of variation in their characteristics. (Later, scientists found the source of variation to be genetics.) They may show differences in size, coloration, strength, behavior, and many other features. Some of these characteristics are passed on from parents to offspring. In fact, for centuries plant and animal breeders have bred organisms to emphasize or increase certain prized characteristics. Plant breeders, for example, try to improve disease resistance by selecting plants with high resistance and mating them together, in the hope of obtaining offspring with even higher resistance. Likewise, horse breeders may select mating pairs to try to increase speed, stamina, or both.

Sometimes these techniques can result in significant differences from one breeding group to another. Figure E2.12 illustrates six different types of vegetables that humans eat. All six plants are of the same species. Humans have bred them for different agricultural uses and to suit different human tastes.

If humans can bring about change in a population of organisms by this type of artificial selective breeding, Darwin reasoned that perhaps selection in nature also could bring about change. Such change

Figure E2.12 Six types of kale These kale plants differ primarily in the part of the plant that stores the most starch.

cauliflower

kale

broccoli

cabbage

Brussels sprouts

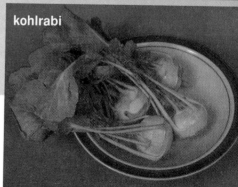
kohlrabi

eventually might result in the production of new species. A species is a group of organisms whose members very closely resemble each other because they share many key characteristics. Despite obvious differences in superficial appearance, all humans are one species because we share many fundamental characteristics. Organisms belonging to different species usually vary from one another in numerous key characteristics. For sexually reproducing organisms, one characteristic that separates species is the ability to mate and produce fertile offspring. A horse and a donkey, for instance, can mate and produce offspring called mules, but because horses and donkeys belong to two different (although very closely related) species, these offspring mules are sterile.

How would the process of **natural selection** take place? One way to trace Darwin's thinking is to consider carefully the information displayed in Figure E2.13. This summary was proposed by Ernst Mayr, a noted biologist at Harvard University, as a way to understand the mix of evidence and inference that underlies the theory of evolution. Let us follow the argument, point by point, and try to understand some of the thinking that led Darwin to his conclusions.

The first three facts in Figure E2.13 describe the situation that inevitably exists in any environment:

• In the absence of limits to population growth, organisms reproduce very rapidly.

• Nevertheless, our observations tell us that most populations in the wild are relatively constant in size.

• Our observations also tell us that the resources available in any natural environment are limited.

List of Mayr's points

Fact 1: All species have such great potential to produce large numbers of offspring that their population size would increase exponentially if all individuals that are born would reproduce successfully.

Fact 2: Except for seasonal fluctuations, most populations are normally stable in size.

Fact 3: Natural resources are limited, and in a stable environment they remain relatively constant.

Inference 1: Because more individuals are produced than the available resources can support, and the population size remains stable, there must be a fierce struggle for existence among the individuals of a population. This results in the survival of only a part, often a very small part, of the offspring of each generation.

Fact 4: No two individuals in a population are exactly the same; rather, every population displays an enormous variety of characteristics.

Fact 5: Much of this variation can be inherited.

Inference 2: Survival in the struggle for existence is not random but depends in part on the characteristics that the surviving individuals inherited. This unequal survival is a process of natural selection that favors individuals with characteristics that fit them best in their environment.

Inference 3: Over the generations, this process of natural selection will lead to a continuing gradual change in populations, that is, to evolution and the production of new species.

Figure E2.13 The logic of the theory of natural selection Ernst Mayr has summarized the logic of Darwin's theory by extracting the major points of evidence and inference. (This list is adapted from E. Mayr, *The Growth of Biological Thought: Diversity, Evolution, and Inheritance,* Cambridge, Mass: Harvard University Press, 1982.)

One important factor that influenced Darwin's views was an essay written in 1798 by Thomas Malthus, a member of the English clergy. Malthus argued that much of the human suffering that we see on earth is the inescapable result of the tendency of the human population to grow beyond the resources available to support it. This capacity for overpopulation seemed to Darwin to be a characteristic of every population. It led Darwin to infer that individual organisms within a population must face a struggle for survival, and that only a few individuals need to survive to pass on their characteristics from one generation to the next. The rest fail to develop; die of starvation, predation, or other causes before they reproduce; or do not reproduce for other reasons. The effect of this loss of reproductive individuals from one generation to the next is that wild populations tend to be relatively stable in size.

The remaining two facts suggested to Darwin how this struggle for survival is played out among the members of a natural population:

• Individual organisms within a population are different in particular characteristics from one another.

• Most, although not all, of this variation can be inherited from one organism to the next.

From these observations, Darwin inferred that the outcome of the struggle for existence depends, to a certain extent, on the characteristics that an organism inherits. Individuals whose inherited characteristics best equip them to survive in a particular environment will be most likely to reproduce and to leave behind offspring with these same beneficial traits. We call these beneficial traits **adaptations** (Figure E2.14). By contrast, those organisms with characteristics that make it more difficult to survive and reproduce in an environment would be less likely to survive and to reproduce. As a result, their particular characteristics would be less likely to be passed on to surviving offspring.

For example, suppose a population of rabbits lived in a moderately warm climate. Within that population, a small proportion of the individuals might have thick fur. If this thick fur resulted from a difference in their genetic material as compared with that of other rabbits living in the same area, then this difference

a.

b.

c.

Figure E2.14 Examples of adaptations a. **snake caterpillar** b. **flicker** c. **red Irish lord fish** Can you identify the adaptations possessed by these organisms? How might each adaptation benefit the organism?

might be passed on to their offspring. Imagine first that the thicker fur is a disadvantage to the rabbits in this environment because they overheat when they are chased by coyotes. As a consequence, these thick-furred rabbits do not survive and reproduce as well as the rabbits with thin fur. You probably can see, in this case, that thick fur would not be a very common trait in this population. Now imagine that, through a period of several decades, the climate in this environment gets colder. In the colder climate, the rabbits with the thicker fur now would have an advantage. That is, the thick fur would be an adaptation that helps them survive the periods of cold weather. All other things being equal, the thick-furred rabbits would be more likely to survive long enough to reproduce and pass on their characteristics than the thin-furred rabbits. As Figure E2.15 shows, after many generations, the number of rabbits with thick fur might increase in the population. This is the process of natural selection.

Although Darwin's understanding of selective breeding led him to propose his theory of natural selection, there is a key difference between selective breeding and natural selection. In selective breeding, humans *choose* how to allow organisms to reproduce in order to pass on desirable characteristics. In natural selection, however, each organism's ability to meet the challenges of survival and reproduction in a natural setting determines the mix of characteristics that will be transmitted to the next generation.

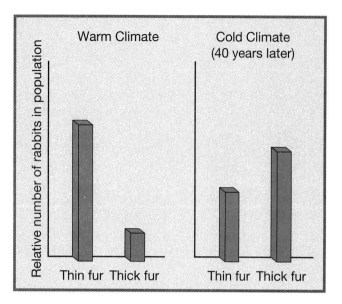

Figure E2.15 Characteristics of a population change due to natural selection

In such a setting, many pressures affect survival and reproduction. As you saw in the rabbit example, changes in the environment can alter the balance of life so that some characteristics are more beneficial than others. Sometimes these changes are catastrophic, like the effect a volcanic eruption might have on both local and very distant populations of organisms.

Competition among organisms for limited resources also acts as a selective pressure. For example, a plant needs sunlight to grow and sufficient soil to support its roots and provide water and minerals. The plant must cope with its environment to survive and produce seeds for the next generation. The plant, however, is not alone. Nearby plants may grow faster than this plant and may block the sunlight, or they may have longer roots that drain off limited water. In this way, these other plants may compete with the first plant for survival. Differences in the ability of various organisms to compete in an environment with limited resources is one of the selective pressures that can bring about evolution.

Predation is another pressure that many organisms face. Living long enough to reproduce requires that the predator be successful in locating and catching prey. The prey, however, must escape predation until it is able to reproduce. In a predator/prey relationship, it may appear that the predator has all the advantages. Not so: think of a lion whose jaw is broken by a well-aimed kick from a gazelle. The lion will die slowly of starvation because it can no longer eat. On the other hand, the gazelle that is not alert enough or strong

Figure E2.16 The predator/prey relationship imposes selective pressure on living organisms.

enough to avoid the lion's jaws also will die, in this case, rather suddenly and dramatically.

What determines an organism's ability to survive and reproduce in the face of these pressures? Generally, it is not one characteristic or two, but a mix of characteristics that helps the organism adapt in a particular environment. Organisms, however, do not *acquire* these characteristics to *help* them survive. Instead, these characteristics occur among individuals in a population as a result of natural, inherited variation. Natural pressures in the environment then act on the differences in the individuals in a population and influence which organisms survive to contribute their characteristics to the next generation.

It is important to realize that individual organisms do not evolve. Rather, populations evolve. Darwin's theory of evolution by natural selection proposes that sometimes the characteristics of a population can change so dramatically as a result of natural selection that eventually the population becomes recognized as a distinct species, different from the ancestral form. Here we return to Darwin's basic idea: descent by modification. Descent from a common ancestor provides a powerful explanation for the similarities that we see among related forms of life, and modification through variation and natural selection provides a powerful explanation for their differences.

It is difficult to overestimate the importance to modern biology of the theory of evolution by natural selection. Perhaps Darwin's greatest contribution was in giving biologists a way to understand the enormous diversity that we see in living systems. The theory of evolution explains the origin of this diversity and helps us understand the relationships that exist among modern species and between modern species and their ancestral forms. As most biologists will tell you, these aspects of biology do not make much sense except in the context of evolution.

Just a Theory?

What do we mean when we say that we have a theory about something? Usually we mean that we have an idea about how to explain something. We may suggest a theory about who killed the butler in the late-night thriller, or we may have a theory about why Ms. Figueroa always looks so tired when she delivers the milk on Monday morning, or we may have a theory about what the coach of the local AAA baseball team said to his pitcher after they lost last night's big game.

These theories that we develop about things often are very tentative. That is, when we say that we have a theory about something, we may mean that we don't have a lot of evidence that it is correct. Because we don't have much evidence, our theories usually aren't shared by very many people. Nevertheless, it is fun to suggest clever ideas.

In contrast, scientists use the word *theory* in a very different way than we do. Whereas the types of theories (or explanations) that we develop tend to be based on little evidence, scientific theories are explanations that are extremely well accepted by the scientific community because they are supported by a variety of strong evidence.

A good example of a scientific theory is the atomic theory, an explanation developed in 1803 by the British teacher and chemist John Dalton. You already may be familiar with one of the most important ideas of Dalton's atomic theory: that all matter (or things) in the universe is composed of tiny particles called atoms. This idea was revolutionary at the time that Dalton first suggested it, but since then physicists have accumulated an enormous body of evidence to support it. Part of that evidence is the result of our modern ability to detect, to manipulate, and even to subdivide atoms. Even though Dalton did not predict that atoms could be subdivided into smaller pieces such as electrons and protons, the general outline of his theory still stands. Cell theory, the understanding that complex living organisms are composed of small building blocks called cells, is an example of a biological theory that is supported by an enormous body of evidence.

Figure E2.17 The word *theory* often means different things to different people.

Despite the evidence that has accumulated in favor of the atomic organization of matter, or of the cellular organization of living organisms, and despite the fact that no modern scientist questions most of these basic principles, we still refer to these ideas as *theories.* Clearly, scientists use the word *theory* very differently from the way most nonscientists use it.

To understand how an explanation becomes known as a scientific theory, we have to consider first the meaning of the term **hypothesis**. You may have learned already that a hypothesis is a trial idea about something. During the course of their research, scientists develop trial ideas, or hypotheses, to explain the reasons underlying an event or a phenomenon or to explain the relationships that they think they see among objects, events, or processes. As you will see in Unit 2, predictions made from such hypotheses are useful because they can be tested. Testing our hypotheses allows us to distinguish those trial ideas that adequately explain the phenomenon under study from those that do not.

You probably can see, then, that the explanations that we develop for things and events around us actually are *hypotheses,* not scientific theories. As hypotheses, these trial ideas or tentative explanations can be tested. In fact, in scientific terms, the police who investigate the death of the butler do so by systematically testing a whole set of hypotheses, not theories.

By contrast, a scientific **theory** is a hypothesis that already has been extensively tested and is supported by a large body of observations and evidence. A good theory explains data that we already know and relates and explains additional data as they become known. In fact, a good theory also predicts new data and suggests new relationships that we may not already have recognized among phenomena.

Why do scientists call Darwin's explanation of descent with modification the *theory* of evolution instead of the *hypothesis* of evolution? First, it is called a theory because of the enormous amount of evidence that suggests that it is a correct explanation. As you already have seen, this evidence spans a wide range of different scientific fields, from anatomy to geology to embryology to physical anthropology. Second, it is called a theory because it explains both the evidence that Darwin saw and recorded and the new data we continue to collect. It also successfully predicts new phenomena. That is, Darwin's explanation for biological change across time continues to be supported, even by evidence collected more than 100 years after his work.

For these reasons and others, this explanation is almost universally accepted by scientists around the world.

Does this wide acceptance of Darwin's work mean that Darwin explained everything there was to explain about evolution or that every part of his explanation is correct? No, not necessarily. Darwin was not able to explain exactly *how* characteristics passed from one generation to the next—he did not include in his theory an explanation of the relationship between natural selection and genetic inheritance. However, this omission does not make Darwin's ideas incorrect. Since Darwin's time, biologists have added information on natural selection and genetic inheritance to the theory of evolution. In fact, the addition of this information to Darwin's basic proposal illustrates the power of a sound scientific theory. As new information about inheritance became available to scientists, it supported, rather than contradicted, the original explanation.

So, when we talk about explanations for things in the world around us, we need to be careful to decide whether we are talking about a theory (as we typically use the term) or we are talking about a *theory* (as scientists use the term). The difference between the two is a big difference, indeed.

Figure E2.18 When used by scientists, the word *theory* means an explanation that is supported by evidence and well accepted by other scientists.

Describing Life: An Impossible Challenge?

An alien spaceship orbiting a planet to make observations ejects a special robotic lander toward the planet's surface. Retrorockets fire to slow the lander's speed, and parachutes open to help it land gently. Now the lander settles down on the planet's surface, amid rocks and soil. Its mission: to determine whether life exists on the planet.

At first, the lander sits motionless, a foreign object on the bleak landscape. Then, slowly, it activates its electronic senses. It rotates its twin cameras to scan the horizon. It measures the local weather: temperature, barometric pressure, wind speed, and direction. It uses special equipment to sniff the air to determine its composition. The atmosphere is thin, but it *does* contain some water vapor, even a little more than observers had expected. This is a good sign . . .

(Continued on page E35)

Figure E3.1 This photograph of the surface of Mars was recorded by equipment on the Viking Lander 2 on 18 May 1979 and relayed to earth in June 1979. You can see it shows a thin coating (1/1000 of an inch) of water ice on the rocks and soil.

Recognizing life on an alien planet such as Mars is indeed a challenge. At minimum, such an endeavor requires that we start with a good idea of how to recognize life on earth. But even this task is not simple. Living systems share many characteristics, but do any of these actually distinguish living things from nonliving things?

You already have an intuitive sense about life. If you were to ask your classmates to identify a tree, a dog, and a rock as living or nonliving, chances are great that all their answers would be the same. But if you then asked your classmates precisely *how* they know what is living and what is not, their responses probably would vary. And suppose you asked them to categorize a less familiar object, such as the scaly, grayish-green lichen on a boulder (like the one depicted in Figure E3.2)? This time, some of your classmates might say that this stuff is not alive.

How can we describe life so that we always can identify it when we see it? Perhaps the easiest way to begin thinking about life is to consider what happens when an organism dies. Think, for example, of a bird that has just died. The bird can no longer move, or eat, or keep itself warm. Even if you touch it, it does not respond. Eventually the dead body will become decayed and disorganized; it will never recover its form or function, nor will it ever again produce offspring.

This simple example provides some clues to the nature of life by showing that certain properties of a living system are lost when death occurs. If, however, you try to write a simple description of life that absolutely *distinguishes* it from nonliving substances, you may find it rather difficult. Rather than trying to describe life precisely, let us first examine the characteristics that we generally observe in living systems. Then, perhaps, we can consider how we might go about recognizing life, both on earth and on another planet.

All forms of life, even vastly different forms such as humans, apple trees, spiders, and microscopic bacteria, share many basic characteristics. This should not surprise you: in fact, it is *because* living things share much in common that you are able to make some judgments about whether an unknown object is likely to be alive or not.

Understanding these common characteristics is fundamental to understanding biology. In fact, these characteristics are so important that we have summarized them as six unifying principles of biology, and we have organized the flow of topics in this course around them. You already have encountered evolution, one of these principles. As you begin to develop a deeper understanding of the remaining five principles, you will be developing a rich understanding of how all living systems, including humans, function.

Evolution: Patterns and Products of Change in Living Systems. As you already have seen in this unit, one significant characteristic of living systems is that they evolve, or change, across time. Through natural selection, the individuals in a population with characteristics that make

Figure E3.2 Lichens are organisms that consist of a close association between a fungus and a photosynthetic organism such as an alga or a blue-green bacterium.

them best-suited for growth to reproductive maturity are those whose particular characteristics become the trademark for the population. In some species, such as the horseshoe crab, characteristics remain unchanged for long periods of time. More often, however, natural selection results in gradual change in populations. These changes eventually lead to distinctly different populations of organisms that display an amazing range of diverse characteristics. Figure E3.3 illustrates how scientists think one type of organism may have evolved. *Evolution* represents the first unifying principle considered in this course.

Homeostasis: Maintaining Dynamic Equilibrium in Living Systems. A second characteristic of life—and a second unifying principle of biology—has to do with a living system's ability to maintain an *internal balance*, referred to as homeostasis. All organisms regulate their internal systems in response to changes in

their surroundings. When you are startled, your heart beats faster, sending blood through your body at a faster rate. This response ensures that your body will continue to have a good supply of oxygen and nutrients (which are carried in the blood) during a possibly stressful or dangerous time. In fact, all organisms show a similar type of internal regulation. Bacteria adjust their production of certain key products in response to changes in the nutrient levels in their environments; plants respond to changes in humidity by opening or closing tiny holes in the underside of their leaves; and some animals can change their coloration in response to their environment (see Figure E3.4).

Energy, Matter, and Organization: Relationships in Living Systems. Another common characteristic of life is *organization*. All living systems are highly organized forms of *matter*. This matter is made from atoms held together in ways that form large, complex molecules.

Although scientists have identified more than 100 different types of atoms, one of the most remarkable similarities among all living things is that they are made predominantly from only a few types of atoms, notably carbon, nitrogen, oxygen, hydrogen, phosphorous, and sulfur.

The molecules of living materials are organized into complex structures known as cells. Cells are the basic structural units of living matter. Because most cells are too small to see with the unaided eye, scientists did not see cells until 300 years ago, after the invention of the microscope. As Figure E3.5 illustrates, cells are baglike structures made of a membrane that encloses and protects the contents.

A related property of all living systems is that they require *energy* to build and maintain their highly organized structures and to carry out all of their activities. Recall that the bird, once dead, eventually will decay and disintegrate, losing its distinctive shape and appearance and becoming more and more indistinguishable from the matter around it. The loss of the bird's characteristically high organization follows the more basic loss of its ability to obtain matter and energy from its environment and to use that matter and energy to keep its body (its matter) repaired and functional. Together, the ideas of matter, energy, and organization represent the third unifying principle of biology.

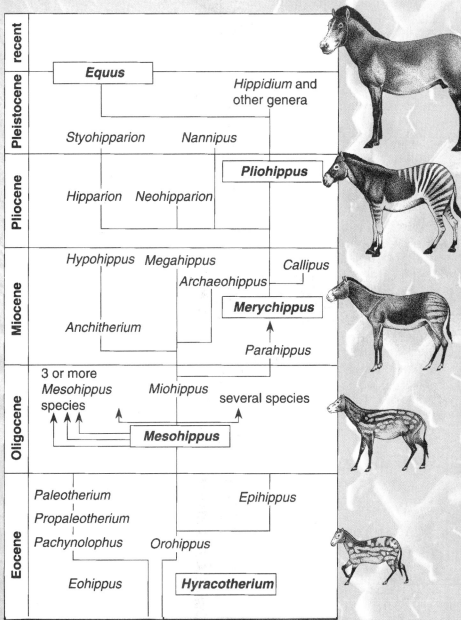

Figure E3.4 This anole lizard can change color in response to its environment.

Figure E3.3 A proposed evolutionary tree What evidence do you think scientists used to develop this explanation?

Continuity: Reproduction and Inheritance in Living Systems.

The organization and the function of living systems depend on specific plans that are encoded in each organism's genetic material, or DNA. For example, maple trees display a characteristic structure and function because they possess DNA characteristic of maple trees,

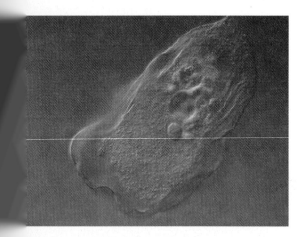

Figure E3.5 The cell is the basic unit of living matter. Most prokaryotic cells are 1–10 μm in diameter. Most eukaryotic cells are 10–100 μm in diameter, but protists can be much larger. Notice the organization of the interior of this *Thecamoeba* cell, which is 10 μm.

and humans grow and function in ways that we recognize as distinct from other life forms because humans possess DNA characteristic of humans. DNA is a long and complex molecule that stores information (see Figure E3.6). One of the most significant characteristics that unifies living systems is the universal nature of this DNA. Although the instructions that direct an organism's cellular activities and developmental events are specific for its species, all organisms, from bacteria to humans, use the same DNA to communicate those

instructions. The ability to transfer those instructions—through DNA—to the next generation during reproduction represents a fourth important unifying principle of life, that of *continuity*.

Development: Growth and Differentiation in Living Systems.

The *ability to grow and develop* represents the fifth unifying characteristic of living systems. Growth is an important activity in the early life of a human. Growing requires the body to assemble new tissue. As the organism's size increases, the way in which the organism's tissue is organized also changes. Human adults not only are larger than children, but they also are shaped differently and can do a variety of things that human infants cannot do, such as walk and talk. Many plants, in a similar manner, begin life as small seedlings that push up through the soil and grow and develop into mature plants that look quite different from the early seedlings. Plant growth also involves the addition of new tissue and the organization of new parts such as leaves and reproductive structures.

Ecology: Interaction and Interdependence in Living Systems.

Finally, all living systems on earth are part of an interactive and interdependent web of life.

Organisms do not exist in isolation, but rather live as one element in a complex community of life (refer to Figure E3.7). Imagine a wooded area alongside a stream on an early summer day. Plants provide shelter and food for a variety of birds; perhaps a rabbit has dug a burrow nearby and now feeds on wild berries growing in the light shade close to the forest; and not far away a fox has just left her den in search of food for her young. This community of different, yet interdependent, living systems illustrates the sixth unifying principle of biology, the *interactive and interdependent nature of life.*

To the extent that these brief descriptions capture the essence of each of the unifying principles, we might say that in this short list of characteristics, we have described life—as it exists on earth. Can we say that any one of these principles *defines* life, in the sense that it alone is necessary for life and that it alone is an indicator of life? Probably not. Just as a *combination* of characteristics identifies you as a human, a *combination* of these principles indicates the presence of life.

To this day, curiosity about life on Mars remains high. Unfortunately, although two of the three tests described in Figure

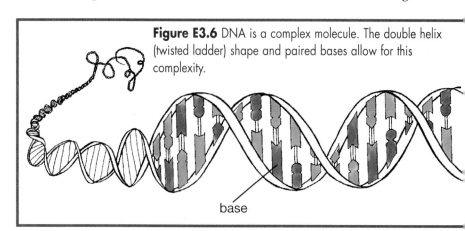

Figure E3.6 DNA is a complex molecule. The double helix (twisted ladder) shape and paired bases allow for this complexity.

base

(Continued from page E31)

Finding evidence of life on Mars would be very exciting, but how should the scientists involved in the effort look for it? The landing craft, though technologically very complex, is small and its designers had to make careful decisions about the equipment that it carries and the activities that it is able to accomplish. Should they look for signs of evolution, of growth and development, of reproduction? All are fundamental characteristics of life, but all probably occur too slowly to be detected by this tiny craft. Instead, designers decided to test the sample for more immediate, more easily recognizable signs of life: those that have to do with a living system's requirement to use matter and energy obtained from the environment to maintain its complex organization.

Suddenly a mechanical arm extends from the strange craft and scoops up some of the Martian soil. At last, humans have collected a sample from another planet. No matter that the craft does not have the physical capability to return to earth. Technology is extending our hands and eyes now.

Slowly the soil sample is deposited in a special chamber. A specially designed apparatus adds a mixture of radioactive gases. Some of the smallest forms of life on earth use light energy and certain gases in the environment around them to build more complex molecules that are vital to their organization and function. Is there anything in the Martian soil that will use these gases as "molecular food"? Will these radioactively labeled molecules slowly begin to accumulate in the soil sample, as living things remove them from the air and use them to maintain and build their internal structures?

E3.8 (refer to page E36) yielded some interesting results, scientists failed to duplicate the results with subsequent samples. This was disappointing and suggested caution in interpreting even the changes that the distant instruments *did* detect. In fact, by 1979, most scientists involved with the project had agreed that although they could not rule out the possibility that life exists on Mars, all the data that they collected in the original experiments could be explained as resulting from purely *chemical* (not biological) causes.

Describing life . . . a difficult, but not an *impossible* challenge. Looking for life, using earth's criteria, in a very different environment more than 40 million miles away . . . more difficult to be sure, but *impossible*?

What do you think?

Figure E3.7 Organisms along and in this stream interact and depend on each other.

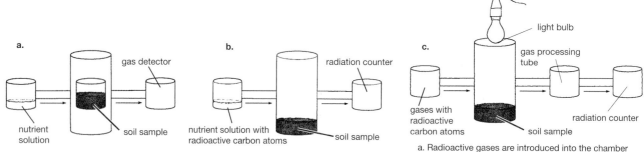

a. Soil sample is suspended in a porous cup.
b. Nutrient solution is added to the soil sample.
c. Changes in gas content are measured by a gas detector.

b. Soil sample is sprayed with radioactively labeled nutrient solution.
b. Any radioactive carbon dioxide that is produced by the soil and released into the air above the sample is detected and counted.

a. Radioactive gases are introduced into the chamber containing the soil.
b. The light is turned on as a source of energy.
c. The chamber is heated to release newly made substances into the air.
d. The air is processed to separate complex substances from the simple gases that had been introduced earlier.
e. Any radioactive carbon that is contained in these complex molecules is detected and counted.

Figure E3.8 a. A gas exchange experiment tested the Martian soil for evidence of organisms that took in gases from the Martian atmosphere and nutrients from the soil and gave off gases as wastes. When this experiment is performed using earth's soil, the experiment indicates the presence of microscopic organisms that take in oxygen and nutrients and give off carbon dioxide. b. A label-release experiment tested the Martian soil for evidence of organisms that could use simple nutrients and give off waste gases. This experiment was similar to the gas exchange experiment and served as an important check on its results. Again, this experiment gives strong, positive results when earth's soil is tested. c. A pyrolytic-release experiment tested the Martian soil for evidence of organisms that might build large, complex substances out of simple gases in the Martian atmosphere. When this experiment is performed on earth's soil, the experiment indicates the presence of microscopic organisms that use the energy of sunlight to help them build sugars and other large, complex molecules.

FIVE KINGDOMS

In which of these pairs of illustrations are the organisms most closely related? Figure E3.9 shows two animals that bear little resemblance to each other. In contrast, Figure E3.10 shows two types of cells, each an individual organism and each looking quite like the other.

Surprisingly, from an evolutionary point of view, the two animals are much more closely related than are the two single-celled organisms. The animals are an African elephant and a close relative, a small mammal known as a hyrax. What you cannot see in Figure E3.9 is all of the ways in which these organisms are similar, from the basic structures of their cells to the structures of their feet and teeth.

Figure E3.9 a. **African elephant (Loxodonta africana)** The average male African elephant is 250 cm high and weighs 5000 kg. b. **Rock hyrax (Procavia capensis)** A rock hyrax may be 30 cm high and weigh 4 kg.

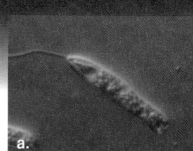
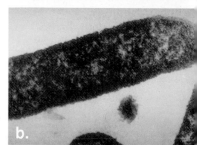

Figure E3.10 a. This *Peranema* is about 40 µm. b. This *Escherichia coli* is 3.5 µm in length (photographed at X35,000).

E36

On the other hand, the organisms in Figure E3.10 are very distant in their evolutionary connection, despite the fact that each is a single cell. If you look closely, you can find one of the characteristics that marks these two organisms as being very different. Notice that *Peranema* has an interior compartment that is missing in the other cell. That compartment is a nucleus, a membrane-enclosed structure in the cell that houses its DNA. The second cell is a bacterium called *Escherichia coli*. Like other types of bacteria, its DNA is not separated from the rest of the cell contents by a surrounding membrane—it lacks a nucleus.

These two cells illustrate the single largest dividing point that biologists recognize among all of the species on earth. The bacterial cell is a very simple type of cell known as a **prokaryote**. It has no nucleus, and the genetic material that it contains is a huge molecule of DNA, without any fancy packaging. In great contrast, the *Peranema* is a more complex type of cell called a **eukaryote**. Eukaryotes have cells with nuclei and DNA that is packaged with proteins to form structures known as chromosomes. Eukaryotic cells also may have other specialized, membrane-enclosed compartments that perform a variety of functions, such as energy transformation and protein storage and packaging. Although many similar processes go on in prokaryotic cells, these cells do not contain such compartments.

The structural differences and the evolutionary distance between prokaryotes and eukaryotes are so great that biologists categorize all organisms on earth on the basis of this distinction. Figure E3.11 illustrates the five major types of organisms recognized by most biologists today. Note that one of the kingdoms includes all of the prokaryotic organisms; in contrast, the organisms in each of the other four kingdoms are eukaryotes.

It would not be surprising if the classification scheme shown in the figure and used in this course seems a bit foreign to you. After all, most of us grow up thinking that the world contains only two basic categories of organisms, plants and animals. We are not alone in this: from the days of Aristotle to the mid-1800s, almost everyone was content with this simple subdivision. We generally have little reason to question it, because we rarely encounter living systems that are so different in external appearance that they don't seem to fit.

By the middle of the nineteenth century, however, some scientists had started to question whether organisms such as fungi and bacteria really fit well into either the plant kingdom or the animal kingdom. Despite these questions, suggestions to increase the number of kingdoms were largely ignored, and it was not until the 1960s that the prevailing attitude in the scientific community began to change. Scientists were discovering new forms of life and were using new microscopic and biochemical techniques to examine cell structure and function in even well-known organisms. This led to an increasing amount of evidence that supported proposals to increase the number of basic categories that biologists recognize. Figure E3.12 illustrates some of these multikingdom schemes and will help you trace the changes that have occurred in scientists' thinking to bring us to the five-kingdom system that is most often used today.

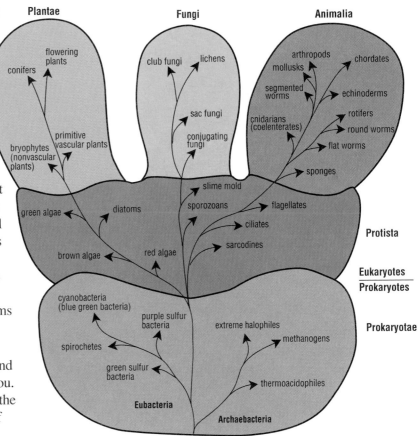

Figure E3.11 A five kingdom scheme The monera kingdom is the prokaryotic group. Plants, animals, fungi, and protists are all eukaryotes. What differences do you think separate the organisms in each kingdom?

As you read the brief descriptions of the five kingdoms that follow, look for patterns in the criteria that determine each group. Look as well for differences that distinguish one basic type of organism from the next. Do you see some of the reasons that biologists can no longer accept a two-kingdom view?

Kingdom Prokaryotae (Monera). The main criterion (or qualification) for membership in this kingdom is the presence of the prokaryotic type of cell (a cell that lacks membrane-enclosed compartments). Prokaryotae are all of the bacteria, which usually are single cells but may occur in groups of cells. Bacteria come in a variety of shapes, as depicted in Figure E3.13, and some can swim by means of long, whiplike tails. Prokaryotes occur in almost every environment, from the inside of the human mouth to nearly boiling hot springs and even in the ice of Antarctica.

Bacteria show a great diversity in the processes that they use to obtain energy. Many bacteria can use the sun's energy directly to power the reactions required for making their own food through photosynthesis. Others use energy derived from the matter (food molecules) that they acquire from their environments. As a group bacteria can digest almost anything—even petroleum. This ability is fortunate for us: bacteria that can recycle matter through decomposition increasingly are being used to help with environmental clean-up efforts. All bacteria reproduce by dividing into two, but some also exchange small amounts of DNA—a form of sexual reproduction.

Kingdom Animalia. Among the four eukaryotic kingdoms is the kingdom in which humans are found, the kingdom Animalia. Animals are multicellular—they have a complex organization of many specialized cells. Animals also are characterized by their ability to bring food into their bodies and digest it. In addition, most animals reproduce sexually and have senses and nervous systems that enhance their ability to move.

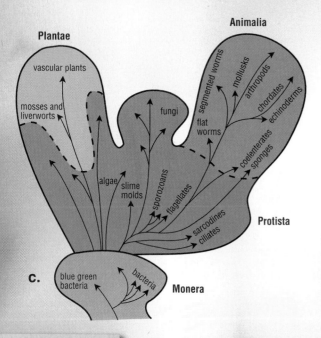

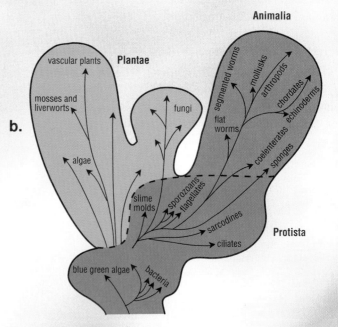

Figure E3.12 Scientific ideas change across time.
a. The first attempts to categorize life resulted in this two kingdom division between plants and animals. b. This model shows three kingdoms: plants, animals, and protists. c. Scientists developed this four kingdom scheme when they realized the great differences between eukaryotes and prokaryotes.

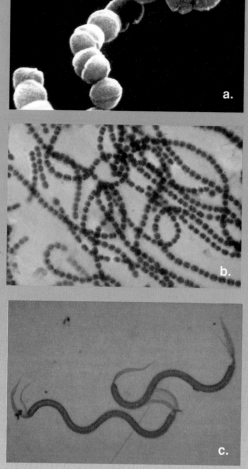

Figure E3.13 Examples of prokaryotes
a. These *Streptococcus* bacteria (photographed at
X40,000) can cause strep throat. b. *Nostoc*
(photographed at X400), a cyanobacterium, is
common in freshwater lakes. c. *Spirella voluntans*
(photographed at X400) is part of a group of bacteria
named for its characteristic spiral shape.

Animals live in marine and freshwater environments,
inhabit the soil, or live on land. In addition, animals come
in a range of sizes, from microscopic worms that live in
human blood to whales that can reach lengths of 27 m
(89 ft). Figure E3.14 shows a diversity of animals.

Kingdom Plantae. Another eukaryotic kingdom, the
kingdom Plantae, includes organisms that acquire their
energy not from eating, but from the sun. Plants carry out
photosynthesis, a process by which cells use energy from
sunlight to produce their own food. Photosynthesis takes
place in membrane-enclosed structures within plant cells
called chloroplasts. Chloroplasts contain chlorophyll, the
light-absorbing pigment that gives plants their
characteristic green color.

Plants are multicellular and their cell membranes are
surrounded by a rigid cell wall that provides support.
Most of them reproduce sexually. Plant forms are diverse
and include mosses, liverworts, club mosses, ferns,

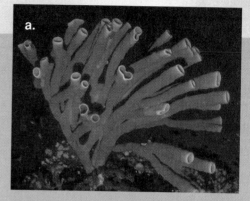

Figure E3.14 Examples of animals a. Tube
sponges from the Red Sea b. A click beetle in Arizona
c. A male hooded oriole from the southwestern region
of the United States

conifers, and flowering plants, as shown in Figure E3.15.
The bulk of the world's food and much of its oxygen are
produced by plants.

Kingdom Fungi. Kingdom Fungi, also a eukaryotic
kingdom, includes organisms that grow directly from

reproductive cells called spores. Fungi, like plants, have cell walls, but they do not carry out photosynthesis. You probably are more familiar with the members of this kingdom than you realize. Fungi such as mushrooms become large, multicellular organisms, with tissues made of slender tubes of cells (hyphae) that may contain more than one nucleus. Other fungi, such as yeasts, live as single cells during their entire life cycle. Still others, such as molds and rusts, live as tiny multicellular structures on the surface of bread that has been sitting around too long and lettuce that is going bad.

Fungi do not digest food inside their bodies as humans do. Instead, they release molecules called enzymes into their surroundings that break down (digest) biological material that other living systems have produced. The smaller food molecules then are absorbed into the cells. Thus fungi, along with many bacteria, play an important role as decomposers in many communities of organisms. The diversity of fungi includes yeasts, molds, morels, mushrooms, shelf fungi, puffballs, and plant diseases such as rusts and smuts (see Figure E3.16). Some fungi also interact closely with green algae or cyanobacteria to form the organisms known as lichens.

Kingdom Protista. Finally, the kingdom Protista is a grab bag of all the remaining eukaryotes that do not belong to the animal, plant, or fungi kingdoms. Protists live in water and in moist habitats, such as in the soil, on trees, and in the bodies of other organisms.

Protists show a remarkable range of diversity in their methods of obtaining food, their methods of reproduction, their life cycles, and their lifestyles. Most protists are microscopic single cells and many grow as colonies—clusters of individual cells. Others, such as brown algae living in the ocean, may form multicellular structures up to 100 m (328 ft) long. Some protists are brightly colored algae that produce their food through photosynthesis, and others are slime molds that obtain their food by decomposing the dead tissues of other organisms. Still other protists are parasites of animals, plants, or fungi. A single droplet of pond water viewed under the microscope reveals a world of protists in their myriad of shapes. Figure E3.17 depicts several protists.

Just as scientists have moved from using two large kingdoms to represent their thinking about earth's living systems to using three, four, and (today) five kingdoms, so they may have six kingdoms in the future. Evidence obtained during the last two decades suggests that a group of organisms called archaebacteria, which currently are classified in Kingdom Prokaryotae (Monera), differ in many respects from other bacteria in that kingdom. The archaebacteria include organisms

Figure E3.15 Examples of plants a. This moss, *Lycopodium*, grows in moist areas. b. A sword fern, *Polystichum munitium*, in Olympic National Park, Washington c. An apple tree, *Malus* spp, in full bloom

that live in environments similar to those that probably existed early in the earth's history, such as hot springs (like those in Figure E3.18), sulfur-containing muds at the bottom of ponds, salt ponds, and salt lakes. For this reason biologists think that the archaebacteria are among the very oldest organisms on earth. Because of their age and their differences from other bacteria, they perhaps merit a kingdom of their own.

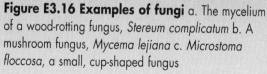

Figure E3.16 Examples of fungi a. The mycelium of a wood-rotting fungus, *Stereum complicatum* b. A mushroom fungus, *Mycema lejiana* c. *Microstoma floccosa*, a small, cup-shaped fungus

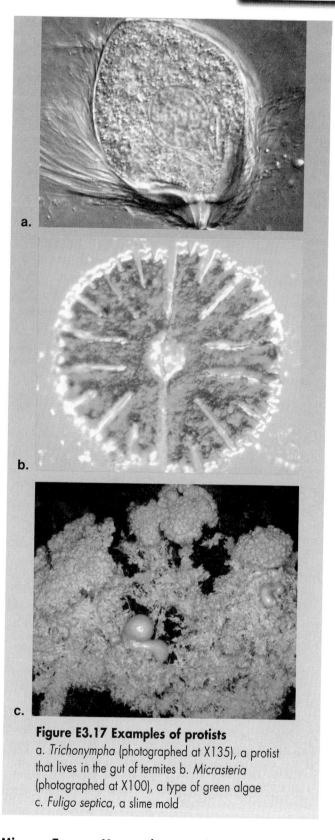

Figure E3.17 Examples of protists
a. *Trichonympha* (photographed at X135), a protist that lives in the gut of termites b. *Micrasteria* (photographed at X100), a type of green algae c. *Fuligo septica*, a slime mold

Figure E3.18 Minerva Terrace, Mammoth Hot Springs, Yellowstone National Park Archaebacteria live in environments like these hot springs.

From Cell to Seed

Have you thanked a green plant today? Plants play a critical role in our existence on earth—they produce the oxygen that we breathe, the food that we eat, and the multitude of materials that we use, from rubber, to lumber, to medicines, to coffee. Perhaps we ought to ask, Have you thanked a 3.5 billion-year-old single-celled organism today? because such an organism likely was the ancestor of all modern plants.

To understand how that could be, we need to trace the history of plant evolution. One of the ways we can begin to understand this history is to recognize that each of the major events of plant evolution that scientists think took place involved the appearance of a major adaptation. These events led to the emergence of the hundreds of thousands of different species of plants that currently inhabit every imaginable place on earth, from the frozen Arctic tundra to lush tropical rain forests.

The ancient seas. We begin our survey at a point about 3.5 billion years ago (see the time line in Figure E3.19). Evidence indicates that plants, like all other modern species, evolved from single-celled organisms that first lived in ancient seas and resembled modern prokaryotes. The atmosphere above these seas is thought to have consisted of a mixture of gases, largely water vapor, carbon dioxide and carbon monoxide, nitrogen, hydrogen sulfide (the stuff that makes rotten eggs smell), and hydrogen. Because this mixture of gases probably contained little or no oxygen, animals and plants, as we know them today, could not have survived.

The first single-celled organisms that lived in these seas most probably used complex molecules in their environment as their source of energy. These molecules likely were formed as a result of chemical reactions that occurred among the various substances present in the seawater. At some point, however, the growing population of living cells probably started using these complex molecules faster than they were being formed. Scientists think that as these primitive food molecules became scarce, the limitation of resources favored the survival of occasional cells that were able to

ERA	MILLIONS OF YEARS AGO	
CENOZOIC	7	apelike ancestors of humans appear
	38	origin of modern mammals
MESOZOIC	66	dinosaurs become extinct
	140	flowering plants appear
	225	mammal-like reptiles appear
PALEOZOIC	395	first amphibians and insects appear
	430	first land plants appear
	520	first vertebrates appear
PRECAMBRIAN	1400	oldest eukaryotic fossils
	2000	significant levels of O_2 in the atmosphere
	3500	oldest stromatolites and prokaryotic microfossils
	3700	oldest earth rocks
	4600	origin of earth

Figure E3.19 Time line of major evolutionary events Why do you think each of these events is significant?

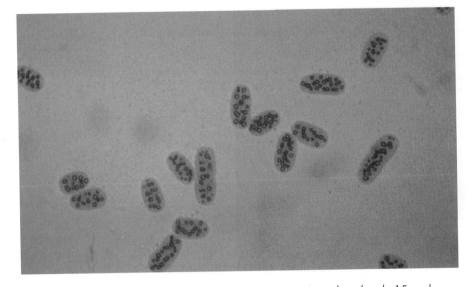

Figure E3.20 Modern phototrophic bacteria, *Chromatium okenii*, though only 15 μm long, are able to build complex molecules. The globules you see are sulfur particles.

UNIT ONE EVOLUTION: Patterns and Products of Change in Living Systems

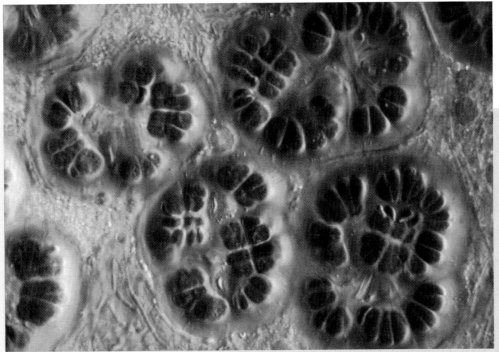

Figure E3.21 Photosynthetic bacteria Bacteria such as these *Gomphosphaeria*, which are each 1 μm in diameter, contributed to the production of oxygen in the earth's early atmosphere.

The appearance of eukaryotic cells. The fossil record indicates that just about this time, another key evolutionary event occurred. By about 1.4 billion years ago, more complex cells—ancestral eukaryotes—had appeared in the fossil record. There is evidence that some of these early eukaryotes incorporated photosynthetic bacteria within their cells and, as a result, were able to carry out photosynthesis.

The appearance of multicellularity. At this point in evolutionary history, some of these eukaryotic organisms consisted of groups of cells rather than a single cell. A multicellular organism would have had a better chance of surviving on land than would a one-celled organism. The outer layer of cells might have protected the inner cells from drying out rapidly, and the inner cells might have been efficient at photosynthesis. Other cells of the same organism might have become specialized in collecting water or nutrients from the environment. The specialized functions of different cells in one organism would have enabled the organism to exploit more of the resources in its new environment. The first plant may have been a specialized, multicellular organism, somewhat like the modern alga *Chara* in Figure E3.22, that was able to live and reproduce on land if ocean spray or tides kept it moist.

Two plant groups apparently evolved from such relatively complex multicellular green algae.

use sulfur compounds, carbon dioxide, and the energy in sunlight to build their own complex molecules. Because these cells possessed a chemical apparatus that was capable of building complex molecules from simple sources, they were largely independent of the dwindling supply of complex molecules that still floated free in the ancient seas.

The appearance of oxygen in the atmosphere. Scientists think that it was these first self-sufficient organisms that gave rise to modern plants. (Bacteria that use this same apparatus, such as those in Figure E3.20, still exist.) Even sulfur compounds, however, were not available in unlimited supply. A substance that was abundant was water, and the appearance of cells that could use water instead of sulfur compounds with which to build complex molecules was a major evolutionary advance. This type of photosynthesis (a process in which water, carbon dioxide, and light energy are used to build complex molecules) releases oxygen gas, a substance that probably had not been present in the atmosphere of the primitive earth. Today, photosynthesis is the major method of supplying energy either directly to living systems or indirectly to organisms that prey on others. Modern photosynthetic cells, such as the blue-green bacteria shown in Figure E3.21 and including those found in plants, contain a green pigment called chlorophyll that absorbs energy from sunlight.

By about 2 billion years ago, significant amounts of oxygen had collected in the atmosphere of the earth and formed a layer of ozone. (Ozone molecules are composed of 3 atoms of oxygen.) The ozone layer blocked out some of the dangerous ultraviolet light from the sun. Because ultraviolet light damages DNA, the establishment of an ozone layer made it possible for organisms to survive on land (water blocks ultraviolet light quite well).

One group is represented today by mosses (refer to Figure E3.15a) and related plants. These organisms possess few adaptations to life on land and require a moist environment in order to live and reproduce. The other group, which includes fossils of the oldest land plants, has many adaptations to life on land.

Adaptations that enhanced survival on land. One adaptation to life on land was the development of a waxy material that reduced water loss by providing a protective covering over the outer plant cells. The development of vascular tissue—a sort of plumbing system that carries water from the ground up through all the parts of the plant—was another important adaptation. Vascular tissue, depicted in Figure E3.23, enables plants to grow to large sizes. The oldest common ancestor of modern vascular plants developed about 420 million years ago.

The earliest vascular plants, however, were limited by their means of sexual reproduction, which required them to live in a moist environment. In early land plants (and their present-day descendants), sperm cells could reach egg cells only if the plant was covered with a film of moisture in which the sperm could swim. As a consequence, these plants were more successful in environments that contained at least a moderate amount of moisture.

Eventually, however, new species evolved that carried out sexual reproduction internally, within specialized reproductive structures that eliminated the need for external moisture. These changes probably occurred about 400 million years ago and enabled these new species to inhabit drier areas. In the most advanced modern vascular plants, the reproductive structures are located in cones or flowers.

Another adaptation that evolved about 350 million years ago was the production of seeds. Seeds are products of sexual reproduction and contain an inactive,

Figure E3.22 *Chara* *Chara* is a modern multicellular alga. The orange and yellow globules are reproductive structures.

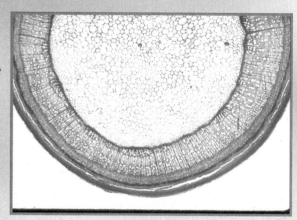

Figure E3.23 This magnified cross-section of a young maple tree stem (photographed at X7) shows the vascular tissues as rings around the outside.

Evolution Produces Adaptations

As you think about all of the organisms that you have observed, you see that despite their common ancestry and despite the common properties that they share as living systems, they really are quite different from one another. Even similar types of organisms—for example, the plants that you examined in the activity *Adaptation, Diversity, and Evolution*—have quite different characteristics. How can we explain the wide range of characteristics among the organisms on earth?

The key to this question is natural selection, the process by which evolutionary change occurs. You examined natural selection in the Chapter 2 activity *Modeling Natural Selection*. To understand the link between natural selection and diversity, consider briefly the sequence of events that might happen to a population of organisms living in a particular environment. A **population** is all of the organisms of one species living together in one area at the same time. As members of the same species, these organisms share certain common characteristics.

E44

a.

b.

Figure E3.24 How are these seeds likely to be dispersed? a. The female cone in this pinon pine will mature and have seeds on the inside of the scales. b. A cantaloupe has seeds inside the fleshy fruit.

tiny plant embryo packaged in material that provides food when the seed germinates and begins to grow. This system allows wide dispersal of new organisms because the seeds are spread by various means (wind, animals, water) and begin to grow in new locations.

In cone-bearing plants such as pines and firs (refer to Figure E3.24a), seeds are dispersed from open structures, but in flowering plants, seeds develop inside a specialized structure that becomes the fruit of the plant (see Figure E3.24b). The flowering plants, products of millions of years of evolution, are the most successful type of living plants. They are represented by more than 250,000 species and have adapted to every habitat on earth except Antarctica. If you doubt their value and importance to us, try these simple tests:

• Ask a friend or a member of your family to quickly name five types of plants.

• Ask the same person to quickly name five important commercial products that come from plants.

Chances are that most, if not all, of the five plants or products that they name are, or come from, flowering plants, some of the modern-day descendants of simple, single-celled ancestors that lived in primitive seas about 3.5 billion years ago.

Nevertheless, within any population there also is some variability in characteristics (just look around you in your classroom). This variability results from the normal events of sexual reproduction as well as from random changes (mutations) that occur in an organism's DNA during reproduction.

What happens to this population if the environment does not change? Does this variability increase? Or do the members of the population continue to look very much like each other and like past generations?

Most likely, future generations of that population will continue to look very much the same. This continuity in adaptive characteristics from one generation to the next occurs because of natural selection. Characteristics that are likely to be most widely represented in later generations are those of individuals that have the best ability to reproduce. Most of these characteristics likely have served as adaptations for that population for some time. On the other hand, new characteristics that may have appeared that *decrease* an organism's chance for survival at this time and in this environment are not likely to be passed on to offspring because the organism often dies before it reaches reproductive maturity.

Does this mean that the population never will change? Not at all. First, it is possible that a new variation might randomly occur that would *increase* an organism's ability to survive and to reproduce. In this case, natural selection would tend to perpetuate that new adaptation and the population might slowly begin to change. Second, it also is possible that the environment might change and that characteristics that were not adaptive in the earlier environment suddenly become beneficial. Remember as well, that characteristics that represent adaptations at one point in the history of a species may cease to do so as the environment changes. Adaptations relate to specific environments, and if the environment changes, then a characteristic that had been an adaptation may no longer provide an advantage. Unless such a characteristic is *harmful* in the changed environment, however, it may not immediately disappear from the species because it will not be selected against. Gradually, the characteristic may become less prominent among individuals of the population in later generations because it no longer is advantageous.

The correlation between surroundings and adaptations is an explanation for the diversity within and among species. The earth has a variety of environments, including arctic tundra, scorching deserts, warm coral reefs, immense ocean depths, alpine forests, rolling grasslands, and rocky coastlines. Each of these environments contains a variety of species, some unique to that environment. Each species has its own distinctive way of surviving and reproducing, which is somewhat dependent on its adaptations. The enormous diversity of species in the five kingdoms is in part a result of the range of adaptations to the enormous variety of environments.

The adaptations that exist in an extreme environment, such as the desert, clearly show the relationship between adaptations and surroundings. Many desert plants have structures such as spines instead of leaves (see Figure E3.25) and are coated with a thick, waxy outer layer that reduces water loss. Most of the mammals are inactive during the day and often hide in the shade or burrow into the sand to escape the drying heat. Life in a hot, dry climate challenges an organism's ability to maintain the body's balance of water, salts, and temperature; many adaptive characteristics have arisen that respond to these environmental conditions.

As organisms compete for food and protection in a living environment, the characteristics that enable them to survive will depend on the current physical conditions, such as amount of rainfall, temperature, and availability of light. They also will depend on the presence of and interaction with other species. For example, a nighthawk has the beak structure, the eyesight, and the capability of

Figure E3.25 Desert adaptations The horned lizard (*Phryhosoma cornutum*) and the cactus (*Opuntia* spp) show several adaptations to the desert environment. How many do you see?

maneuvering quickly in flight that enable it to catch flying insects. The nighthawk is in a good situation to compete for food, as long as there are flying insects to be had. In an unusually dry season, insects may not reproduce in large numbers and the adaptive characteristics of the nighthawk would be less useful. If this climatic change persisted for many seasons, it would have long-lasting effects. The nighthawk population would decline.

Sometimes characteristics such as those of the nighthawk can be adaptive and yet somewhat misleading in terms of an evolutionary pattern. Insect-eating bats, for example, have some of the same characteristics as the nighthawk—they are well adapted for quick turns in flight and have a way to sense prey at night (they rely mainly on sound-based sensory perception). Both species have adapted to compete for the same food source at the same time of night, yet the nighthawk is a bird and the bat is a mammal. In order to trace the true relatedness of these two species, we must consider a *combination* of traits rather than one or two.

Through the long years of evolutionary history, new species have arisen as new adaptive traits appeared within some subgroup of a population. These adaptations must appear at a level sufficient to make this group distinct. Thus a new species at first is related very closely to the remaining members of the species from which it was derived. As more time passes, generally more differences appear between the new and old species. Thus classification criteria, which reflect the pattern of evolutionary change, often are characteristics that are, or once were, adaptations.

Organizing Diversity

Whenever people collect information, they develop systems for organizing it. Think of the ways people organize the following: notes for research papers, computer files, recipes, or CD collections. Scientists organize information in specific ways for specific purposes. Health care professionals, for instance, organize information about blood types in a way that differs from how geologists organize information about soil types. Biologists, likewise, have developed their own systems for organizing the information that they accumulate about different types of living systems. One important way to categorize this information is by how long ago certain organisms shared a common ancestor. Thus, biologists have developed classification schemes that reflect our understanding of the evolutionary relationships that exist among the millions of different known species.

One of the criteria that biologists use to construct these schemes is structural similarities among organisms. A pattern of similar characteristics or homologies (refer back to the essay *Modern Life: Evidence for Evolutionary Change* on page E20) may suggest evolutionary relatedness. Appearances alone, however, can be misleading when it comes to recognizing biological relatedness. For example, not all traits are significant when considering questions of relatedness. Some characteristics are *acquired* during an organism's lifetime, such as bigger muscles built up by weightlifting. Because these characteristics cannot be inherited, they cannot be used as clues to evolutionary relationships. The characteristics that represent biological relatedness are those that are **heritable**, that is, that can be passed on

from parents to offspring by way of DNA. True homologies always involve heritable characteristics.

Similarly, not all organisms that look different are necessarily unrelated. Think back to the elephant and the hyrax that you saw in the essay *Five Kingdoms* (page E36). Or consider organisms that were commonplace some 150 million years ago—the dinosaurs. Long before computers and other special effects provided the technology for creative movie re-enactments, films depicted dinosaurs by superimposing close-up images of lizards against a backdrop that suggested enormous size. These images were not very convincing to those who had visited a museum and seen fossilized skeletons of the extinct giants or reconstructions of them based on scientific data. Nevertheless, most people probably reacted more favorably to images of fearsome reptiles (dinosaurs) than they would have reacted to close-ups of fearsome songbirds. Yet birds also are fairly close relatives of the dinosaurs, perhaps even closer than modern reptiles, despite the fact that birds look less like their prehistoric dinosaur ancestors.

Another type of evidence that biologists use to establish evolutionary relatedness and to organize meaningful classification schemes comes from biochemical examination of the proteins or DNA found in each organism. Biochemical homologies have become increasingly important in determining relationships. Sometimes comparisons of DNA sequences have changed our understanding of the relationship between organisms. The greater the similarities in DNA sequences, the more closely related two organisms are thought to be (refer to Figure E2.7, *Comparing DNA Across Primates*, on page E21).

Biochemical techniques have helped to clarify some classification problems. For example, for many years experts could not agree on the classification of the giant panda (see Figure E3.26). Some experts

Figure E3.26 Giant panda (*Ailuropoda melanoleuca*) The giant panda is closely related to bears.

grouped pandas with bears; others grouped them with raccoons. New techniques for studying the homologies in DNA have led to a greater understanding of the evolutionary relationships between bears, raccoons, and pandas. Now, the giant panda is classified with the bears, and the evidence suggests that the raccoon and bear families diverged from a common ancestor between 35 and 40 million years ago.

As new information about organisms becomes available, scientists may alter their opinions about how they should group species to reflect these evolutionary relationships. Although evolutionary history does not change, the classification schemes that scientists use to reflect that history improve with new knowledge and may change a great deal.

Biological classification schemes. Modern biological classification schemes generally contain a number of categories, each category representing a group of organisms with a particular degree, or level, of relatedness to each other. Organisms that have the greatest number of shared characteristics are grouped together in the category of **species**. You are a member of the species known as *Homo sapiens*, which is Latin for "knowing man." The first word, *Homo*, is the name of a group of species that share many homologies and as such form a larger category, known as a **genus**. The second name, *sapiens*, is the descriptive specific name within the genus group.

Although common names for a species may vary in different regions or different countries, scientific names do not vary. The use of scientific names is very important to accurate communication and efficient research. For instance, in California a gopher is a small, burrowing rodent with the scientific name *Thomomys bottae* (Figure E3.27a), whereas in Florida a

gopher is a type of tortoise whose scientific name is *Gopherus polyphemus* (Figure E3.27b). Imagine how difficult it would be for people who used the same common name for different organisms to communicate without confusion. Using scientific names avoids this problem.

As important as the concept of a species is, the category itself is sometimes hard to define. As a human, you share the physical characteristics of bipedalism, a precisely opposable thumb, and a relatively large, complex brain with other organisms in your species. Recall from Chapter 1, however, that

Figure E3.27 a. *Thomomys bottae*, a burrowing animal called a gopher in California b. *Gopherus polyphemus*, a tortoise called a gopher in Florida

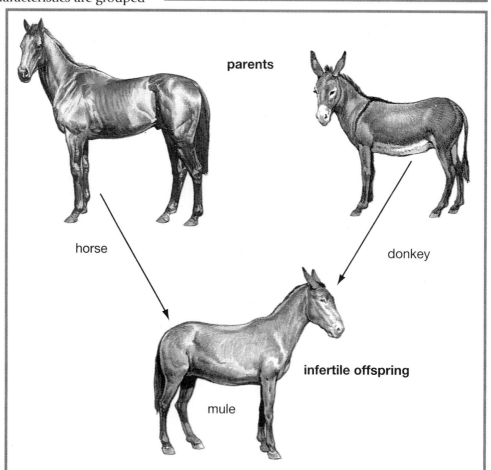

Figure E3.28 Mating between a horse and a donkey results in a mule, which usually is infertile. The donkey and horse are considered to be separate species. All animals are shown about 1/45 life-size.

these characteristics also are shared to some degree by certain other primates. Decisions about which organisms are the same species and which organisms constitute a different species are not always easy to make. For some organisms that reproduce sexually, clues to where species boundaries occur can be gained from determining whether or not two organisms can, and do, interbreed to produce offspring that also will be able to reproduce. Production of fertile offspring is important if species characteristics are to be passed on to future generations (refer to Figure E3.28).

The criterion of interbreeding is not strict because many organisms that reproduce sexually do not, however, interbreed in natural conditions. For example, many plants, including dandelions and peas, reproduce sexually by self-fertilization. Other plants, such as strawberries, reproduce mainly by means of shoots or roots that grow into new plants, as shown in Figure E3.29a. And many organisms, such as bacteria and other microbes, rarely or never reproduce sexually. Instead they divide into two cells or a new cell buds off (depicted in Figures E3.29b and E3.29c). In these cases biologists must rely on shared characteristics to define the boundaries between species.

Humans are one species among 1.7 million that biologists have described so far, but scientists estimate that the earth's total number of species may be 30 to 40 million. How can such a huge number of species be organized according to their evolutionary relationships? Recall that biologists define different *levels* of relatedness. A species represents the closest level of relatedness in biological classification schemes; that is, members of the same species are considered to be the same type of organism. At the next level, the genus is a group of related species. Members of the same genus are very similar (the genus Felis includes the domestic

cat and several other species, such as the ocelot). Groups of similar genera (plural of genus) form a **family** (the family Felidae includes many genera, such as cats and panthers) and so on. Families are organized in a large group known as an **order**, orders are grouped together in a **class**, and classes form still larger categories known as **phyla** (singular, phylum) or, for plants, as **divisions**. Figure 3.4 in the activity *Using Unity to Organize Diversity* shows the relationships of several organisms from the species level to the **kingdom** level, the category in biological classification systems that includes the largest number of related organisms.

Biological classification offers not only a way to organize almost 2 million different species of organisms into categories based on their evolutionary relationships, but also a way to organize what we know about these species. Its power and usefulness in this regard is easily illustrated. What could you tell someone about the organism named *Gyrodon merulloides*? Not much, probably. If you were told, however, that this organism is classified in the same major category as a mushroom, an image of its general characteristics suddenly comes to mind.

The usefulness of the species-to-kingdom scheme also is illustrated by its enormous lasting power. This system developed gradually over about 100 years from the system of naming species that a Swedish botanist, Carolus Linnaeus, established in 1753. With some changes, the scheme still is in use today. It has accommodated a tremendous volume of new knowledge that has been added in the past 200 years. And, if biologists are correct in their estimates of the number of species that remain to be discovered and described, we can expect it to accommodate the data yet to be examined and added.

 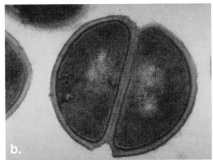 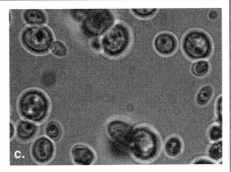

Figure E3.29 Not all organisms reproduce sexually or interbreed. a. Strawberries develop new plants at the ends of runners. b. Bacteria (*Staphylococcus*), photographed at X10,000, divide into two new bacteria. c. Yeast bud off new cells (photographed at X600).

COMPARTMENTS

How do you feel on a very cold day? If it is wet and windy outside, you certainly notice a cold sensation on your face, and your fingers and toes may become chilled or even a little numb. Yet if you pick up some snow and hold it in your bare fist, the snow melts. Obviously your body is warmer than its snowy surroundings, even though your fingers and toes may feel cooler than the rest of you.

By contrast, what happens to your body on a very hot day? While you may *feel* hot and sweaty, your internal temperature does not vary much. In fact, if you compared your body's temperature under these two quite different external conditions, you probably would discover that your body has maintained a fairly constant internal temperature that is different from either the freezing winter air or hot summer air surrounding it.

How is it possible for your body to maintain an internal temperature that is very different from the temperature of its environment? Part of the reason that your body can do this is that it is a *compartment* that is separated from the air around it. In fact, you might think of all living systems as containers that hold very specialized contents. Containers have walls, or external boundaries, that hold the contents and separate them from the surroundings. Living systems also have external boundaries; in humans, the skin separates the inside of the

Figure E4.1 What do this sports fan and her thermos bottle have in common? How are they different?

body from the outside environment. In single-celled organisms the cell membrane (and cell wall, if there is one) forms the boundary between the inside and the outside environments. By forming a special compartment, the external boundary of the living system allows conditions to differ greatly on the inside and the outside, just as the wall of a thermos bottle helps keep a hot beverage warmer than the outside air on a winter day (see Figure E4.1).

Temperature is not the only way in which the internal environment of an organism may differ from its surroundings. All living systems require specific internal conditions for life processes to take place. Some of the conditions other than temperature that have to be controlled include the amount of water, the concentration of specific minerals, the levels of various nutrients, the level of oxygen, and the pH. For example,

marine fish are surrounded by water yet they must separate their internal water from the water that surrounds them, because seawater is much saltier than the fluids inside their bodies. As you might expect, it is partly the boundary that is formed by a fish's scaly, waterproof skin that allows this difference in salt levels to exist and, in so doing, allows a variety of critical life processes to take place.

Complex organisms also may contain many smaller compartments, each with its own internal environment. Just as your skin separates the environment inside your body from the environment outside, the wall of your stomach separates the environment inside your stomach from the environment of the rest of your abdomen.

The same is true of other living systems. Close examination of a leaf reveals that it is covered by a waxy

UNIT TWO HOMEOSTASIS: Maintaining Dynamic Equilibrium in Living Systems

material that forms an external boundary that protects the living tissue within (refer to Figure E4.2a). Closer examination of the leaf reveals a series of veins embedded under the surface (see Figure E4.2b). These veins form separate compartments within the leaf, compartments that are distinct from the surrounding leaf tissue and that connect to other, tubular, compartments in the stem. As Figure E4.2b shows, microscopic examination of the veins and the surrounding leaf tissues reveals even smaller compartments, each enclosed within the leaf's overall structure, yet each a distinct compartment in itself.

Like humans and plants, most other complex organisms also have several different levels of compartments. At the microscopic level, the basic

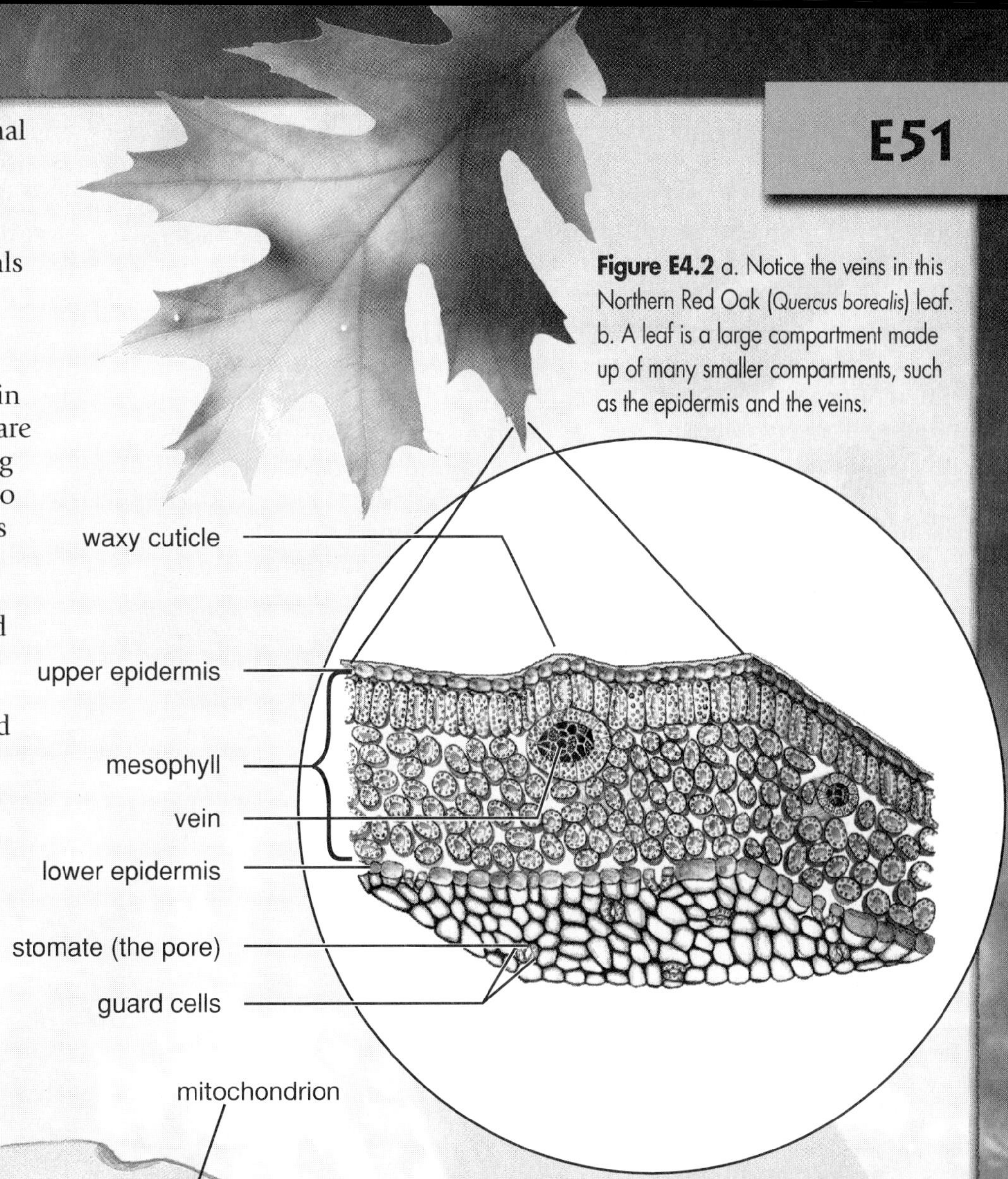

Figure E4.2 a. Notice the veins in this Northern Red Oak (*Quercus borealis*) leaf. b. A leaf is a large compartment made up of many smaller compartments, such as the epidermis and the veins.

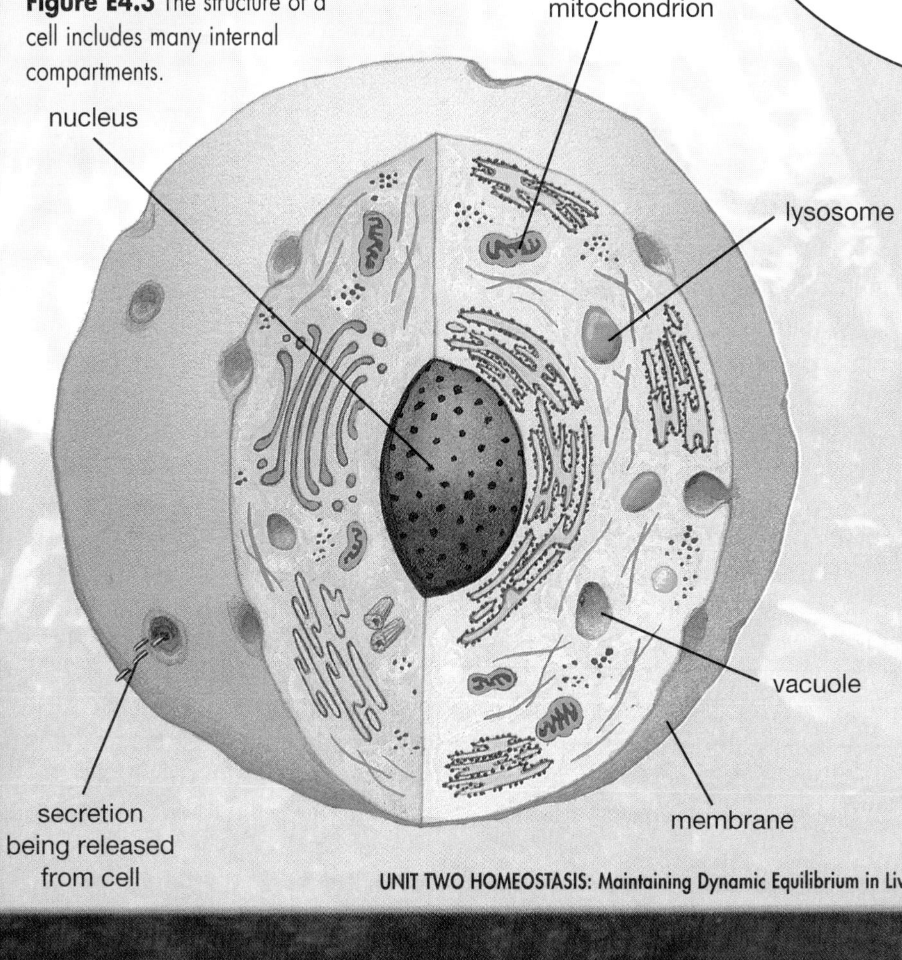

waxy cuticle
upper epidermis
mesophyll
vein
lower epidermis
stomate (the pore)
guard cells

Figure E4.3 The structure of a cell includes many internal compartments.

nucleus
mitochondrion
lysosome
vacuole
membrane
secretion being released from cell

unit of life is the cell, a compartment whose boundary is formed by one or more membranes. (Plant cells and some bacteria also have rigid cell walls that provide strength and protection.) Prokaryotic organisms consist of only one compartment, the cell itself. In contrast, eukaryotic organisms, even those whose bodies also consist of only one cell, contain smaller compartments. Several of these compartments are illustrated in the cell shown in Figure E4.3. Foremost among such subcellular compartments is the **nucleus**, which contains DNA.

What is the advantage of having so many compartments in living systems? Remember that all living systems require specific internal conditions for life processes to occur. The presence of many different compartments within one cell or one organism allows for the presence of many different internal environments. These environments provide the different conditions that may be required for very different functions to occur. The environment within the nucleus, for example, provides the conditions required for a variety of processes involving the cell's DNA to take place. In contrast, the environment within the lysosome (see Figure E4.3) provides the conditions required for a variety of very different functions to occur. In humans the pH of the stomach contents is much lower than the pH inside the small intestine. This difference in pH is related to differences in the types of digestive processes that occur in each organ and, in most of us, is very carefully controlled.

What would happen if the conditions within these various compartments or within your body were to undergo a significant lasting change? As you will see in Chapter 6, changes beyond certain rather narrow limits can be dangerous to the organism involved, mostly because these changes disrupt the normal processes of life. Remember that the structure of compartments, both large and small, allows for a range of very specific internal conditions to exist, and that each condition is necessary in its own way for these processes to take place.

Think back for a moment to the sports fan and her thermos pictured in Figure E4.1. The caption under the photograph asks you to suggest how the fan and her thermos are similar. Probably, you recognized right away that both the fan and her thermos are *containers*, or compartments, with outside boundaries that allow the conditions of the contents (in this case, their temperatures) to be different from the condition in the environment. But the caption also asked you how the fan and her thermos are different. For example, what do you think will happen over time to the temperature of the coffee in the thermos? In contrast, what prediction can you make about the temperature of the fan's body? Clearly there is more to this challenge of maintaining specific internal conditions than just the presence of a boundary. In this chapter and the next, you will examine in detail how living systems respond to this challenge. And next time you melt a snowball by gripping it tightly in your bare hand, perhaps you will think about your body as a biological compartment.

Membranes

For a cell to form a compartment, it must have a boundary around its contents. You might compare the membrane around a cell to the boundary walls of a beverage container because the membrane separates the *inside* of the cell from the *outside*. A cell is like a beverage container because a cell's contents include a liquid-like substance that contains a variety of different molecules, and its membrane forms a barrier around these molecules.

The cell membrane is more complicated, however, than the walls of most containers because it lets some things pass through it, while preventing the movement of other things. If you think about this for a moment, you should be able to see how important this is. To act as an effective boundary, the membrane must prevent the free movement of molecules into or out of the cell; in general, it must be *impermeable* to most substances. Yet molecules that the cell needs for its activities must be able to pass through the membrane. Because of differences in their permeabilities to various molecules, biological membranes are said to be **selectively permeable** to the variety of molecules that they encounter.

The selective permeability of a membrane (its ability to regulate the passage of molecules) depends on the structure of the membrane, in particular on the organization of the large molecules of which it is made. As Figure E4.4 shows, a typical cell membrane is made up primarily of lipid (fat) molecules. An important property of lipid molecules is that they do not mix well with water. In the cell membrane, the lipid molecules are arranged in two layers that are oriented so that the parts of the lipid that have the lowest tendency to interact with water form the *interior* of the membrane. This hydrophobic, or water-repelling, part of the membrane prevents molecules that are very water-soluble from passing easily through the membrane barrier. By contrast, fat-soluble molecules dissolve in this interior part of the membrane and move easily in and out of the cell. Water also moves easily through the membrane because water molecules are small and have no electric charge.

Notice also in Figure E4.4 that the membrane contains many protein molecules among the lipids. Proteins are large molecules made up of many small molecules (called amino acids) that are linked together to form a long folded chain. Some of these protein molecules, particularly those on the outer surface of the membrane, act as receptors, structures to which other molecules bind and which facilitate communication and physical interaction among cells. Many of these membrane proteins are involved in regulating the passage of substances into and out of the cell. Some act as carriers that specifically transport certain molecules such as glucose across the cell membrane. Others form special gates or channels through which various water-soluble

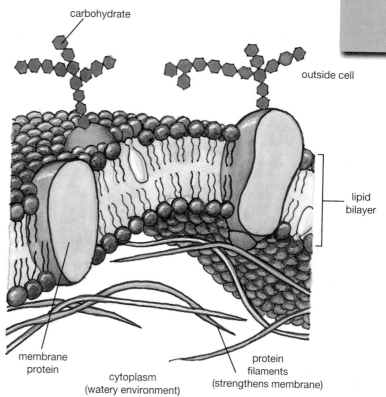

Figure E4.4 Diagram of a section of the lipid bilayer of a cell membrane
Lipids are oriented in such a way that the interior of the membrane repels water. Protein molecules may span the membrane or be exposed on the inner or outer surface. What structures facilitate selective permeability?

substances such as **ions** (molecules that have an electrical charge) are able to pass into and out of the cell. For example, when table salt (NaCl) is dissolved in water, it forms charged ions (Na^+ and Cl^-). Because sodium and chloride ions can cross the cell membrane only through special protein channels, the cell can regulate the movements of these ions very precisely.

Thus, biological membranes really are more complex than the walls of an insulated beverage container. They also perform a far more complex and important role; as you will see, cell membranes are critical to controlling conditions inside cells.

Molecular Movement

Someone in the next room opens the oven and takes out a pan of hot brownies. Almost immediately your mouth starts to water as you smell the chocolate. How did the odor get to your nose? A few molecules from the chocolate were released by heat and traveled through the air, and you detected them with your sense of smell. In

fact, the protein receptors on cells in your nose detected these chocolate molecules.

This simple example demonstrates an important property of molecular movement. Molecules disperse in random directions, but overall, they tend to spread from an area where they are more concentrated to an area where they are less

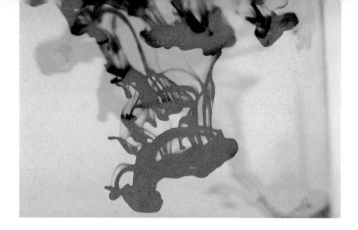

Figure E4.5 Diffusion Diffusion is the movement of molecules from an area where they are more concentrated to an area where they are less concentrated, eventually resulting in an even distribution throughout.

concentrated. This dispersion continues until the concentration of molecules is the same everywhere (see Figure E4.5). This type of molecular movement is called **diffusion**. If you could sample the concentration at various points from the most concentrated area to the least concentrated area, you would find that the concentration decreases gradually. This gradual decrease forms a distribution known as the concentration gradient. Because molecules diffuse from areas of higher concentration to areas of lower concentration, they are said to move *down* the concentration gradient.

Diffusion occurs because the universe, and all the molecules in it, drifts into a less ordered state unless some energy input keeps it orderly. Think, for example, about your room. It takes your energy to maintain order—to keep your clothes off the floor, your CDs in one place, and your papers on the desk. Without a continual input of energy, the organization of your room soon disintegrates and begins to look like the room in Figure E4.6. The degree of disorder in a system is called **entropy**. Living systems use energy continuously to maintain their own order and unique internal environment.

The diffusion of molecules from higher to lower concentration occurs in solutions as well as in the air. For example, a solute, such as table salt (NaCl), dissolved in a solvent, such as water, undergoes diffusion until it is uniformly distributed. The

movement of molecules, such as salt or sugar molecules, across a biological membrane depends in part on the relative concentration of the solute in the cell's internal and external environment. The movement also depends on special membrane proteins that act as carriers or channels.

One type of diffusion, called **osmosis**, can result in the buildup of significant pressure inside a membrane-enclosed compartment. Osmosis is the movement of water from an area of greater concentration to an area of lesser concentration of water. For example, "pure" water has a higher concentration of water (100 percent) than a 10 percent salt solution has, which contains only 90 percent water. Thus, water will move into a compartment containing a salt solution. When the initial concentration of water is greater outside a cell than inside, water rushes into the cell. In this way,

Figure E4.6 Entropy, or the tendency toward disorder, in a system What evidence of entropy do you see in this drawing? What is needed to put this room into order?

UNIT TWO HOMEOSTASIS: Maintaining Dynamic Equilibrium in Living Systems

osmosis can result in such a buildup of pressure inside the cell that the cell swells and bursts. Conversely, if the concentration of water is greater inside than outside a cell, water leaks out and the cell shrinks. Figure E4.7 shows a range of cell responses in both animal and plant cells.

Osmotic pressure plays a special role in plant cells because of their rigid cell walls. These cells have an internal water-filled compartment known as a **vacuole**. The water in a vacuole exerts osmotic pressure on a cell's contents, pressing the cell membrane tightly against the cell wall. This pressure gives plant structures such as leaves their firmness. What do you think is happening in a vacuole when a leaf wilts?

Diffusion, including osmosis, accounts for much of the exchange of materials between the internal and external environments of individual cells. Now recall that cell membranes are only *selectively* permeable. Thus, only certain molecules can diffuse freely through a cell membrane. Large molecules or molecules with electrical charges cannot cross lipid membranes efficiently by diffusion alone. However, as Figure E4.8 illustrates, membrane proteins have methods of transporting these molecules through membranes.

In **passive transport**, no energy is expended by the cell, and transport proteins move substances *down* their concentration gradient, either into or out

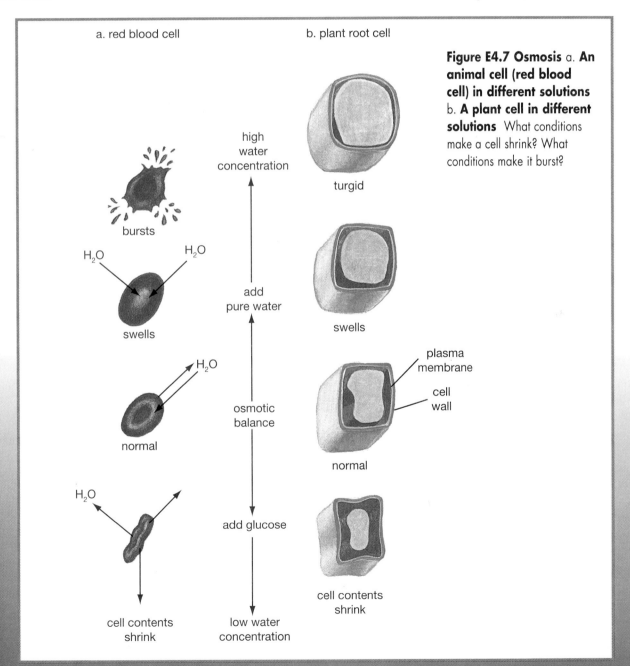

Figure E4.7 Osmosis a. **An animal cell (red blood cell) in different solutions** b. **A plant cell in different solutions** What conditions make a cell shrink? What conditions make it burst?

of the cell. Each type of protein can transport only specific substances. Cells use passive transport to move many essential ions.

In **active transport**, cells use energy to move substances with the help of transport proteins. Through active transport, substances can move across a membrane *against*, or up, the concentration gradient. For example, the soil around a plant may contain low amounts of elements necessary for the plant's growth. Ordinary diffusion would not provide the plant with enough of these elements. By means of active transport and other processes, however, the plant's root cells can accumulate these elements in relatively high amounts. This example is a little like collecting all of the clothes scattered about a room and placing them in a laundry basket. Once inside the plant, the elements then can be transported to all parts of the plant.

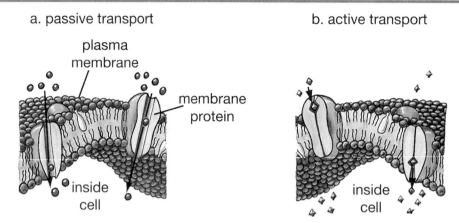

Figure E4.8 Passive and active transport a. In **passive transport** substances move across membrane proteins, down their concentration gradient. b. **Active transport** uses energy and membrane proteins to move substances against their concentration gradient. Active transport also can move substances with their concentration gradient.

Making Exchanges throughout the Body

Imagine spending the day in a one-room studio apartment. You would have easy and direct access to all that you needed. Now imagine spending the day in a 20-story office building with your office on the 15th floor, the restrooms on the 10th floor, and the cafeteria on the 3rd floor. The same daily tasks that you might perform in the studio apartment suddenly become more complicated. You would not have direct access to all that you needed. Instead, you likely would require more complex systems (for example, an elevator and hallways) to help you move from place to place and to complete your daily tasks.

In a similar way, single-celled organisms such as bacteria have their entire living compartment directly in contact with the environment that surrounds them, and they can interact directly with this environment. For multicellular organisms, however, life is more complicated.

In humans and other multicellular organisms, most of their cells are buried deep inside their bodies. These cells are not in contact with the external environment from which they must obtain oxygen and food and into which they must release wastes. Despite this lack of contact with the external environment, the levels of oxygen and nutrients must stay sufficiently high inside the cell and waste products must be removed from the cell

continuously if the cells are to survive. For multicellular organisms, as for all life forms, survival depends on the body's ability to create an acceptable set of internal conditions in the face of an external environment that is constantly changing.

In humans such conditions are maintained through the interaction of a number of interrelated organ systems. An organ system is a group of organs that work together to perform a common function. Some are delivery systems that carry vital materials to different parts of the body. Other organ systems are regulators of specific conditions and respond to various signals from different parts of the body.

The gas exchange system is a delivery system that provides air to the lungs, where oxygen is exchanged for carbon dioxide and delivered to the bloodstream. The circulatory system transports oxygen, nutrients such as glucose, and a variety of other substances to cells throughout the body and carries waste products such as carbon dioxide from the cells. The urinary system eliminates excess water, salts, and wastes that contain nitrogen. The activity of all of these systems depends on the exchange of materials across boundaries. Many of these exchanges occur by the processes that move molecules—diffusion, osmosis, and passive and active transport. These processes depend on sensitive communication between tissues and organs, between the internal and external environments, and between the brain and the rest of the body.

Let's look closely at one delivery system. The circulatory system is essential for maintaining appropriate internal conditions because it forms an extensive network that distributes materials throughout the body. At the center of this network is the heart, which is a muscular organ that pumps blood through the entire circulatory system. Branching out from the heart is a series of blood vessels. These blood vessels are interconnected tubes that carry blood and help control its flow. The blood itself is a complex mixture of water, dissolved gases, nutrients, cells, large molecules, such as proteins, and a variety of other substances. Blood moves away from the heart through arteries and returns through veins.

aortic arch
carotid artery (to head)
jugular vein (from head)
b.
right ventricle
lung
left ventricle
descending aorta (artery to body)
vena cava (vein from body)
liver
kidney
intestine
a.

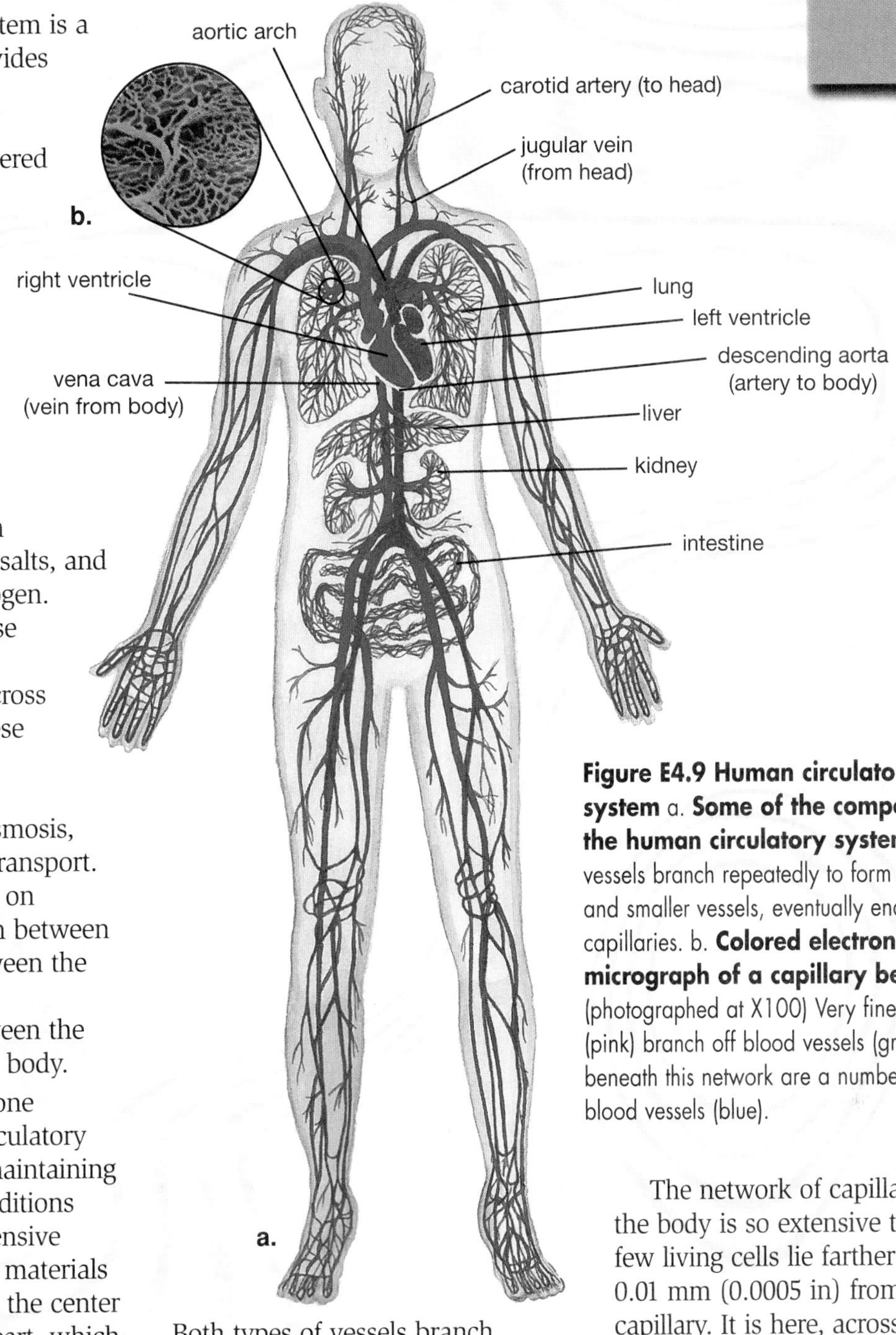

Figure E4.9 Human circulatory system a. **Some of the components of the human circulatory system** Blood vessels branch repeatedly to form smaller and smaller vessels, eventually ending in capillaries. b. **Colored electron micrograph of a capillary bed** (photographed at X100) Very fine capillaries (pink) branch off blood vessels (gray); beneath this network are a number of large blood vessels (blue).

Both types of vessels branch repeatedly to form hundreds of millions of smaller vessels. These are connected to each other by even smaller vessels called arterioles (tiny blood vessels that carry blood *to* capillaries), venules (tiny blood vessels that carry blood *away* from capillaries), and capillaries. In fact, the human body has more than 10 billion capillaries, thin-walled vessels that are barely as wide as the diameter of one cell. Figure E4.9 illustrates the human circulatory system.

The network of capillaries in the body is so extensive that very few living cells lie farther than 0.01 mm (0.0005 in) from a capillary. It is here, across the thin capillary walls, that the actual exchange of materials between the blood and the body's cells occurs. As blood passes through a typical capillary, oxygen, glucose, and other substances move out into the fluids that surround the cells. These substances then move from the fluids into the cells. Simultaneously, carbon dioxide and other wastes enter the blood.

This is a good example of how intricate the systems of exchange in the human body can be.

Disposing of Wastes

At one time or another you probably have been responsible for taking out the trash. If so, you have a pretty good idea of how much waste people produce each day. The body produces waste because it does not use all of the matter that it takes in. Also, biological processes that are going on inside you all the time produce waste materials. These wastes could become toxic if they were to build up in the body so the human body, like any living system, needs a way to dispose of wastes.

As cells produce waste products, they deposit them in the blood stream. These wastes are carried away from cells by the blood stream and are eliminated from the body in a variety of ways. One waste product, carbon dioxide, is carried back to the lungs and eliminated by exhaling. Excess salts and some other waste products can be eliminated through perspiration. Metabolic wastes—chemical wastes produced by biological processes—are eliminated primarily by the urinary system.

Have you ever smelled the pungent odor of glass cleaners? This odor comes from ammonia, which is a toxic substance that is produced by cells as waste when proteins are broken down. If ammonia accumulates in the blood and tissues, it causes life-threatening illness. Humans, however, possess an organ system that efficiently removes toxins such as ammonia. As ammonia is formed, the liver immediately converts it to urea, a relatively nontoxic substance. The kidneys remove urea, including it in the urine that is excreted from the body. The urinary system performs this essential function continuously and also carries out other important roles, such as regulating blood pressure and adjusting fluid volumes in extracellular compartments.

The urinary system is one of the body's regulatory systems that operates automatically, but it is directly influenced by behavior, such as drinking water and other liquids. The regulatory organs of the urinary system are the two kidneys, which are illustrated in Figure E4.10. The kidneys adjust the content of blood by filtering it to remove waste substances and then restoring the correct balance of water, salts, and other key compounds to the blood. Tubes called ureters connect each kidney to the bladder, a sac that holds the waste-containing urine until it is released from the body.

The kidneys are complex and active organs. Each kidney consists of more than 1 million microscopic **nephrons**,

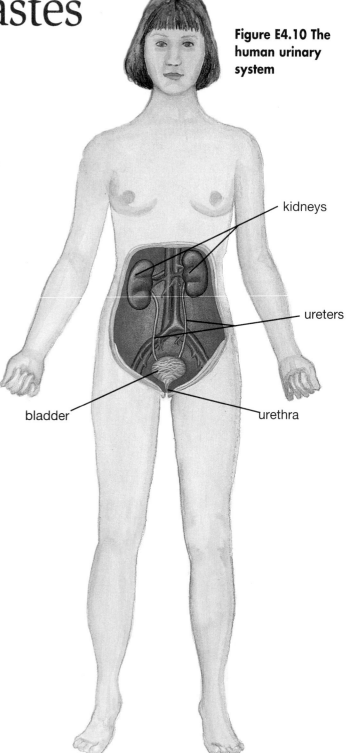

Figure E4.10 The human urinary system

kidneys

ureters

bladder

urethra

which are long coiled tubes that act as filtering units (see Figure E4.11). One end of the nephron forms a capsule that surrounds a specialized capillary bed that is condensed into the shape of a knot. This capillary bed, called a **glomerulus**, is part of the blood vessel network. The glomerulus differs from other capillary beds because it is supplied by an arteriole and drained by another arteriole, rather than by a venule. The arteriole that drains the glomerulus transports the blood to a second capillary bed that surrounds the rest of the nephron. This second capillary bed returns the blood to the venous side of the body's vascular system for its continued circuit back to the heart. It is in and around these two capillary beds and their nephrons that the kidney cleans and regulates the composition of the blood.

Although the kidneys are small compared with other organs (together they weigh only about 300 g or 10 oz), the kidneys and their supply of arteries normally carry almost 20 percent of the heart's output of blood at any time. This amount can vary, depending on circumstances; as it varies, so does the rate of urine production.

To understand how the kidney regulates the body's internal environment, think about its shape and its components, and recall what you have learned about membranes and compartments. Now, look at Figure E4.11 and, as you read the description of the kidney as a filter, try to trace what happens. Begin by finding the three major tubes of the kidney, located on the concave side of the kidney. These tubes are an artery, a vein, and a ureter. By visualizing what happens in these three tubes, you can begin to see what takes place inside the kidney.

We know that arteries carry blood from the heart to the rest of the body, so it is not surprising that blood enters the kidney by way of the renal artery. Likewise, we know that blood returns to the heart by means of veins, so it follows that the renal vein

carries blood out of the kidney. We also know that arteries and veins are connected to each other by smaller vessels called arterioles, capillaries and venules.

All of the substances in the blood entering the kidney do not leave through the renal vein, however. Some substances are removed from the blood and become urine, which then leaves the kidney by way of the ureter. Urine collects in the nephrons in the kidney. The nephrons interact with the capillary beds in a special way, helping to produce, collect, and transport urine. In the process, they clean the blood and regulate its composition. The cleaning and regulation of blood composition occurs in three phases along the nephron. These are **filtration**, **reabsorption**, and **secretion**.

Filtration takes place in the glomerulus, a specialized structure that carries blood under much higher pressure than other capillary beds. Together the two kidneys filter as much as 125 mL of plasma *every minute*. If that entire amount were excreted, the results would be pretty severe. To offset this fluid loss, you would have to drink 7.74 L of liquids *every hour*. That's 180 L every day! You'd be spending all your time drinking and urinating.

Shortly after the kidneys filter 125 mL of fluid per minute, however, 99 percent of the fluid is reabsorbed and returned to the blood. **Reabsorption** takes place between the nephron tubule and the second capillary bed. It occurs by means of active

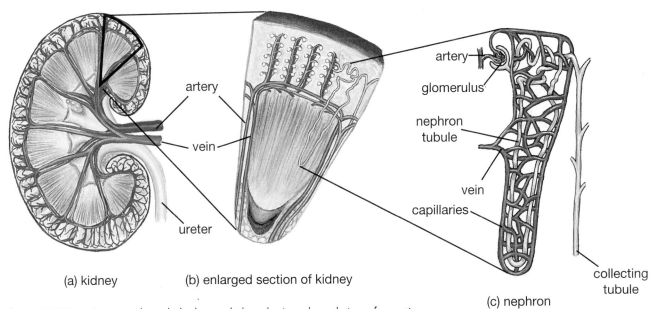

Figure E4.11 a. A section through the human kidney. b. An enlarged view of a section of a kidney. c. An enlarged view of one nephron with its surrounding capillaries. The urine leaving the collecting tubule eventually enters the bladder, where it is stored temporarily.

transport of sodium ions from the filtrate into the capillary blood, creating an osmotic pressure that moves water in the same direction. Water moves back into the blood according to the degree to which the cells in the wall of the tubule are permeable to it (that is, according to how easily the water can pass through the tubule wall). The rate of water movement depends on internal conditions in the body. If blood volume is low, then water is reabsorbed very efficiently back into the blood stream. If, in contrast, blood volume is high (perhaps you just drank a couple of tall glasses of water), then much of the extra water passes quickly through the tubule and enters the bladder. Thus the rate of urine production is directly related to water balance in the body.

The third phase of kidney function, **secretion**, occurs farther along the nephron. Secretion provides a way for wastes that were not filtered from the blood in the glomerulus to be excreted in the urine. These wastes move out of the capillaries around the tubule and then are actively transported by the cells of the tubule wall into the interior of the tubule. From the tubule these wastes move through the kidney and on to the bladder as urine.

As you can see, together these processes allow very precise adjustments in the composition of the blood and removal of waste products. The vital importance of these processes is illustrated by the observation that acute kidney failure that is not treated immediately rapidly leads to death from fluid imbalances, incorrect ion levels in the blood, and the accumulation of metabolic wastes.

Homeostasis ● ● ● ● ● ● ● ● ●

Have you ever wondered why you don't faint every time you stand up? Does it surprise you that even if you skip lunch you still can walk and talk? Perhaps you never thought about these questions. Although it's obvious that you can stand up without fainting and continue to function without eating lunch, explanations for these occurrences really are quite complex. For instance, the cells in your brain all are exceedingly sensitive to tiny changes in the levels of oxygen and sugar in their immediate environment; even small decreases in these critical substances can cause fainting. One reason that the effects of gravity don't cause a drop in the blood and oxygen flow to the brain when you stand up is that your blood pressure immediately rises to maintain adequate oxygen flow. Likewise, you can get away with skipping lunch because a declining level of sugar in your blood stream triggers your liver to begin releasing the sugar it holds in storage.

Your body must continuously make these adjustments and many others to create and maintain an environment within which your brain can function. These adjustments are made *automatically*. Taken together, these complex internal adjustments assure that conditions within your body remain within rather narrowly defined limits, a condition of balance called **homeostasis** (see Figure E5.1).

Humans are not the only organisms that maintain homeostasis. In fact homeostasis is a fundamental characteristic of *all* living systems. In many animals internal organs that are similar in function to those in humans help to maintain homeostasis. In plants, specialized structures, such as those illustrated in Figure E5.2, have evolved that enable plants to maintain balanced conditions.

In simpler organisms, however, much simpler mechanisms operate to maintain homeostasis. In a single-celled organism such as an amoeba, the removal of toxic waste products is accomplished without complicated internal organs and with very few specialized structures. Instead, basic processes, such as diffusion and osmosis, are sufficient. Even though these mechanisms for maintaining homeostasis may seem simple when compared with an entire circulatory system, they are critical for maintaining the amoeba's life. Ultimately different organisms must balance different conditions and may use different mechanisms to do so. For all organisms, however, maintaining the delicate homeostatic balance means life, and losing this homeostatic balance for an extended period of time means death.

In order to maintain homeostasis, two things are required: an organism must be able to *sense* when changes have occurred in the external and internal environment, and it must be able to *respond* with appropriate adjustments. For example, humans can monitor **stimuli** or external signals such as cold because we have sensory neurons in our skin that allow us to feel the outside temperature. Once the message "cold" is received in the brain, our body can respond by changing blood flow. This change is involuntary, or *automatic*, because we do not consciously control the physiological processes that cause the heart rate to increase or

UNIT TWO HOMEOSTASIS: Maintaining Dynamic Equilibrium in Living Systems

decrease or the blood vessels to dilate or constrict. In other words, we do not have to *decide* that the body should attempt to keep the brain, heart, and liver at a nearly constant temperature even if that means sacrificing tissues at the surface of the body; the body will attempt to keep the core warm anyway.

The human body's response to change is quite specific as well as involuntary. For example, the body responds to cold temperature by diverting circulation to keep the most important internal organs warm. This type of response is appropriate for the external conditions. If the body becomes too hot, however, the circulatory system diverts blood flow away from the internal organs to protect them from damage caused by excess heat.

Even though these examples are rather dramatic, the human body routinely senses and responds to thousands of small changes each day, most of which you are not even aware. It is through many small, specific, and automatic changes that living organisms are able to sense and react to an environment that is ever-changing and sometimes hostile. Luckily, the mechanisms responsible for maintaining balance are always on the job.

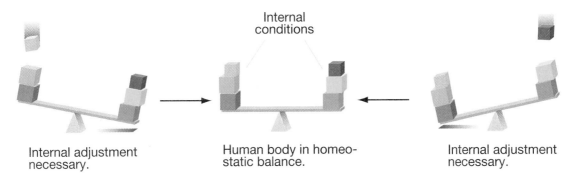

Internal conditions

Internal adjustment necessary.

Human body in homeostatic balance.

Internal adjustment necessary.

Figure E5.1 Homeostasis The human body is maintained in a state in which the internal conditions are balanced. When the balance is disturbed, the body adjusts its internal conditions to restore balance.

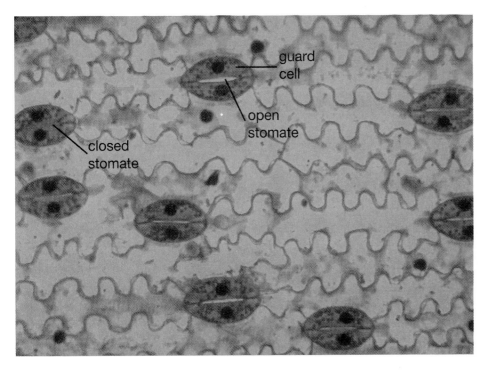

guard cell

open stomate

closed stomate

Figure E5.2 Guard cells control the rate of water loss in plants. Water loss is controlled by the condition of special cells, called guard cells, that regulate the size of microscopic pores in leaves. These pores are called **stomates**. When the plant has sufficient water, the guard cells swell and the stomates open. Water then evaporates through the stomates. When the plant is low on water, the guard cells shrink, and the stomates remain closed, which preserves water. Based on the appearance of stomates in this leaf, how would you describe this plant's water balance?

Careful Coordination

As you learned in Chapter 4, multicellular organisms face a great challenge in maintaining internal conditions. Most of their cells are buried deep inside their bodies, far removed from the external environment from which they must obtain their oxygen and nutrients. They, therefore, are not able to rely on simple processes to maintain balance. In humans, acceptable internal conditions are created through the interactions of a number of organ systems. These interactions allow cells deep in the body to maintain contact with the outside, even though this contact may be indirect. Organ systems contain many organs that communicate with each other. Continuous, coordinated adjustments made by all of the body's organs help to *regulate* homeostatic balance.

Continuous automatic adjustments are possible because most body systems are directed by the nervous system and the endocrine system—two organ systems that reach every part of the human body. The nervous system is known for directing rapid, short-term, and very specific responses in the body. A reflex, such as the fortuitous jerk that pulls your hand back when you accidentally touch something hot, is an example of a rapid nervous system response. This reflex is a homeostatic response to a potentially dangerous rise in skin temperature, and it illustrates the interaction of sensory, nerve, and muscle systems. In this case, receptors in the skin send nerve signals to nerve cells in the spinal cord. These in turn stimulate muscles in the arm to contract suddenly, and the hand withdraws from the hot surface.

In contrast, the endocrine system, which is illustrated in Figure E5.3, usually directs slower and longer-lasting changes. In cases of dehydration, for instance, sensors in a specialized part of the brain called the

hypothalamus (Figure E5.4a) detect a shortage of water in the body. These sensors in turn respond by signaling the pituitary, an endocrine gland also located in the brain, to release chemical messengers called **hormones** into the blood stream (Figure E5.4). Hormones trigger changes in the activities of a wide variety of cells and organs throughout the body. (See the table *Origins and Effects of Hormones*, Figure E5.5, on pages E63-E64 for more information about the endocrine system.) At this point, the endocrine system interacts with the circulatory system because the blood stream carries the hormone vasopressin to the kidneys. The kidneys, which are part of the urinary system, respond to this signal by retaining more water. The net effect of all this sensing, responding, and interacting is a regulated decrease in urine output and an increase in the desire to drink water, as illustrated in

Figure E5.6. The general process by which the body automatically senses changing conditions and responds to them is called **feedback**.

Feedback plays a significant role in homeostasis because it is the mechanism that regulates how the body responds to changing conditions when an imbalance is detected. Feedback can operate in one of two ways. It can direct organ systems to interact in ways that change internal conditions *away from* the initial conditions or *toward* the initial conditions. The regulation of blood pressure, that is the pressure exerted by blood against the walls of circulatory vessels, serves as a good example of changing conditions away from initial conditions. Like many other internal conditions, blood pressure must be maintained within very defined limits. If the pressure becomes too low, nervous system sensors called baroreceptors, located on large arteries in the chest and neck, send a signal through the nervous system to the brain. As

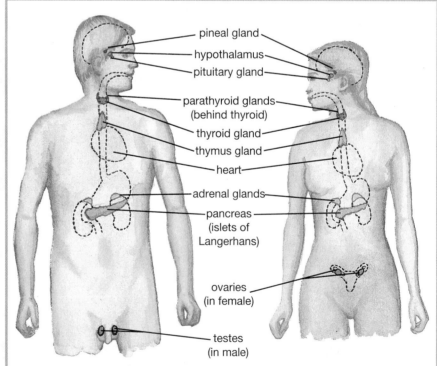

Figure E5.3 The endocrine system The dotted lines show non-endocrine organs, such as the stomach and kidneys.

- pineal gland
- hypothalamus
- pituitary gland
- parathyroid glands (behind thyroid)
- thyroid gland
- thymus gland
- heart
- adrenal glands
- pancreas (islets of Langerhans)
- ovaries (in female)
- testes (in male)

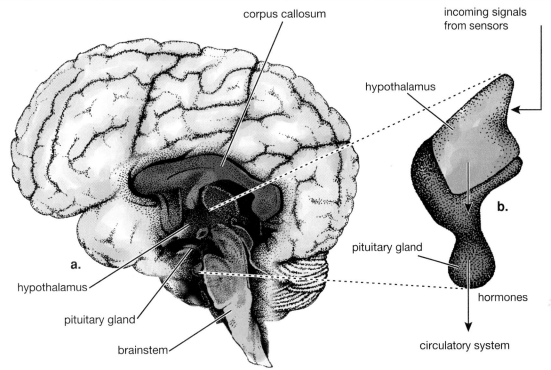

Figure E5.4 The hypothalamus a. The hypothalamus is a specialized part of the brain that is part of both the endocrine system and the nervous system and is involved in detecting changes in internal conditions. b. The hypothalamus responds to feedback by signaling other endocrine organs. In response to dehydration, the hypothalamus can release hormones that act on the pituitary gland, which lies just underneath the hypothalamus in the brain. The pituitary gland then releases hormones into the circulatory system.

Gland/Organ	Hormone	Target	Principal Action
Pineal	melatonin	unknown in humans—perhaps hypothalamus and pituitary	regulates circadian rhythms (day/night cycles)
Hypothalamus	corticotropin-releasing (CRH)	anterior pituitary	stimulates secretion of ACTH
	gonadotropin-releasing hormone (GnRH)	anterior pituitary	stimulates secretion of FSH and LH
	prolactin-inhibiting hormone (PIH)	anterior pituitary	inhibits prolactin secretion
	somatostatin	anterior pituitary	inhibits secretion of growth hormone
	thyrotrophin-releasing hormone (TRH)	anterior pituitary	stimulates secretion of TSH
Hypothalamus (via posterior lobe of pituitary)	antidiuretic hormone (ADH or vasopressin)	kidney	controls water excretion
	oxytocin	breasts, uterus	stimulates release of milk; contraction of smooth muscle in childbirth

Figure E5.5 Origins and Effects of Hormones

Figure E5.5 Origins and Effects of Hormones (Continued)

Figure E5.7a shows, the brain then responds by sending a signal to speed the heart rate, which in turn increases the blood pressure. In addition, the brain also signals small arteries and veins to constrict, or narrow, thus further increasing blood pressure. In this case, the initial condition was low blood pressure and the response was to change conditions to increase the blood pressure. Such changing conditions represents a type of feedback known as **negative feedback**.

Figure E5.7b shows an example of **positive feedback**, which is a type of regulation in which the body responds to changes by adjusting internal conditions toward the initial condition. In the example shown, the initial condition is a small clot that begins to develop in response to a bleeding wound. Positive feedback triggers a regulatory response in which still more clotting fibers accumulate at the site of injury. This has the effect of increasing the size of the clot, which helps to reduce the loss of blood.

By delicately balancing positive and negative feedback mechanisms, the body is able to regulate all of the changing internal conditions that humans typically experience. Indeed, the process of regulating blood pressure in the brain is so finely tuned that even when the body's blood pressure is slightly high, such as when you exercise, the blood pressure in the brain remains normal. This is because instantaneous adjustments are made to maintain normal pressures as well as the necessary amount of blood flow that is so important to the brain.

Gland/Organ	Hormone	Target	Principal Action
Pituitary (anterior lobe)	adrenocorticotropic hormone (ACTH)	adrenal cortex	secretes steroid hormones
	follicle stimulating hormone (FSH)	ovarian follicles, testes	stimulates follicle; estrogen production; spermato-genesis
	growth hormone (GH)	general	stimulates bone and muscle growth, amino acid transport, and breakdown of fatty acids
	luteinizing hormone (LH)	mature ovarian follicle, interstitial cells of testes	stimulates ovulation in females, sperm and testosterone production in males
	prolactin	breasts	stimulates milk production and secretion
	thyroid stimulating hormone (TSH)	thyroid	secretes thyroxine
Parathyroid	parathyroid hormone (parathormone)	intestine, bone	stimulates release of calcium from bone; decreases excretion of calcium by kidney and increases absorption by intestine
Thyroid	calcitonin	intestine, kidney, bone	inhibits release of calcium from bone; decreases excretion of calcium by kidney and increases absorption by intestine
	thyroxine	general	stimulates and maintains metabolic activities
Thymus	thymosin	lymphatic system	possibly stimulates development of lymphatic system
Heart	atrial natriuretic factor (ANF)	blood vessels, kidneys, adrenal glands, regulatory areas of the brain	regulates blood pressure and volume; excretion of water, sodium, and potassium by the kidneys
Adrenal cortex (outer portion)	aldosterone	kidneys	affects water and salt balance
	androgen, estrogen	testes, ovaries	supplements action of sex hormones
	cortisol	general	increases glucose, protein, and fat metabolism; reduces inflammation; combats stress
Adrenal medulla (inner portion)	epinephrine norepinephrine	general (many regions and organs)	increases heart rate and blood pressure, activates fight-or-flight response
Pancreas (endocrine tissues)	glucagon	liver	stimulates breakdown of glycogen to glucose
	insulin	muscle, liver cells	lowers blood sugar level; increases storage of glycogen
Reproductive organs			
Ovaries	estrogen	general	stimulates development of secondary sex characteristics; bone growth; sex drive
	progesterone	uterus (lining)	maintains uterus during pregnancy
Testes	testosterone	general	stimulates development of secondary sex characteristics; bone growth; sex drive

Figure E5.6 Feedback and regulation work to reverse the effects of dehydration and restore balanced internal conditions.

1. **Change in internal condition.** During exercise, water is lost through heavy breathing. This can lead to dehydration.

2. **Feedback from sensors.** Sensors in the hypothalamus and near the heart detect a loss of water. The concentration of certain components in the blood is an important signal.

3. **Regulation.** Hormones such as vasopressin are released into the circulatory system in response to the feedback signals. Vasopressin acts on the kidneys, causing them to retain water. You also feel thirsty.

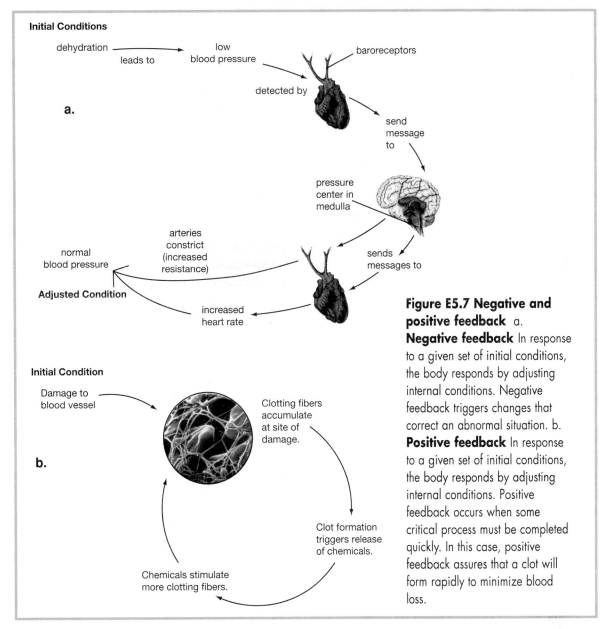

Initial Conditions

dehydration ——— leads to ——→ low blood pressure ← baroreceptors

detected by

a.

send message to

pressure center in medulla

sends messages to

arteries constrict (increased resistance)

normal blood pressure

Adjusted Condition

increased heart rate

Initial Condition

Damage to blood vessel

Clotting fibers accumulate at site of damage.

b.

Clot formation triggers release of chemicals.

Chemicals stimulate more clotting fibers.

Figure E5.7 Negative and positive feedback a. **Negative feedback** In response to a given set of initial conditions, the body responds by adjusting internal conditions. Negative feedback triggers changes that correct an abnormal situation. b. **Positive feedback** In response to a given set of initial conditions, the body responds by adjusting internal conditions. Positive feedback occurs when some critical process must be completed quickly. In this case, positive feedback assures that a clot will form rapidly to minimize blood loss.

Regulation and Homeostasis

by Knut Schmidt-Nielsen

Source: From *Developing Biological Literacy: A Guide to Developing Secondary and Post-secondary Biology Curricula* by BSCS. © 1993 by Kendall/Hunt.

Living organisms tend to maintain a relatively constant state that is optimal for survival. This constancy is well regulated and is called homeostasis. Deviations from the usual state or composition of the organism are met with corrective action. Deviations that are too great cause serious difficulties and eventually death. Homeostasis is important at all levels, from the single cell through the level of organs to the entire organism. In ecology the term applies to the self-regulating maintenance of population densities.

Living organisms have elaborate mechanisms to resist changes and maintain constant conditions. Regulatory mechanisms respond to deviations or departures from the normal; corrective action returns the system to the normal range. Mechanisms to maintain constancy involve **feedback** control. A familiar example of feedback control is the heating system of a house that is regulated by a thermostat. If the temperature falls below the desired temperature (the setpoint), it serves as a signal that causes corrective action (more heat), thus restoring the system to the desired temperature. When the deviation (in our example a decreased temperature) is offset by a corrective action in the opposite direction (increasing the temperature), it is called **negative feedback**.

A biological example of feedback control is the maintenance of a relatively constant body temperature. If our body temperature increases, we normally increase heat loss by sweating; if the body temperature drops, we respond by shivering to increase heat production. The responses thus tend to maintain the body temperature at the normal level.

An example of homeostasis in physiology is the aquatic animals that maintain salt concentrations in their bodies different from those in the surrounding water. Marine invertebrates maintain the same total osmotic concentration as in the surrounding seawater, but the concentrations of various salts in their blood invariably differ from those in the surrounding water. Similarly, the concentrations of various salts inside the cells differ from those in the blood and extracellular body fluids. These differences are characteristic of each animal species, and their lives depend on the maintenance of these differences. Marine fish, in addition to maintaining internal concentrations different from those in seawater, maintain their overall osmotic concentration at about one third of that in seawater.

Freshwater has very low concentrations of salts, and both invertebrates and vertebrates that live in fresh water (fish, some amphibians, and reptiles) maintain internal concentrations that far exceed those in the very dilute water. The maintenance of differences between the organism and the surrounding water depends on elaborate control mechanisms, a subject that generally is referred to as osmoregulation.

Many of the processes that serve to maintain constant conditions in higher vertebrates, and especially in mammals, are reasonably well known and understood. Such functions include the maintenance of constant body temperature, constant heart and breathing rates, constant blood pressure, and constant level of blood sugar. For example, the mechanisms involved in the maintenance of a constant water content of the body, in spite of changes in water intake, are well known. If water intake is restricted, the kidneys respond by reducing the volume of urine, eliminating waste products in high concentrations with a minimal use of water. If water intake is large, the kidneys respond by producing large volumes of dilute urine. These changes in renal function are under hormonal control; if there is water shortage, the pituitary gland produces antidiuretic hormone ADH (also known as vasopressin); if there is a surplus of water to be eliminated, the production of the hormone ceases and the urine volume increases. In mammals it is the retention and conservation of water that is promoted by the antidiuretic hormone ADH. In other animals it may be water elimination that is augmented by a hormone. Consider a mosquito, which in a single blood meal may ingest twice its body weight, a load that greatly impairs its flying and maneuvering ability unless it is rapidly eliminated. A blood-sucking mosquito, before it has even completed its meal, begins to urinate at a high rate. The sudden increase in urine production is regulated by a diuretic hormone that stimulates secretion of fluid by the Malpighian tubules.

We are familiar with the responses to acute changes in physiological demands such as exercise. The increased oxygen demand during exercise is met by increased rate and depth of breathing, increased blood flow to the working muscles, and increased heart rate (plus a moderate increase in stroke volume). These changes are carefully regulated to meet the demands and, when exercise ceases, the rates return to the resting levels.

Also familiar to us are the slow changes that occur in response to increased physical demands on the organism during athletic training. The changes involve well-controlled reorganization and remodeling of physiological systems in response to the increased demands. Muscles subjected to increased use respond with growth and with less obvious changes, such as increases in number of mitochondria, in Krebs cycle enzymes, and in cytochromes—responses that augment the performance of the muscles. Training also augments the heart's capacity for pumping blood and the lungs' capacity for oxygen uptake.

If the bones are subjected to long-term changes in forces exerted on them, they undergo well understood changes. Bones are far from dead and inactive mineral tissue; living cells in the bones (osteoclasts and osteoblasts) respond to changes in the forces exerted on the bones by resorption and deposition of mineral substances (crystals of inorganic calcium phosphates). A major question that today we are unable to answer fully is how the forces on the bones are communicated to the cells responsible for the remodeling of the bones.

The immediately preceding sentence points to an unsolved problem that is of the greatest importance. Although the mechanisms involved in the maintenance of steady state responses to short-term changes such as exercise are reasonably well understood, the mechanisms involved in the regulation of long-term changes are not. For example, consider the ability of the liver to regenerate. A mammal can tolerate surgical removal of a large portion of its liver. The remaining liver will begin to regenerate the lost tissue. The growth of new tissue continues until the liver reaches its initial size, and then it ceases. Little is known about the signals that initiate the regeneration except that certain growth-promoting substances are involved, and nothing is known about the signals that make the growth cease when the regenerating liver reaches the "correct" size.

Many related problems pertain to other areas of the growth and development of an organism. We know, for example, that the normal growth of a human infant depends on the release of pituitary growth hormone. The sudden spurt in growth of the human male in early puberty is initiated by the influence of gonadal hormones and their effect on the production of growth hormone, but what in turn regulates this increase in production of male hormone and makes it cease when the body has reached the "correct" size is poorly understood.

Consider a fertilized mammalian egg (refer to Figure E5.8) that has undergone two divisions and hence has reached the four-cell stage. One of the four cells can be removed and then transplanted into the uterus of a suitably prepared female of the same species. This cell, only one fourth of the size of the original egg, is now capable of growing into a normal embryo that eventually develops into a normal, full-sized fetus that is delivered at term. We do not understand how a single cell that represents only a fraction of the mass of the fertilized egg grows into a normal fetus, what signals control its growth to the right size, and what makes growth cease at the exact correct size. How does the organism "know" what its right size is?

Today, we are seeking answers to these and similar questions pertaining to the long-term control of growth and development, as well as long-term remodeling in response to changing demands on the organism. Our current knowledge is inadequate, and the many unsolved problems in these areas are foremost in the minds of many contemporary biologists.

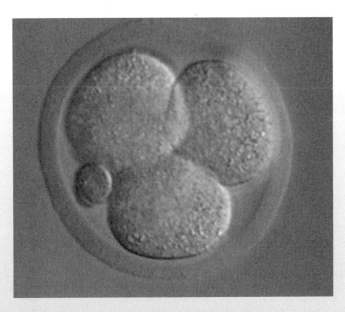

Figure E5.8 Mouse embryo at four-cell stage of development Each of the large mouse cells (only three are visible) will develop into a normal mouse if separated from each other. How does the mouse embryo "know" the right size of a mouse?

The Breath of Life

Take a nice, deep breath, and let it out slowly. What do you think happens in your body with each breath that you take?

Each time you breathe in, you draw air into your lungs. This action is important for your survival because air contains oxygen—a substance that every cell of your body needs to maintain normal conditions. Each time you breathe out, you expel air out of your lungs. This action is also important because it helps rid your body of carbon dioxide—a substance that is produced in cells as a byproduct of energy metabolism.

Transporting oxygen from the outside environment to the internal cells of your body, and carrying carbon dioxide from these same cells back to the outside environment, requires a finely regulated interaction of a number of organ systems. The organ system most directly involved in regulating your body's interaction with the atmosphere is the **gas exchange** system. The central organs of the gas exchange system—the lungs—form two compartments that connect to the outside environment through your trachea (windpipe) and your nose. The air inside these lung compartments is not actually inside the internal environment of your body. Instead, the membranous tissues of the lungs themselves separate this air from the rest of the cells of your body.

How then, does oxygen move from your lungs into the internal environment of your body? And how does carbon dioxide move from this internal environment back into your lungs (actually, back into the external environment)? The answers involve a combination of simple chemical processes and complex homeostatic regulation.

As you draw another deep breath, think about the path that the air must travel. The air passes through the nose, where it is warmed, moistened, and cleaned, or through the mouth. Then it enters the trachea, passes the vocal cords, and enters a branching system of bronchial tubes in each lung compartment. The surfaces of these breathing tubes are lined with mucus and cilia, which are tiny hairlike structures that move in a wavelike manner and sweep debris out of the passages. When the air finally reaches the ends of the passages in the lungs, it enters smaller compartments that are made up of many tiny air sacs called alveoli (singular, alveolus). This pathway of air entering the lungs is shown in Figure E5.9.

Once the oxygen in the air has reached the alveoli, it is in the smallest lung compartment, but it still has not passed into the body's internal environment. To reach this environment, the oxygen must diffuse across the alveoli's thin membranous walls (the alveolar membranes). The large number of alveoli tremendously increases the surface area of lung tissue that oxygen molecules have to cross. In fact, the surface area of these air sacs is 40 times larger than the entire outer surface of the human body. This very high surface area increases the amount of oxygen that can move into the body's internal environment and, as you will see, also increases the amount of carbon dioxide that can leave the body and enter the lungs to be exhaled.

The movement of oxygen across the alveolar membranes to enter the body involves the interaction of the gas exchange system with another organ system, the circulatory system. As shown in Figure E5.10, a dense network of capillaries filled with blood surrounds each cluster of alveoli. This blood comes into such close contact with the thin membranes of the alveoli that simple diffusion makes the final entry of oxygen into the body possible. The diffusion of oxygen depends on its relative concentrations in the air sacs and in the blood inside the capillaries that surround them. If the concentration of oxygen is lower in the blood than in the air sacs, the oxygen diffuses from the air sacs into the blood. In the blood the oxygen binds to the protein hemoglobin, which is present in red blood cells, and then is carried to all parts of the body. In this way two systems work together to deliver oxygen to cells deep inside the body that have no direct contact with the outside environment.

At the same time that oxygen is diffusing into the blood, the waste gas carbon dioxide is diffusing out of the blood and into the alveoli. Recall that carbon dioxide is a by-product of certain energy-yielding processes that take place in cells. Carbon dioxide is transported by the blood from cells all over the body. When it arrives in the capillaries surrounding the alveoli, it diffuses across the alveolar membranes into the air inside the lungs. The relative amount of carbon dioxide in the blood and in the air inside the alveoli determines the direction

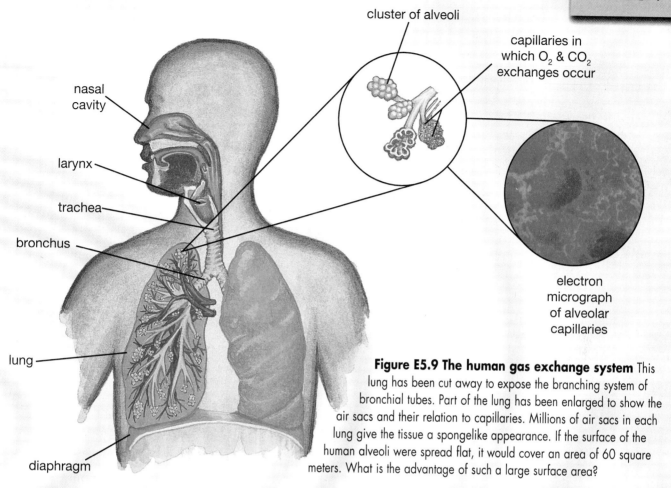

Figure E5.9 The human gas exchange system This lung has been cut away to expose the branching system of bronchial tubes. Part of the lung has been enlarged to show the air sacs and their relation to capillaries. Millions of air sacs in each lung give the tissue a spongelike appearance. If the surface of the human alveoli were spread flat, it would cover an area of 60 square meters. What is the advantage of such a large surface area?

of diffusion. Because its concentration usually is higher in the blood than inside the alveoli, carbon dioxide usually diffuses out of the blood and into the air inside the lungs. As you might expect, the enormous surface area that is available in the lungs for diffusion speeds the release of carbon dioxide from the blood into the lungs. When you exhale, you release this carbon dioxide from your lung compartments into the external environments.

Like many other homeostatic processes, breathing involves a precisely regulated set of events that requires continuous monitoring by feedback systems in the gas exchange, circulatory,

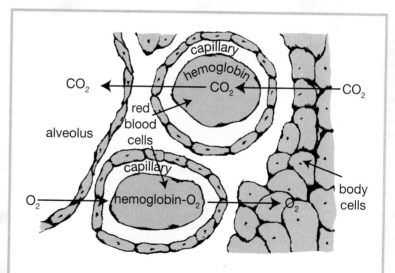

Figure E5.10 The gas exchange and circulatory systems work together. Oxygen from the lungs is transported by red blood cells to all body cells. Carbon dioxide produced in body cells moves in the opposite direction, from body cells to the lungs.

and nervous systems. Consider, for example, what happens to breathing rate during a period of rapid exercise. As energy-yielding processes in the body speed up, the production of carbon dioxide also increases, causing the blood to become more acidic. Sensory receptors in the aorta, in the brain, and in arteries that lead to the head detect this increased acidity and send a signal to the respiratory centers in the brain. These centers respond by stimulating the diaphragm and rib muscles to contract more rapidly, increasing the breathing rate. This faster breathing rate increases the rate at which oxygen is brought into the body and the rate at which carbon dioxide is released. As the exercise period ends and the rate of carbon dioxide production in the body declines, the blood becomes less acidic and this change is detected by the sensory receptors. This information also is relayed to the respiratory centers that in turn signal the diaphragm and rib muscles to contract more slowly.

This regulatory system works automatically, which explains why you do not have to consciously control your breathing rate for flow. The signals involved are very powerful: although you can exert some conscious control over your breathing rate, you cannot hold your breath indefinitely. Once the carbon dioxide level in your blood reaches a critical level, the homeostatic signals override your efforts to keep your muscles rigid, and you are forced to take a breath.

Take one last deep breath. Can you describe what is happening in your lungs as you inhale and exhale? Can you remember how the rate of your breathing is normally controlled? Now consider this: because of the gas exchange system and many other equally complex homeostatic systems, many important adjustments that you never have to think about take place in your body.

What is the evidence that this is going on? Think of all the little breaths you took between those nice deep breaths.

Behavior and Homeostasis

What made Ralph, the character in *A Pause That Refreshes*? (Chapter 4), head to the refrigerator for a cool drink? Why does a lizard move toward a heated rock when its external environment cools off? What makes you reach for a sweatshirt when you enter an air-conditioned movie theater? These questions all are focused on observable behaviors that appear to help maintain homeostasis. But what are the signals that trigger an organism to respond to each set of changing conditions in a specific way?

All these examples of behavior, or observable responses, have a physiological basis. In other words, homeostasis is maintained by internal processes. Sometimes these internal processes manifest themselves in visible behaviors. While it is tempting to assume that the behavioral response is the process that maintains homeostasis, you should remember to ask this question: What is happening on the inside? Your body's internal conditions, and those of other organisms, are controlled by a variety of interconnected monitoring and feedback systems. All organisms continuously receive stimuli that trigger their monitoring and feedback processes. These stimuli arrive in many forms: light, temperature, sound, water, chemicals. Living systems vary greatly in the type of response they have for each type of stimuli. The feedback processes involve physiological responses that may include observable behaviors.

Internal conditions such as the level of carbon dioxide, body temperature, and salt concentration are examples of conditions that are controlled by physiological processes. Some organisms, however, have interesting behavioral responses that augment (enhance) the internal homeostatic responses. You have learned that carbon dioxide plays an important role in regulating breathing rate. In general, your breathing rate is determined physiologically by the acid-base balance of the blood. Panting is a typical behavioral response to increased exercise that restores acceptable carbon dioxide levels. Under unusual conditions, such as fever, aspirin poisoning, or anxiety, the body responds with the atypical behavior of hyperventilation. In this potentially dangerous situation, the body "overbreathes" or increases the breathing rate above the body's need to blow off carbon dioxide. Consequently, carbon dioxide is lost more rapidly than it is produced in the tissues, and your brain does not get the feedback to signal breathing. Eventually, you may pass out from lack of oxygen. Humans can reduce the danger of this response with the conscious behavior of placing a paper bag over the victim's mouth and nose. This trick increases the level of carbon dioxide in the air being breathed in, so the body attains the minimum carbon dioxide level consistent with a normal breathing rate.

Scientists categorize the mechanism animals use to regulate body temperature regulation into two major groups: endothermic (those that generate heat internally) and ectothermic (those that collect heat from outside the body). Mammals and birds are endothermic. Animals such as fish, reptiles, and insects are ectothermic. Regardless of which type of animal an organism is, temperature regulation is a critical survival tool. Many fundamental cell processes depend on enzymes that function best in very narrow temperature ranges. This is why doctors are concerned when their patients run high fevers. A modest increase in temperature can help kill pathogens, but a large increase or sustained increase will destroy vital cell functions and can put the patient at risk.

Mammals and birds maintain relatively constant temperatures by balancing heat production with heat loss. For example, as you, or other endotherms, digest food, you generate heat, you lose heat through the evaporation of water (by sweating, panting, breathing) and the transfer of heat. You can increase the production of heat by eating more, exercising, or shivering, and you can decrease the heat loss by adding insulation. You can put on a sweater; a bird fluffs up its feathers. These behaviors are all responses to changes in the external conditions. In each case, the organism used feedback to regulate a response that started physiologically and was coordinated with an observable behavior.

Evolutionary processes have resulted in a variety of interesting behavioral adaptations to the challenge of maintaining a constant temperature. Some of the most notable are the adaptations that mammals show in extreme climates. For example, small desert mammals may dwell underground or be active at night to minimize the impact of the hot, dry days. Similarly, small mammals in very cold environments will live in tunnels under the snow. The temperature in these tunnels does not drop below -5°C (-23°F), even when outside air temperatures fall below -50°C (-58°F).

Ectotherms do not have internal processes that help regulate their internal temperature. Instead, they have internal receptors that trigger specific behavioral responses when their internal temperatures rise or fall out of

Figure E5.11
Violent shivering can increase the body's heat production by as much as 18 times normal.

Figure E5.12
Collared lemming Collared lemmings live along the arctic edges of northern Europe, Siberia, and North America. They maintain their internal temperatures in a very cold climate by living in tunnels. What behaviors do you use to stay warm?

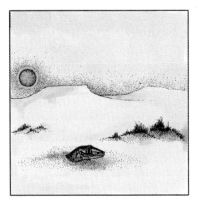

Figure E5.13 The lizard is able to maintain a fairly constant body temperature by changing its body position relative to the position of the sun.

a safe range. This type of behavior in desert lizards has been studied extensively. Biologists have found that lizards' responses are so finely tuned that they are able to maintain a body temperature between 36°C (97°F) and 39°C (102°F) by moving in and out of the sunshine and by adjusting their orientation to the sun.

Reptiles, birds, fish, and humans show a variety of physiological adaptations for regulating their salt concentration. Marine reptiles such as turtles have special salt glands above their eyes that excrete the excess salt they take in with their food. Birds have a similar adaptation except that the salt solution drains out of their beaks. In humans and other mammals, excess sodium ions are removed by the kidneys and excreted in the urine. Some animals, such as the spider crab that lives in estuaries (a saltwater environment), can sense changes in the ambient salt level but do not have a physiological mechanism for removing the salt. Instead, the spider crab moves to areas of lower or higher salt concentrations as necessary to maintain its internal balance.

Many animals have observable behaviors that are related to maintaining homeostasis. But the ability of humans to think about their behavior, make choices about behaviors, and access technological solutions increases our range of responses in comparison to other organisms. We can cool and heat our external environment; we have developed sport drinks to restore our electrolyte balance after we sweat; and we have access to a wide range of foods, beverages, and drugs that can restore or destroy a homeostatic balance. For people with hypertension, certain medications are life-saving. On the other hand, the tendency to consume processed or packaged foods that

have excessive quantities of sodium can aggravate the same hypertensive condition.

So although Ralph's physiological responses signaled him to restore the water balance in his body, his conscious behavior determined whether that balance would be restored. By choosing an alcoholic beverage, which acts as a diuretic, Ralph made his internal condition worse instead of improving it. As you learned in Chapter 1, humans are distinguished from other animals by a set of characteristics, especially the capacity of our brain. That powerful brain gives us the capacity to override certain physiological responses with our behavior and the capacity to make informed decisions about our behavior that help maintain homeostasis.

Figure E5.14
Western gull (*Larus accidentalis*)
Western gulls have a wingspan of 30–40 cm. Note the drop of saltwater at the tip of this bird's beak. Salt glands help sea birds eliminate excess sodium. How does your body control its salt concentration?

Beyond the Limits

The human body has a remarkable ability to adjust to changes in the external environment and maintain a stable internal environment, but it has its limits. When people are in a harsh external environment such as the desert the body's continuous challenge to maintain internal balance is even greater than usual. Small errors in judgment such as not drinking enough water can have serious consequences in such situations. Even under pleasant conditions, homeostasis may be disrupted if one or more of the body's regulatory systems breaks down. In either case symptoms of illness may appear.

The factors that disrupt homeostasis and stress the body are called **stressors**. Sunlight can be a stressor and so can a physical injury or an infection.

You already are familiar with some of the disruptions that can take place when heat is accompanied by dehydration. If stressors result in a mild or temporary disruption of healthy internal conditions, the problem may pass, and the body may be able to recover normal balance. For example, a deep cut that bleeds profusely can temporarily disrupt the body's fluid balance, but if the bleeding is stopped by applying direct pressure to the wound, the body may be able to regain a normal balance. In slightly more serious circumstances, a transfusion of blood may be necessary to help restore homeostasis. If the disruption of internal conditions is very extreme, the regulatory systems may be permanently damaged, and the body may not be able to re-establish homeostasis. For instance, the extremely rapid blood loss that occurs when a person's arm or leg is severed can outpace the body's ability to adjust to the changing conditions. If medical help is not immediately available, serious illness or even death can result (Figure E6.1).

Even if a disruption of homeostasis is less severe than the previous example, the disruption still can pose a serious threat if it lasts for a long time. Chronic diseases such as diabetes (Figure E6.2) or heart disease are examples of this type

Figure E6.2 Chronic disruption of homeostasis Diabetes mellitus is a chronic disease that can cause tissue damage throughout the body. The disorder is caused by inadequate amounts or functioning of the hormone insulin, which regulates levels of glucose in blood, liver, and muscle cells. If this disease is not well-controlled, diabetes can result in poor circulation and can damage tissues such as a. eye, b. heart, c. kidneys, and d. peripheral tissue.

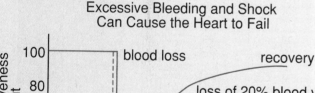

Excessive Bleeding and Shock Can Cause the Heart to Fail

Figure E6.1 Blood Loss The body adjusts for blood loss by clotting the blood and increasing the heart rate, which slowly restores the effectiveness of heart output (see 20% blood loss curve). There is a limit to how much blood can be lost and still have homeostatic mechanisms restore function. In the event of a 40% blood volume loss, the effectiveness of heart output continues to drop, and eventually, the heart stops.

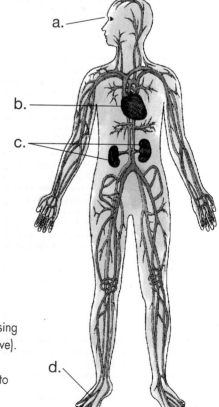

UNIT TWO HOMEOSTASIS: Maintaining Dynamic Equilibrium in Living Systems

of disruption. In these long-term disruptions organ systems or tissues often are damaged slowly over time, making the body's attempt to regain the balanced conditions of homeostasis progressively more difficult.

Other stressors that could overwhelm homeostatic regulation include a lack of nourishment (starvation), a lack of oxygen (suffocation), the presence of toxins in the air being breathed, damage to the gas exchange system (such as that caused by smoking), a large dose of toxic compounds (such as a drug overdose), or a serious infection. An extensive bacterial infection called septicemia, which affects many aspects of the human body, can induce an extremely dangerous form of

shock. In the early stages of the infection, toxins produced by the bacteria enter the blood stream and cause symptoms such as fever, flushed skin, and rapid heartbeat. The patient develops local hemorrhages and tiny blood clots. Eventually, the patient can go into irreversible shock and die. Possible sources of the infection include an untreated bladder infection, a skin infection caused by *Streptococcus* or *Staphylococcus* bacteria, an obstructed part of the intestine, or an unsanitary abortion.

The examples of disruptions of homeostasis just described are somewhat familiar because they are human injuries or illnesses. The danger, however, of homeostatic disruptions affects

all living systems. Every species on earth is adapted for survival in a specific habitat and under specific conditions, and if an organism is put in very different situations, its regulatory mechanisms may be pushed beyond acceptable limits. Consider the very different situations faced by fish living in freshwater and those living in saltwater (see Figure E6.3.) Both types of fish must adjust their internal concentrations of water and solutes such as sodium chloride (salt) so that a proper balance is maintained. Mechanisms have evolved that are effective at accomplishing this task for each type of fish, just as kidneys help regulate these same conditions in humans. The adjustment mechanisms,

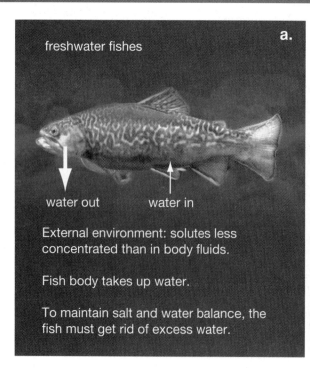

freshwater fishes

a.

water out   water in

External environment: solutes less concentrated than in body fluids.

Fish body takes up water.

To maintain salt and water balance, the fish must get rid of excess water.

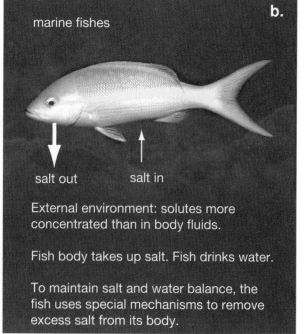

marine fishes

b.

salt out   salt in

External environment: solutes more concentrated than in body fluids.

Fish body takes up salt. Fish drinks water.

To maintain salt and water balance, the fish uses special mechanisms to remove excess salt from its body.

Figure E6.3 a. Freshwater fish must continually remove large amounts of water so that the solutes in their internal environment are not diluted. b. Marine fish must keep removing solutes such as sodium chloride via their gills. What do you think would happen if each of these organisms were placed in the other's environmental setting?

however, are sufficient only for the usual environmental variations in water and salt balance. This is why we define homeostasis as maintaining conditions *within certain limits*. If either type of fish suddenly is put in the other's external conditions, the fish would be *beyond its limits* to such an extent that it most likely would die. The effect of extreme stressors on any organism can be the loss of ability to restore normal internal conditions, a state of homeostatic imbalance that is often fatal.

Coping with Disruptions: The Role of Medicine in Homeostasis

What happens if internal conditions are pushed beyond the limits of our body's ability to compensate? In societies with advanced technology, scientists and physicians have developed medical technologies that provide temporary help and, in some cases, long-term help when normal homeostatic mechanisms fail. Medical techniques often are aimed at adjusting internal conditions or temporarily taking over the role of one of the body's regulatory systems. In some cases adjusting internal conditions requires surgical intervention, an invasive process in which a physician cuts into the body, often to repair or remove the tissues or organs that are impairing the body's ability to heal itself. For example, surgeons can repair some heart defects by opening a patient's chest and replacing valves or rerouting arteries that are important to maintaining proper circulation. In addition, surgical procedures are used to insert technological devices permanently in the body. A pacemaker is an electronic device that can improve circulation by providing a regular heartbeat, which in turn provides the stimulations necessary for heart contraction and pumping.

Non-invasive medical techniques also use technology to help restore the body's internal balance, but they do so without the need for surgery. For instance, if a person's lungs have collapsed or if the muscles that control the lungs are paralyzed, physicians often use a mechanical ventilator to make certain that oxygen is delivered to the blood stream and that waste products such as carbon dioxide are removed from the patient's body. Dialysis machines are capable of filtering blood and assisting with fluid regulation in patients whose kidneys have failed. Less permanent procedures such as electrical muscle stimulation, which helps retrain muscles to contract properly after long periods of inactivity, often are used in physical therapy to help patients regain the ability to function independently.

Health care professionals also use medical technology to gather information about the nature of various injuries or illnesses. The information contained in vital signs is the first step in **diagnosis**, which is the process of determining a patient's condition. Medical workers can make accurate measurements and collect extensive information by using technological tools that extend the senses, tools such as stethoscopes, blood pressure cuffs, and thermometers. For example, the analysis of

X-ray images provides medical experts with a way to assess internal structures such as the bones of the foot shown in Figure E6.4. An electrocardiograph electronically maps the rate and rhythm of the beating heart and prints out a record such as the ones shown in Figure E6.5. This record, an

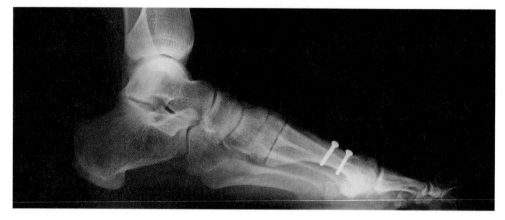

Figure E6.4 Imaging technology reveals internal structures Diagnosis of disruptions such as broken bones or tumors is made more efficient through analysis of X-ray and other images that allow physicians to view internal structures. Physicians used this X-ray of a foot to determine how well the patient's bones were healing after surgery.

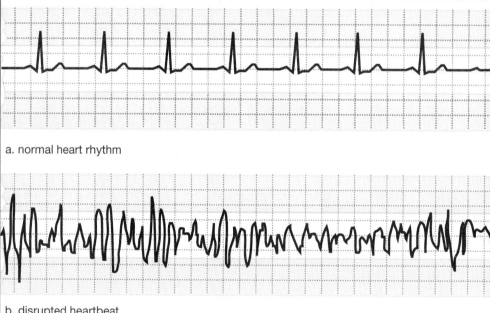

a. normal heart rhythm

b. disrupted heartbeat

Figure E6.5 Electrocardiograms help physicians assess heart function. Electrocardiograms reveal a. normal and b. disrupted heart function. This particular disruption in heart function is called ventricular fibrillation, the contracting of the muscles of a ventricle rapidly and continuously in an uncoordinated manner. How would a disrupted circulatory system affect gas exchange?

electrocardiogram, helps a cardiologist (a heart specialist) interpret the effectiveness of a patient's heart and circulatory system.

Technological advances in the analysis of blood, another important indicator of the internal conditions in the body, also have resulted in improvements in diagnosis. The biochemical and cellular composition of blood often reveals clues about the health of internal organs. In one type of blood test, technicians count blood cells of many different types and use the results to evaluate whether the patient has a blood disorder such as leukemia or simply an infection. The ratios between different blood cell types and the total blood volume also are important indicators of internal health. In a healthy person's blood, red blood cells generally are present in a much higher concentration than white blood cells, at a ratio of about 700:1. An abundance of white blood cells usually indicates an infection. The components of blood plasma (the noncellular, liquid part of blood) also may vary with disruptions in homeostasis so their measurement is useful in diagnosis as well. Plasma components include cholesterol, lipids, protein, glucose, and electrolytes (that is, solutes such as chloride, potassium, calcium, and sodium).

In cases where an internal imbalance is detected by diagnostic tests such as blood tests, physicians often turn to drug technologies for assistance in restoring a patient's homeostasis. Drug therapies are used to treat high blood pressure, heart problems, diabetes, psychological disorders, cancer, and many other potentially serious ailments. For instance, the widespread use of antibiotics to treat bacterial infections has resulted in a dramatic decline in the number of deaths due to infections.

Antibiotics and many other drug technologies are recent developments, and they provide an example of how technology has changed our cultural view of medicine during the past 70 years. Before the discovery of the most common antibiotic, penicillin, in the mid-twentieth century, people in the United States often contracted life-threatening infections in hospitals and thus tended to think of hospitals as treatment facilities of last resort. They would risk a hospital stay only if all in-home medical treatments had failed to heal them. Today people generally view hospitals as places to go when they need to be cured. Unfortunately the widespread use of antibiotics provides a selective pressure that has resulted in the evolution of bacterial strains that are resistant to certain antibiotics, and bacterial infections once again are becoming more difficult to treat. Figure E6.6 depicts an experiment that tested the resistance of two types of bacteria to an antibiotic.

The tremendous increase in the use of medical technology has resulted in an increase in the quality of life for many people, and researchers are continuing to make further advances. A number of ethical questions have been raised by some of these advances, however, and health care professionals and the public must struggle to resolve them. For instance, our increased use of life-support technology, not only for temporary support during major surgery but also for the long-term maintenance of the terminally ill, forces us to think about how we define quality of life. New technology also raises financial concerns because the costs for many advanced medical treatments are extremely high. As our medical knowledge increases and as we develop more tools to assist medical professionals, society must learn how to balance the costs and benefits of the treatments.

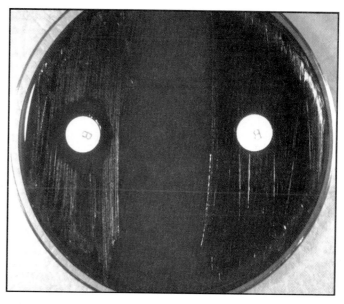

a.   b.

Figure E6.6 Bacterial resistance to antibiotics On this petri dish, two strains of bacteria were spread evenly across a nutrient material. Disks soaked in the antibiotic bacitracin were then added to the plates, and the two bacterial strains were observed after further growth. The clear area in (a). indicates a lack of growth near the antibiotic disk. In (b). there is no clear area around the disk, indicating that the bacteria grew there. What evidence do you see that indicates that bacteria can be resistant to the effects of antibiotics?

Avoiding Disruptions: The Immune System

Regulatory mechanisms that help correct homeostatic disruptions in the circulatory, digestive, gas exchange, and other systems are critical to the health of living organisms. Just as important are the mechanisms that *prevent* these disruptions from occurring. In humans the system that provides powerful protection against disruptions caused by infections and foreign toxins is called the **immune system**. Like a fortress, your body's immune system offers many defenses to keep out or kill invaders.

The body's largest organ, the skin, is considered a *nonspecific* barrier or defense; it does not have to recognize an invader to keep it out. As the body's first line of defense, the skin helps to guard the body's internal environment from many types of outside attack. The defensive role of the skin does not imply, however, that the skin is a sterile environment on which nothing can live. In fact millions of harmless bacteria live on the skin (refer to Figure E6.7). In most cases these harmless bacteria are beneficial because they compete effectively against other, potentially harmful, microorganisms, which collectively are known as **pathogens**. Pathogens are disease-causing microorganisms. Through time humans and harmless bacteria have benefited from their interactions. Humans have benefited because these bacteria

help to protect us from pathogens. Harmless bacteria have gained a resource-rich environment, the human skin.

Even if pathogens occasionally outcompete the skin's harmless bacteria, they still must pass through the barrier of the skin if they are to thrive at the human host's expense. Like the walls of a fortress, the skin helps to guard your internal environment from outside attack. This wall, however, is not perfectly secure. Natural doors to the internal environment, such as the mouth, ears, eyes, nose, genital, urinary, and anal openings, provide opportunities for pathogenic invasion. Additional nonspecific defenses have evolved to help protect these natural openings; these include tears, saliva, mucus, and sweat. Some of these bodily secretions contain an antibacterial

enzyme called lysozyme. Cuts in the skin also provide avenues for infection because these openings give invading pathogens direct access to the tissues of the body's internal environment.

Protection against pathogens is not unique to humans; all living systems have some resistance to invasion by foreign material. Bacteria, for example, produce special enzymes that destroy foreign DNA, such as the DNA of invading viruses, but preserve the DNA of the bacterial cell. Plants produce chemicals that kill areas of plant tissue infected by a fungus so that the fungus does not spread. The tissue breakdown means that part of the plant is sacrificed, but this loss is less dangerous to the plant than is an infection that could spread throughout the organism.

If pathogens successfully pass the barrier of the skin and reach the inside of an organism, other components of the immune system take over to provide protection. In humans and other mammals, the invading

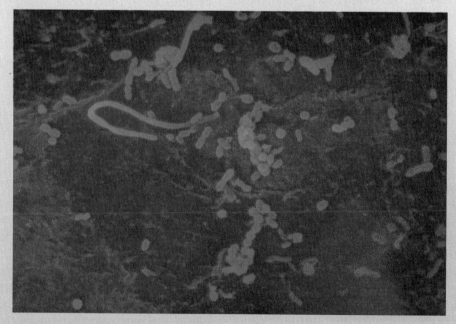

Figure E6.7 Scanning electron micrograph of bacteria on skin, magnified X24,000

pathogens still must face an army of cellular and molecular defenses. For example, within many body compartments such as the circulatory system and lungs, there are nonspecific defense cells that recognize and scavenge *many* types of invading organisms and toxins. Among these scavenger cells are **macrophages**, one of which is illustrated in Figure E6.8. In the lungs macrophages help protect the body against pathogens that humans breathe in with the air they inhale.

Macrophages also act as generals in the immunity army; they activate, or signal, helper T-cells, which *help* recruit more immune cells to defend the body. Helper T-cells coordinate the immune system's *specific* barriers by directing groups of specialized cells to attack only *certain* invaders. (This response is very different from the nonspecific defenses provided by macrophages and the skin, which act to repel *all* invaders.) For instance, helper T-cells can activate another class of immune cells, the killer T-cells, and direct them to kill specific infected cells. This line of specific defense is called the **cell-mediated response,** and Figure E6.9 summarizes some of the complex interactions involved in its regulation.

Cell-mediated responses are an important part of the defense against viral infections because viruses hide *inside* cells. Once viruses enter the cells of their

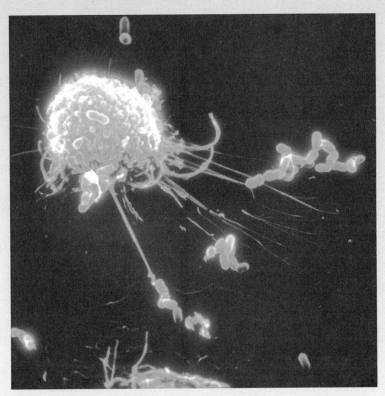

Figure E6.8 Macrophage engulfing bacteria A macrophage is a nonspecific scavenger cell that helps protect the body against pathogens. In this photo, cellular extensions from a macrophage have captured several bacteria. Macrophages also play a role in the specific immune response.

host, the infected organism, they no longer can be scavenged by macrophages. The viruses then use the cell's own biochemical machinery to make macromolecules and to reproduce. With this effective adaptation the virus uses the host cell as a shield. Antibiotics, which kill bacteria reproducing *outside* cells, generally are not effective against viral infections because the viruses are hidden inside cells. Most chemicals that would kill a virus-infected cell also would kill healthy cells. Killer T-cells, on the other hand, can distinguish virus-infected cells from uninfected cells by the unique molecular signals on the infected cells' surfaces. Based on these molecular signals, the killer cells can destroy the infected cells and leave the uninfected cells unharmed.

Macrophages also help activate another set of specialized immune cells, called B-cells. Once these cells are activated, some of them can produce molecular defenses known as circulating **antibodies**, which enter the blood stream and move throughout the body. Antibodies are protein molecules that recognize chemical signals from specific pathogens; these chemical signals are known as **antigens**. When circulating antibodies come in contact with the specific pathogens for which they are targeted, the antibodies stick to, or bind, the antigens of the invaders; this leads to the destruction of the invaders. This line of specific defense is called the **antibody-mediated response**, and it works together with the other components of the immune system to help protect the body against damage from invading pathogens or the toxins they produce (Figure E6.9 illustrates this response).

The rapid and effective response of the immune system sometimes works against the host. Antigens in insect venom, pollen, animal dander, or food protein are not necessarily a serious threat to human homeostasis, but the immune system still recognizes them as foreign invaders. Thus exposure to these antigens can produce

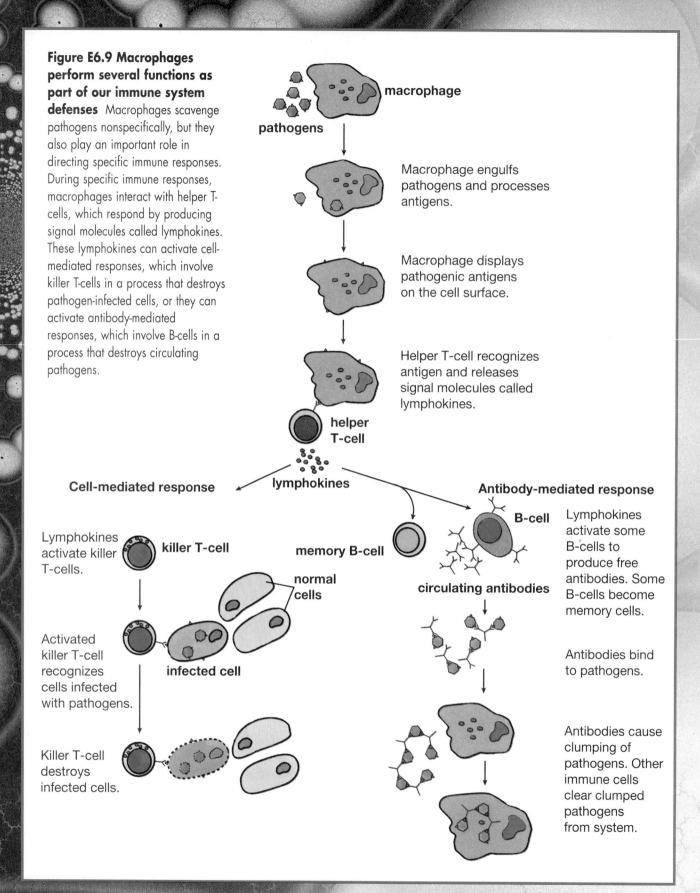

Figure E6.9 Macrophages perform several functions as part of our immune system defenses Macrophages scavenge pathogens nonspecifically, but they also play an important role in directing specific immune responses. During specific immune responses, macrophages interact with helper T-cells, which respond by producing signal molecules called lymphokines. These lymphokines can activate cell-mediated responses, which involve killer T-cells in a process that destroys pathogen-infected cells, or they can activate antibody-mediated responses, which involve B-cells in a process that destroys circulating pathogens.

pathogens

macrophage

Macrophage engulfs pathogens and processes antigens.

Macrophage displays pathogenic antigens on the cell surface.

Helper T-cell recognizes antigen and releases signal molecules called lymphokines.

helper T-cell

lymphokines

Cell-mediated response

Antibody-mediated response

memory B-cell

B-cell

Lymphokines activate killer T-cells.

killer T-cell

Lymphokines activate some B-cells to produce free antibodies. Some B-cells become memory cells.

circulating antibodies

normal cells

Activated killer T-cell recognizes cells infected with pathogens.

infected cell

Antibodies bind to pathogens.

Killer T-cell destroys infected cells.

Antibodies cause clumping of pathogens. Other immune cells clear clumped pathogens from system.

unpleasant hypersensitivities such as bee-sting reactions, hay fever, and asthma. Such responses occur because the antigens trigger the body's production of antibodies that selectively bind to a certain type of cell found in blood and tissues. These cells have granules that contain many biologically active substances, including chemicals called histamines. When the antibodies trigger the granule cells to release histamines into the surrounding tissues, the result is usually sneezing, itchiness, and teary eyes. In more serious reactions the histamine can cause smooth muscles to contract and blood vessels to swell. That reaction, in turn, can cause the airways in the lungs to constrict, which can result in severe breathing difficulty. The antihistamines found in allergy medicines often are effective in reducing these symptoms because they counteract the histamines released by the granule-containing cells.

The defensive forces of the immune system not only react to and destroy foreign material and cells infected by foreign material, they also identify and destroy human cells that have changed into the abnormal cells characteristic of cancer. Cancerous cells arise occasionally when individual cells become faulty and no longer respond to the

homeostatic signals that regulate their behavior. One faulty response is for these abnormal cells to begin dividing and producing new tissue. If enough abnormal cells accumulate, a cancerous tumor forms. If the cancer progresses, it may invade other areas of the body and eventually cause death. To prevent this, an active, healthy immune system is an essential protection. The relative rarity of cancer, particularly among young people, is a sign that the immune system usually attacks and destroys abnormal cells before a cancer develops.

With so many lines of defense, including an immune system that is constantly destroying foreign material, why does anyone ever get sick? The answer is that the immune system, like other systems that regulate homeostastis, has limits. There are limits to how many pathogens or abnormal cells the immune system can control. If infections become too widespread or destroy too much tissue, or if an invader can escape detection, the body may not be able to protect itself.

People also become sick when their immune systems weaken. Many stressors, such as

inadequate sleep, smoking, drug use, and anxiety, can impair the immune system's ability to protect the body. With depleted natural defenses pathogens can take over, leading to many infectious diseases. Acquired immunodeficiency syndrome, or AIDS, offers an extreme example of just how important the immune system is to good health. In AIDS the human immunodeficiency virus, or HIV, directly attacks cells of the immune system, particularly the helper T-cells, disabling most of the immune system's specific responses. As a result the immune defenses of the infected person are weakened tremendously, and the victim often is unable to fight off even minor infections. In addition, the victim is left vulnerable to many serious diseases, including pneumonia and cancer. Ironically, the tragic consequences of AIDS actually emphasize the impressive ability of the body to defend itself in most situations.

Self and Nonself

The immune system uses a complex network of defensive forces to eliminate invaders that threaten the human body, but such protection is beneficial only when the defenders can distinguish between the host and the invaders. In other words, an effective immune system is able to distinguish between cells of the body ("self") and foreign cells such as pathogens ("nonself"). The human immune system is so well-adapted to making these

distinctions that it even can distinguish between cancerous cells, which are a threat, and our normal cells. This important distinction usually keeps the immune system from attacking the body it is supposed to defend. Bacteria are protected in a similar way; they have protective enzymes that degrade foreign DNA such as the DNA found in bacteria-attacking viruses, but not the bacteria's own DNA.

Blood transfusion is another example of the body's ability to distinguish self from nonself. The surface of red blood cells contains molecules that identify the blood group to which the cells belong; this is the basis for distinguishing blood as a certain *type*. These molecules can vary among individuals and, as Figure E6.10 shows, are identified by the immune system as belonging to either self or nonself. If blood of an incompatible (nonself) type is transfused into the body of a patient by mistake, the patient's immune system attacks the blood, making the patient so dangerously ill that death may occur.

Blood Type	Molecule on Red Blood Cell	Antibody in Plasma
O	none	anti A, anti B
A	A	anti B
B	B	anti A
AB	AB	none

Figure E6.10 ABO blood types Blood types are inherited. A patient with type A blood will have antibodies against type B molecules; that is, the antibodies will recognize type B blood as foreign, and the patient's immune system will attack the type B blood. People with blood type O can donate to people with other blood types without danger because there are no blood-type molecules on type O cells to which antibodies can bind. The blood recipient's immune system will not attack type O blood. People with type O blood, however, can only accept blood transfusions from type O blood. Can you explain why?

When the human immune system cannot distinguish between self and nonself, it attacks and damages tissue in the body. In fact, many diseases are disruptions of immune function. These diseases are called **autoimmune diseases**, which means simply that a malfunction of the immune system causes the body to damage itself. Autoimmune diseases include rheumatoid arthritis, in which the immune system causes inflammation and damage to joints, and multiple sclerosis (MS), in which the immune system slowly destroys the nervous system. Intensive research into the causes of the immune system's failure to distinguish between self and nonself is currently underway. Scientists hope that a better understanding of this important function will aid in the development of new therapies and, perhaps, cures for autoimmune diseases.

Immune System Memory ● ● ● ● ● ● ● ● ●

The human immune system is amazingly efficient at protecting against outside attacks. A powerful feature of the immune system is the ability of certain immune cells to retain a **memory of infection**. Due to this memory, your immune system can produce a more powerful attack against pathogens the second time your body is exposed to them. This response is possible because a few of the immune cells that recognized and fought a particular invader in the past remain in your body; they store a memory of the previous infection in the structure of their molecules. In other words, these immune cells are already programmed to respond quickly if the same pathogen tries to invade a second time. With this programmed response, your body may fight off the infection without ever having symptoms of illness. Figure E6.11 compares the immune system's responses after the first and second exposures to the antigens of one pathogen.

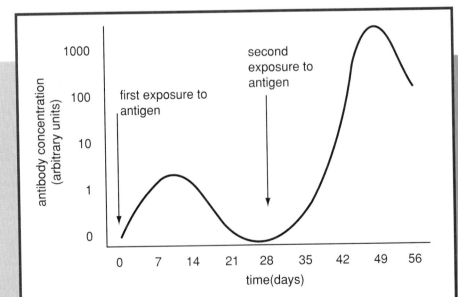

Figure E6.11 Response of the immune system to antigens In the primary response (after the first exposure to a pathogen), the peak concentration of antibodies is reached in 7 to 14 days. In the secondary response (after a second exposure to the same pathogen), a much higher concentration of antibodies is reached in less time.

UNIT TWO HOMEOSTASIS: Maintaining Dynamic Equilibrium in Living Systems

The rapid and potent response to a second exposure, however, does not guarantee that your immune system will fend off a second infection before you become ill. This is particularly true if the pathogen has mutated, or changed, slightly since the first infection. Rapidly evolving pathogens can stay one step ahead of the immune system, an adaptation that explains how slightly different strains of the same cold virus can circulate through a school, continually reinfecting students and staff.

The memory feature of the body's immune system not only explains why you recover more quickly from infections caused by related pathogens; it also explains how individuals become immune to particular illnesses. For example, if you have had measles, mumps, or chicken pox once, you generally do not contract that illness a second time, even if you are exposed to the pathogen that causes it.

Medical researchers have exploited immune memory by developing vaccines. Vaccines trick the body's natural defenses into reacting against a pathogen that is not attacking the body. Thus the vaccinated individual can acquire immunity to a disease without ever having the disease. A common type of vaccine consists of inactivated pathogens, such as viral or bacterial particles, that are injected into the body. The vaccine can activate a specific immune response because it includes antigens for a particular pathogen. The vaccine does not cause a primary infection and illness, however, because the pathogen is either dead or disabled or because only some of its antigens (such as proteins) were introduced into the body. The newest approach to vaccines involves inserting the genetic material (DNA) of a virus into human cells to produce viral antigens; these antigens then trigger an immune response.

As Figure E6.12 shows, vaccines can have a dramatic, positive impact on the spread of disease. There are vaccines for certain types of influenza, tetanus, rabies, measles, mumps, smallpox, chicken pox, hepatitis B, and many other diseases. Some diseases, however, lack effective vaccines because dozens of varieties of similar viruses can cause the same symptoms. For example, a vaccine developed against one variety of the virus that causes the common cold would not provide protection against all of the other varieties of cold viruses. Certain other diseases are difficult to prevent through the use of vaccines because they mutate so rapidly. In the case of human immunodeficiency virus (HIV) a vaccine would protect the body against only one strain of the virus; it would not protect against infection by each new mutant.

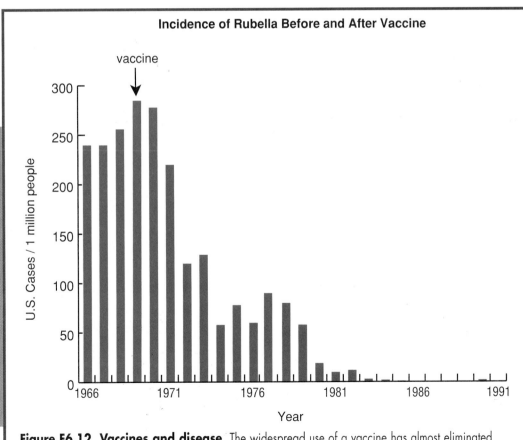

Figure E6.12 Vaccines and disease The widespread use of a vaccine has almost eliminated rubella, or German measles, in the United States.

Avoiding Disruptions: Behavior, Choices, and Risk

What human behaviors reduce the risk of disrupting homeostasis? Simple reflexes such as closing the eyelids in response to a sudden, threatening movement or retracting the fingers quickly from a hot surface, are protective behaviors. Humans, with their capacity for complex thought, also have the option of choosing behaviors that may prevent disruption of bodily functions by reducing risks. For example, you can avoid being around someone who has the flu, you can choose always to buckle your seat belt while in a car, and you can adopt the habits of getting adequate sleep and eating nutritious food. None of these behaviors eliminates the risk of injury or illness, but they represent some factors under your control that may greatly reduce the risk of either mild or major disruptions.

As we consider risk, it is important to distinguish controllable from uncontrollable factors. For instance, do you think you are in danger of developing lung cancer? Some factors such as a genetic tendency to develop certain forms of cancer are not under your control because you may have inherited this genetic tendency. Exposure to toxic substances in polluted outdoor air or in an unsafe work place increases the risk of certain cancers. Would this factor be under your control?

One of the most controllable factors related to lung cancer is cigarette smoking. Cigarette smoke damages the lungs' protective mechanisms and leaves a smoker more vulnerable than a nonsmoker to infection or to damage from other pollutants. Smoking does not guarantee that a smoker will get cancer, but it greatly increases the risk of lung cancer. It also increases the risk of heart disease from damaged blood vessels. A further disadvantage of smoking is that it damages the elasticity of lung tissue with each inhalation of smoke. This damage is progressive and results in the slow, and often painful, fatal disease known as emphysema. Figure E6.13 compares healthy lungs with diseased lungs.

Many smokers find it extremely difficult to quit smoking because the nicotine contained in tobacco is one of the most addictive chemicals known. What is more interesting is why people who have access to accurate data about the effects of smoking choose to

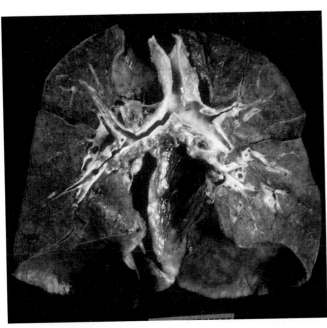
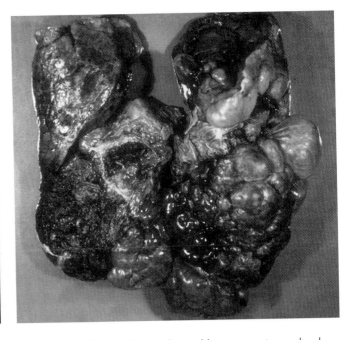

Figure E6.13 Smoking destroys healthy lung tissue Healthy lungs [on the left] are light in color and have a consistent alveolar structure. Lungs damaged by emphysema (on the right) are dark in color and have bulbous growths where alveoli have been destroyed. These lungs lose their elasticity through time, making breathing difficult.

start smoking. With smoking, as with other controllable and risky behaviors such as skiing, drinking alcohol, or driving without a seat belt, an individual must weigh the risks against the benefits to make an informed decision.

A number of behaviors and risks, some controllable and some not, affect the efficiency of the immune system. It is not easy to test what effect each factor has on immune function because the immune system is so complex. However, the factors listed in Figure E6.14 do appear to have an effect on the efficiency of the immune system.

Even if the immune system is strong, it may not successfully combat an infection if the number of pathogens is very large or if the invading organism damages enough tissue to weaken the body and lower the immune response. Once again, there are ways to reduce the risks and give the immune system a good chance to work adequately. If you avoid exposing yourself to sources of infection, including polluted water, contaminated food, and animals or people with contagious diseases, you will have reduced your own risk with regard to infection. Keep in mind that exposure may not mean simply being near the source of infection. Sitting in the same room with contaminated food will not make you ill, but eating it may do so. Being sneezed on could give you a sore throat or a common cold. Simply shaking hands with a person who is infected with HIV will not cause you to become infected, but coming in contact with his or her body fluids by touching a bleeding wound or having sexual intercourse certainly can do so. Overall, the immune system like the body itself needs to function in a balanced manner, as indicated in Figure E6.15.

Enhances Defense	Impairs Defense
• adequate rest	• fatigue and lack of sleep
• moderate exercise	• extreme exercise (marathons, cross-country ski racing)
• good nutrition	• poor nutrition
• positive mental attitude	• anxiety or depression
	• smoking
	• excessive alcohol use
	• excessive antibiotic use
	• certain infections (such as mononucleosis or HIV)

Figure E6.14 Factors influencing immune function

Level of Immune Function	Response of Immune System to Stressors	
	Internal	External
overactive	autoimmune diseases	allergies
normal	immune system removal of abnormal cells	immune system removal of toxins and successful fighting of infection
underactive	cancerous cell growth	susceptibility to infections

Figure E6.15 Relationship between immune function and various stressors
Like other homeostatic systems, the immune system can become improperly regulated, resulting in disruptions caused by over- or underactivity.

Individual Behavior Can Affect Larger Groups

Not all the factors that influence the risks to your health are under your control, but some of them are controllable by others in your society. For example, you may make the choice never to drive a car recklessly or while under the influence of alcohol, yet you still could be injured in an accident caused by someone else who drove recklessly or while intoxicated. In this case the source of danger (drunk or reckless driving) is controllable, but not by you, the victim.

There are even more complicated situations in which your risk as an individual is partly related to the group of friends or the larger society of which you are a member. If your society has laws that restrict the amount of air pollutants that can be released from cars or factories, then society's overall risk, as well as your individual risk, of getting lung cancer will be reduced. Is this a controllable or uncontrollable factor? If you avoid excessive use of a car or vote to support pollution restrictions, then as an individual you will be contributing to the reduction of risk. But your contribution alone will not have a very large effect. It takes many personal decisions to direct public policies that can affect a group or population.

Every organism is constantly at risk of disrupting its homeostasis through injury or illness, yet the planet is populated with an enormous number of organisms. How do they all survive? The answer is that they don't. Still, despite the risks a sufficient number of organisms survive long enough to reproduce, thus continuing the existence of the species as they slowly evolve. In this way the immune system plays a role in protecting organisms from being killed by disease and infections before they can reproduce. Behavioral adaptations also help reduce the risks that individual organisms face and thus increase the species' chances for survival.

Through a combination of immunological and behavioral adaptations, all living organisms have evolved some mechanism that helps them cope with potential disruptions of homeostasis. Certain mammals such as rabbits have two primary behavioral responses to danger; they hold very still, reducing the chances of being spotted by a predator; if the danger is more imminent, they run away quickly. Many flowering plants protect themselves against freezing by losing their leaves in the winter cold. Even bacteria can detect toxic substances and respond by moving away. Humans, therefore, are not the only organisms that have behavioral and physical means to maintain homeostasis and to survive. Humans, however, may be the only organisms with the ability to make conscious decisions about behaviors that can affect not only their own species but every other species on the planet.

Figure E6.16 Uncontrollable risks Even though the driver of this car was driving carefully and wearing a seatbelt, she was severely injured by a reckless driver.

Ethical Analysis

Scientific inquiry can tell you *how* things work and can help answer questions about whether something *could* happen in a given situation. Science, however, cannot tell you what you *should* do in a given situation. Questions that involve issues of should are ethical questions. In an ethical dilemma your choices depend not only on facts about the world, which science may help explain, but also on values—on ideas that are important to you, your family, or your society. Individual, familial, or societal values often are determined by complex interactions between religion, sociology, and philosophy.

The thinking and evidence-collection processes that scientists use to study questions about the natural world also are valuable for studying ethical questions. Just as scientists must gather data to support or disprove their hypotheses, ethicists must work hard to develop strong arguments that support their positions, and evidence is as crucial to the ethical process as it is to science. An opinion that lacks the weight of critical thinking and supporting information likely will not be a persuasive opinion. People who approach ethical choices in a careful way, using analysis and reasoning to make their decisions, are using the process of ethical analysis.

What are some steps in the process of ethical analysis?

First, *identify the question* of interest clearly and precisely. It is difficult to construct strong arguments unless you are very clear about the question you wish to address. For example, you might ask: Is it ethical to have a law that *requires* car occupants to wear seat belts?

Second, *gather information* about the issue in question. This information might include expert opinions about the issue from various perspectives, such as those of philosophers, historians, theologians, economists, and scientists. The information also must be accurate so that you can use it effectively to support your arguments.

Third, *evaluate the information* to understand how it applies to the issue you are facing and to the individuals and groups of individuals involved. It is important to evaluate the information as it pertains to all groups that may be affected. You must consider how the issue affects the *interests* of each individual, of particular groups of people, and of society as a whole. It also is important to consider these interests in the light of both the *consequences* of any actions and any *rights*, or freedoms, that might be denied. For example, consider again the issue of requiring all people to wear seat belts while traveling in a vehicle. Such a law promotes the best interests of the individuals and of society because it reduces the

Figure E6.17 Complicated issues have no simple solutions
The facts of such issues must be analyzed in the light of the interests of everyone affected.

chance of injury in a car accident. Some argue, however, that it infringes on the rights and freedom of individuals.

Fourth, use your reasoning and your data to *form well-reasoned arguments* that support one or more solutions to the ethical issue or conflict. For instance, evaluate the health care costs that society must pay when car occupants who don't use seat belts are injured in accidents. A cost might be an increase in insurance premiums for other car owners. You might reason that the rights of individuals not to wear seat belts are less important than the interest that society has in reducing health care or insurance costs.

Fifth, you and the people to whom you present your arguments and conclusions must critically *analyze your case* to determine its validity. For instance, an economist might challenge the assertion that car occupants without seat belts contribute significantly to overall health costs. In such a case, you could analyze the new cost data (which the economist is using to support the position that individuals' rights should not be denied). Then you could determine whether it is more or less reliable than the data used to support the original argument (that society's interest in reducing health costs is of greater importance).

Sixth, *make a recommendation* about what *should* be done about the issue. Use these well-supported arguments to help decide how you, or groups of people (society), plan to take action to address a particular issue.

Human Performance: A Function of Fitness

For Yates and Sullivan, the characters in the story *The Sky Awaits*, as for most athletes, being fit means performing a vigorous physical routine without debilitating fatigue. Is the level of fitness that is required to be an athlete a good general definition of fitness? To be fit, do you need to have the strength required to put a supersonic jet into a 90° bank at six Gs? Do you need to have the speed and endurance necessary to place in the top 10 percent in a 41.9-km (26.2-mile) marathon to consider yourself fit? On the other hand, if you are not ever going to participate in such activities, does this remove the need for you to be fit?

If world-class athletic performance is the standard, then very few of us could consider ourselves fit. Even Yates and Sullivan, prepared as they are for

Figure E7.1 Human performance Observe the different levels of performance in this variety of complex human activities. Some activities are a part of daily life, whereas others are highly specialized. Each has its own special requirements, although general fitness for life helps in all cases.

their type of work, would not necessarily be well prepared for all types of athletic activity. To be fit for their jobs, Yates and Sullivan must have a special quality—the ability to withstand sudden, strong accelerative forces without blacking out or losing their concentration. That does not necessarily mean that they would be capable of functioning effectively as a sprinter or a weight lifter. Likewise, a runner ready to

compete in a 100-meter race may be able to sustain short-term bursts of speed, which require strong leg muscles and good heart capacity, but may not necessarily be fit for the special tasks of high-performance pilots.

Clearly, an athlete's body must be prepared for the specific demands of a specific activity, and the fitness requirements differ, depending on the nature of the activity. Outstanding performance in one sport does not guarantee outstanding performance in another.

Also, a person who lacks the special skills required for any form of athletics may be very fit for life itself. To function normally in the world, our bodies must be capable of performing certain basic physical activities. We walk; we talk; we gesture; we sometimes run. Even when we are sitting still, our bodies continually maintain a basic level of internal function. Fit-for-life individuals usually are in good physical condition and health, and they find that they are able to perform routine activities of life easily, as illustrated in Figure E7.1.

What, then, is a useful general definition of fitness? Some authorities define fitness as the ability to perform routine physical activity with enough energy in reserve to meet an unexpected challenge. A fit individual not only gets through each day but is able to run up three flights of stairs carrying a heavy book bag and still answer the question, "Why are you late?" when she reaches class. A less fit person may have to stop to catch her breath at the top of the second flight and still may be unable to answer for several seconds after collapsing into a chair.

If being fit means having the physical capacity to perform particular functions, athletic and non-athletic, what does the body require to be fit? Whether flying jets, swimming, running to catch a plane, or just walking to school, the body requires two basic resources: matter and energy. All biological activity requires a regular and sufficient source of matter. Matter, as you recall, is what makes up the universe.

Consider the common factors in these activities: a ballet dancer executing a complex spin, a cheetah racing at high speeds to attack a gazelle, and a lily bud opening its petals. Each of these activities relies on the organization of matter into the needed structures and on the regular repair and replacement of those structures as necessary. For the human and the cheetah, the muscles of the limbs and back act against the skeleton, coordinating movements that are directed by sensory perceptions such as sight and touch and that are processed by the brain. For the lily, day length stimulates blooming and coordinates the opening of the petals. In each case a highly coordinated interaction of structural systems must take place.

The physical activities you can perform are determined largely by your level of fitness. Of course, performance ultimately is limited by how your body is built. If your body can perform the basic functions of life, then clearly you possess a basic fitness for life. If

you also can bound up those three flights of steps so quickly that you beat the school bell, then you also possess a level of athletic fitness.

Biological activity requires energy. Your body needs energy to organize matter, that is, to repair, maintain, and build the body. Energy also provides the power that is required for that organized matter to perform work. It takes energy, for example, to build the proteins required for muscle growth. It also takes energy for the muscles to move. In fact, your body requires energy not only to drive its external functions such as walking and running but also to drive a variety of internal cellular and subcellular processes that are vital for life.

If all human performance requires matter and energy, what is the difference between a person capable of winning a grueling 160-km (100-mile) bike race and another capable only of watching the race on TV? One way to

Figure E7.2 What structural systems must interact for this activity to take place?

understand this difference in fitness is in terms of their bodies' relative abilities to apply matter and energy to the task at hand. The racer's matter is organized more effectively than that of a less fit person: although the two individuals may weigh the same, the racer's muscles are larger and stronger, so he is better able to apply the needed power to the pedals. Likewise, the racer's body supplies energy more effectively than that of a less fit person: although the two individuals both have circulatory and gas exchange systems, the racer's heart pumps blood more efficiently, which in turn supplies more oxygen and nutrients to the cells. The racer, thus, is better able to keep his legs pedaling and his brain working as he nears the finish line.

Figure E7.3 Bicycle racer and couch potato: a comparison a. Bicycle racer: Weight: 70 kg (little fat, much muscle); Exercise level: high; Diet: well-balanced. b. Couch potato: Weight: 70 kg (much fat, little muscle); Exercise level: low; Diet: high in fats and sugars.

FOOD: OUR BODY'S SOURCE OF ENERGY AND STRUCTURAL MATERIALS

They sat in the cold mess-hall, most of them eating with their hats on, eating slowly, picking out putrid little fish from under the leaves of boiled black cabbage and spitting the bones out on the table. . . . The only good thing about skilly was that it was hot, but Shukhov's portion had grown quite cold. However, he ate it with his usual slow concentration. . . . Sleep apart, the only time a prisoner lives for himself is ten minutes in the morning at breakfast, five minutes over dinner and five at supper.

The skilly was the same every day. Its composition depended on the kind of vegetable provided that winter. Nothing but salted carrots last year, which meant that from September to June the skilly was plain carrot. This year it was black cabbage. The most nourishing time of the year was June: then all vegetables came to an end and were replaced by groats.* The worst time was July: then they shredded nettles into the pot.

The little fish were more bone than flesh; the flesh had been boiled off the bone and had disintegrated, leaving a few remnants on head and tail. Without neglecting a single fish-scale or particle of flesh on the

*Groats are hulled and crushed oats or wheat.

brittle skeleton, Shukhov went on chomping his teeth and sucking the bones. He ate everything—the gills, the tail, the eyes when they were still in their sockets . . . .

A spoonful of granulated sugar lay in a small mound on top of [his bread-ration]. . . he sucked the sugar from the bread with his lips . . . and took a look at his ration, weighing it in his hand and hastily calculating whether it reached the regulation fifty-five grammes. He had drawn many a thousand of these rations in prisons and camps, and though he never had an opportunity to weigh them on scales . . . he, like every other prisoner, had discovered long ago that honest weight was never to be found in the bread-cutting. There was short weight in every ration. The only point was how short. So every day you took a look to soothe your soul—today, maybe, they won't have snitched any.

Source: From *One Day in the Life of Ivan Denisovich*, by Alexander Solzhenitsyn, translated by Ralph Parker, Translation © 1963 by E.P. Dutton and Victor Gollancz, Ltd. Copyright renewed © 1991 by Penguin USA and Victor Gollancz, Ltd. Used by permission of Dutton Signet, a division of Penguin Books USA Inc.

In its most basic sense, **food** is any substance that your body can use as a raw material to sustain its growth, repair it, or provide energy to drive its vital processes. In extreme situations such as Shukhov's, food is whatever will keep you alive.

A complete analysis of most of the substances we call food would show that much of what we eat consists largely of water. A tomato, for example, is about 95 percent water. Water is an important nutrient that we often take for granted, but our bodies need enormous amounts of it in comparison to other nutrients. An average American diet includes about 2 L, or 2000 g, of water each day. In contrast, most of us eat only about 50 g of protein in a day and only milligrams of many vitamins and essential elements.

The bulk of our food is composed of the three major classes of nutrients: carbohydrates, proteins, and fats. The tiny remainder consists of vitamins, essential elements, and an assortment of other substances. An important function of the digestive system is to break down large nutrient molecules into molecules small enough to pass across the lining of the digestive tract into the body. For example, complex carbohydrates such as starch and glycogen are broken down into simple sugars such as glucose. Proteins are broken down into their amino acids and fats are broken into an array of slightly simpler molecules.

If food technically can be anything that keeps you alive, what constitutes good nutrition? The phrase *good nutrition* means ensuring that your body receives what it requires to remain healthy and functional and avoiding those things that may cause it harm. Exactly what constitutes *good* may vary somewhat with the circumstances. For example, although not consistent with the general guidelines for a healthy diet, Jennifer Yates's breakfast was appropriate for her unusual type of physical activity. With its relatively high-fat foods, her breakfast contained sufficient stored energy and low bulk. A high-bulk breakfast such as pancakes and cereal would have put considerable physical strain on her digestive system as G forces in flight multiply the mass of any stomach or intestinal contents.

Determining what constitutes a level of good general nutrition is the job of biochemists, nutritionists, and other health professionals. Their understanding of how the body uses energy and matter is critical to their work. As their understanding of the way the body uses nutrients improves through research, the nutritional guidelines that professionals suggest often change. Figure E7.4 outlines some nutritional guidelines for general fitness. (Note that the "calories" listed on food labels are actually kilocalories, or kcals, a measure of the energy in food.)

There are, of course, many ways to achieve the balance of nutrients outlined in Figure E7.4. In fact, an individual's *diet*—the types of food that he or she eats on a regular basis—may reflect a number of influences such as personal preference, cultural background, and the varieties of food that are available. Even Shukhov's diet provided *some* of these nutrients. The "putrid little fish," for example, probably provided the prisoners with a critical source of protein. Although dietary proteins can supply some energy, they are far more important for the repair and maintenance of body tissues and for normal growth and development.

Humans obtain needed protein from a variety of sources. Different protein sources are important because, for our bodies to remain healthy, our diet must include the essential amino acids—those that our bodies cannot produce. High-protein food from animal sources (meat, milk, eggs) has the proper balance of amino acids for the human diet, as does the plant source soybeans. Other plant-derived foods, such as grains, nuts, and seeds, are good sources of protein as well, but most plant-derived foods lack

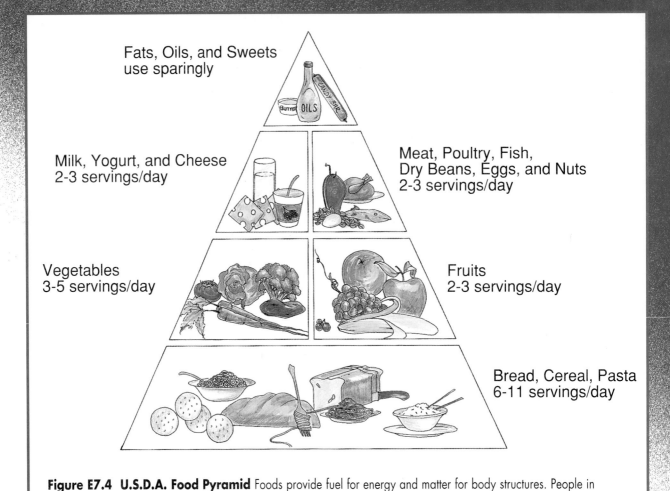

Fats, Oils, and Sweets
use sparingly

Milk, Yogurt, and Cheese
2-3 servings/day

Meat, Poultry, Fish,
Dry Beans, Eggs, and Nuts
2-3 servings/day

Vegetables
3-5 servings/day

Fruits
2-3 servings/day

Bread, Cereal, Pasta
6-11 servings/day

Figure E7.4 U.S.D.A. Food Pyramid Foods provide fuel for energy and matter for body structures. People in different cultures select and prepare foods in a variety of ways. What foods would you choose to meet the U.S.D.A. recommended guidelines?

one or more of the essential amino acids. It is possible, however, to obtain completely balanced amino acids by combining these plant-derived foods in the diet. The combination of legumes (beans, peas, and peanuts) with grains or nuts can provide balanced protein diets.

The fish probably also served as Shukhov's only significant source of lipids. In this regard the prisoners' diets might have been healthier than our own. Although fats are important nutritionally for making hormones and cell membranes and for storing energy, a majority of Americans consume too many fats, and many of these fats are the wrong type. Of particular concern are cholesterol and saturated fats, which are present in animal products such as meat, cheese, and butter. Because of the apparent link between the intake of these lipids and cardiovascular disease, physicians now recommend that fats should make up less than 25 percent of the daily kcals. (Some physicians recommend diets much

lower in fat, particularly fat from animal sources, to avoid or reverse blockage of blood vessels and accompanying damage to the heart. These low-fat diets also work well for weight reduction.)

The carrots, cabbage, and groats in Shukhov's diet likely provided an important source of fiber and carbohydrates. Complex carbohydrates such as starch generally are regarded as the best source of energy for physical activity because they supply glucose, the fuel that the body can use most readily. Carbohydrates are present in fruits, vegetables, and grains such as wheat, oats, and groats. In addition to providing carbohydrates, these foods also provide fiber, vitamins, and essential elements. Dietary fiber (roughage) comes mainly from cellulose in plants and cannot be broken down in the human digestive tract. Fiber absorbs water and toxins, contains no usable kcals, and reduces the time food spends in the digestive tract, thus ensuring regular elimination.

Important Vitamins for Human Health

Vitamin Fat-Soluble	Sources	Functions	Deficiency Symptoms
A (retinol)	Liver, green and yellow vegetables, fruits, egg yolks, butter	Forms eye pigments; helps cell growth, especially of epithelial cells	Night blindness, flaky skin, lowered resistance to infection, growth retardation
D (calciferol)	Fish oils, liver, action of sunlight on lipids in skin, fortified milk, butter, eggs	Increases calcium absorption from gut; is important in bone and tooth formation	Rickets (defective bone growth)
E (tocopherol)	Oils, whole grains, liver, mayonnaise, margarine	Protects red blood cells, cell membranes, and vitamin A from destruction; helps maintain muscles; synthesizes DNA and RNA	Fragility of red blood cells, muscle wasting, sterility
K (menadione)	Synthesis by intestinal bacteria; green and yellow vegetables	Assists liver in synthesis of clotting factors	Internal hemorrhaging (deficiency can be caused by oral antibiotics, which kill intestinal bacteria)
Vitamin Water-Soluble	**Sources**	**Functions**	**Deficiency Symptoms**
B_1 (thiamine)	Whole grains, legumes, nuts, liver, heart, kidney, pork, macaroni, wheat germ	Facilitates carbohydrate metabolism, nerve transmission, and RNA synthesis	Beriberi, loss of appetite, indigestion, fatigue, nerve irritability, heart failure, depression, poor coordination
B_2 (riboflavin)	Liver, kidney, heart, yeast, milk, eggs, whole grains, broccoli, almonds, cottage cheese, yogurt, macaroni	Forms part of electron carrier in electron transport system; aids production of FAD; activates B_6 and folic acid	Sore mouth and tongue, cracks at corners of mouth, eye irritation, scaly skin, growth retardation
Pantothenic acid	Yeast, liver, eggs, wheat germ, bran, peanuts, peas, fish, whole grain cereals	Facilitates energy release and biosynthesis; stimulates antibodies and intestinal absorption	Fatigue, headaches, sleep disturbances, nausea, muscle cramps, loss of antibody production, irritability, vomiting
B_3 (niacin)	Yeast, liver, kidney, heart, meat, fish, poultry, legumes, nuts, whole grains, eggs, milk	Serves as coenzyme in energy metabolism; is part of NAD^+ and NADP	Pellagra, skin lesions, digestive problems, nerve disorders, diarrhea, headaches, fatigue

Figure E7.5 Important Vitamins for Human Health

Figure E7.5 Important Vitamins for Human Health (Continued)

Important Vitamins for Human Health			
Vitamin	**Sources**	**Functions**	**Deficiency Symptoms**
B_6 (pyridoxine)	Whole grains, potatoes, fish, poultry, red meats, legumes, seeds	Serves as coenzyme in amino acid and fatty acid metabolism and in the synthesis of brain chemicals, antibodies, red blood cells, and DNA; is essential to glucose tolerance	Skin disorders, sore mouth and tongue, nerve disorders, anemia, weight loss, impaired antibody response, convulsive seizures
Biotin	Cauliflower, liver, kidney, yeast, egg yolks, whole grains, fish, legumes, nuts, meats, dairy products; synthesis by intestinal bacteria	Serves as coenzyme in fatty acid, amino acid, and protein synthesis; promotes energy release from glucose; facilitates insulin activity	Skin disorders, loss of appetite, depression, sleeplessness, muscle pain, elevated blood cholesterol and glucose levels
Folate (folic acid)	Liver, yeast, leafy vegetables, asparagus, salmon	Serves as a coenzyme in nucleic acid synthesis and amino acid metabolism; is essential for new cell growth	Failure of red blood cells to mature, anemia, intestinal disturbances, diarrhea
B_{12} (cobalamin)	Liver, organ meats, meat, fish, eggs, shellfish, milk; synthesis by intestinal bacteria	Serves as coenzyme in nucleic acid synthesis; helps maintain nervous tissue; plays a role in glucose metabolism	Pernicious anemia, fatigue, irritability, loss of appetite, headaches
C (ascorbic acid)	Citrus fruits, tomatoes, green leafy vegetables, peppers, broccoli, cauliflower	Is essential to formation of collagen, an intercellular substance that holds cells together; protects against infection; maintains strength of blood vessels; increases iron absorption from gut; plays important role in muscle maintenance and stress tolerance	Scurvy, failure to form connective tissue, bleeding, anemia, slow wound healing, joint pain, irritability, premature wrinkling and aging

Finally, the prisoners' diet must have included *some* vitamins and essential elements, or they would have been considerably less fit than they seemed to be. Usually needed only in very small amounts, vitamins work as partners with enzymes in cellular activities and are necessary for normal growth and maintenance of life. Thiamine and riboflavin, for example, are B-complex vitamins, which play an important role in the chemical processes by which energy is released from food. Deficiencies of thiamine can lead to muscle atrophy, paralysis, mental confusion, and even heart failure.

Elements are the basic components of matter. Essential elements such as sodium and calcium are important to maintaining homeostasis. Figure E7.5 describes the sources, functions, and deficiency symptoms of vitamins. Figure E7.6 provides similar information about the major essential elements.

Important Elements for Human Health

Name	Food	Function	Deficiency Symptoms	Excess Symptoms
Calcium (Ca)	Dairy products, green vegetables (broccoli, greens), legumes, tofu (bean curd), small fish (with bones)	Helps in bone and tooth development; facilitates muscle contraction, blood clotting, nerve impulse transmission, and enzyme activation	Osteoporosis, stunted growth, poor quality bones and teeth, rickets, convulsions	Excess blood calcium (rare), loss of appetite, muscle weakness, fever
Chlorine (Cl)	Table salt, soy sauce, processed foods	Helps maintain acid-base balance; assists hydrochloric acid formation in stomach; promotes bone and connective tissue growth	Metabolic alkalosis (rare), constipation, failure to gain weight (in infants)	Vomiting
Magnesium (Mg)	Whole grains, liver, kidneys, milk, nuts, dark green leafy vegetables, seafood	Serves as a component of chlorophyll, bones, and teeth, and as a coenzyme in carbohydrate and protein metabolism	Infertility, menstrual disorders	Loss of reflexes, drowsiness, coma, death
Phosphorus (P)	Soybeans, dairy foods, egg yolks, meat, whole grains, shrimp, peas, leafy green vegetables	Serves as component of bones, teeth, nucleic acids, phospholipids, proteins, and ATP	Bone fractures (rare), disorders of red blood cells, metabolic problems, irritability, weakness	Decreased levels of calcium, muscle spasms, jaw erosion
Potassium (K)	Whole grains, meats, fruits, vegetables, milk, peanut butter	Helps maintain body water and pH balance; plays important role in nerve and muscle activity, insulin release, glycogen and protein synthesis	Muscle and nerve weakness, poor digestion	Abnormalities in heartbeat or heart stoppage, muscle weakness, mental confusion, cold and pale skin (all are rare)
Sodium (Na)	Table salt, soy sauce, processed foods, baking soda, baking powder, meat, vegetables	Helps maintain body water and pH balance; plays role in nerve and muscle activity and glucose absorption	Weakness, muscle cramps, diarrhea, dehydration, nausea	High blood pressure, edema, kidney disease
Sulfur (S)	Dairy products, nuts, legumes, garlic, onions, egg yolks	Serves as component of some amino acids; plays role in enzyme activation and in blood clotting	None known; protein deficiency would occur first	Excess sulfur-containing amino acid intake leads to poor growth

What Happens to the Food You Eat?

Have you ever watched a pizza commercial on television and heard your stomach begin to growl as the actor pulls up a slice with melted cheese stringing behind? In response to your thoughts of pizza or to an actual need for nutrients to maintain your body's supply, hormonal signals begin to prepare the digestive system for action. For example, a decrease in nutrient levels in your blood sends a signal to the hunger center in the hypothalamus of your brain. The hypothalamus responds by triggering the release of digestive juices such as gastric juice into the stomach. A feeling of hunger then motivates you to find food. These initial responses to hunger stimulate the secretion of hormones such as gastrin, which stimulate further secretion of digestive juices in the stomach. Thus, a feedback system alerts the body that it needs food, and you change your behavior to obtain it.

Although fungi digest their food outside the organism and then absorb the simple nutrient products, humans and many other animals must bring food inside their bodies to provide proper conditions for digestion. (Figure E7.7 shows the components of the human digestive system.) The first steps are familiar: obtain food, break it down mechanically by chewing, and then swallow it.

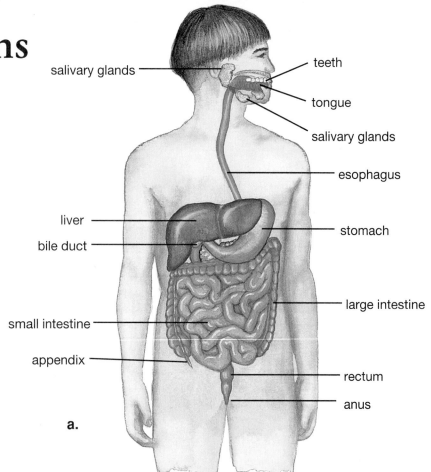

a.

Figure E7.7 The digestive system of the human body a. Location of organs and tissues involved in digestion b. Examples of enzyme action in breaking down food. Notice the compartments in which the enzymes for specific substrates act.

b.

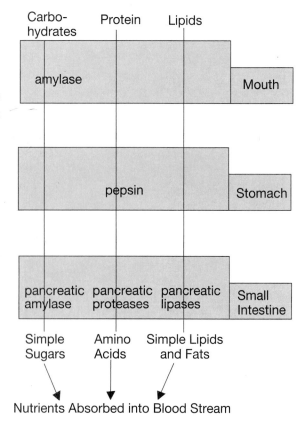

UNIT THREE ENERGY, MATTER, AND ORGANIZATION: Relationships in Living Systems

The mechanical breakdown of food is not enough. No matter how finely you chew bits of steak, the proteins remain intact. If your body is to use the proteins, it must break them down all the way to amino acids. Similarly, bread, like the finely powdered flour from which it is made, still has its carbohydrates intact, mainly in the form of long storage molecules known as starch. If your body is to use such carbohydrates, your body must break them down to sugars. Starch is a molecule manufactured in plants that is made up of large numbers of sugars bonded together in a large branching molecule, as shown in Figure E7.8.

No amount of chewing can break down proteins and carbohydrates to these smaller molecules, yet starch does begin breaking down to sugar while still in the mouth. The chemical action of special proteins known as **enzymes** is critical to this step in digestion. Enzymes are protein molecules that catalyze, or speed up, specific molecular reactions that otherwise would take place only very slowly. The molecules to which enzymes bind are called **substrates**; particular protein and carbohydrate molecules are examples of substrates. When an enzyme catalyzes a substrate reaction, the substrate is converted into a different molecule. For example, when you eat a steak, the stomach enzyme pepsin binds to steak proteins and, with other protein digestive enzymes in the small intestine, breaks down the proteins to their component amino acids, as illustrated in Figure E7.7b. Enzymes that help accomplish the breakdown of food are present

Proteins

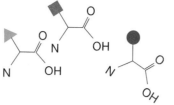

amino acids: carbon, nitrogen, hydrogen, and oxygen

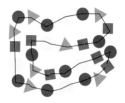

proteins: many amino acids bonded together in a chain that folds into a precise shape

Carbohydrates

simple sugar: carbon, hydrogen, and lots of oxygen

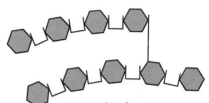

starch: many simple sugars bonded together in branching structures

Lipids

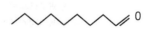

fatty acid: carbon, hydrogen, and a small amount of oxygen

simple fat: consists of three fatty acid molecules joined to a molecule of glycerol

Figure E7.8 Examples of macromolecules and their components
Macromolecules, such as proteins, carbohydrates, and lipids, are chains of smaller molecules. For example, proteins are made up of long chains of many different amino acids, and complex carbohydrates such as starch and glycogen are made up of long chains of simple sugars. (Starch and glycogen are storage molecules in plants and animals, respectively.) The lipids known as simple fats are made up of three long chains of fatty acids combined with glycerol, a small alcohol.

in your mouth, your stomach, and your small intestines.

In general, digestive enzymes break down complex molecules into their simple components. Figure E7.9 lists some examples of digestive enzymes, the reactions they carry out, and the

pH conditions under which the enzymes function.

Even though chewing by itself is not enough to break down the macromolecules in food to small molecular components, it does perform a very important digestive function. As you chew, the surface area of your food increases greatly, and the food becomes moistened with saliva, a fluid that is secreted by salivary glands located under your tongue. Saliva contains the enzyme called salivary amylase, which begins breaking down the macromolecule starch into the sugar molecule maltose. (Maltose consists of two glucose molecules joined together.) The increased surface area means that more starch molecules are accessible to amylase, and thus the conversion to maltose can be accomplished more quickly.

After you swallow, it takes less than 10 seconds for the chewed and moistened food to pass through the esophagus into the stomach. The food is moved by the coordinated waves of contraction known as peristalsis.

Peristalsis is produced by the muscles that encircle the digestive tract. In the stomach, partially digested food is churned and mixed with more digestive fluids for about three or four hours. The stomach stretches during this process and makes you feel full. This sensation reduces your motivation to continue eating (unless the food is so tasty that you ignore these signals and continue eating anyway). This is another example of feedback, which you learned about in Chapter 5.

The pancreas and the liver contribute to the digestive process. The pancreas is an organ in the abdomen that produces 1.4 L per day of fluid that contains enzymes that contribute to the final digestion of the remaining macromolecules. The

liver produces 0.8 to 1.0 L of bile per day, which is stored in the gall bladder until it is delivered to the small intestine after a meal. Bile contains bile salts, which act in a manner similar to detergents by breaking fat into tiny droplets, increasing the surface area of the fats so that enzymes can work on them easily. The digested nutrients, now present as small sugar, amino acid, and fatty acid molecules, are absorbed across the wall of the small intestine, where they enter the blood stream. Undigested remains of food cannot cross this wall. Instead, this part of the food enters the large intestine, where water is removed from the remains and is returned to the blood. The body then eliminates the compacted solid wastes.

Role of Some Digestive Enzymes		
Type of Enzyme	General Reaction	Optimal pH
amylase	starch → double sugars (maltose)	pH 6.9–7.0
proteases	protein → amino acids	pH 2.0 (pepsin) pH 7.0–8.0 (most others)
lipases	fats → fatty acids + glycerol	pH 8.0

Figure E7.9 Role of Some Digestive Enzymes

Anorexia Nervosa: Dying to Be Thin

She has been getting increasingly moody, she hasn't menstruated in three months, and the circles under her eyes suggest to others that she has not been sleeping well. The cold she caught

two weeks ago has lingered, despite her efforts to shake it. Though she denies feeling tired, she seems to have less energy each day, and yesterday, her father noticed her swaying a bit—dizzy, perhaps?—when she jumped up to answer the phone. Yet, after

dinner (an unhappy meal in which her parents pushed her to eat and Christine insisted she was not hungry), she went out to run her customary 3 km (1.9 miles).

Although Christine doesn't know it and probably wouldn't

admit it, she has anorexia nervosa, an eating disorder that affects an estimated 1 million individuals, mostly teenage girls, in the United States. Christine doesn't see that she is undernourished; in fact, she will insist against all evidence to the contrary that she is fat and needs to lose weight. Her self-discipline is the envy of all her friends. She denies her hunger and she exercises relentlessly until she is convinced that she has burned off any "excess" calories she might have consumed.

Left untreated, Christine likely will continue starving herself—possibly to death. As her nutritional base deteriorates, she will experience hormonal changes, her heart muscle will become weak and thin, and her digestive system will begin to function less and less efficiently. Electrical activity in her brain may become abnormal, and electrolyte imbalances in her body will cause her to become more and more at risk for sudden heart failure.

Because the underlying causes of anorexia nervosa are complex and involve self-image and mental attitudes, treatment of the disorder also is complex. Successful treatment must take into account the whole individual: the physical self, the cultural self, and the psychological self. Not surprisingly, treatment is most successful when the entire family is involved and participates honestly in the process.

The Structural Basis of Physical Mobility

Mary, James, Lolita, Madonna, Rodriguez . . . Writing your name seems simple enough, doesn't it? To do even this simple task, however, requires a highly coordinated series of muscle movements in your arm, hand, and fingers. Energy is necessary for all of these movements. Likewise, to transmit a nerve impulse from one part of the body to another, a neuron must be able to channel some of its available energy into specific, regulated movements of particular ions across its cell membrane.

All physical activities require some type of structure that can translate the energy of food into useful biological work. The muscles and skeleton of your arm, hand, and fingers as well as the membrane of a neuron are biological structures. The functions of these structures are quite specific. A membrane cannot move your fingers, nor can a muscle carry nerve impulses. These examples illustrate that there is a close relationship between a physical structure and its function.

Consider, for example, the organization and function of skeletal muscle, the type of muscle that produces the movements of your limbs. To generate most types of movement, muscles must work in opposing groups against the skeleton. Figure E7.10 illustrates the organization of muscles in your arm. Notice that the biceps and triceps are attached to the bone of the upper and lower arm by tendons, flexible cords of connective tissue. The biceps' tendon is attached to the bone of the lower arm on the *inside* of the elbow joint so that, when you contract the biceps, your arm bends. By contrast, the triceps' tendon attaches to the bone of the lower arm on the outer *side of the elbow* so that, when you contract the triceps, your arm straightens.

Muscles in a vertebrate limb work on the bone to which they are connected just as two sets of pulleys would work on a lever. A relatively small amount of shortening of either produces a large movement of the bone. Even though this is the case in a wide variety of organisms, the details can vary greatly. These differences mean that different organisms are capable of different types of movements.

This variation is especially evident in organisms that have the same overall body plan but have adapted to different situations. Compare, for example, the forelimb of the cheetah to that of the mole shown in Figure E7.11. Both are vertebrates, and their limbs work according to the same principles (and even use the same muscles) as the human arm. The primary functions of these limbs differ greatly, however; the cheetah's limbs are adapted for running after fleet-footed prey, whereas the mole's limbs are adapted for burrowing in the ground. What structural details underlie these adaptations?

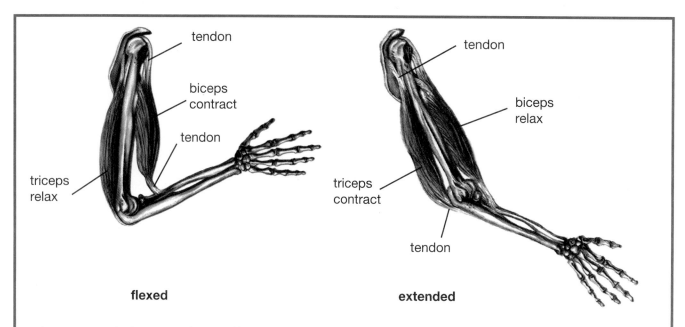

Figure E7.10 The human arm functions like a lever through the action of biceps and triceps muscles on a fulcrum point, the elbow. Although these muscles are attached to the bones in the upper and lower arms, muscle contraction causes only the lower bone to move.

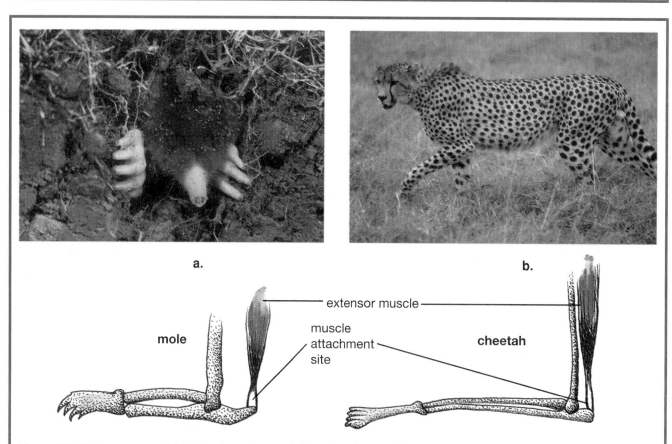

Figure E7.11 Structure and function in moles and cheetahs The type of movement possible in an organism depends on the precise arrangements of skeleton and muscle. a. Short, heavy bones like those of the mole are typical of skeletons of animals that require power. b. Thin, light bones like those of the cheetah favor speed. Think about the effect of applying forces to the two muscle attachment sites.

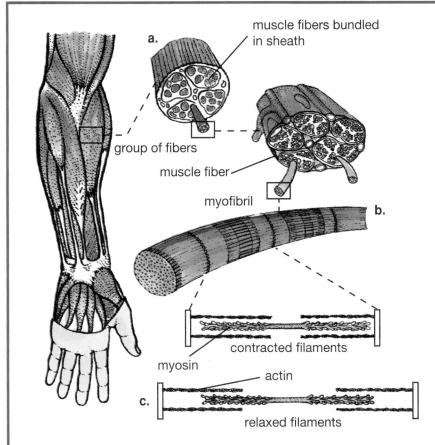

muscle fibers bundled in sheath

a.

group of fibers

muscle fiber

myofibril

b.

contracted filaments

myosin

actin

c.

relaxed filaments

Figure E7.12 a. A muscle is composed of many muscle fibers bundled in a sheath. b. Each muscle fiber is made up of many parallel myofibrils. c. Each myofibril is made up of protein molecules organized into thick and thin filaments.

many bundles of muscle fibers (Figure E7.12a). The thin and thick lines visible in Figure E7.12c are filaments, specialized structures within muscle fibers that are composed of two types of long, thin protein molecules. When you contract a muscle, an expenditure of energy enables the filaments within each fiber to slide past each other, much as your interlocked fingers can slide past each other when you move your hands together. This sliding of the filaments shortens the muscle fiber; together the shortening of many muscle fibers shortens the whole muscle. When you relax your muscle, these filaments return to their original positions, and the muscle regains its natural appearance and shape.

Studying muscle fibers at a subcellular level explains why it is important that muscles work together. While the movement of the molecular filaments past each other can shorten the muscle, it cannot lengthen the muscle again. That is, when a muscle relaxes, it cannot return to its normal length by itself. This inability to lengthen means that the muscle by itself cannot push on anything but can only pull. This is why it is important that muscles in the limbs of vertebrates are arranged so that for every set of muscles that pulls the bone of a limb in one direction, there is a set that pulls it back the other way. Were that not the case, many movements would be impossible.

The mole's digging forelimbs must generate power rather than speed. As the diagram in Figure E7.11a shows, the bones of such limbs are short and thick, and the projection at the elbow to which the extensor muscles attach is quite long in proportion to the lower limb bone. Because of this structural arrangement, the extensor muscles generate great power in the lower limb when they contract. In the limb of a running animal such as the cheetah, however, the extension at the elbow to which the extensor muscles attach is very short in proportion to the long lower limb bone (Figure E7.11b). As a result the same amount of contraction by the extensors moves a cheetah's foot much farther than a mole's foot, enabling a cheetah to run at great speeds.

The importance of structure to the function of muscles is not only apparent in the size and shape of the limbs to which they are attached, it also is evident if you look at the microscopic internal organization of skeletal muscle (see Figure E7.12). As you can see, a muscle is a complex structure that is made up of

The functional advantages of different structures also is evident in organisms that have quite different body plans. In vertebrates groups of muscles work in antagonistic pairs and act against an internal support system, the bony skeleton. Invertebrates have quite different support systems. For example, many soft-bodied invertebrates have an efficient support system composed of a surprising substance: water. A water-based support system, or hydrostatic skeleton, is not as odd as it might sound. Water, like other liquids, is not compressible. This characteristic means that although a flexible container filled with water may

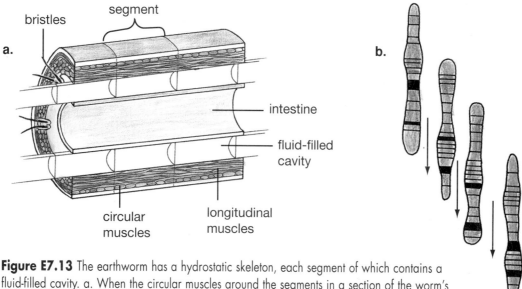

Figure E7.13 The earthworm has a hydrostatic skeleton, each segment of which contains a fluid-filled cavity. a. When the circular muscles around the segments in a section of the worm's body contract, the fluid in those segments is squeezed, and the segments become longer and thinner. b. When the longitudinal muscles contract, the segments become shorter and thicker. The earthworm moves by anchoring one part of its body with its bristles while it extends another part.

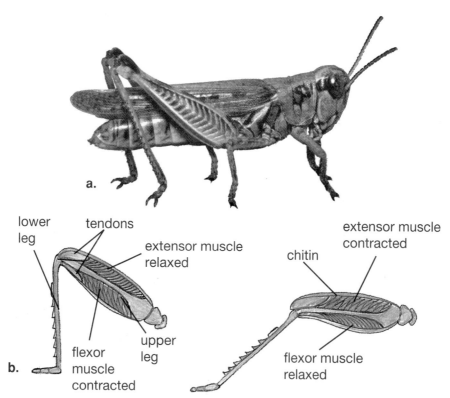

Figure E7.14 Exoskeletons have muscles attached to the inside of the skeleton, but these muscles still work in opposing pairs.

change shape in response to pressure, its volume remains constant. As Figure E7.13 shows, the contraction of the circular muscles around the segments in a particular part of the worm's body squeezes on the watery fluid in those segments, causing them to become longer and thinner. Dynamic contraction of the antagonistic longitudinal muscles, on the other hand, causes the segments to become shorter and thicker. When the worm crawls along, it alternately extends and contracts different parts of its body in this way and uses stiff bristles on each segment to anchor some sections while extending others. Many soft-bodied animals such as slugs and jellyfish have variations on this system of an internal hydrostatic skeleton surrounded by antagonistic groups of muscles.

Another common type of invertebrate support system is the **exoskeleton**—a hard skeleton on the outside of the body. (The internal support system in vertebrates is called an **endoskeleton**.) The grasshopper shown in Figure E7.14a is a good example of an animal that has an exoskeleton. Figure E7.14b shows diagrams of a grasshopper's leg. Note the antagonistic set of muscles. When a grasshopper draws up its leg, the lower muscle contracts while the upper muscle remains relaxed. When the grasshopper hops, the upper muscle contracts while the lower one relaxes. Although these muscles attach to the *inside* of an external skeleton, the mechanical aspects of the system are very similar to those of the human arm. In addition to being quite strong for its mass, an exoskeleton also provides a layer of armor that protects the soft parts of the animal's body. The essay *The Ant That Terrorized Milwaukee* considers what happens if an animal with an exoskeleton gets too big.

The Ant That Terrorized Milwaukee

The huge, black thorax towered over Damien, blocking the light. The enormous insect was rearing on its back two pairs of legs; its front legs pawed at the air like a huge stallion or, more accurately, like a menacing and hideous monster. The sensitive antennae quivered and twisted through the air, searching, searching for anything that might challenge it. The air was heavy with the animal's stench.

Now that he was this close, Damien could understand why his pitiful attempts to bring down the beast had failed so miserably. The animal's body was encased in a shiny, hard, black substance, which appeared to be impenetrable to any weapon Damien might be able to get his hands on. Through his fear he vaguely remembered something important he knew about insects . . . yes, that was it! A hard outer skeleton—what did Miss

Brumstitzler call it?—but, no, too late . . . . As the moving mouth parts drew nearer, Damien's last, desperate thought was, "But she *insisted* they couldn't get this big. . . . "

You probably have heard statements such as: "An ant can carry 10 times its own weight, so an ant the size of a person could lift a car," and "A grasshopper the size of a horse could jump the length of a football field." An overused plot in horror films has insects or other minute creatures become gigantic. You have seen that multicellular organisms exhibit a wide range of body plans that work by the same basic structural and mechanical principles. With such a range is there any limit to how big, how fast, or how strong an organism might be?

For physical reasons giant creatures usually are not possible because basic structural and mechanical considerations limit the situations and range of sizes for which particular body plans are suitable. For example, growth is one limitation to animals with exoskeletons. Because the exoskeleton encases the whole body in armor, it must be shed completely every time there is significant growth. The animal is relatively helpless and vulnerable while the new exoskeleton hardens. Another major limitation of such a body is due to the material out of which exoskeletons are made—a complex carbohydrate called chitin. Hollow tubes of chitin are very strong for their mass, as long as they remain small. At larger sizes, however, the mass of the chitin exoskeleton that would be needed to support the body against gravity and to withstand the force of opposing muscle systems increases to impossible levels. An ant the size of a person probably could not even pick itself up, let alone wreak havoc on Milwaukee.

ENERGY'S ROLE IN MAKING STRUCTURES FUNCTIONAL

The detailed structure of muscle fibers explains how muscles contract, but it does not explain where a muscle gets the enormous amounts of energy required for vigorous contraction. Scattered among muscle fibers are many cylindrical **mitochondria**, subcellular compartments in which oxygen is used to release energy. The energy release that takes place largely within mitochondria is the major source of energy for low-intensity physical activity. This form of energy release is called **aerobic energy release** because it requires oxygen.

Although aerobic energy release is the source of a great deal of energy, limits to the rate of the underlying cellular reactions mean that it cannot keep up with the energy requirements of all types of human performance. When large amounts of energy must be available very quickly (for example, during a sprint to catch a rapidly disappearing bus), another type of energy release can supplement the aerobic process in muscles for short periods of time. This process does not require oxygen and is therefore called anaerobic energy release. Anaerobic energy release can deliver a significant amount of energy quickly, but it can be sustained for only a few minutes. A disadvantage of anaerobic energy release is that it produces by-products such as lactic acid that can build up in the muscles and interfere with their ability to contract. Recovery from extreme anaerobic exercise requires approximately 1 to 1.5 hours of rest, with adequate oxygen delivery.

Because of the energy demands of contracting muscles, strenuous human performance requires a significant increase in circulation, and the blood flow to exercising muscles may reach 15 times the normal levels. The increased blood flow ensures adequate oxygen delivery for aerobic exercise and also removes by-products. Even so, vigorous exercise eventually results in muscle fatigue, a condition in which the glycogen supplies of the muscles are so depleted that the energy-releasing mechanisms no longer can function. The only solution for extreme fatigue is rest; with sufficient time and proper food, the glycogen stores are replenished and normal function can resume.

Factors Influencing Performance

Genetic and Gender Differences. Are great athletes or great dancers *born* or *made*? Probably the answer is that great athletes are partly born and partly made. On the one hand, as humans and individuals we are born with a certain basic set of physical capabilities. On the other hand, many things that we do and that we don't do affect how successfully we are able to use our inborn capabilities.

One reality that we all deal with is that we are humans, not cheetahs or moles or ants. As humans we are capable of performing the functions that we can because our bodies (basically, our organized matter) are able to acquire and use energy in particular ways. At a species level the basic structure and function of our bodies determines how we move, and certain restrictions limit our capacity to perform. Our height, for example, is determined largely by inheritance and may affect whether or not we are likely to be first-class basketball players. Inheritance appears to be important in gymnastics, as most successful gymnasts tend to have small, compact bodies. Our gender and genetic makeup also influence other physical factors such as skeletal mass, maximal lung capacity, and the rate at which our bodies use energy; these may influence the types of physical activity that we can perform best.

We can see the effect of gender on the body's physical development and activity in two simple examples. One result of the increase in testosterone levels that occurs in most young men during puberty is an increase in the cross-sectional area of their muscles. Because a muscle's strength increases in proportion to its size, this increase in muscle mass means that males at puberty and older tend to be stronger than females of the same age and height. Anabolic steroids are popular among some athletes in training because they mimic some of the effects of testosterone on muscle mass. Although steroids may improve athletic performance, such substances have a number of serious and sometimes irreversible effects. These effects include high blood pressure, alterations in heart muscle, and reduced fertility, all reasons for the Olympic Committee's ban on steroid use as a performance-enhancing technique.

Similarly, a key issue to consider when evaluating your dietary habits and their relationship to your fitness is the manner in which your body uses energy. Basal metabolic rate, or **BMR**, is the approximate rate at which your body uses energy and it is expressed as kilocalories used per kilogram of body mass per day. BMR differs for different individuals and is an indicator of how much energy your body uses. Rapidly growing children tend to have a high BMR, whereas people with an underactive thyroid gland have a low BMR. In general, females tend to have a lower BMR than males, due to higher percentage of body fat. This means that females use energy more efficiently than males and tend to require less food for the same amount of physical activity. The combination of your BMR, your diet, and your level of exercise determines your mass. With an intake of food that balances the body's nutritional demands against the BMR and activity level, a person's mass remains nearly constant. Taking in even slightly more food than the body needs for energy can, through time, cause the body to store the excess kcals as fat. In moderate, controlled dieting, the number of kcals taken in falls slightly short of the body's energy demands, which forces some of the matter and energy for life processes to come from stored fat. With a more severe or sustained reduction in kcals (such as in fasting or starving), even the body's stored fat supplies eventually do not provide adequate matter and energy. In that case, the body breaks down its own structural components, such as muscle, to keep itself alive.

Conditioning. Despite these gender and genetic differences, human performance also is based in large part on general physical fitness, and we can improve our fitness by conditioning. Consider what happens, for example, with a group of healthy hikers. Those who normally engage in an exercise program soon take the lead, while others lag behind, breathing heavily and suffering from aching muscles. One basis for these differences is the way the body changes during a regular exercise program.

Conditioning can improve both general and athletic fitness in several ways. The major effect of regular exercise is to bring about changes in the structure and function of the body. Muscles enlarge and become stronger as a result of repeated work, and mitochondria increase in number and size. The muscles' capacity for glycogen storage increases as does its blood supply. The net effect is a greater ability to convert fuel into useful energy.

What is a reasonable amount of exercise for staying generally fit? As little as 20 to 30 minutes of moderate exercise two to three times a week can help your heart, lungs, and circulatory system to work more effectively. Exercise tends to decrease your resting heart rate (not your heart rate while you are exercising). In addition, exercise increases the heart's output with each beat, thus requiring it to work less. Long-term conditioning also builds muscle mass and tone (firmness); strengthens the skeleton by maintaining, or increasing, bone mass; and improves the communication between nerves and muscle. Better strength, coordination, and endurance are the rewards. Aerobic activities, such as jogging, brisk walking, bicycling, or swimming, can accomplish these goals.

Behavior. In a similar fashion, the lifestyle that an individual adopts influences fitness. One important element of lifestyle is what and how much a person eats and drinks. For example, people who are interested in maintaining their weight must keep the number of kilocalories that they take in equal to the number of kilocalories that they expend. On the other hand, people who are interested in losing weight must take in fewer kilocalories than they expend. Athletes who are training for a marathon might want to increase the amount of carbohydrates that they eat in order to increase the amount of glycogen available to muscles. This is called carbohydrate loading. Figure E7.15 describes diets in each of these categories so that you may compare the relative amounts of servings in each food group.

A good mental attitude also can lead to behaviors that promote good general fitness, just as various mental and emotional disorders can lead to behaviors that endanger fitness. An estimated 2 million individuals in the United States suffer from eating disorders, conditions in which chronic irregularities and abuses in an individual's eating patterns can endanger life itself.

Toxins. In addition to behavioral disorders, **toxins** also influence performance. Toxins are substances that, when taken into the body, ultimately cause diminished performance or impairment of health. Even medications such as anti-inflammatory drugs (ibuprofen, for example) may be toxins under certain circumstances, especially if they are taken at higher doses than directed.

Illegal, or so-called "street drugs," diminish performance as do legal toxins such as alcohol and tobacco. Alcohol initially may feel like a stimulant because it produces a temporary sense of well-being. Thus, under the influence of alcohol, we may think of ourselves as feeling and performing better. Actually, alcohol depresses the central nervous system, which causes loss of coordination and impaired performance. Ultimately, alcohol causes cells and tissues to use oxygen less efficiently and to produce less energy. In large amounts over long periods, it damages brain and liver tissue. Even in small quantities, it impairs judgment, hence the intoxicated individual's misleading perception of stimulation.

Tobacco contains several substances that can adversely influence performance. Carbon monoxide from burning tobacco binds to hemoglobin, the oxygen-carrying molecule in red blood cells; this binding reduces the hemoglobin's ability to carry oxygen. A two-pack-per-day cigarette smoker generates enough carbon monoxide to reduce his or her blood oxygen level to that of a nonsmoker who is experiencing the thin air of high mountain altitudes (3048 m or 10,000 feet). Both tobacco smoke and unburned tobacco such as chew or dip produce high levels of nicotine in the blood stream and can result in addiction. Nicotine affects performance directly because it constricts blood vessels, thus impairing oxygen delivery.

Diet Comparison Food Groups (Daily Servings)						
	Milk	Meat	Fruit	Vegetable	Grain	Fats, Oils, and Sweets
Maintenance Diet	3*	2–3	2–4	3–5	6–11	**
Weight-Loss Diet	3*	2	2	3–5	6	**
Carbohydrate-Loading Diet	3*	3 or more	7 or more	5 or more	11 or more	**

* teenagers and young adults; 2 for older adults

** You can select foods from the Fats, Oils, and Sweets category only if you can afford the kcals after eating the recommended servings from the essential food groups.

Note: The diets listed in this table are approximations based on information from the U.S. Department of Agriculture's Daily Food Guide. *These do not constitute dietary recommendations*. Individuals should check with their physician before going on any diet.

Figure E7.15 Diet Comparison

companies continually use technological advances to produce, for example, lighter-weight tennis rackets and more flexible vaulting poles that give athletes a competitive edge.

Athletes also use devices that simulate competitive conditions and allow them to build their skills. At the Olympic Training Center in Colorado Springs, Colorado, swimmers test their fitness in a device called a flume, shown in Figure E7.16. A flume is a simulator containing water that runs at gauged speeds. The swimmer can swim in place against moving water under controlled conditions to increase speed and endurance. Ski team members can practice all year long on roller skis that duplicate the feel of cross-country skis. Therefore, snow is not always necessary for an athlete who is trying to build the endurance and coordination skills necessary for competitive skiing.

The use of such technologies may raise ethical questions. For example, do these technologies confer an unfair advantage on competitors who can afford them? Wealthy nations that are able to pay the cost of such devices may produce superior performers. Are these uses of technology fair? To address these and other issues, regulatory agencies exist for each major sport and for large sporting events such as the Olympics. Many of these agencies are international in scope.

We have seen that several factors affect human physical performance. These include genetic, behavioral, and technological factors, and they all are related to how effectively we use our body's energy supplies.

Figure E7.16 Swimmers training for the Olympic Games can test their speed and endurance by swimming in a flume. This equipment controls the speed and direction of water flow through the swimming tank.

Other important aspects of lifestyle that affect fitness include the amount of sleep a person gets and how one handles stress. By the choices that we make, humans can control many, though not all, of the factors influencing any type of fitness.

Technology. Technology can help us both measure and improve our individual fitness.

Athletes use weight machines and computerized aerobic exercise machines to build and to measure fitness. Specialized clothing and equipment enhance performance by increasing comfort level and efficiency. Computers can model the stresses produced by various activities and help researchers design athletic shoes for specific sports. Sports equipment

Matter and Energy Are Related

Humans are very good at identifying differences between things that we encounter in our physical world because many distinguishing features are directly observable through our senses. If someone asked you to describe how the following four examples of matter are different—skin, plant roots, water vapor, and gears in a watch—you could do so easily. Identifying similarities, however, can be quite a bit more challenging, so it would not be surprising if you were stumped by the riddle: What do skin, plant roots, water vapor, and gears in a watch have *in common*? In this case a common feature of all these items is that they are made of atoms, the basic building blocks of all matter from the smallest virus to the stars in the largest galaxy (see Figure E8.1). Because atoms are microscopic and cannot be observed directly through our senses, the importance of atomic interactions generally is not appreciated. As you will see, however, these interactions are critical to understanding the relationship between matter and energy.

Individual atoms may be hidden from our view because of their small size, but if many atoms are assembled together as an *organized collection*, they may form unique structures that are large enough to be clearly visible. It is the organization of many types of atoms that explains why skin is distinctly different from the gears of a watch. As Figure E8.2 shows, the source of this important atomic organization is chemical bonds. **Chemical bonds** hold atoms together in very predictable ways to form molecules; energy is stored within the structure of a molecule's bonds and atoms.

It is difficult to describe exactly how energy exists in the structure of chemically bonded atoms, but try to imagine it by thinking about the following example in which magnets represent atoms. Consider what happens when you arrange two round, flat magnets so that opposite poles are close to each other. The forces of attraction between the north and south poles pull the individual magnets tightly together. If you arrange several magnets in this way, instead of having several flat magnets, you have a solid, cylindrical stack. The energy stored in this new large magnet is much greater than the energy stored in any of the small

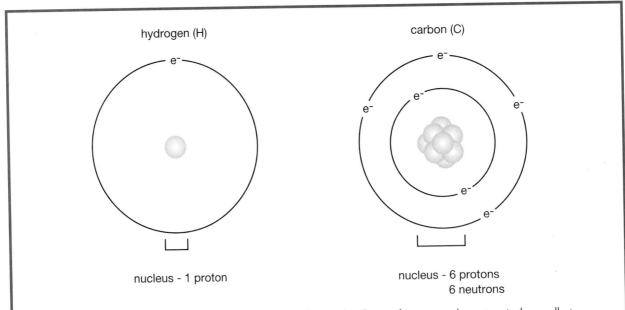

Figure E8.1 Atoms All matter is composed of elements, the simplest forms of matter, and an atom is the smallest particle of an element that still has the properties of that element. The unique properties of an element are the result of the number and type of subatomic particles present in its atoms. For instance, six electrons, six protons, and six neutrons are characteristic of the element carbon (symbol, C). Electrons are extremely lightweight, negatively charged subatomic particles that orbit rapidly around the nucleus of an atom. The nucleus, which is located at the center of an atom, is composed of two heavy subatomic particles: protons, which are positively charged, and neutrons, which have no charge at all and thus are neutral. Scientists have identified more than 100 different types of atoms, or elements.

magnets. The storage of energy in molecules is somewhat similar.

Magnetic energy is one familiar type of energy, but energy comes in a variety of other forms, including heat, light, electrical, solar, nuclear, mechanical, and chemical energy. Heat energy is the energy of movement. Most people are familiar with this type of energy through their experiences with friction. A rug burn is the result of the friction between a carpet and someone's moving skin, and bald tires are the result of friction between a road surface and rotating rubber treads. Movement also occurs at the molecular level. When molecules move, they too produce heat, and the greater the motion, the greater the heat that is released. For example, the coils on an electric stove heat up because an electrical current in the metal coils causes the metal atoms to move very rapidly.

When molecules slow down, heat decreases and objects feel cold. Many substances can absorb heat, that is, they can remove the heat energy caused by the molecular motion of nearby substances. A hot pan placed in cold water becomes cool as the hot metal transfers its heat energy to the surrounding water molecules, which respond by moving faster and becoming warmer.

Not all heat is the result of friction between a material and its surroundings. Chemical reactions also can release heat to the surroundings; such reactions are called **exothermic** (*exo* = out, *thermic* = heat) reactions. In an exothermic chemical reaction, heat is released when the atoms in molecules are reorganized, that is, when the chemical bonds between atoms are broken and new ones are formed. If a chemical reaction results in the production of heat, then the reaction is releasing energy. The explosive combustion of grain dust is a chemical reaction in which molecules are quickly reorganized in the presence of oxygen (an element present in the air) and a spark. Such a reorganization of matter can cause the uncontrolled release of a tremendous amount of energy.

Processes that absorb, or take in, heat are called **endothermic** (*endo* = in, *thermic* = heat); such processes require an *input* of energy before they can occur. For example, ice does not melt unless it receives an input of heat energy from another source, such as the glass and surrounding liquid in a container of iced tea. When ice does absorb heat, which occurs any time the ice is in contact with something warmer than itself, the organized water molecules in the

ice begin to move and become disorganized. In other words, an input of energy alters the molecular organization (but not the chemical bonds), and the result is that the ice melts. Many complex factors determine whether a particular molecular reorganization will be endothermic or exothermic.

As Figure E8.3 illustrates, the melting of ice is an endothermic *process*, but it is not an endothermic chemical *reaction* because no breakage or formation of chemical bonds occurs, only a change in state. (The *solid* water molecules become *liquid* water molecules.) Chemical reactions, however, also can be endothermic if an input of energy is necessary to reorganize atoms in the molecules. For instance, the oxygen atom and the hydrogen atoms in a molecule of water are quite stable when bonded to each other. It is necessary to add energy, in the form of a spark or ultraviolet light, to reorganize the atoms and form a molecule of oxygen (O_2) and two molecules of hydrogen (H_2).

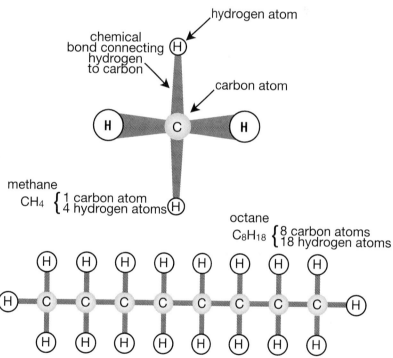

Figure E8.2 Molecules Molecules are formed when atoms are organized by chemical bonding. Combinations of different atoms, organized in unique ways through chemical bonds, account for the unique properties of macroscopic materials. For instance, in one ordered arrangement, carbon and hydrogen form a smelly gas called methane. In another arrangement, carbon and hydrogen form octane, a flammable liquid used in gasoline. Note that in both molecules hydrogen always has one chemical bond and carbon always has four chemical bonds. The number of bonds that a particular atom has is determined by the number of electrons orbiting around it.

The formation of water from oxygen and hydrogen is typical of how matter and energy interact in chemical reactions. This reaction also shows how chemical bonds are a link between matter, in the form of atoms and molecules, and energy. Understanding this link is important in biology because all living systems require energy and matter for survival. In the case of humans, the thousands of different chemical reactions that occur in our cells each day depend primarily on chemical energy, and evolutionary adaptations enable these reactions to take place in a controlled manner—without sparks and explosions.

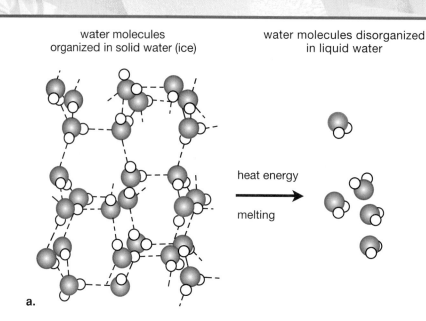

water molecules organized in solid water (ice)

water molecules disorganized in liquid water

heat energy

melting

a.

Figure E8.3 Endothermic processes and chemical reactions take in heat. a. An input of heat energy is necessary to melt ice, but note that there is no change in the chemical bonds of the water molecules that make up the ice. During the endothermic *process* of melting, the solid water molecules that are held rigidly in place in ice simply change state and become liquid water molecules, which are free to move around. b. In an endothermic *chemical reaction*, an input of energy causes the chemical bonds to break and reform. The addition of electricity to water molecules breaks and reforms chemical bonds and leads to the formation of two new molecules, hydrogen and oxygen.

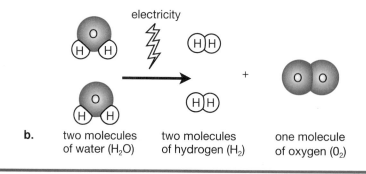

electricity

b.

two molecules of water (H_2O)

two molecules of hydrogen (H_2)

+

one molecule of oxygen (O_2)

Energy Is Converted and Conserved

When electricity flows through the metal coils on a stove burner, making the coils red hot, it does more than ready the teapot for tea. It illustrates an essential property of energy: energy can be **converted** from one form into another. In this case electrical energy is converted into heat energy. This property has enabled scientists and engineers to develop techniques and tools for making many types of energy accessible to humans. For example, power-generating dams harness the mechanical energy of water by passing the water over turbines, which rotate and convert the movement of the water into electrical energy.

Without energy conversion, gasoline would not be much more than a smelly, toxic liquid. In fact, gasoline is a useful tool in technologically advanced societies because in its liquid form gasoline stores a large amount of chemical

Figure E8.4 Energy conversion Molecules of the chemical gasoline store energy. This energy can be released by combustion and converted into mechanical energy in an automobile. What other energy conversions can you think of?

energy in its molecular structure. Gasoline, however, can be a dangerous substance because it is highly flammable, but it is exactly this property that makes it valuable as a fuel for industry and transportation. As Figure E8.4 suggests, we have learned how to control the explosive property of gasoline by designing engines that convert gasoline's stored chemical energy into a more useful, mechanical form—a form that can be used to power machinery.

Exploding grain elevators are another example of the conversion of energy from a stored form (the molecules of grain dust) to heat (the energy of motion). Such explosions also are an example of the fact that energy exists in two forms: an inactive form and an active form. The inactive form is called **potential energy**. This is energy that is available for use and is stored in the structure of matter. Grain dust contains a great deal of potential energy, but when this energy is released, it becomes active. Active energy is called **kinetic energy**.

A boulder sitting at the top of a hill contains a great deal of

potential energy simply because of its position, as Figure E8.5 shows. When the boulder falls off a cliff, however, its potential energy is released in the form of kinetic energy, or the energy of movement. Although the potential energy poses no threat to you—you can sit on the boulder as it rests at the top of the

hill—you would not want to come into contact with this same boulder as it careens down a hill, displaying its kinetic energy. Likewise, there is little to fear from pumping tremendous amounts of potential energy into a car's gas tank, but don't light a match or you'll see that potential energy converted into kinetic energy very quickly.

Once potential energy has been converted into kinetic energy, one of two things happens to the kinetic energy. It may be captured and made useful, or it may be wasted. If, for example, the boulder were attached to a rope and, as it rolled down the hill, it turned a wheel that raised water from a well, the kinetic energy would be captured and made useful. If, on the other hand, the boulder just rolled and rolled until it stopped, the kinetic energy would be wasted. The fact that the boulder stops at all indicates that yet another energy conversion has occurred. Each time the boulder turned, the earth beneath it was slightly deformed. As this happened, the kinetic energy of the boulder

a.   b.

Figure E8.5 Potential energy and kinetic energy a. The boulder resting at the top of the cliff has a great deal of potential energy and no kinetic energy. b. The falling boulder has a great deal of kinetic energy but less and less potential energy. How much kinetic or potential energy does the boulder have after it comes to rest at the bottom of the hill?

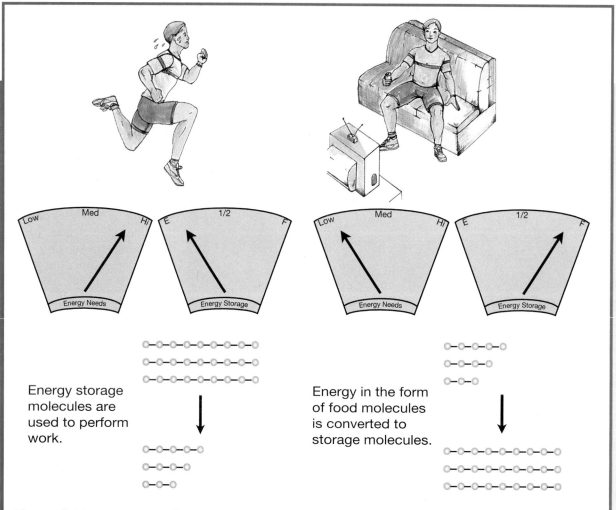

Figure E8.6 Living systems regulate energy storage. When the energy needs of humans are high, our bodies convert storage molecules such as glycogen into a form of chemical energy that can be used to perform work. When energy needs are low, our bodies convert food molecules into storage molecules that can be used to meet future needs. Are your energy needs high or low right now? Is your body trying to store energy, or is it using previously stored energy to perform work?

Energy storage molecules are used to perform work.

Energy in the form of food molecules is converted to storage molecules.

caused molecular movements in the grass, soil, and stones and was released as heat. In this case the release of heat is essentially wasted energy because the energy was not captured and used. The end result is that no energy is created or destroyed; it is simply *transferred* from the boulder to the surrounding grass, soil, and stones.

Energy transfer also takes place in your body, where food is supplied to the digestive system, nutrients and energy are removed and supplied to muscles and other

tissues, and heat is produced. Even through all of the intermediate steps necessary for this transfer, no energy is lost or created. This phenomenon, called the **conservation of energy**, means that even though the amount of energy at one location can change, the total amount of energy in the universe remains the same.

In living organisms this phenomenon is important because energy is conserved at the level of molecules. For instance, if one large storage molecule contains eight units of energy and it is broken down into two smaller molecules, then the two smaller molecules

together (plus any small amount of heat or side reactions) will contain eight units of energy as well. The conservation of energy and the ability to transfer energy from one molecule to another play key roles in maintaining the energy balance of living systems. The energy needs of organisms vary from time to time, and the ability to convert energy from one form to another means that organisms can use the form of energy that best matches their current energy needs. As Figure E8.6 illustrates, when energy needs are low—for example, during sleep— living systems store potential energy in large storage molecules, which contain many bonds and atoms and, thus, much energy. When energy needs are high–for example, during exercise–living systems convert some of this stored potential energy into mechanical, or kinetic, energy such as in the contraction of muscles.

The source of energy in molecules. What is special about molecules that allows them to store energy? The arrangement of atoms in molecules (like the arrangement of magnets) affects the energy properties of those molecules, and the arrangements of atoms are determined by chemical bonds. There are several types of chemical bonds, and most of the differences between the molecules' energy properties are due to the type of chemical bonds between the molecules' atoms. The forces of attraction in chemical bonds occur for different reasons, and they vary in their strength. The stronger the bond, the greater the energy needed to break it.

Some bonds occur between atoms that have lost or gained electrons. These positively or negatively charged particles are called ions, and oppositely charged ions attract each other to form **ionic bonds**. Ionic bonds are responsible for holding together the sodium and chloride ions in a crystal of NaCl, common table salt. Ionic bonds are relatively weak, a fact that you can demonstrate for yourself by dissolving some salt in ordinary tap water. As the salt dissolves, the ions (Na^+, Cl^-) separate from each other and associate with water molecules.

Stronger bonds result when the atoms that make up a molecule *share* their electrons. These bonds are called **covalent bonds**. Figure E8.7 illustrates the electrons that are shared between hydrogen and oxygen in a water molecule. The carbohydrate molecules glycogen (in muscles and liver) and starch (in plants) are complex molecules and rich sources of potential energy;

each is a macromolecule with large numbers of covalent bonds. A great deal of energy often is needed to make or break covalent bonds, and in most cases living systems use enzymes to help with these reactions.

The role of enzymes. Enzymes are critical because they can reduce the amount of energy necessary to start a chemical reaction. (This energy is called the energy of activation.) If a chemical reaction involving enzymes is represented by a stone propping up a boulder on a hillside, then think of the boulder as the molecules involved in the reaction and the stone that is propping up the boulder as the energy barrier that must be overcome to make or break covalent bonds, that is, to get the chemical reaction going (Figure E8.8).

Clearly, there is a lot of potential energy in the boulder, but without a strong push (much energy), the boulder cannot get past the stone and move down the hill. Likewise, without a lot of energy, many chemical reactions cannot take place. In this analogy enzymes work to reduce the stone to the size of a pebble. Now a gentle push is all that is needed to get the boulder rolling. Similarly, in the presence of enzymes, a small amount of energy can start a chemical reaction.

This analogy should immediately suggest one advantage of a lower energy requirement: living systems can avoid the danger associated with the release of large amounts of energy. In fact, high-energy chemical reactions do not occur in living systems because specific enzymes have evolved that accomplish necessary reactions in a precise, controlled, and very rapid manner. The conversion of starch to simple sugars by the digestive system enzyme amylase is an example. The ability to *accelerate* reactions at relatively low energies is why enzymes are so important to the functioning of living systems.

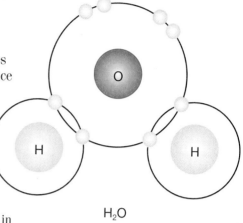

H_2O

Figure E8.7 Covalent bonds in molecules A water molecule consists of one oxygen atom (which has six electrons in its outer orbit) sharing two of its electrons with two hydrogen atoms (which have only one electron each). These shared electrons represent a force of attraction known as a covalent bond.

Energy for cellular activity. Although enzymes reduce the energy needed to start chemical reactions, nearly all biological reactions require the input of some energy before they can proceed. The source of this energy is long-term storage molecules such as glycogen. These molecules are so large and inaccessible, however, that a cell cannot use the energy in storage molecules directly. To make this energy useful, cellular reactions convert some of the energy in large molecules into another form of chemical energy that can be used *directly* by cells. A special type of molecule called **a**denosine **tri**phosphate, or **ATP** (see Figure E8.9), accomplishes this by carrying energy (in its molecular structure) between reactions within a cell. ATP is a **carrier**–a molecule that carries or transfers something (the way a mail carrier carries the mail). In fact, ATP also is an energy storage molecule, but it is a short-term storage molecule; it carries energy

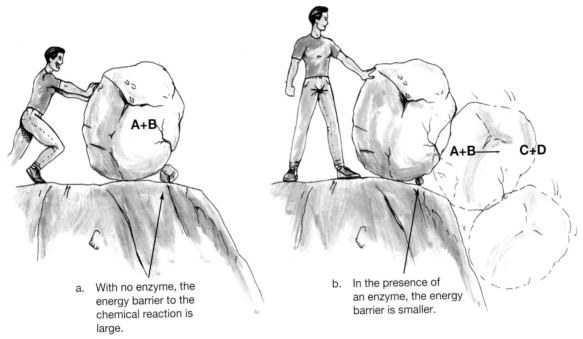

a. With no enzyme, the energy barrier to the chemical reaction is large.

b. In the presence of an enzyme, the energy barrier is smaller.

Figure E8.8 Enzymes reduce the amount of energy needed to start a reaction. a. Without enzymes the covalent bonds of many molecules (A and B) cannot be reorganized by chemical reactions in living systems. Starting such reactions (like pushing the boulder down the hill) requires too much energy. b. With enzymes, the barrier to starting the reaction is much smaller, and less energy is needed to start the reaction.

to places in the cell where the energy is used directly to fuel chemical reactions. In these reactions the energy is used to help perform cellular work; the energy is not all wasted as heat.

Even though the direct source of energy for cellular work is ATP, this valuable energy source is many steps removed from the original energy sources. For example, tracing the path of energy flow backward makes it clear that the energy in ATP comes from the energy in long-term storage molecules, and the energy in long-term storage molecules comes from food molecules. Thus, any reaction that requires ATP for energy has an indirect link to the food eaten by living organisms. The transfer of energy from matter in the form of food to matter in the form of ATP is, therefore, one of the most important processes in all of biology.

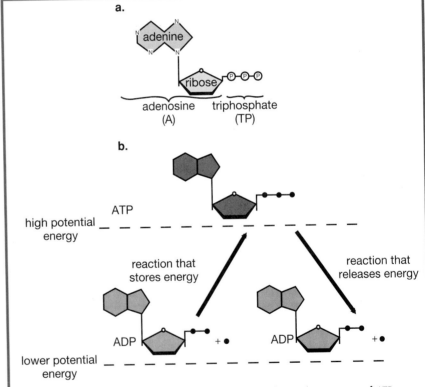

Figure E8.9 Structure of ATP a. This diagram shows the structure of ATP (adenosine triphosphate), which is one of several energy carriers that are found in all living organisms. Each ATP molecule is made up of a main section (the A of ATP) to which are attached three identical groups of atoms called phosphates (the TP of ATP). b. Energy is stored in ATP until it is released by reactions that remove the third phosphate group, forming a molecule of ADP (adenosine diphosphate). ATP acts as an energy carrier by alternately storing and releasing energy.

Historical Connections between Matter and Energy

Two thousand years ago the Greeks had a thriving society that focused on the pursuit of knowledge and fitness. These people understood that there is a connection between the fitness of the mind and the fitness of the body. They started the Olympic Games to let men show their athletic skill in different events. (Women were not permitted to participate in the ancient Olympics, although the Greeks had a contest for women called the Heraea. The Heraea was held every four years and had fewer events than the men's games.) One of the principal events of the games was the marathon, a foot race that commemorated the feat of a Greek messenger who ran from the city of Marathon to the city of Athens, a distance of 26 miles 385 yards.

While some Greeks were showing what their bodies could do, some were extending the limits of their minds by developing explanations for the natural world. A key assumption of their explanations was that the world was composed of four primary elements: fire, water, earth, and air. They described the composition of everything in the world by some combination of these four elements. These ideas were held to be true until the beginning of the eighteenth century.

By the mid-1700s, scientists were using new explanations to describe the natural world, and one of the ideas they focused on was the composition of fire. What was it made of? Unlike the Greeks, who assumed that fire was a primary element, eighteenth-century scientists experimented with fire and relied on observations to guide their explanations. They noted that when something like coal was burned, it gave off an oily substance. They named this substance *phlogiston*, meaning "fatty earth." The better something burned, the more phlogiston they assumed it contained; coal apparently contained much phlogiston.

Joseph Priestley used the idea of phlogiston to explain the results of one of his experiments: a mouse in a glass container with a burning candle dies quickly because phlogiston is poisonous. The more the candle burns, the more phlogiston is produced.

A few years later, in 1772, a Frenchman named Antoine Lavosier generated an alternative explanation for the death of the mouse in Priestley's experiment. Lavosier reasoned that the mouse died not because

Figure E8.10 Discus thrower (*Diskobolos*) This marble statue is a Roman copy of a bronze made by the Greek sculptor Myron. The original statue was life-size and sculpted about 450 B.C.

phlogiston was *added* to the air but because the fire *removed* something from the air. On the basis of additional experiments, he concluded that to burn, the candle flame required an element in the air. Lavosier called this element *oxygen*. When all of the oxygen was consumed, the mouse would die. Thus, not only does a flame require oxygen to burn, but mice require oxygen to live.

Lavosier's ideas form the foundation for today's understanding of fire and the release of energy when matter is burned. As later scientists examined these processes on a cellular level, they realized that energy is released systematically and, for most organisms, oxygen is required for the systematic release of energy from food.

Controlling the Release of Energy from Matter

In the grain storage explosion, energy was released from matter suddenly and dramatically. When you eat bread or cookies, however, you do not explode. Clearly, the body must release the energy stored in grain in a way that is more controlled. When you eat flour made from grain, your body's cells release the energy stored in this form of matter, but it takes place one small step at a time. Enzymes make this stepwise release of energy from the starch molecules possible by promoting chemical reactions that can proceed safely inside cells. The process by which enzymes convert the energy stored in macromolecules (such as starch and glycogen) or in the smaller molecules (such as glucose) that make up these macromolecules into readily available energy in the form of ATP is called **cellular respiration**.

The steps of cellular respiration can be divided into three main stages, as illustrated in Figure E8.11. The first stage, known as **glycolysis** (*glyco* = sugar, *lysis* = to split), occurs in nearly every living cell. During glycolysis, glucose in the cytoplasm of the cell is split into two smaller molecules, and a small amount of energy is converted to ATP. The small molecules from glycolysis still

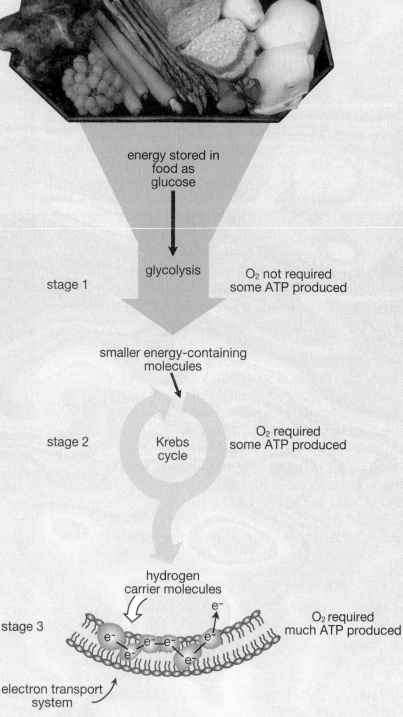

energy stored in food as glucose

stage 1 — glycolysis — O_2 not required some ATP produced

smaller energy-containing molecules

stage 2 — Krebs cycle — O_2 required some ATP produced

hydrogen carrier molecules

stage 3 — electron transport system — O_2 required much ATP produced

Figure E8.11 The three stages of cellular respiration Glycolysis, which occurs in the cytoplasm of the cell, breaks glucose into smaller molecules. These molecules are transported into the mitochondrion and further broken down to carbon dioxide in the Krebs cycle. In the electron transport system, the hydrogen atoms released from glycolysis and the Krebs cycle are used to form many molecules of ATP.

contain much stored energy. In the second stage of cellular respiration, known as the **Krebs cycle** or the tricarboxylic acid (or TCA) cycle, these molecules are broken down into carbon dioxide in the mitochondrion. This further breakdown releases additional energy, which is converted to several more ATP molecules.

In addition to the small amount of energy transferred from glucose to ATP in these first two stages, energy is transferred in the form of hydrogen atoms. These hydrogen atoms are important because they are responsible for producing the greatest amounts of ATP in cellular respiration. Special hydrogen carrier molecules make this possible by transporting these hydrogen atoms to the third stage of respiration, which involves the **electron transport system**. This stage also occurs in the mitochodrion. In this system the hydrogen atoms first are separated into electrons (e^-) and protons (H^+), and the electrons are transported along a series of transfer molecules until finally, after many small steps, they combine with oxygen and protons to form water.

As the electrons move through this series of transfers, a complex osmotic process transfers the relatively large amount of energy originally present in the starting glucose molecule into many smaller ATP molecules. This end product of cellular respiration is essential because all living systems can use ATP *directly* for cellular work such as movement.

Cellular Respiration: Converting Food Energy into Cell Energy

One of the unifying principles of biology is that all living systems require a source of energy for survival, but for what specifically is this energy needed? And does the potential energy in the molecules of food satisfy these energy needs? The specific energy needs of organisms include mechanical work, such as the contraction of muscles; active transport; and the building of new molecules, cells, and higher structures, all of which are necessary for tissue repair and growth. Indirectly, the potential energy in molecules of food satisfies the energy needs of organisms, but the structure of these molecules does not store energy in a form that can be released and used *directly* for cellular work. Thus, some method of converting the energy of food into more usable energy seems essential.

Through the selective pressures of evolution, such metabolic processes exist. Not surprisingly the simplest of these metabolic processes is remarkably similar in all living organisms, whether prokaryotic or eukaryotic. Eukaryotic organisms as different as yeast cells and humans use nearly identical energy conversion steps, a fact which suggests that these biochemical pathways evolved a very long time ago, before simple organisms became more complex. To see exactly *how* organisms convert food energy into cell energy, let us follow the path of a glucose molecule to see how and where its potential energy is harvested by the cell. The process begins in the cytoplasm when glucose first enters glycolysis, a process that does not require oxygen (see Figure E8.12). A glucose molecule has six carbon atoms. It is quite stable; that is, the bonds holding its atoms together are not easily broken. Because of this stability, the cell must use a small amount of energy to begin the glucose-splitting reactions (Step a). This is similar to lighting a match to start a fire.

In a series of steps that require enzymes, the glucose then is broken down into two molecules that have three carbon atoms each (Step b), and the atoms in these molecules are rearranged to form two molecules of pyruvate (Step c). In the final steps a small amount of ATP is produced, but most of the energy of the original glucose remains in the two pyruvate molecules.

In addition to the ATP, two molecules of NADH (**n**icotinamide **a**denine **d**inucleotide with a **h**ydrogen) are produced. NADH is an energy carrier, but it also carries hydrogens, which are composed of an electron and a proton. The role of NADH as a hydrogen carrier is especially important because the electrons and protons from the hydrogens are used in the electron transport system to produce large amounts of ATP.

In humans and most complex organisms, the pyruvate molecules produced by glycolysis now enter

glucose

Figure E8.12 A molecular view of glycolysis In glycolysis glucose is broken down into two molecules of pyruvate in many enzyme-catalyzed steps. In this process the energy from some ATP is needed to begin the glycolysis reactions (see (a) where ATP gives up some of its energy and becomes ADP). ATP also is produced in later steps, as is NADH, a hydrogen carrier molecule.

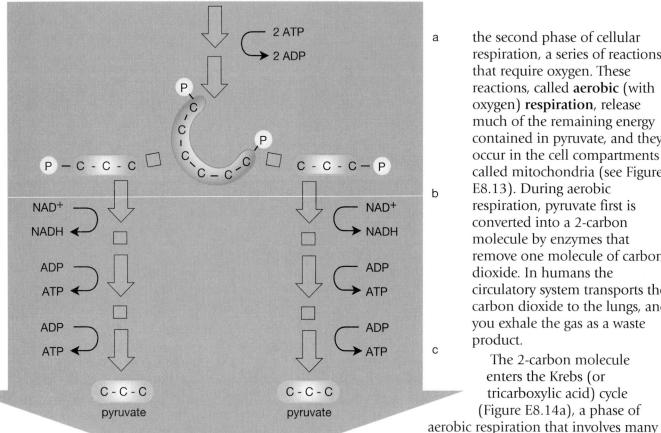

a

the second phase of cellular respiration, a series of reactions that require oxygen. These reactions, called **aerobic (with oxygen) respiration**, release much of the remaining energy contained in pyruvate, and they occur in the cell compartments called mitochondria (see Figure E8.13). During aerobic respiration, pyruvate first is converted into a 2-carbon molecule by enzymes that remove one molecule of carbon dioxide. In humans the circulatory system transports the carbon dioxide to the lungs, and you exhale the gas as a waste product.

b

c

The 2-carbon molecule enters the Krebs (or tricarboxylic acid) cycle (Figure E8.14a), a phase of aerobic respiration that involves many enzymes and molecular rearrangements. This cycle completes the release of energy from pyruvate by breaking it down to carbon dioxide (Step b). The net result of the cycle is that the potential energy trapped in the structure of pyruvate is used to

Figure E8.13 A mitochondrion The reactions of aerobic respiration take place in this cellular organelle. Mitochondria have two membranes, an inner and an outer. The highly folded inner membrane forms the inner compartment, and the space between the two membranes forms the outer compartment of a mitochondrion. The enzymes involved in the oxygen-requiring steps of cellular respiration are located inside the mitochondrion.

outer membrane
inner membrane
outer compartment
inner compartment

UNIT THREE ENERGY, MATTER, AND ORGANIZATION: Relationships in Living Systems

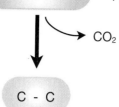

C - C - C   pyruvate

CO₂

C - C

a.

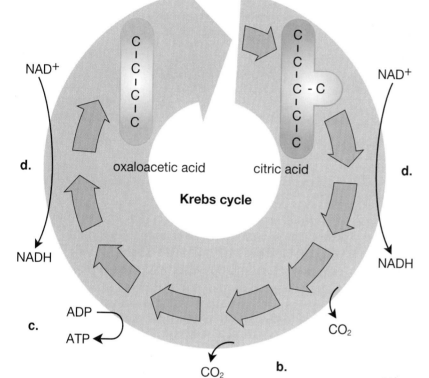

Figure E8.14 The Krebs cycle The starting molecule for aerobic respiration is a 2-carbon molecule that is derived from the pyruvate that was generated by glycolysis. The 2-carbon molecule combines with a 4-carbon molecule to produce a 6-carbon compound. Then many enzyme-catalyzed reactions occur that release two molecules of carbon dioxide (b) and produce some ATP (c). In addition, several NADH molecules are produced (d).

NAD⁺

NAD⁺

d.

d.

NADH

NADH

oxaloacetic acid   citric acid

Krebs cycle

ADP

c.

ATP

CO₂

CO₂

b.

Figure E8.15 Electron transport
The final step in aerobic respiration is the transfer of electrons from NADH to a chain of electron carriers embedded in the inner membrane of the mitochondrion. As the electrons pass from one electron carrier to the next, energy is released and protons are pumped into the outer compartment. The resulting proton concentration gradient drives the production of ATP as protons flow back into the inner compartment of the mitochondrion, through an ATP-producing enzyme complex. At the end of the electron transport chain, the electrons join with oxygen and protons and form water.

produce a little more ATP (Step c) and a lot of hydrogen carrier molecules, including NADH (Step d).

Figure E8.15 outlines the third major stage in cellular respiration, the electron transport system. This stage, which also is aerobic, is the point at which the energy stored in hydrogen carrier molecules such as NADH is used to produce large amounts of ATP. Each electron transport system consists of a series of electron carrier molecules that are embedded in the inner membrane of a mitochondrion.

During this third stage of respiration, the hydrogen atoms carried by NADH are separated into electrons and protons; the electrons

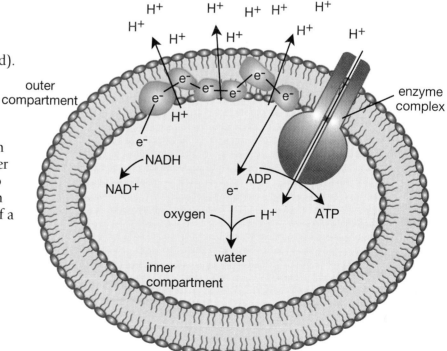

are transported along the chain of electron-carrying proteins. As the electrons move from one carrier to the next, they release energy. Some of the energy is used to pump protons from the inner compartment to the outer compartment of the mitochondrion. This causes protons to accumulate in the outer compartment until they are at a higher concentration than the concentration inside. The difference in the concentration inside and outside of the compartments, a concentration gradient, is a source of potential energy because diffusion exerts a force to equalize the proton concentration on both sides of the membrane. As the protons diffuse toward the area of lower concentration, they pass through an enzyme complex in the membrane. Like a waterwheel capturing the potential energy of a flowing stream to grind wheat, this enzyme complex uses the potential energy from proton diffusion to make ATP from ADP and phosphate. In this way much of the energy from the electrons of NADH, energy that originated with glucose, is transferred to ATP, and the ATP can serve as the energetic *push* that starts chemical

Figure E8.16 How is perspiration related to electron transport?

reactions in the cell. Ultimately, the transferred electrons end up combining with protons and molecular oxygen, which results in the formation of water. The requirement for oxygen at this final transfer explains why humans and other animals as well as plants, fungi, protists, and many prokaryotes require oxygen for their survival. As it turns out, the presence of oxygen at this stage is tightly linked to the regulation of ATP levels.

Regulation and Energy Production

Cellular respiration, like nearly all important processes in living systems, is carefully regulated to maintain the homeostatic balance of the organism. In this case the condition that the organism must sense and respond to is the supply of available energy (see Figure E8.17). In general, living organisms need a supply of ATP at all times, even when an organism is relatively inactive. For instance, individual cells must maintain an osmotic balance with their surroundings

by continually transporting sodium (Na^+) and potassium (K^+) ions across their selectively permeable cell membranes. This process requires energy. At certain times, however, an organism's energy needs rise beyond the levels of ATP needed for routine cellular activity. When this happens, the cells detect increased levels of ADP, a signal that ATP is being used up rapidly. The mitochondria respond by increasing the rate of electron transport and proton pumping and by producing more ATP from ADP.

The level of ADP is not the only signal that influences energy production in cells. Notice in the essay *Controlling the Release of Energy from Matter* on page E116 that only the very last step in cellular respiration involves oxygen directly, yet the Krebs cycle and electron transport system cannot proceed unless oxygen is present. Glycolysis, on the other hand, is an anaerobic process: it does not require oxygen. The presence or absence of oxygen in the cell, however, affects energy production because it dictates the fate of the pyruvate

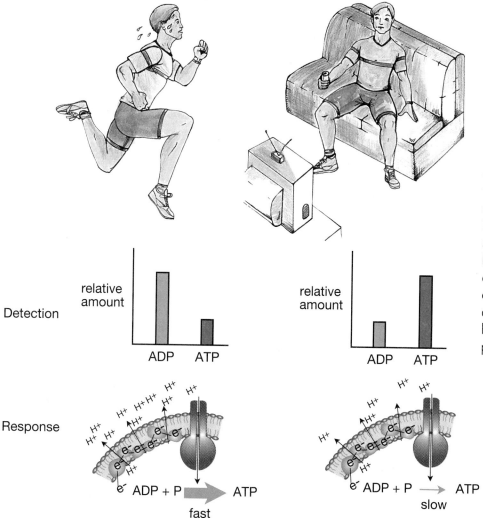

Figure E8.17 Regulation of energy Cells can detect and adjust the supply of available energy. When energy needs are low, there are low levels of ADP, and ATP is made at a slow rate. When energy needs are high, there are high levels of ADP, and the electron transport phase of cellular respiration speeds up. How does this system help maintain homeostasis? Is this an example of positive or negative feedback?

formed in glycolysis. Figure E8.18 shows these possible fates.

Under aerobic conditions, enzymes convert pyruvate into a 2-carbon molecule that enters the Krebs cycle and the other stages of cellular respiration. Under anaerobic conditions, pyruvate can be converted into lactic acid or alcohol in a process called **fermentation**. For example, animal cells can use NADH to convert pyruvate into lactic acid.

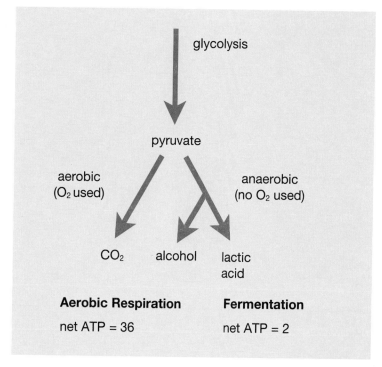

Figure E8.18 Aerobic and anaerobic energy release When oxygen is present, pyruvate enters the Krebs cycle and is converted into carbon dioxide. Hydrogen carriers transport electrons to the electron transport system, which produces much ATP. In the absence of oxygen, pyruvate can be converted into alcohol or lactic acid. What are the advantages of each process?

In this way the cell can continue to produce a small amount of ATP until oxygen becomes available again. During vigorous exercise, extremely active muscles do not receive enough oxygen for cellular respiration, and the cells must rely on anaerobic energy production. Certain bacteria, such as those responsible for souring milk or making yogurt, also produce lactic acid. Yeast carry out a type of fermentation that results in the production of alcohol. Fermentation is an inefficient energy-production process; it releases only a small portion of the energy available in a molecule of glucose. Despite this inefficiency fermentation was the only method of energy release until about 2.5 billion years ago, when the first oxygen-releasing photosynthetic organisms evolved and changed the earth's atmosphere forever.

Just how inefficient is anaerobic energy production? The amount of ATP produced during aerobic respiration in the mitochondria is about 18 times greater than that made in anaerobic glycolysis. This difference in the efficiency of energy release from glucose is significant because it made possible the evolution of complex organisms. Remember that entropy means all things move toward disorder *in the absence of energy*, but it is possible with an input of energy to increase order and the organization of complex living systems. The greater efficiency of aerobic respiration also explains why aerobic conditioning gives humans an advantage in performance. During aerobic conditioning, muscle tissues produce more mitochondria, providing the potential for greater energy release. In addition, circulation to the muscles improves, providing a greater supply of oxygen. As a result of these two changes, aerobic respiration can convert a greater proportion of the energy in glucose into usable cellular energy, making an exercising muscle work more efficiently.

Whose Discovery Is This?

Two students working at the same table in the library keep getting told to be quiet due to their frequent talking.

"What's your report on?" asks Inez.

"The guy who discovered photosynthesis," Fernando replies. "Well, actually the guy who first figured out that plants use light and carbon dioxide from the air to make their own food. Lots of guys have worked out the details."

"Not *just* guys," Inez points out. "Women scientists, too."

"Yeah, but this stuff was in the eighteenth century. It was all guys then," he laughs. "What's yours about?"

"Same thing."

"So you're looking up stuff on this Dutch guy, Ingen-Housz?" Fernando asks.

"Ink and who? No, a Swiss guy, Senebier. He was the first to figure out photosynthesis," Inez corrected him.

"No, he wasn't. It says right here that Ingen-Housz was a medical doctor and a friend of Benjamin Franklin. In 1779 he published a book describing how green plants use the visible part of sunlight to restore air. That means put oxygen back into it, so an animal could breath it and live. So he discovered photosynthesis."

"But my reference book says that Senebier was a minister who did scientific experiments. He published a paper describing the exchange of gases by plants. He said they take up carbon dioxide and give off oxygen while making substances containing carbon," Inez exclaims. "Yours must be wrong."

"Let's see that." Fernando grabs the thick book and reads the passage that Inez had been studying.

"Hmm. You're right. It does say Senebier did it. But my book says Ingen-Housz. How are we supposed to know who was the discoverer?"

The librarian leaves her desk and walks toward them. "What *is* the disturbance here?" she inquires sternly.

"We're trying to figure out who discovered photosynthesis," Fernando explains.

"Oh, everyone knows that was the English minister and scientist Joseph Priestley, who was also the first to describe oxygen in air," the librarian replies. She is surprised when both students start laughing.

Ingen-Housz, Senebier, and Priestley all experimented with photosynthesis during the latter part of the eighteenth century. Priestley and Senebier are often given credit for the discovery of photosynthesis because of their debates on the topic, but Ingen-Housz's publication of 1779 predates the others' work.

Getting Energy and Matter into Biological Systems

There is a saying that you must have money to make money. A similar loop exists in biological processes: you need energy to get energy. For instance, cellular respiration releases usable energy from storage molecules, but a certain amount of energy is needed to start this process. Likewise, food supplies the energy and matter requirements of many organisms, but the food itself usually is derived from other living systems. How, then, do energy and matter get into living systems in the first place? The main way is through photosynthesis, the series of reactions by which plants, algae, and some bacteria use the sun's energy, in the form of light, to synthesize macromolecules from smaller, simpler molecules.

You might think of a plant as a solar-powered factory that converts the radiant energy of sunlight (solar energy) into potential energy stored in chemical form. Solar energy varies in its strength, from the warming rays of infrared to the damaging rays of ultraviolet. Visible light, which is only a small fraction of the total energy coming from the sun, consists of a spectrum of colors, each with a different wavelength and energy content, as shown in Figure E8.19. Photosynthesis uses only certain wavelengths, or colors, of visible light. This fact is

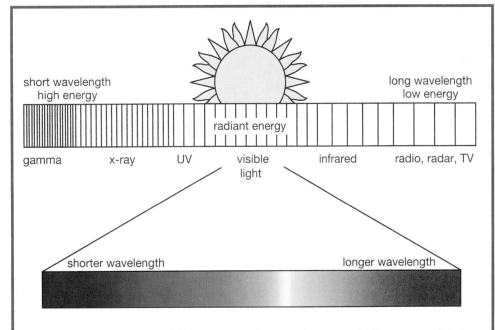

Figure E8.19 Spectrum of light energy The sun is the source of different types of radiant energy, including damaging ultraviolet (UV) light, light that can be detected by the human eye (visible light), and warming infrared light. Photosynthesis uses only a small portion of this spectrum of light energy.

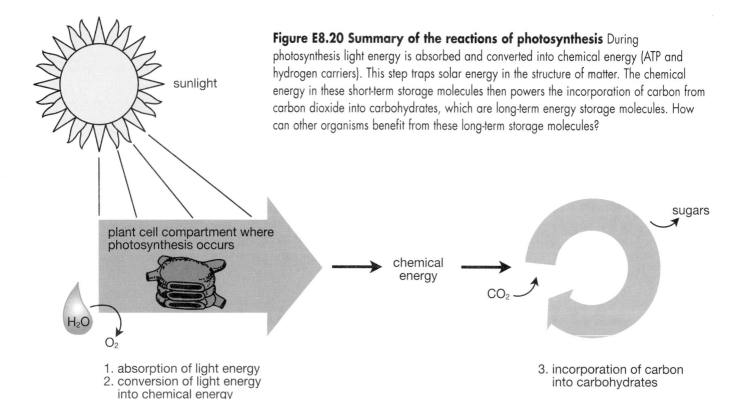

Figure E8.20 Summary of the reactions of photosynthesis During photosynthesis light energy is absorbed and converted into chemical energy (ATP and hydrogen carriers). This step traps solar energy in the structure of matter. The chemical energy in these short-term storage molecules then powers the incorporation of carbon from carbon dioxide into carbohydrates, which are long-term energy storage molecules. How can other organisms benefit from these long-term storage molecules?

sunlight

plant cell compartment where photosynthesis occurs

H_2O

O_2

chemical energy

CO_2

sugars

1. absorption of light energy
2. conversion of light energy into chemical energy

3. incorporation of carbon into carbohydrates

emphasized by the green color of plants, most of which appear green because their pigments reflect green light, rather than absorb it. Not only does photosynthesis depend on particular wavelengths of light, it also works more or less efficiently depending on the intensity of this usable visible light. The ideal intensity of light varies for different plants. Of course, many factors that have nothing to do with light also affect photosynthetic efficiency, factors such as availability of water and nutrients in the soil.

Three major events occur in plant cells during photosynthesis: (1) absorption of light energy, (2) conversion of light energy into chemical energy, and (3) storage of potential energy in carbohydrates. These three events occur in two distinct but interdependent sets of reactions, which are summarized in Figure

E8.20. In the first two steps of photosynthesis, light energy is absorbed and converted into chemical energy in the form of ATP and hydrogen carrier molecules. These steps bring the energy of the sun into living systems by trapping it in matter. These energy-carrying molecules then supply the energy necessary to drive the third step of photosynthesis, which combines carbon from carbon dioxide in the atmosphere with other atoms to form carbohydrates.

Photosynthetic organisms use these carbohydrates for long-term energy storage (much as humans use glycogen), and other living systems use them indirectly when they eat the photosynthetic organisms.

Bringing Solar Energy into Living Systems. The reactions of photosynthesis take place in **chloroplasts**, small compartments inside certain plant cells. Figure E8.21 shows the location of the

chloroplast-containing cells in a plant leaf, and the enlargement illustrates the organization of a chloroplast. Within the chloroplast is a highly ordered system of membranes, the thylakoids, which are folded so that they form small compartments somewhat like the infoldings in a mitochondrion. Surrounding the thylakoids is a colorless fluid known as the stroma. The internal structure of the chloroplast is very important in the process of converting light into chemical energy.

Embedded in the thylakoid membranes are organized arrangements of pigment molecules, including the green pigments called **chlorophylls**, which give plants their color. Chlorophyll absorbs light energy in the visible wavelengths, and this absorption is how potential energy is brought into living systems. The absorbed energy sets up a flow of electrons, as shown

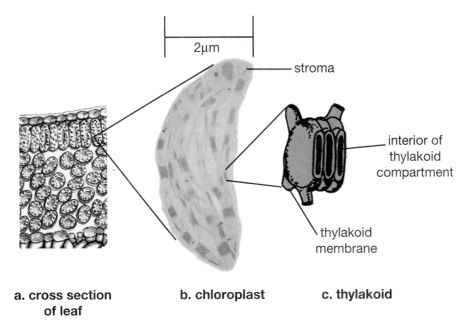

a. cross section of leaf   b. chloroplast   c. thylakoid

Figure E8.21 The conversion of light energy into chemical energy occurs in subcellular compartments known as chloroplasts. a. The green disks in the cells in this diagram of a leaf cross section are chloroplasts. b. This electron micrograph of a chloroplast shows the layers and stacks of thylakoid membranes. c. The thylakoid membranes contain chlorophyll and other pigments and form subcompartments within the chloroplast in which the light-trapping reactions occur. What other subcellular compartments are in a plant cell?

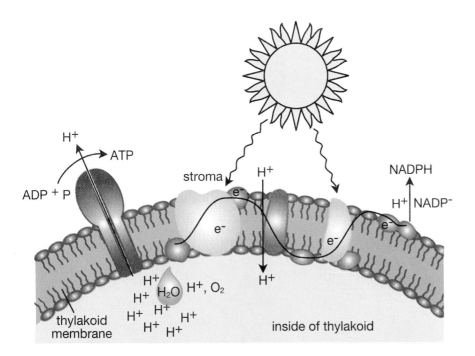

Figure E8.22 ATP production in the chloroplast during photosynthesis
Absorption of light energy sets up a flow of electrons from water through pigments and other molecules in the thylakoid membranes. Protons accumulate on the inside of the thylakoids. The resulting proton gradient functions much like that in a mitochondrion. The gradient supplies potential energy that enables a membrane-spanning enzyme to synthesize ATP from ADP and phosphate.

in Figure E8.22, which begins when certain chlorophyll molecules lose energy-rich electrons. These electrons move through an electron transport system in the thylakoid membranes and, as this happens, some of the energy originally captured from the sun is used to power the transport of protons across the thylakoid membranes. The protons accumulate inside the thylakoids and form a concentration gradient. As in cellular respiration, the protons then diffuse down their concentration gradient through an enzyme complex in the membranes, and ATP is synthesized.

In addition to producing ATP, the flow of electrons set up by the initial absorption of light energy gives rise to NADPH, an energy and hydrogen carrier similar to the NADH formed during cellular respiration. The electrons lost by chlorophyll that start the electron flow ultimately are replaced by electrons removed from water. As a result water is separated into oxygen, electrons, and protons. The oxygen is released by the plant as oxygen gas, a waste product of photosynthesis. The waste product is the source of the oxygen on which animals, plants, and all aerobic organisms rely for respiration.

The first two events of photosynthesis (the absorption of light energy and its conversion into chemical energy) form three products: oxygen gas, ATP, and NADPH. The products ATP and NADPH now are available to provide energy for the third event in photosynthesis, the reactions that form carbohydrates. This important process involves converting the energy trapped in short-term storage molecules into a form of long-term energy storage.

Bringing Carbon into Living Systems. The final reactions of photosynthesis are important because they use the chemical energy trapped in ATP and NADPH to bring matter into living systems and organize it in a way that stores energy for long periods of time. These reactions are called **carbon fixation reactions** because they incorporate, or *fix*, carbon into carbohydrates: they bring atmospheric carbon dioxide into the cell and combine it with other molecules to form carbohydrates. Although carbon fixation reactions do not directly involve the absorption of light energy, they do require the products of that absorption, ATP and NADPH. Figure E8.24 is a simplified diagram of these carbohydrate-producing reactions, which take place in the stroma of the chloroplast.

Carbon dioxide from the air enters the plant through the stomates (look back at Figure E5.2 on page E61). In the carbohydrate-producing reactions, the carbon dioxide is added to an existing 5-carbon sugar, creating a 6-carbon sugar. This increases the total number of carbon atoms in

Figure E8.23 This *Anacharis* plant is actively engaged in photosynthesis. What gas is present in the bubbles?

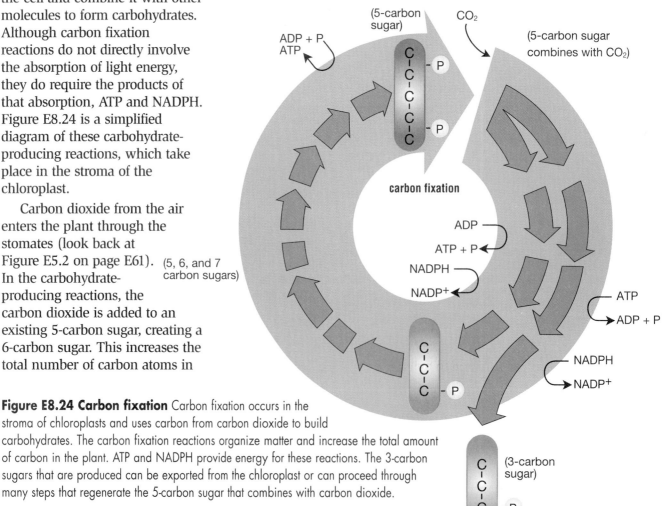

Figure E8.24 Carbon fixation Carbon fixation occurs in the stroma of chloroplasts and uses carbon from carbon dioxide to build carbohydrates. The carbon fixation reactions organize matter and increase the total amount of carbon in the plant. ATP and NADPH provide energy for these reactions. The 3-carbon sugars that are produced can be exported from the chloroplast or can proceed through many steps that regenerate the 5-carbon sugar that combines with carbon dioxide.

the plant and is the critical point at which matter is brought into living systems. ATP and NADPH provide the energy for these reactions. The 6-carbon sugar quickly splits into two 3-carbon sugars, which have several possible fates.

In a cycle of reactions, these 3-carbon sugars are rearranged into a variety of other sugars, including the 5-carbon sugar that first combines with carbon dioxide. Some of the 3-carbon sugars are exported from the chloroplast and used to form other carbohydrates such as sucrose and starch as well as lipids, amino acids, proteins, chlorophyll, and other molecules needed by the plant cell. Sucrose can be transported to nonphotosynthetic tissues of the plant such as the roots, where it is a source of matter (carbon) and energy for the cells in those tissues.

Metabolism Includes Synthesis and Breakdown

Why is it that growth seems to be characterized by inefficiency and wasted effort? Think about the latest highway expansion project. Lanes are closed, the old surface is torn up and hauled away, and only after much time and inconvenience is new asphalt laid to create a wider, more efficient roadway. Likewise, to build a large office building in a densely packed city, older, smaller buildings first must be razed and the debris cleared away. In both cases the raw materials from the outdated structures are broken down, and often recycled, before new materials can be used to make newer, larger, or more useful structures.

Living systems function in much the same way except they continuously break down and build up *molecules*, and the reactions responsible for both processes are linked by energy. For example, when you eat potatoes, your body breaks down the potato starch to the small glucose molecules of which it is made. The glucose then can be broken down further during cellular respiration to provide immediate energy in the form of ATP, or it can be transported to your liver or muscles and combined with other glucose molecules to form glycogen, which is a large, energy-storage molecule. Your body uses some of the energy that is released during cellular respiration to build, or synthesize, the glycogen.

The sum of all the chemical activities and changes that take place in a cell or an organism is known as its **metabolism**. Generally, breakdown reactions (such as cellular respiration) release energy, and synthesis reactions (such as photosynthesis) require energy. The ATP that is produced by breakdown reactions then becomes the source of energy for many cellular activities, including processes like muscle contraction.

Thus, ATP, as well as other energy carriers, provides a critical *link* between reactions that produce energy and those that require it. Without such links the energy released from breakdown reactions would be wasted, and no energy would be available for biosynthetic reactions, which are necessary for growth and repair.

The breakdown processes in cells produce a variety of smaller molecules that can be converted into the intermediate compounds that are formed in glycolysis and the Krebs cycle. This makes metabolism more efficient because these intermediate molecules can be used in cellular respiration just as glucose is used. For example, fats can be broken down into glycerol, a 3-carbon molecule, and fatty acids, long chains of carbon and hydrogen. Glycerol can be converted into one of the 3-carbon intermediates of glycolysis and can enter the energy-releasing process of cellular respiration. Fatty acids can be converted into the same 2-carbon molecule that enters the Krebs cycle. Proteins can be broken down to amino acids, and after the nitrogen-containing group has been removed, the remaining carbon skeleton can be broken down to intermediates that also can enter glycolysis and the Krebs cycle, as shown in Figure E8.25.

These intermediates can be used in synthesis reactions as well (see Figure E8.25). Organisms require many different types of macromolecules for their structures and activities. For example, each of the many chemical reactions that takes place in an organism requires a specific enzyme, and the organism must synthesize these enzymes. Cell membranes require specific lipids, and these too must be synthesized. Whether protein, carbohydrate, or fat, each molecule carries out a function that is made

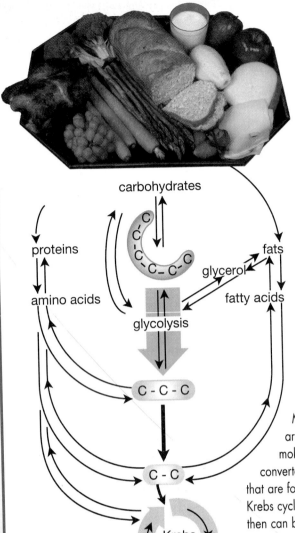

carbohydrates

proteins

amino acids

glycerol

fats

fatty acids

glycolysis

C - C - C

C - C

Krebs cycle

**Figure E8.25
Metabolic pathways**

Macromolecules in food are broken down to smaller molecules that can be converted into the intermediates that are formed in glycolysis and the Krebs cycle. These intermediates then can be used in cellular respiration to produce ATP. The same intermediates also are a source of carbon skeletons for the synthesis of macromolecules.

possible by its molecular organization.

Although the tissues of organisms in most species are composed of only a few elements and the same types of macromolecules are found in a diverse range of living systems, the particular molecular arrangements in a given organism are different from all others. That is why the protein that you consume in your diet, which has been made by a plant or animal of a different species, cannot be used directly as a prefabricated protein in your body. Instead, the dietary protein is broken down to its component amino acids, and your cells assemble the amino acids into the specific protein patterns that your body requires.

The synthesis of new proteins is the body's most efficient use of amino acids. In situations such as starvation or the extreme stages of anorexia nervosa, however, the body's cells can compensate for the lack of carbohydrates and fats in the diet by breaking down protein in an effort to maintain homeostasis. Unfortunately, when this occurs the muscles of the body are consumed as fuel. Although the breakdown of protein is a last resort, it indicates how flexible cells can be in their metabolism.

Garbage among Us—from Then until Now!

Matter—lots of it. Wood. Paper. Glass. Metals. Plastics. Rubber. Cloth. Food. Yard waste. All of these types of matter end up in our landfills. In fact, did you know the following?

- In the United States each person throws away an average of 1400 pounds of garbage each year.
- A convoy of garbage trucks long enough to encircle the

earth's circumference six times would be required to carry all the municipal waste generated in the United States in one year.

- As of 1990 more than 50 percent of the states in the

Figure E9.1 Overloaded with waste and no place to go.

United States were having problems finding a place to dump trash; sometimes trash travels great distances before it is dumped (Figure E9.1).

- Lettuce buried in a landfill may take more than seven years to decompose completely.

- A hot dog can last more than 10 years in a landfill.

- A steak buried in a landfill can retain its fat 15 years after burial.

- In spite of recycling efforts, waste paper is still a major component of landfills nationwide, making up 40–50 percent of the waste in landfills.

You throw away things every day, but did you ever wonder what people could tell about you if they sorted through your trash? Garbage and waste can tell us quite a bit about how organisms acquire and use the matter and energy in their community. Let's consider a few communities of

different types and reflect on the cycling of matter through them. How is this process similar and different in these communities? How does matter move from one organism to another? How much waste and what type of waste does each community produce? What happens to the waste in each community?

Bats in Mammoth Cave National Park, Kentucky. Two bats flit about in the sky at twilight, barely visible in the last light of evening. They dart about catching insects; one bat descends briefly, flying just above the surface of a pond to take a drink, while the other nips a moth in midair. Soon, the bats dart into the entrance of a cave that is partially hidden by evergreens (see Figure E9.2). Several fleas, ticks, and mites have latched onto the bats' coats during the evening. As the bats sit on a ledge in the cave, resting and cleaning themselves, some of the insects fall to the cave floor,

where they become food for the organisms that dwell there. During the night one of the bats excretes its wastes—a substance called guano. Guano is not solid like human feces, but rather thick and mudlike; it covers the floor of the cave like carpeting. Bacteria, fungi, and other one-celled organisms grow on the guano and use it as their food—their source of energy. In turn many of these small organisms are destined to become the food for insects, and some of these insects may become food for the bats.

The Earliest Anasazi Indians. The year is 950 A.D. As a hint of daylight appears in the east, a young Anasazi woman awakens and checks on her two young children still asleep beside her in their cliff dwelling (Figure E9.3). They live along a basin in the high-plateau region of the American Southwest. As the woman rises and starts a fire, her mate begins to stir. She retrieves

Figure E9.2 A gathering of little brown bats Little brown bats (*Myotis lucifugus*) such as these live in caves throughout Mammoth Cave National Park, Kentucky.

Figure E9.3 Cliff dwellings The Anasazi lived in cliff dwellings such as this one photographed by Rich Buzzelli. These structures can be found throughout the Four Corners region of the American Southwest.

some kernels of corn from a slab-lined hole and begins to grind them into a coarse meal. It is late spring, and the couple hopes to finish planting the corn today. They want to work in the morning before the day becomes hot. To plant, they use a stick to make holes in the ground, drop kernels of corn into each hole, and then cover them with soil.

Later the woman and her children join other women and children to gather yucca from the plateau. The Anasazi families will not only eat the fruit and seeds of the yucca, but they will use the roots for soap and shampoo, and the strong, sturdy fibers from the leaves for making intricate baskets, sandals, aprons, mats, and cradle boards. Small pieces of the yucca and other plant material are discarded in a pile along with corn husks and cobs, worn-out mats, and broken tools made out of bone.

In memory of Rich Buzzelli

Some mornings the woman's mate joins other men in the community to hunt for rabbit and deer. In addition to preparing and eating the meat, the Anasazi make clothing for the winter months from the pelts of rabbits and tools and utensils such as needles from the bones. Bones that are not used for tools or utensils are discarded in the small but growing pile of waste. At the bottom of this pile, the organic material has decayed

a.

b.

c.

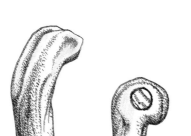

d.

e.

f.

Figure E9.4 Artifacts from Pueblo Bonito Archaelogists found these artifacts and more at the Pueblo Bonito site on the Chaco River Valley. They are dated at about 950–1000 A.D. a. black on white bowl, 2¾ in. high b. turquoise pendants, 1 in. high c. shell ornament, 2 in. long d. bone scrapers, 6½ and 5 in. long e. bone awls, 4½ to 7¾ in. long f. bone needle, 2¾ in. long g. projectile points, 1½ in. and 2 in. long h. basket, 15½ in. high.

g.

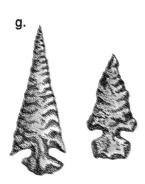

h.

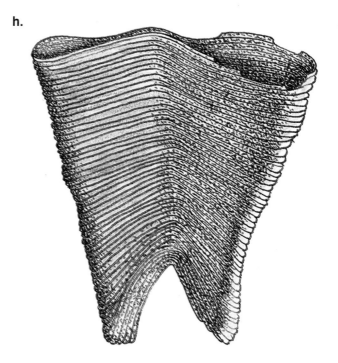

enough so that it is almost indistinguishable from the soil beneath it.

Family in Maineville, Ohio. A single mother looks in on her 8-year-old daughter Sonya, who has the flu. She has a big box of tissues beside her bed and a pile of used ones in the wastebasket nearby. Sonya also has finished one carton of juice and has begun another. One of the family's three cats is resting at the foot of the bed. (Someone needs to clean the litter box today.)

This family lives in a three-bedroom, two-bathroom house at the end of a quiet street in a small community north of a major city. Sonya has a 15-year-old brother, Matt, who is keeping his distance because he doesn't want to get sick. He has a huge history report due Monday; he works away at the computer most of the afternoon and prints the entire report four times before he is satisfied. He places the rejected copies in the recycle bin in the kitchen.

The mother is an architect and has been working at home all day on a balsa-wood model of a hospital addition. (Scrap wood seems to be everywhere.) In the evening on the way home from the grocery store (with 10 bags of groceries), the mother stops at a fast-food restaurant and picks up some chicken dinners, which everyone enjoys—even Sonya who is beginning to feel better. As Matt tosses the last dinner carton into the garbage, he notices that it's full again—Whose week is it to take out the garbage, anyway?

Matter in Nature Is Going Around in Cycles . . . What Next?

Imagine a riverbed similar to the one where the Anasazi Indians lived 1000 years ago. It might look something like the illustration in Figure E9.5. In such a riverbed ecosystem, matter is on the move from one place to another or from one organism to another. Water runs through it; nutrients move from the soil into plants rooted along the river; beavers gnaw through branches and twigs and use them to build their dams; and a great blue heron nabs an unsuspecting fish from the middle of the river. When the types and amounts of matter in such an ecosystem remain essentially the same across time, the ecosystem remains in balance. When the types or amounts of matter in an ecosystem change, the ecosystem changes as well.

As we look more closely at the riverbed ecosystem, we realize that there are two types of material components—biotic and abiotic. **Biotic** matter is living matter, and **abiotic** matter is matter that is not living. Both types of matter cycle within an ecosystem.

Figure E9.5 A beaver busy at work What effect might this beaver (*Castor canadensis*) have on the movement of matter in its ecosystem?

Beavers are part of the biotic matter along this riverbed. In the spring two to four young beavers are born into each beaver family along the river. During the months while young beavers are growing and becoming more self-sufficient, the local beaver community can withstand the temporary increase in its population.

E132

Eventually, however, many of last year's young will move out of the community to establish their homes elsewhere along the river. A few of last year's young may remain, replacing the beavers that die, but the total population of beavers in this area will remain relatively stable. If the number of beavers in this area increased significantly, this increase might place a stress on the available resources—the food supply and the sites and materials for building dams, for example. If a situation such as this continued for several seasons without a similar increase in the resources that beavers need, the resources might be depleted to the point that the environment no longer would be able to support any beavers. They then would die off or move elsewhere. Such a scenario would cause the community to change.

Not only does matter move through an ecosystem, but some of it makes a complete cycle within it. For example, when beavers die along the river, their bodies gradually decay, and the nutrients derived from their bodies eventually mix with the soil. Plants then acquire some of these nutrients from the soil, and in turn various animals (including beavers) feed on these plants. In this way nutrients that once were part of an animal's body cycle through various types of matter and are taken up by another animal's body.

Water is an example of abiotic matter that moves through and cycles within an ecosystem. Water is constantly moving through this river; the water flowing through the river at a certain spot today is different from the water that flowed through this same spot yesterday. Water also is part of a cycle. Some is added to the river by rain or snow and from tributaries and ground water reserves in the plateau; some water leaves the river through evaporation (see the drawing of the water cycle in Figure E9.6).

Through time, when the net inflow is the same as the net outflow, the river remains essentially the same. When the inflow is either increased or decreased significantly, the river changes. A significant increase in the water flow would cause the banks to flood and many plants along the riverbed to be destroyed. A sustained increase in the amount of water, however, may support a different number and variety of plants and animals in the community and consequently change its makeup. Similarly, a decrease in the water flow may cause the disappearance of certain plants and animals as competition for water increased. Again, if such a change were sustained, the makeup of the community would change as well.

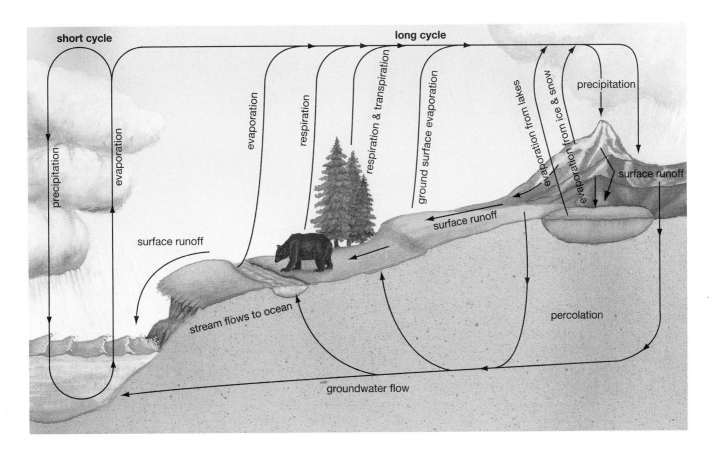

Figure E9.6 The water cycle The water cycle collects and redistributes the earth's water supply.

These same principles of movement and cycling of matter apply to the many other, less obvious components that make up the riverbed ecosystem. Consider, for example, a single atom of carbon. If you could follow a single atom of carbon in the riverbed community through time, you would see it cycle through many different molecules. At one point it may form part of a protein molecule in a floating leaf of duckweed. At another time the same carbon atom may become one of the atoms within a DNA molecule in the genetic material of a fish or a frog that ate the duckweed. At still another point it may remain for a long period within dead plant or animal material in the mud of the river until it is finally used by bacteria or fungi and rejoins the biotic community. Figure E9.7 illustrates some of these interactions and relationships in the carbon cycle. Again, when the total number of carbon atoms in an ecosystem remains approximately the same along with the proportion of carbon atoms to other atoms, the ecosystem remains essentially stable. When the number or proportion of carbon atoms changes significantly, the community changes.

Other less obvious cycles of matter that occur in ecosystems depend on the existing environmental conditions. Essential elements, such as calcium, potassium, and phosphorous, are available to the biological community only after they dissolve in the ground water and are taken up by the roots of plants. Through experience people have discovered that many desert lands will produce large crops, at least for a time, if they are irrigated. Often, essential elements are abundant in these soils, but they are not readily available to plants due to the shortage of water.

A contemporary issue that is important in the tropics involves a change that humans have introduced. Humans have been removing native communities of

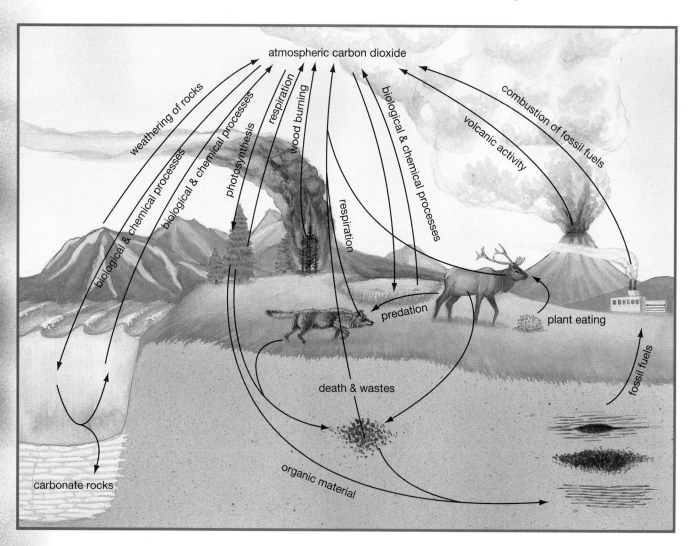

Figure E9.7 The carbon cycle In a stable ecosystem, the total number of carbon atoms will remain approximately the same.

UNIT THREE ENERGY, MATTER, AND ORGANIZATION: Relationships in Living Systems

plants and animals and growing crops in their place. In addition to the loss of the native plants and animals, a change such as this disrupts the cycling of elements and may result in an additional loss of soil fertility.

Another issue that is related to the cycling of matter is that various organisms in a community often cannot efficiently metabolize compounds such as pesticides that are foreign to the community but have been introduced for one reason or another. As a result, foreign compounds may accumulate at toxic levels in the tissues of some organisms. They also may build up in the environment and persist there for long periods of time before they are returned to the cycle.

Taken as a whole, when a community is stable within an ecosystem such as the riverbed, it exhibits a type of large-scale dynamic balance that resembles the homeostasis of individual organisms. The same is true in other communities such as those found in a desert, a temperate pond, a pine forest, or the arctic tundra. On a larger scale, the same is also true for the entire **biosphere**. The biosphere includes all the organisms as well as the soil, water, and air that surround and support them. As the environment changes, communities may change as well. Homeostasis continues around a balance point, but that point of balance can change through time.

Worms, Insects, Bacteria, and Fungi—Who Needs Them?

Did you know that 100 million bacteria can live in a single gram of fertile soil? Did you also know that 250,000 earthworms can live in a 1¼ acre field of rich topsoil?

This is one reason why earthworms are considered a farmer's best friend. Earthworms, organisms that decompose decaying organic matter, can work through 10 tons of topsoil a year, aerating it and increasing its fertility.

The quality of the topsoil is important because topsoil serves as a link between the living and nonliving world. Abiotic nutrients enter the living world when they are absorbed by plants and are returned to the nonliving world when they are excreted by animals as waste. The waste ends up in the topsoil, where it is broken down into simple nutrients by the soil's inhabitants. At this point the cycle can begin again as these nutrients are reused by plants.

Earthworms play a vital role in keeping the nutrient levels in the soil high because they consume partially decomposed organic matter such as dead leaves and roots. They then excrete nutrient-rich waste, which mixes with the soil and creates humus. Other organisms that perform a similar function in the soil are insects, bacteria, and fungi. Together these organisms help return vital elements, such as phosphorous, calcium, and nitrogen, to the environment where other organisms can use them.

The importance of recyclable elements to the health of most communities emphasizes the role of organisms that decompose organic matter. This is particularly apparent in the tropics because the layer of topsoil is thin, the temperature is high throughout the year, and rains are frequent. Under such conditions nutrients break down rapidly and those that are not returned quickly

to the living part of the ecosystem flow away in ground water and are lost to the community. Figure E9.8 illustrates this situation for a tropical rain forest. The community appears extremely lush because most of the nutrients are held in living material almost continuously, and matter recycles rapidly from one organism to another. The critical location for this rapid recycling is the relatively thin layer of dead plant material that forms the topsoil on the forest floor. In the constant heat and moisture of the tropics, a group of small organisms in this layer quickly breaks down much of the fallen leaves and other dead organic matter. The nutrients they release into the ground water then move downward toward the sterile clay beneath and are reabsorbed rapidly by the rich network of tree roots. The nutrients move up through the tree's roots, trunk, and branches to its leaves, where the nutrients are

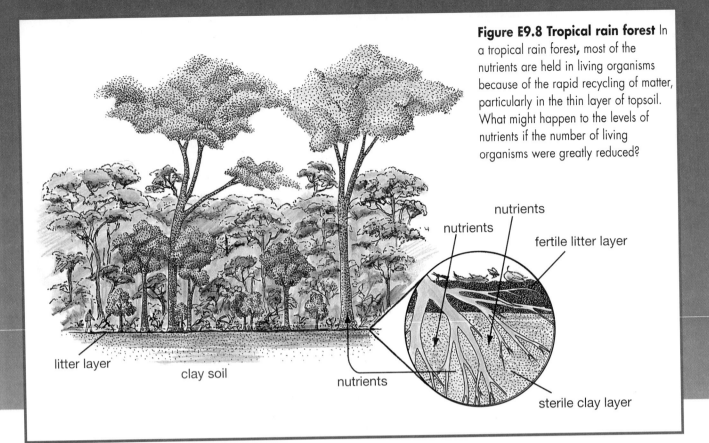

Figure E9.8 Tropical rain forest In a tropical rain forest, most of the nutrients are held in living organisms because of the rapid recycling of matter, particularly in the thin layer of topsoil. What might happen to the levels of nutrients if the number of living organisms were greatly reduced?

nutrients

nutrients

nutrients

fertile litter layer

litter layer

clay soil

nutrients

sterile clay layer

recycled into living matter through photosynthesis.

In environments with less extreme conditions, decomposition is much slower and virtually stops in the winter. In these environments, dead organic material can remain in the topsoil much longer without losing its nutrients. If the community is to persist, however, the processes of recycling eventually must return the nutrients in fallen leaves and other dead organic matter to the plants.

Let's Ask Drs. Ricardo and Rita

Drs. Ricardo and Rita, long-time colleagues at the same university, answer questions and concerns about relationships in communities.

Dear Dr. Ricardo,

I am a biology student who has just completed drawing a food web for class. I noticed something as I was making my food web: I learned that the sum of all feeding interactions is called a **food web**. I've included a sample food web (Figure E9.9).

A food web involves feeding interactions among **producers** (organisms that are able to make their own food by using matter and energy from the nonliving world), **consumers** (organisms that feed on other organisms), and **decomposers** (organisms that feed on decaying organic matter).

I understand this concept, but I still need your help. In our food webs, we listed our producers on the bottom. Then we listed the animals that eat plants, the herbivores. Above these we included

UNIT THREE ENERGY, MATTER, AND ORGANIZATION: Relationships in Living Systems

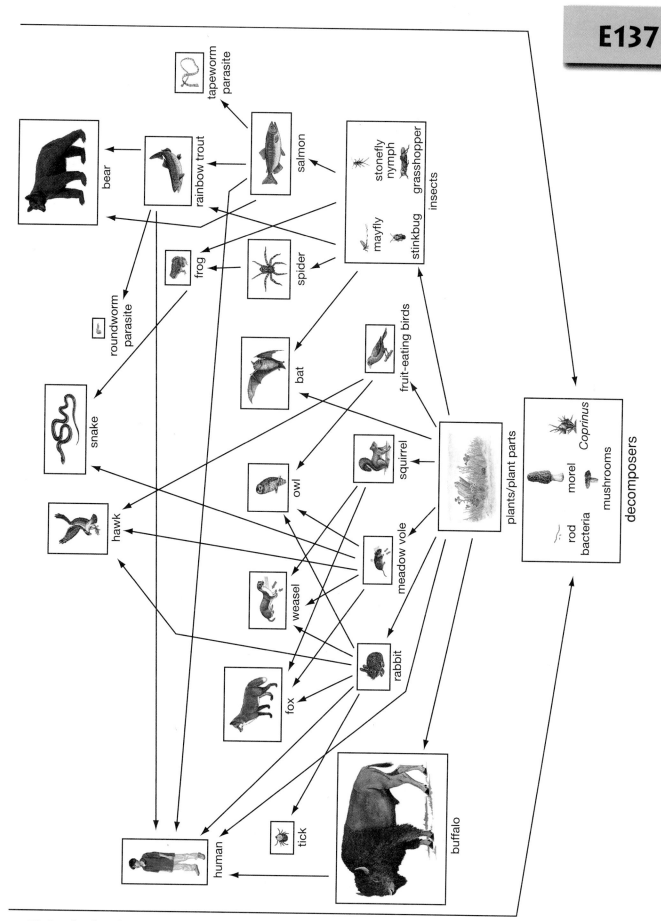

Figure E9.9 A food web Food webs in a community can be very complex. Can you find any more relationships?

the omnivores and carnivores. We also included decomposers in our webs. Because we have included virtually every level of interaction and type of relationship, I was wondering if this is basically what a community is?

—Elsie Conrad, Eco High School

Dear Elsie,

How observant you are! You have done a terrific job of describing a community, which is a collection of organisms that live and interact with each other in a given area. You also described what biologists call **biomass**. Biomass refers to the mass of all living organisms in a given environment. As you probably know, some herbivores may not eat an entire plant. For example, a rabbit may eat only part of a violet. A carnivore such as a mountain lion may not eat the bones of the rabbit. What these creatures eat is considered the **consumable biomass**.

What happens to the remaining parts of the violet and the bones of the rabbit? These types of biomass, the leftovers, often represent a substantial portion of the total amount of energy and biological material in a community. This resource does not go to waste, however. Remember the decomposers in your food web? These organisms (for

example, fungi and bacteria) specialize in using the matter that other organisms do not consume. Though generally not as visible as the other groups, these decomposers serve a vital function in communities; they reduce dead biomass, such as the partially consumed violet and rabbit, into molecules that the producers can reuse. So, you see, you could have drawn arrows from all the creatures to the decomposers.

Dear Dr. Rita,
I drew a food web in my class, but I still don't understand what it means in terms of the real world. Please bring food webs to life for me.

—Charles Webb

Dear Charles,

Thanks for your comment. I would love to bring food webs to life for you! Think of a riverbed that the Anasazi Indians might have lived near, and study the detailed drawing of such a riverbed (Figure E9.10).

In the open water of the river, algae and microscopic water plants are the main producers. These producers are

cottonwoods
sycamores

willows
service-
berries
dogwoods

monkshood
elephantellas
violets
sedges

great blue herons
dippers
racoons

tiger salamander
beavers
algae
watercress

Figure E9.10 Cross section through the edge of a river bed in Southwestern United States The interactions among organisms in an ecosystem involve the cycling of matter and the flow of energy. What is the ultimate source of energy?

E138

consumed by insects and by small aquatic animals such as the tiger salamander and various fish. Higher-level consumers such as the great blue heron live by eating either first-level consumers (organisms that have eaten the producers) or smaller predators that have fed on them. Note that at every step in the process, some biological matter in the form of inedible plant or animal parts, organic waste products, or whole dead organisms, passes to the decomposers. The decomposers then break down the matter to simple molecules that can be used by the producers.

The food web of the river ecosystem is based on solar energy that is converted into food by producers that live in this ecosystem. Not all ecosystems, however, support themselves in this manner. A good example of a rich natural ecosystem that is *not* self-supporting is the seashore. Virtually all of the organisms that inhabit the zone between high and low tides are consumers. They ultimately depend on plant material and other living or dead organic matter that is brought in by each high tide. Many of the seashore creatures feed on this matter directly—for example, mussels and clams filter seawater for microscopic bits of food. Many other species are predators, species such as the sea star, which preys on clams and mussels. The organisms in this community also depend on solar energy, but most of that energy is converted into food by producers that live in deeper ocean communities.

Dear Dr. Ricardo,

    I keep hearing about trophic levels when people talk about food webs. What are trophic levels? Are they some sort of tool used to measure food webs? Are they important? Thanks for helping me.

    —Lynn Gifford

Dear Lynn,

    I used to be confused about trophic levels too. Let me explain. A food web almost looks like a layered wedding cake with each higher level having fewer organisms. The feeding level occupied by an organism is its **trophic level** (which is depicted by the pyramid in Figure E9.11).

Herbivores occupy the first trophic level in a food web. Small predators occupy the second trophic level, and larger predators occupy the third trophic level.

Producers do not belong to any trophic level because they do not eat; they are capable of using solar energy directly to fuel their own metabolism and to produce new biomass. All other organisms are consumers. These consumers cannot make all of the biological molecules they need to build tissue and support their metabolism. After all, when was the last time you raised your arms to the sun, captured solar energy, and converted it into glycogen? We consumers can obtain chemical energy only by consuming biomass, that is, by eating other organisms. We can use a second pyramid to demonstrate the number of consumers that occupy each trophic level (Figure E9.12).

What about omnivores? What about decomposers? What trophic levels do they occupy? They can occupy more than one. A decomposer can exist in the first trophic level when it breaks down the remaining biomass of a plant that was partially eaten by a herbivore. The same decomposer also might exist in the third trophic level when it breaks down the bones of a carnivore. Omnivores also occupy different trophic levels. When you eat a salad and a steak for a meal, you are occupying two trophic levels. You see, organisms of a community interact with

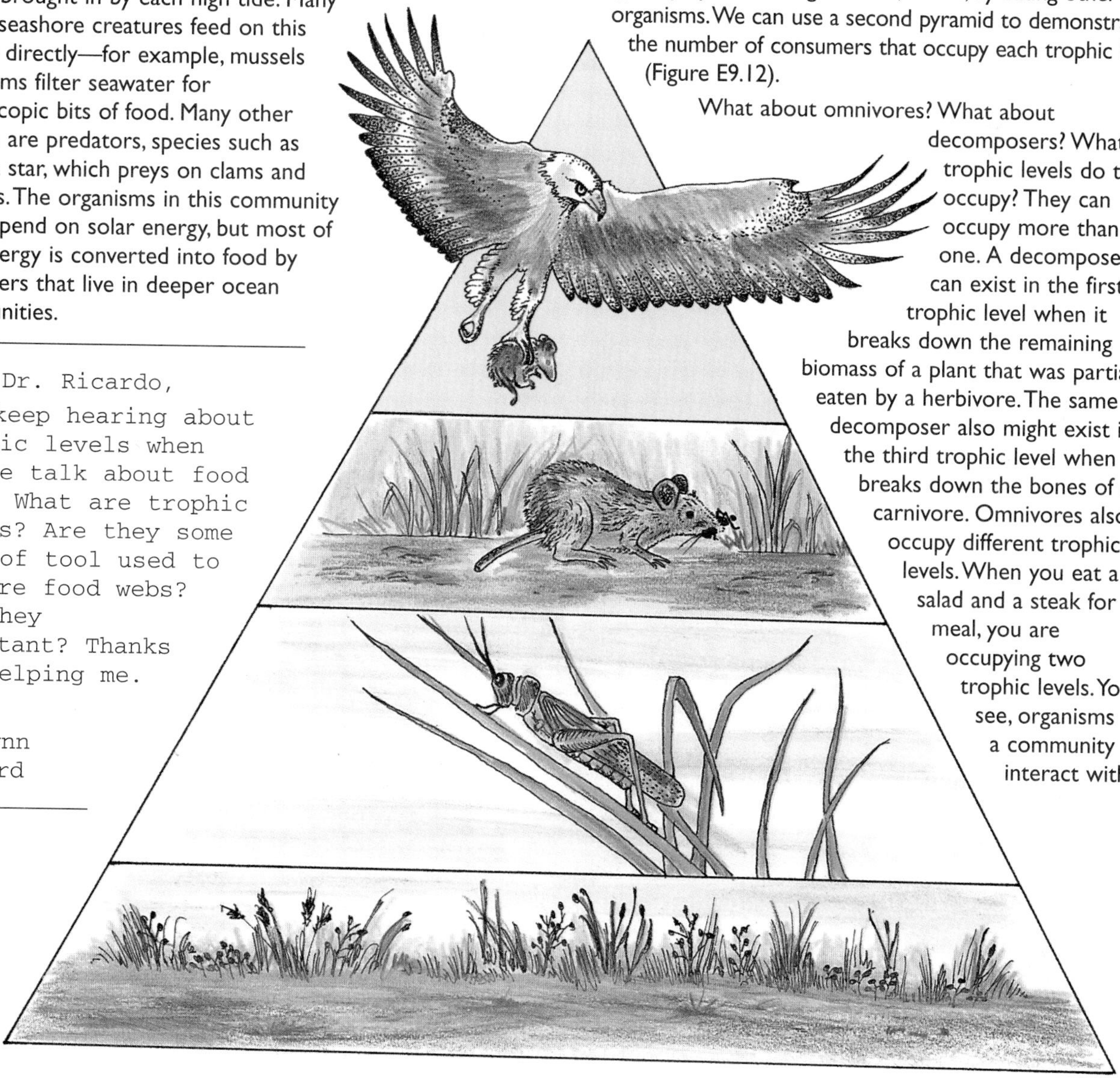

Figure E9.11 A pyramid of trophic levels Trophic levels provide one way to represent the interactions of the organisms in a community.

each other in many ways. One way is through acquiring food, as I have described here.

Dear Dr. Rita,

    I am convinced that the food web that I drew is all wrong. When I drew arrows between organisms, I pointed the arrow toward the organism that does the eating. Shouldn't the arrow be pointing toward the thing being eaten?

    —Connie Allen

Dear Connie,

    You are not alone. Many times students who draw food webs for the first time want to draw the arrows so they point toward the organisms being eaten. Remember, though, the arrows represent *energy flow* and not the act of eating. For example, energy from grass is passed to a rabbit. The energy from the rabbit is passed to a mountain lion. Think of the arrows as meaning "the energy is passed to." A food web helps us understand the flow of energy through a community. Ultimately, the sun is the source of energy for nearly all earth communities.

Did you know that...

☛ Thirty-three percent of all the grain harvested on earth each day is fed to livestock and poultry.

☛ It takes 16 pounds of grain and soybeans to produce just one pound of beef. It takes six pounds of grain and soy to produce a pound of pork, four pounds for a pound of turkey, and three pounds for a pound of eggs or a pound of chicken.

☛ There are currently 1.28 billion cattle populating the earth.

☛ Cattle graze on nearly 24 percent of the landmass of the planet.

☛ Cattle consume enough grain to feed hundreds of millions of people.

☛ Cattle grazing contributes to the increasing numbers of deserts on the earth.

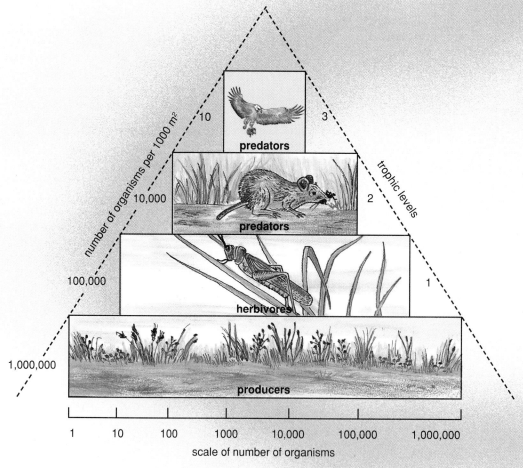

Figure E9.12 A pyramid of trophic levels showing numbers of organisms This idealized pyramid shows the number of organisms per 1000 square meters of grassland habitat.

UNIT THREE ENERGY, MATTER, AND ORGANIZATION: Relationships in Living Systems

Losing Heat

Each time that an organism uses energy, it loses part of the energy in the form of released heat. This means that only a portion of the solar energy that producers take up is stored in biomass that herbivores can eat. In turn only a portion of the plant material that herbivores eat is converted into body parts that could become food for predators. Each transfer of energy from organisms at one trophic level to organisms at the next trophic level results in a decrease in the amount of energy that is available.

Ecologists are able to estimate the amount of energy that is stored in the biomass at each trophic level by taking a sample from a community and harvesting the total biomass represented by the producers, herbivores, and higher-level consumers. They then determine the caloric value of this organic matter.

If we begin with 1 million kcals of solar energy entering the ecosystem, producers convert only a small fraction (0.8 percent) of this solar energy into plant biomass (see Figure E9.13). This biomass represents 8000 kcals of stored chemical energy. The herbivores, such as the grasshoppers that feed on these plants, incorporate only 800 kcals into their biomass, which is only 10 percent of the 8000 kcals of plant energy available to them. Similarly, the predators, such as the mice that feed on the herbivores, incorporate only 80 kcals into their biomass, which is 10 percent of the 800 kcals available to them. Finally, the secondary predators, such as the hawk, that feed on the first level of predators, incorporate only 8 kcals into their biomass. Again, this is 10 percent of the 80 kcals available to them. This pattern in reduction of the amount of energy at successive trophic levels means that the secondary predators acquire only 8 kcals of the 8000 kcals of energy that became part of the food web at

What Happens	Percent
Converted to heat	46.0
Reflected	30.0
Evaporation/precipitation	23.0
Photosynthesis	0.8
Wind, waves, currents	0.2
Total	100.0

Figure E9.13 What happens to the solar energy that reaches earth? Only 0.8 percent of the solar energy reaching the earth is used directly in the production of food.

the producer level. We can use the pyramid in Figure E9.14 to illustrate the reduction in the amount of energy available.

The flow and accompanying loss of energy from one trophic level to the next is the basis for the suggestion that people should eat lower on the food web. For example, a person eating as an herbivore can create 1 kg of human biomass by eating 10 kg of plants. It takes about 100 kg of plants, however, to create 1 kg of human biomass if the person eats as a primary carnivore, that is, if they eat the cow that ate the plants instead of eating the plants directly (see Figure E9.15).

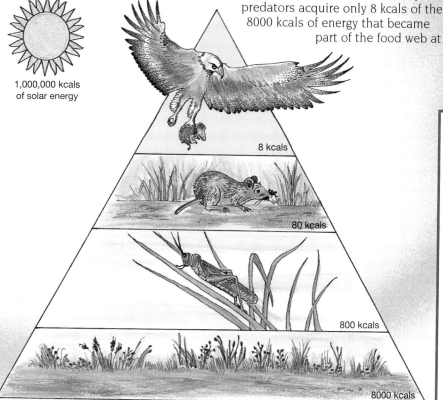

1,000,000 kcals
of solar energy

8 kcals

80 kcals

800 kcals

8000 kcals

Figure E9.14 Energy pyramid This idealized energy pyramid illustrates that only a portion of the energy available at one trophic level is available at the next higher trophic level.

10 kg of plants → 1 kg of human biomass

100 kg of plants → 10 kg of cow → 1 kg of human biomass

a. person eating as herbivore

b. person eating as primary carnivore

Figure E9.15 Human as herbivore and human as primary carnivore The energy relationships between trophic levels are the basis for the suggestion that people eat low on the food chain.

Continuity through Reproduction

Four thousand new bacteria in six hours, six kittens in two months, one elephant in a little under two years (see Figure E10.1). More, more, more. What biological function is served by making more organisms? Whether it be a bacterium, an insect, a tree, a bird, or a human, every individual organism arises as a result of some type of reproductive process. Reproduction is the production of offspring by one or more parents, and all species depend on this process for survival.

We are accustomed to thinking about entering the world somewhat abruptly (through birth) and leaving abruptly (through death). For this reason we sometimes need to remind ourselves that, for a species, life is a continuous process. Our individual existence came about through the fusion of our parents' egg and sperm cells, but genetic blueprints for those egg and sperm cells were passed along through millions of generations. The evidence for evolution indicates that we are part of a journey of life that began on this planet more than 3 billion years ago. Reproduction is the process that made such a journey possible. As we trace our ancestry back through the millennia, using fossil records and biochemical methods, we uncover evidence that links us to other species and that helps us understand how new species slowly emerge. Continuity of genetic information also allows us to trace the short-term history of some genetic factors. For instance, patterns of risk factors for heart disease or cancer may occur within a family, and if we trace the lines of reproduction through family trees, we often can discover these and other patterns.

When you think of your own life, you probably think back to memories of early childhood or wonder about your future; you might speculate about how long your life will be and what experiences it will hold. How important to you are the people who lived before you were born or those who will be here after you die? As humans we place great importance on individual lives and note the length of individual life spans with interest. We tend to view most other species differently, however, and focus on the species as a whole—not on individuals. We think of blue jays or crows as living species because they have living members that continue to produce offspring, whereas we think of the dodo bird as an extinct species that no longer contributes to the living world. We do not mark to any great extent the death of a single blue jay as we would that of an individual human or perhaps a pet. For wild organisms we see continuity as the ongoing presence of the species. Because all individuals eventually die, the ongoing presence of any species depends on a process that ensures new organisms; this process is reproduction.

Figure E10.1 African elephants (*Loxodonta africana*) Elephants have the longest pregnancy of any mammal in the world—close to 22 months. The young nurse for three or four years.

No matter who currently acts as your parent or guardian, you, like all mammals, had two biological parents, each of whom contributed to the blueprint that makes up your own genetic plan. That combination of information from two sources is possible through **sexual reproduction**. Plants that grow from seeds are the product of sexual reproduction, as are cockroaches and elephants. To combine information and produce a new individual, sexual reproduction requires specialized reproductive cells called **gametes**. They differ from all other cells of the parents in that they have only half of the genetic information that other cells have. In humans the female parent produces the female gametes, which are known as eggs or **ova** (singular, **ovum**). The male gametes produced by the male parent are known as **sperm**. Figure E10.2 shows ova and sperm of several species. When gametes

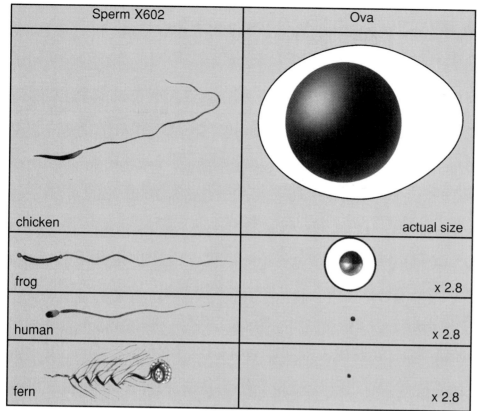

Sperm X602	Ova
chicken	actual size
frog	x 2.8
human	x 2.8
fern	x 2.8

Figure E10.2 Gametes (ova and sperm) from several species In each species sperm are much smaller than the ova. This difference is apparent when you notice the numbers above each column, which show that the illustrations of sperm have been enlarged more than those of the ova. Chicken eggs are among ova that are incubated outside the body of the mother, and thus have a hard protective shell.

come together during sexual reproduction, their nuclei fuse. The product of that fusion is a **zygote**, which will develop into an organism of the next generation. The genetic blueprint in the zygote is the combination of the information that has been contributed by each gamete and in this way represents a new pattern.

Although you may be most familiar with sexual reproduction, not all organisms result from two parents. A hydra, for example, is a tiny animal that lives in water. One way in which it reproduces is by forming a bud on the side of its body, as illustrated in Figure E10.3b. This bud enlarges until it grows into a new hydra that breaks off from the parent. Unless a mutation alters some particular trait, the offspring will have the same genetic pattern as its

parent. When identical offspring are produced from a single parent, the process is called **asexual reproduction**.

Some species can reproduce sexually and asexually, as is the case with the hydra and aspen trees. Grass also may grow from an underground runner or make new individuals from seeds. Yeast, too, can produce sexual gametes or an asexual bud that results in offspring. This characteristic of being able to reproduce either sexually or asexually is just one of many aspects of reproduction. When you consider the diverse species in the five kingdoms of organisms, you find a wide variety of reproductive strategies that differ not only in whether they are sexual or asexual but also in the number of offspring they produce, in the length of time the strategy requires, in the ability of the offspring to survive, and in many other characteristics. In each case, however, the species is perpetuated by the continual production of new organisms that replace those that die.

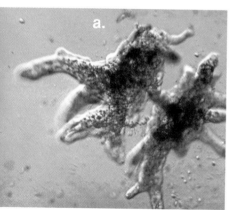
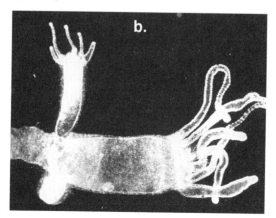
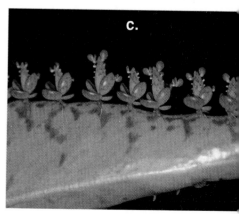

Figure E10.3 Three examples of asexual reproduction a. Amoeba (x100) reproduce by simple cell division (binary fission). b. Hydra (x100) can reproduce by forming a bud that grows into a new individual. c. The *Bryophyllum* leaf (x1) sprouts small plantlets that may grow into a new *Bryophyllum*.

Making More People

A woman living on Trobriand, a tiny island north of New Guinea, completes a long labor and delivers a healthy boy. Although the woman has a husband, she and the others present at the birth do not place much importance on the biological link between the baby and his father. According to Trobriand beliefs, it was important, however, for the father to continue to have intercourse with the mother while she was pregnant so that the fetus would grow. The Trobriand Islanders emphasize the role of the mother in the development of a child and believe that ancestral spirits help plant the seeds that enable a fetus to grow. This belief establishes a link between the past and the future.

One aspect of this view, that children are the products of an ancestral line, fits with the scientific view that offspring acquire their genetic heritage from both parents and, consequently, are related to their ancestors. Scientifically, we know that pregnancy is not caused by spirits planting seeds in the mother but instead requires that a female ovum and a male sperm be brought into contact with each other. We also know that reproduction in humans is highly regulated, as it is in other sexually reproducing species. To have a thorough knowledge of how your own species reproduces, you need to know how the reproductive structures function and how they are regulated.

The Female Anatomy. Before we study the details of human fertilization, it is useful to be familiar with the reproductive structures that produce male and female gametes and how these structures function to bring the gametes together. In humans fertilization normally occurs inside the female body. A woman is born with all the ova she will ever have, around 2 million, which are located in her **ovaries** (see Figure E10.4). As a 7-month-old fetus, a female actually has about 7 million ova, but about 5 million

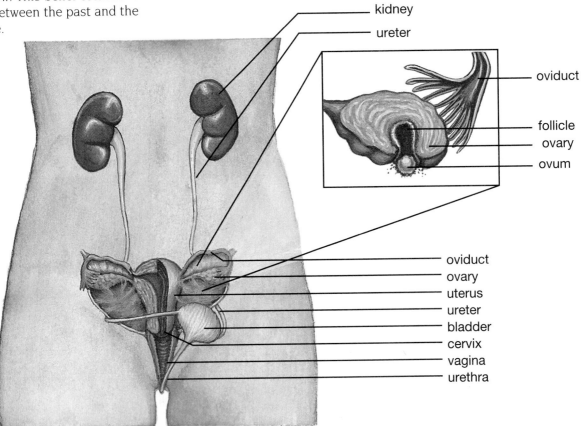

kidney
ureter
oviduct
follicle
ovary
ovum
oviduct
ovary
uterus
ureter
bladder
cervix
vagina
urethra

Figure E10.4 The human female reproductive system This drawing shows the internal organs (ovaries) that produce gametes and the pathway (called the oviduct) that leads from each ovary to the uterus. The uterus expands to accommodate a developing embryo if fertilization occurs. An opening (cervix) at the base of the uterus leads to the vagina, the birth canal, which opens to the external environment. Notice the long path a sperm must travel to join an ovum in one of the oviducts.

of these die before birth. Only about 400 of these will ever mature into healthy ova that are capable of being fertilized. Ova begin maturing in a female's ovaries after she has reached **puberty**, or sexual maturity, which usually occurs between 9 and 14 years old. Although there are two ovaries, usually only one of them releases a mature ovum each month. When an ovum is ready, it bursts out of the ovary into the body cavity and enters a long tube called an **oviduct** that is located nearby. As you see in Figure E10.4, there are two oviducts, one near each ovary.

When fertilization occurs, it usually occurs in the oviduct. Even if fertilization does not take place, the ovum moves through the oviduct to the **uterus**. It is in this muscular, pear-shaped organ that the embryo develops if fertilization occurs and a pregnancy ensues.

The lower, narrow portion of the uterus, known as the **cervix**, provides an opening to the uterus. The cervix extends part way into the **vagina**, which is a muscular tube that connects the uterus to the outside of the body. The vagina is the birth canal through which infants pass as they are born. If fertilization does not occur, the ovum disintegrates within 24 hours after ovulation.

The Male Anatomy. The male body begins producing gametes (sperm) at puberty, around 13 years old. Sperm form in the **testes**, which are part of the male reproductive system (see Figure E10.5). Sperm take about 74 days to develop, and each day millions mature in a healthy, young adult male. The contrast in the numbers of male and female gametes may surprise you, but it becomes more understandable when you know that each sperm has a fairly low

chance of surviving long enough to reach an ovum. As you saw in Figure E10.2, the sperm is much smaller than the ovum. In fact, the sperm cell is one of the smallest cells in the body, and it is very specialized; it has a head that contains the genetic material, a midsection with many mitochondria for energy production, and a long tail that provides motility. After sperm are produced by a male's testes, they are stored in a long, coiled duct called the **epididymis**.

Sperm remain in the epididymis until **ejaculation**, the process in which sperm move through a tube called the vas deferens and are released from the body through the **urethra**. The urethra passes through the **penis**, which is the male organ of both sexual intercourse and urine

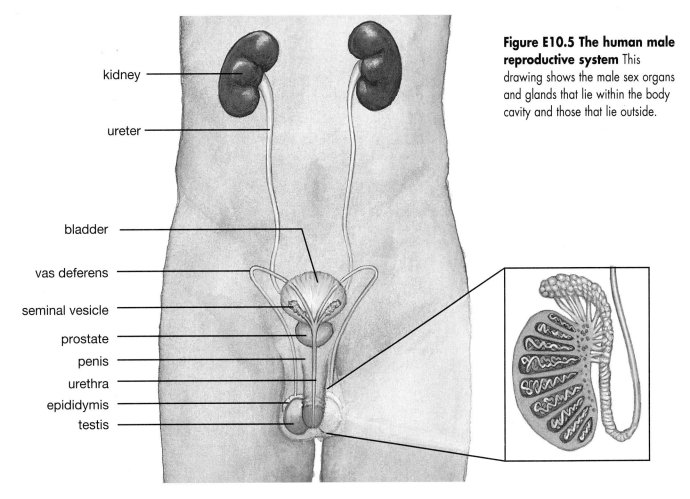

Figure E10.5 The human male reproductive system This drawing shows the male sex organs and glands that lie within the body cavity and those that lie outside.

excretion. A muscle near the prostate gland contracts during ejaculation so that no urine can flow through the urethra during that time. The sperm are carried out of the body in a thick white fluid called **semen**, which also includes secretions from several accessory organs. These accessory organs include the prostate gland, which contributes to the semen a secretion that helps neutralize the acidity of the vagina, and the seminal vesicles, which produce a secretion that nourishes the sperm. During ejaculation, up to 5 mL of semen is released, which contains approximately 300 million sperm cells.

Sexual Intercourse. Sperm cells enter a female's body through the vagina. This usually happens during sexual intercourse. As a male becomes sexually aroused, the blood flow to the penis increases, causing it to become rigid. After the penis becomes rigid, it can more easily enter the female's vagina. As the female becomes sexually aroused, the vagina becomes lubricated, which eases the entrance and movement of the male's penis and also contributes to the mobility of the sperm. An ejaculation results from involuntary muscle contractions that follow the stimulation of sensitive nerve endings in the penis. These involuntary muscle contractions occur during male orgasm. Ejaculation releases sperm-containing semen through the penis. Females have analogous orgasmic contractions, but these contractions are not necessary for the release of ova or for fertilization.

Although millions of sperm may be deposited in the vagina from one ejaculation, probably only a few thousand survive the journey to the oviducts, which takes less than an hour. If an ovum happens to be present in one of the oviducts, many of the surviving sperm swarm around it and release enzymes that promote changes in the outer layers of the ovum so that the sperm can penetrate these layers. Only one sperm actually penetrates the ovum during fertilization. As soon as penetration occurs, changes take place in the cell membrane of the ovum that prevent the entry of any other sperm.

The single cell that results from the joining of the ovum and sperm is referred to as a zygote, and it develops into a human embryo. The zygote moves from the oviduct and almost always implants in the wall of the uterus. After implantation the developing embryo will obtain the nutrients and oxygen it needs from the mother's blood supply.

Hormones and Sexual Reproduction

For both human females and males, the joining of an ovum and a sperm is highly regulated. Hormones are the molecules that delicately tune the reproductive system during development and after sexual maturity. This regulation mainly involves communication between endocrine glands in the brain and the reproductive organs.

The onset of puberty in both males and females is controlled by changes in hormonal levels that influence the reproductive organs. These hormones help bring about some of the noticeable indicators of sexual maturity—changes in body proportions, the appearance of pubic and underarm hair, and increased interest in sex. In males hormonal signals at puberty often cause the vocal cords to lengthen, which lowers the voice. Also, in both males and females, the anterior pituitary in the brain releases two hormones that play important roles in sexual reproduction. These hormones are **luteinizing hormone (LH)** and **follicle stimulating hormone (FSH)**. The release of these hormones is regulated by another part of the brain, the hypothalamus. In males FSH stimulates maturation of the testes, which respond by manufacturing sperm. The main function of LH is to stimulate the release of the major male sex hormone, **testosterone**, from the testes (see Figure E10.6).

Even when a human has matured physically, hormonal regulation does not stop. Hormones continue to signal the events that lead to the production of gametes and promote sexual behavior. In a mature male testosterone stimulates continuous sperm production. Erection of the penis and ejaculation also depend on sufficient levels of testosterone in the blood stream. As shown in Figure E10.6, in addition to the stimulatory signals from the brain to the testes, there also are negative feedback loops. For instance, the levels of testosterone can be reduced if necessary because there is a connection between LH and testosterone secretion. If too much testosterone is in the blood, the hypothalamus signals the anterior pituitary to release less LH. The lowered LH level in turn decreases the amount of testosterone released from the testes.

In females hormonal interactions generally are more complex although the basic types of communication are

E146

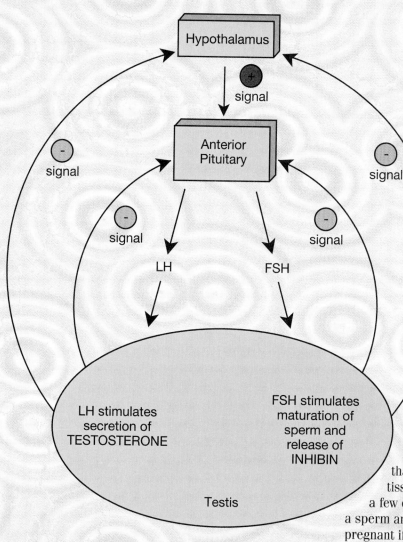

Figure E10.6 Interaction of hormonal signals between the brain (hypothalamus and anterior pituitary) and the testes in the human male Testosterone is the major male sex hormone, but production of viable sperm depends on the action of LH and FSH as well. The involvement of the hypothalamus suggests a mechanism for sexual arousal in response to sexual thoughts.

LH = luteinizing hormone
FSH = follicle stimulating hormone

similar. The major female sex hormones, **estrogen** and **progesterone**, are produced by the ovaries. Estrogen stimulates changes in a female's body at puberty that are analogous to changes in males. After puberty hormones continue to regulate reproductive function in females in a cyclic fashion. In contrast to the continuous production of sperm by males, the female ova mature and are released from the ovaries as part of a monthly cycle of events. This ovum-releasing cycle is part of the **menstrual cycle**, which is illustrated in Figure E10.7. Typically, only one ovary releases an ovum during the menstrual cycle. The release of an ovum is known as **ovulation**. FSH causes the ovum to mature inside a sac, or follicle, on the surface of the ovary. The follicle affects reproductive regulation by releasing estrogen. The presence of estrogen then stimulates the secretion of LH, which in turn stimulates ovulation. Ovulation occurs around the middle of the cycle. The menstrual cycle lasts about 28 days, although it may range from 24 days to more than 35 days.

The female hormones also regulate changes in the uterus that prepare it for the possible implantation of a zygote. During the first half of the cycle, estrogen stimulates a thickening of the inner lining of the uterus

that includes an increased blood supply to this tissue. If sperm enter the vagina anytime between a few days before ovulation to a day or so afterward, a sperm and ovum may join, and the female will become pregnant if the resulting zygote implants in the uterus. Although an ovum can be fertilized only during a 10- to 15-hour interval following its release from the ovary, sperm can survive up to 72 hours, so sexual intercourse several days before ovulation still may result in pregnancy. Because individual menstrual cycles vary so much, it is extremely difficult to predict exactly when a female is likely to become pregnant. As a result, pregnancy often occurs when it is not expected if individuals do not take preventive measures.

During the second half of the menstrual cycle, the levels of progesterone increase (see Figure E10.7). If a zygote reaches the uterus, it may implant itself in the lining of the uterus and begin a pregnancy. A placenta then develops from embryonic and maternal tissue. The placenta nourishes the growing embryo and secretes hormones, including progesterone, that help maintain the uterine environment needed to sustain the pregnancy.

If fertilization does not occur, hormone levels in the female decrease. This decrease causes the blood supply to the inner lining of the uterus to diminish. Part of the lining then disintegrates and passes from the uterus through the vagina to the outside of the body. This shedding of the lining is known as **menstruation**. During

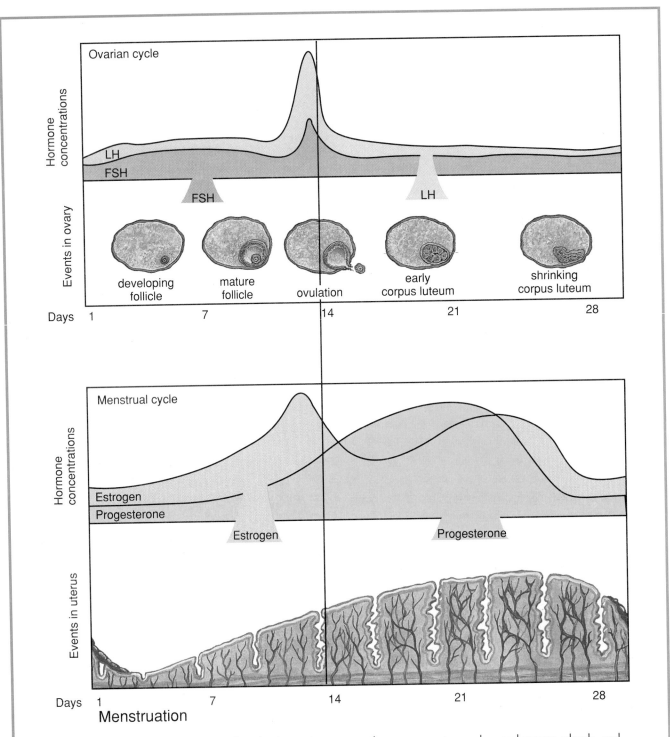

Figure E10.7 The human menstrual cycle Interactions among the nervous system and several organs, glands, and hormones regulate the cycle. Note how the hormone levels change at ovulation, marking two very different phases to the cycle.

monthly menstruation, women lose only a small amount of blood, between 50 mL and 150 mL (about ¼ cup to a little more than ½ cup).

Menstruation is a characteristic of the human female and some other female primates. Most other mammals, including other primates, have analogous cycles known as estrus,

or "heat," that occur only a few times a year. These female mammals ovulate and are receptive to males only during these times. Animals such as deer, elk, and moose enter into estrus once a year, and most dogs go into estrus twice a year. In some mammals, such as mice and rabbits, ovulation is stimulated by intercourse.

In humans females do not remain fertile for their entire lives. Just as menstruation marks the beginning of sexual maturity, **menopause**, which is the cessation of the menstrual cycle, indicates that a female is entering the phase in which she no longer is capable of reproducing. The hormonal effects of menopause are just as profound as those of the menstrual cycle itself and vary widely from individual to individual. Menopause is a normal part of the life of a female and usually occurs between the ages of 45 and 50. Males do not experience the equivalent of menopause, although the number of healthy sperm a male produces may decline with age. If enough sperm are present, however, a male of advanced age is still capable of impregnating a female.

INFERTILITY

Couples trying to conceive children are not always successful. In approximately 15 percent of couples in the United States, either the male or female is infertile. A variety of problems may be the cause. If the male's sperm count is very low, the chances of fertilization decrease dramatically. If the female does not ovulate or ovulates at irregular times, pregnancy may be difficult to achieve. In some women the oviducts become blocked, creating a physical barrier between ovum and sperm.

Some of these problems can be overcome so that natural fertilization takes place. For example, a woman who does not ovulate often may take a pharmaceutical drug that stimulates ovulation. Drugs such as these often stimulate the release of more than one ovum, which increases the likelihood of multiple births. Low sperm count in males, in some cases, results from simple metabolic deficiencies, such as inadequate levels of zinc. Such conditions can be resolved easily.

In more serious cases technology may be used to carry out **in vitro** (literally, "in glass") fertilization. Under laboratory conditions an ovum is removed from a female's body and mixed with sperm. If fertilization takes place, the zygote is allowed to develop for a few days in this specialized environment. If the zygote appears to be healthy and developing normally, it is implanted into the woman's uterus to continue development. In some cases the embryo from one woman may be inserted into another woman during early stages of embryonic development.

Sexual Activity and Health Hazards

Sexual behavior involves some risks. Childbirth itself carries risks to the mother, although modern medical technology has reduced enormously the risk of maternal death during childbirth. Some birth control methods also may have health risks. For example, some men or women become allergic to spermicidal preparations and experience inflammation of the affected tissues. Intrauterine devices (IUDs) can produce cramps and increased menstrual flow in some women and may increase the risk of pelvic inflammatory infections. Oral contraceptives slightly increase the chance of stroke, although this risk is dramatically lower than the risk of a mother dying during childbirth. The risk of side effects from oral contraceptives increases markedly among women who smoke. For example, in the age group 15 to 24 years old, women smokers are seven times more likely to die from the side effects of oral contraceptives than nonsmokers are. This risk is high enough that oral birth control pills are not recommended for women who smoke.

A different type of risk associated with sexual behavior is the danger of contracting a sexually transmitted disease (STD). Sexual behavior usually involves the direct contact of tissues that are enclosed by mucous membranes. Such contact involves the exchange of body fluids, which provides an excellent route of infection for pathogens, such as the viruses that cause herpes, AIDS (HIV), and hepatitis A and B. Figure E10.8 lists some of these pathogens and the symptoms of the diseases they may produce.

AIDS is of particular concern, because it is, at present, a fatal disease. Casual contact, such as being in the same room or shaking hands, with an HIV-infected person will not transfer the virus. Infection requires a transfer of body fluids, particularly blood, semen, or vaginal secretions. Both heterosexual and homosexual intercourse and the sharing of needles by drug users are the most common routes of HIV infection.

Figure E10.8 Examples of Sexually Transmitted Diseases

Disease	Pathogen	Description
Acquired Immunodeficiency Syndrome (AIDS)	HIV (virus) Human Immunodeficiency Virus	Usually no symptoms of infection with HIV, although virus can be transferred to others. The disease AIDS, which usually takes years to develop, has symptoms that include extreme fatigue, weight loss, reduced resistance to other infections, and cancer, and eventually leads to death.
Genital Herpes	Herpes simplex (virus) (Usually type 2, HSV-2)	Causes ulcers on male's penis or female's vulva, vagina, or cervix. (Type 1 virus most often causes fever blisters on the mouth and is also contagious.) These ulcers continue to erupt throughout life.
Syphilis	*Treponema pallidum* (bacterium)	This disease occurs in several stages. Early: Chancre (a hard, painless lesion) on male's penis or on or near female's vagina, or on lips and hands Mid Stage 1: No external signs; blood test would be positive Mid Stage 2: Symptoms come and go: rash on palms and feet, white sores in mouth (which can be spread by kissing) Final Stage: Widespread tissue damage, including damage to the nervous system; mental illness; death
Chlamydia	*Chlamydia trachomalis* (bacterium)	May not show symptoms. If symptoms are present, they are as follows: in females, inflammation of opening to vagina or cervix; in males, watery discharge from penis and inflammation of epididymis. Infection may be transferred to the eye. Females may pass it on to child during delivery. May cause blindness.
Gonorrhea	*Neisseria gonorrhoeae* (bacterium)	May not show external symptoms. If symptoms are present, they are as follows: in females, infection of cervix with discharge; in males, pus from tip of penis (urethra). May become chronic if not treated; may cause infertility in females.

Mating Behaviors of Nonhuman Animals

A Pacific salmon leaves the ocean and swims hard against the current of a freshwater river, driven upstream by his instinct to return to a specific location to reproduce. Sexual drive is a physiological mechanism that leads to a wide variety of mating behaviors in different species. Because reproduction is essential for continuity of a species, those behaviors that promote mating and reproduction are advantageous. Behavioral and physical traits that give individuals a competitive edge in mating and in producing offspring contribute to sexual selection—a process of natural selection that involves specific characteristics or features that one sex prefers in the other. The animal world provides many striking examples.

Figure E10.9 Male bighorn sheep (*Ovis canandensis*) in a competitive sparring match

Male bighorn sheep, which live in the mountainous regions of western North America, often engage in competitive sparring matches in which they bash their heads together, as shown in Figure E10.9. These encounters occur only during the winter mating (rutting) season, a time when males fight to control the areas where females in estrus gather. The massive horns of a particular male, as well as its uniquely characteristic sparring movements, may prove advantageous because the winners of these matches tend to mate more frequently than the losers. Fighting, however, can be expensive in terms of expended energy, lost time, or injury; the losers do not challenge the larger, stronger males once the stronger males have established their dominance. Sometimes young males will challenge the dominance of older males in subsequent years.

In many species actual fighting may not be part of the competitive strategy to attract or gain access to a mate. Simple behavioral displays may suffice. Males of certain species of grouse may gather on a lek, or dancing ground, and begin a synchronized, elaborate dance display for the females. The males spread their tail feathers, inflate their neck pouches, and produce a booming call to accompany the dance. Females select a mate from the group of displaying males. A male's plumage and the subtle features of its dance may improve the chances that a female will select him as a mate. Males of some species such as the bird of paradise have extremely elaborate plumage that further enhances their courtship display. Male songbirds frequently express sexual readiness in the form of songs and calls. These vocalizations may announce the male's defense of his territory or his invitation to mate.

In some species of fish, such as African cichlids and American sticklebacks, ritualized courtship and mating movements that mimic aggressive patterns have evolved. Figure E10.10 illustrates courtship and mating in the American stickleback. In wetland areas spring is heralded by the incessant evening chorus of frogs announcing their readiness to mate.

In many species in which the females provide primary care for the young, being conspicuous can be a disadvantage. Bright colors and elaborate plumage, for example, may increase the danger of predation to the mother and her young. Frequently, striking external features are reserved for males. The elaborate plumage of male birds may become a disadvantage, however, because it makes them more visible to potential predators.

When we observe the various forms of behavior and dress among human males and females trying to attract each other's attention at school, at the beach, at dances, and at other social gatherings, it is tempting to draw parallels between human behavior and display and the patterns we see in other animals. In fact, some human cultures, such as certain Native American tribes, have studied the patterns and movements of particular animals and incorporated them into their dances and rituals. Even though cultural behaviors are not transmitted biologically, many such practices may be adaptive and may enhance reproduction.

In humans sexual drive has a genetic and physiological basis. Certain physical attributes that may attract a member of the opposite sex, attributes such as facial features, body shape, or body proportions, also may have a hereditary basis. Some of the physical and behavioral attributes that make us attractive to a potential mate (such as strength, good health, and intelligence) also may be advantageous to survival or success in life.

a. Male (right) threatens other males.

b. Male's red underside attracts female.

c. Male starts zig-zag dance toward nest; female follows male.

d. Male guides female into nest.

e. Male taps female near tail; female lays eggs.

f. Male bites female; female leaves nest; male enters nest and fertilizes eggs.

Figure E10.10 The courtship and reproductive behavior of sticklebacks

Figure E10.11 During a Turkana wedding ceremony in northwestern Kenya, the bride's family brings sheep and goats as part of the bridewealth. Men and women dance in the background.

Cultures and Mating Patterns

If you take a few minutes to think about the ceremonies in your community that surround the marriage of a young couple or the birth of a child, it is easy to see that in human societies cultural patterns are woven into the fabric of human reproduction and provide a range of expressions. In general, human cultural patterns influence the way that each human society approaches events such as birth, marriage, and death, as well as how it approaches common daily events, such as gathering and preparing food and raising children.

Unlike the roles parents play among most other organisms, raising human children is a long-term commitment. Human societies around the world often mark sexual maturity with great ceremony and approach marriage as a significant event.

In a tribe in the Amazon forest, adolescent boys undergo an initiation rite that marks them as young men. Two older men make incisions in the boys' skin and fill the wounds with hot, liquid tar. The tar prevents infection and leaves a pattern of scars on their bodies. Although the ritual is painful, the boys are proud to be entering puberty. Similar scarification techniques are used by the Aborigines of Australia's Northern Territory and by the Abelam of Papua New Guinea.

Among the Abelam, a female's first menstruation is a time of great celebration. At this time other women cut designs in the skin of the breasts, stomach, and upper arms of the girl and shave her head. Such rituals are conducted with great celebration. In some cultures marriage takes place long before the young pairs are ready to reproduce, as the following story indicates.

Ashok's family was very pleased with the dowry that they would receive after the wedding ceremony from his bride's family. As is often the custom among Hindu families in the rural Haryana state of India, the groom's family had visited the family of the bride-to-be prior to the wedding to inspect the gifts. Now the evening of the wedding feast had arrived, and all preparations for the occasion were complete.

Kamala watched with anticipation as her husband-to-be arrived on horseback from a neighboring village. The approaching darkness enhanced the effect of the fireworks display that heralded Ashok's arrival. A small band played, and villagers in elaborate costumes danced in front of the horse and rider.

At the wedding feast, Ashok's family was served first while Kamala and her family waited patiently. The food included a variety of tasty vegetable curries with roti, a flat, unleavened bread. The festivities continued throughout the evening and into the early morning hours. The wedding ceremony could not conclude until 1:27 a.m., the precise time determined by the family astrologer.

Kamala had celebrated her 11th birthday just two weeks earlier, and Ashok would be 14 next month. The bride and groom would not live together for several more years. It is common in this part of rural India for a father to arrange the marriage of his daughters well before they are old enough to bear children. By tradition it also is very important that they be virgins when they marry.

Kamala's father had spared no expense for this wedding. He had accomplished his duty now that the last of his three daughters was successfully married. Kamala would remain with her parents until one day, perhaps four or five years later, Ashok would return, with ceremony, to claim his bride.

Cultural ceremonies such as marriage promote bonds between males and females; bonds such as these are often referred to as pair bonds. In many cultures the bond formed between individuals is a monogamous bond, which is a bond between one male and one female. Practices such as polygamy, in which a man or a woman takes multiple spouses, also occur across cultures. When you consider the length of time that parents must care for their offspring, one of the biological bases for pair bonding among humans becomes apparent. Similar patterns exist in other primate species, such as gibbons and marmosets. In these groups of primates, adults mate for life and raise their offspring to maturity.

Sexual Behavior, Ethics, and Public Policy. When two people have a child, their genetic information is combined in their offspring. This basic biology seems to underlie the marriage patterns that are practiced in different cultures. In certain periods during the history of ancient Egyptian cultures, for example, the brothers and sisters in ruling families might have married to increase the purity of the family line and avoid the genetic influence of outsiders. More often, however, cultures have developed strong taboos against marriage or sexual activity between close relatives. In the Middle Ages in parts of Europe, for example, ruling families were forbidden to marry anyone to whom they were related in the seven previous generations. With such a strict rule, they ran out of eligible marriage partners. In Navajo traditions young people are not to marry anyone from either their own clan or the clan of their father. Clans are subgroups within a culture that trace their descent either through their mother's line or through their father's line. Taboos against what most people in the United States consider to be incest— sexual relations between members of the immediate family—have provided a certain amount of protection against inbred diseases. In some societies first cousins may marry, but this is not allowed in others.

When a society makes a law that defines how closely related two individuals may be and still marry, the society is proclaiming a public policy. Some decisions such as these are based on scientific information, as in the case of parental contribution to genetic patterns in children. Even so, the differences in the laws in modern societies show that there is more than one cultural approach for addressing the same set of information. Other public policies are based on ethical considerations or cultural values, and in some societies in which the religious practices are not separate from the government, public policy reflects religious beliefs.

Sexual practices among humans are influenced by cultural values and public policy as well as by choices that individuals make. Sexual activity can have serious physical, emotional, and financial ramifications. Indeed, not only would sexual partners be responsible for another human they produce through their union, but they expose themselves to potential health risks. The physical contact between membranes of the reproductive organs and the exchange of body fluids that occurs during sexual intercourse provide an excellent avenue for microbes to be exchanged. The most serious of these pathogens is HIV, the virus that damages the immune system and causes AIDS. Like genital herpes, there is no cure for an HIV infection; unlike herpes, AIDS is fatal.

As a thinking organism, can a human adopt sexual behaviors that consciously avoid these dangers? There is no completely safe sex, only safe abstinence. Individuals who choose to become sexual partners, however, can choose to decrease the risk of infection. For instance, if each person has sex with only one sexual partner and each person has been checked by medical tests to ensure that he or she is free of HIV, the risk of infection is greatly reduced.

Asking questions and making smart choices based on the answers help to reduce the risk of disease and unwanted pregnancy. For example, the use of a physical barrier such as a condom during sexual intercourse reduces the chance of fluid (and pathogen) exchange. The risk is somewhat less if a spermicide is used along with a condom because spermicides may kill HIV. In each case, however, the practices described only *lessen* the odds of infection or pregnancy; they are *not a guarantee* against infection or pregnancy.

How are these personal choices related to public policy? For one thing, federal regulations determine how various methods of birth control are made available. And, as a society, we all bear the price of high-risk behaviors that result in disease because these behaviors increase the pressure on medical services, medical costs, the cost of insurance, and so on.

A conflict of values can become an ethical debate behind the scenes of public policy decisions. One of the best examples of such a debate is the issue of abortion, which is the physical or chemical removal of an embryo from the uterus after implantation. When the abortion is performed under medical supervision during the very early stages of embryo growth (the first three months of pregnancy), the health risk to the mother is extremely slight. However, there is considerable disagreement as to whether the law should allow even early-stage abortions. Some individuals believe that society should not tell a woman what she may or may not do with her own body. Other individuals feel that abortion is unethical and that the embryo is a human being with rights as soon as the zygote is formed.

Phenotype and Genotype

The physical traits that we observe in humans or other organisms, such as blood type, ear shape, and petal color, result when the information in an organism's genetic plan, its DNA, is expressed biologically. These observed traits are referred to as the organism's **phenotype**, and this term can be used to refer to specific traits or the collection of traits that characterize an entire organism. For instance, we could say that a collie has a long hair phenotype rather than a short or medium hair phenotype, or we could say that a collie has a very different phenotype from a Great Dane (the collie is smaller, has longer hair, and shorter legs). The inheritance of a genetic plan from parent to offspring, which occurs during reproduction, provides the offspring with a blueprint for its phenotype. This is the reason that offspring inherit a phenotype similar to their parents' phenotype.

Despite the obvious importance of the genetic plan in determining the phenotype of an organism, the phenotype is not determined solely by this plan. Environmental factors play a critical role as well. For example, the average height of the population of Japan increased by several inches during the early and middle twentieth century (see Figure E11.1). This relatively rapid increase in average height was not caused by changes in the genetic plans of the whole population. Rather, it was due to improvements in the environment of each successive generation of Japanese children during their growth years. Previously an environmental limitation such as poor diet had prevented the members of this population from reaching the maximum size permitted by their genetic plan. Interactions between the inherited genetic plan and the environment are particularly evident for complex phenotypes,

such as height and behavior, because so many genetic and environmental factors are involved.

Scientists in the 1930s conducted a direct test of the hypothesis that environmental factors influence how genetic information produces a phenotype. In a simple yet ingenious experiment, scientists took cuttings from genetically identical plant species and grew them in plots at different altitudes. The cuttings grew extremely well at sea level but barely grew at all at higher mountainous altitudes. Thus the environment affected the phenotype that was produced from a given genetic plan.

Genotype. If phenotype is the result of a genetic plan that is inherited, expressed, and influenced by the environment, what is the genetic plan itself? The genetic plan, or **genotype**, is all of the genetic information in an organism—genetic information that

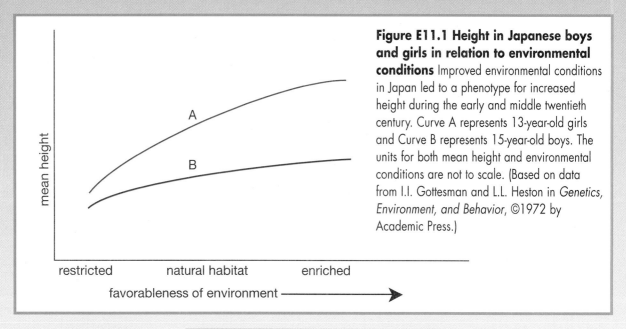

Figure E11.1 Height in Japanese boys and girls in relation to environmental conditions Improved environmental conditions in Japan led to a phenotype for increased height during the early and middle twentieth century. Curve A represents 13-year-old girls and Curve B represents 15-year-old boys. The units for both mean height and environmental conditions are not to scale. (Based on data from I.I. Gottesman and L.L. Heston in *Genetics, Environment, and Behavior*, ©1972 by Academic Press.)

is stored in DNA molecules. In humans and other eukaryotes, the DNA is organized around proteins into distinct structures called **chromosomes**. Each human chromosome contains just a small part of our total genetic information, yet each contains enough information for many phenotypic traits. For example, the very tip of just one human chromosome contains the piece of genetic information that causes blood to clot following an injury. We refer to that particular piece of genetic information as a **gene**—in this example, a gene that produces a blood-clotting protein in healthy individuals. (Alexis's hemophilia A resulted from a defective gene. As a result his body could not produce the clotting protein called Factor VIII.)

Each human chromosome contains thousands of genes, and our entire collection of 46 chromosomes is our complete genotype. Figure E11.2 illustrates some of the many different genes that are present on two of our chromosomes and how each gene

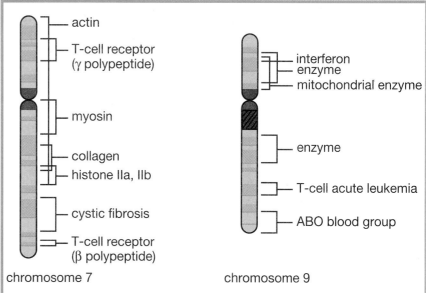

Figure E11.2 Each chromosome carries many genes. This illustration shows a few of the genes that have been mapped to two human chromosomes, chromosome 7 and chromosome 9. Humans have two copies of each of 23 different human chromosomes, which carry an average of 3000 genes each.

occupies a specific location. Not only are there many *different* genes at different locations on a chromosome, but there often is more than one possible version of a particular gene. Some individuals may have a *version* of a blood type gene that specifies type A blood whereas others may have a version of the same gene that specifies type B blood. Different versions of the

same gene are called **alleles**, and the particular combination of alleles in an organism determines the organism's phenotype. In general, every organism that reproduces sexually receives two alleles for every gene, one set of alleles (on one set of chromosomes) from the mother's ovum and one set of alleles (on a second set of chromosomes) from

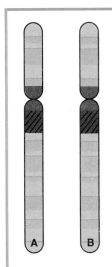
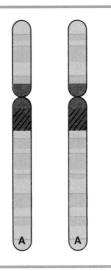

Figure E11.3 Two alleles for a blood type gene A gene can exist in more than one version; each version is called an allele. The combination of alleles in an organism, along with the environment, produces that organism's phenotype. The chromosome pair on the left produces the phenotype of type AB blood. The chromosome pair on the right produces type A blood.

two ways: either (1) the allele for type A was inherited from the father, through his sperm, and the allele for type B was inherited from the mother, through her ovum, or (2) allele A came from the mother and allele B came from the father. Either way this combination of alleles results in a genotype that is heterozygous for the blood type alleles—that is, the alleles for the blood group gene are different (*hetero-*). Alternatively, a person who inherits two identical alleles would have a genotype that is homozygous for the blood type alleles—that is, the alleles are the same (*homo-*). Figure E11.3 shows these combinations.

Some human traits have very simple inheritance patterns. Cleft chin, a phenotype that ranges from a small dimple to a full crease in

the father's sperm. A large variety of alleles is possible when gametes combine to form a new organism, and this variety helps explain why there is so much inherited phenotypic variation in populations of organisms.

The inheritance of genes (and many traits) is not difficult to follow if one remembers how genetic information is organized and that the two alleles for any given gene in a zygote can be either the same or different. For example, a person who has type AB blood has one allele for type A blood and one allele for type B blood. This could happen in one of

Parents

♀ straight ears, white fur   X   ♂ floppy ears, brown fur

Possible offspring

straight ears, white fur   straight ears, brown fur   floppy ears, white fur   floppy ears, brown fur

Figure E11.4 Independent assortment of two traits in rabbits Because the same ear type and fur color do not always stay together, these results demonstrate that the traits for ear type and fur color are assorted independently of each other.

the middle of the chin, is one such example. Geneticists who have studied many human families in which this trait appears have determined that an individual who inherits just one allele for this trait (from either parent) will show the phenotype. The phenotype of cleft chin will result even if the allele inherited through the other parent does not specify cleft chin. A trait in which the phenotype is the same whether the individual is homozygous or heterozygous is called a **dominant trait**. A trait that is exhibited only if the individual is homozygous is a **recessive trait**.

Two alleles in combination can determine one trait of an individual organism. Every organism has many traits, however, and quite often different traits seem to be inherited in a random fashion. For example, baby rabbits inherit genes for the traits of ear type and fur color. They may be born with straight ears, like their mother, and brown fur, like their father, even though the father may have floppy ears and the mother may have white fur (see Figure E11.4). On the other hand, it is quite possible that rabbit offspring from these same parents may be straight-eared and white or floppy-eared and brown. These different combinations are possible because the transmission of an allele for the fur-color gene does not affect (is independent of) the transmission of an allele for the ear-type gene. In other words, the genes governing those two traits undergo **independent assortment**. In this example independent assortment occurs because the genes for ear type and fur color are located on different chromosomes.

This principle of independent assortment was discovered more than 150 years ago in a small European monastery garden. A scholarly monk named Gregor Mendel used pea plants to study patterns of inheritance. In the monastery garden, Mendel experimented with generation after generation of pea plants. His insights later became the cornerstone for explaining many patterns of inheritance.

Case Studies of Two Tragic Genetic Disorders

Huntington disease. Rita is a 30-year-old woman, and she and her husband would like to have a baby. Her father died a few years ago of Huntington disease (HD), a dominant genetic disorder that causes degeneration of the central nervous system. Symptoms of the disease do not appear until age 35 or older. There is no treatment or cure for HD. The disease simply gets worse for five to 15 years before the patient dies. Because Rita's father was a victim of HD, she knows that her chance of developing the disease is 50 percent. Through a genetic procedure that requires a small blood sample, it is now possible to determine whether an individual carries the allele for HD. Is it better for a person to wonder whether she or he has a 50-percent chance of developing the disorder and passing it on to a child, or is it better to know with certainty whether she or he carries the gene? HD causes tremendous psychological stress because its effects often are not noticed until after the individual has had children. Many affected parents feel guilty because they may have passed the allele for this dominant trait to some of their children. These children also face ethical dilemmas about whether to have children.

Cystic fibrosis. Almost from the time his parents brought him home from the hospital, Richard seemed weaker than other infants, and he frequently fell ill with coughs and colds. Then at nearly a year old, and after many trips to the doctor, he was diagnosed with cystic fibrosis (CF). CF is a lethal disease that is characterized by digestive and respiratory problems, and it is the most common genetic disorder among children of northern European descent. Richard's grief-stricken parents were confused and angry when they heard the diagnosis.

"How could our son have such a terrible disease when no one in either of our large families has ever had CF?" asked the distraught father.

"The allele for CF is quite common among Caucasians," explained Dr. Cooper, their pediatrician. "In fact, about one in 25 Caucasians carries an allele for cystic fibrosis. Unlike the situation with Huntington disease, however, a person must inherit an allele for cystic fibrosis from both parents to be affected."

"So you mean that both of us could get sick as well?" asked Richard's mother.

E158

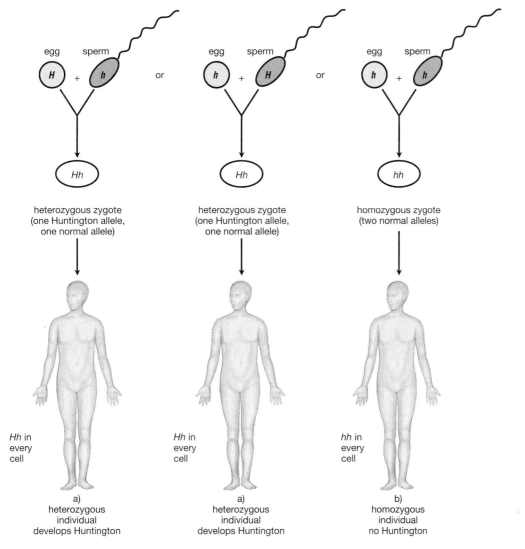

egg sperm or egg sperm or egg sperm

Hh   Hh   hh

heterozygous zygote (one Huntington allele, one normal allele)

heterozygous zygote (one Huntington allele, one normal allele)

homozygous zygote (two normal alleles)

Hh in every cell

Hh in every cell

hh in every cell

a) heterozygous individual develops Huntington

a) heterozygous individual develops Huntington

b) homozygous individual no Huntington

Figure E11.5 Genotypes in Huntington disease Geneticists use a shorthand way of writing genotypes that assigns a capital letter to the allele for the dominant form of a trait and a lowercase of the same letter to the allele for the recessive form. The allele for Huntington disease is symbolized by a capital *H*, and a lowercase *h* is used to indicate the normal allele. a. The zygotes that gave rise to the first two individuals received the *H* allele in the gamete that came from one parent and the *h* allele in the gamete from the other parent, resulting in the heterozygous genotype *Hh*; these persons eventually will develop the disease. b. The zygote that gave rise to the third individual received *h* alleles in the gametes from both parents (homozygous *hh*), so this person will remain free of the disease. An individual who received *H* alleles from both parents (a relatively rare situation) would have the homozygous genotype *HH* and would develop the disease.

"No. Neither of you can get CF because it is a recessive trait—you only have the disease if you are homozygous. The fact that Richard is sick indicates that he is homozygous and that each of you is a **carrier**—you are heterozygous and you each carry one allele for CF. For example, if we indicate the allele for cystic fibrosis with a lowercase *c* and the normal allele with a capital *C*, only persons with the genotype *cc* will develop the cystic fibrosis phenotype, whereas persons with a *Cc* or *CC* genotype will have a normal phenotype. Because the allele for CF is so widespread in the population, the chance that two heterozygotes will marry and have children with the disease is quite high as well. The high frequency of the CF allele in the population explains the tragic fact that cystic fibrosis is so common."

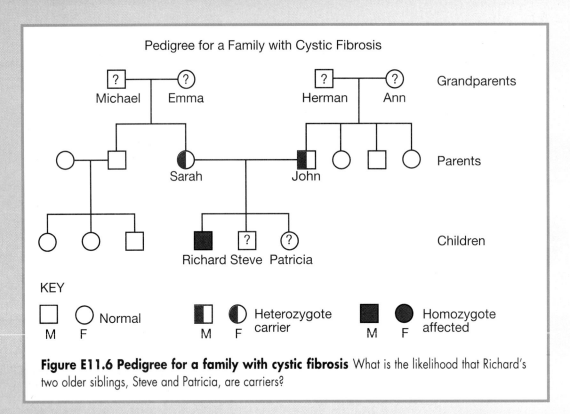

Pedigree for a Family with Cystic Fibrosis

KEY

☐ ○ Normal
M F

◧ ◐ Heterozygote
M F carrier

■ ● Homozygote
M F affected

Figure E11.6 Pedigree for a family with cystic fibrosis What is the likelihood that Richard's two older siblings, Steve and Patricia, are carriers?

Understanding that the alleles for diseases such as CF can be hidden because the trait is recessive, Richard's parents realized that they could not have known they were carriers. They provided the best care possible for Richard, including the administration of antibiotics to help his body fight lung infections and a daily routine of chest-thumping therapies to clear the thick mucous deposits from his lungs. Even though his illnesses continued to be frequent and the unpleasant treatments were a constant part of his life, Richard celebrated his fourth birthday much like any other child—with balloons, cake, and lots of friends. He was a happy child most of the time despite his restricted lifestyle and reduced life expectancy. Fortunately, recent medical advances offer the promise of a healthier life for CF patients.

Meiosis: The Mechanism Behind Patterns of Inheritance

Meiosis is the special type of cell division that leads to the production of gametes in sexually reproducing organisms. The key to understanding how certain phenotype patterns are inherited is being able to follow the chromosomes during meiosis. In nearly all cells of sexually reproducing organisms, chromosomes occur in matching pairs. Body cells such as skin and nerve cells, which contain matching pairs of chromosomes, are said to be **diploid**. For each chromosome pair, one chromosome came from the mother's ovum and one from the father's sperm. The maternal and paternal chromosomes of each pair usually have genes that affect the same traits, although the alleles for each gene may be different.

Gametes (ovum and sperm) contain only one chromosome of each pair. Gamete cells are said to be **haploid**; they have only half the number of chromosomes that the body cells of the organism have—*one* chromosome from each matching pair. To understand the origin of these two very different cell types, consider sexual reproduction at the cellular level. When an ovum and a sperm combine to form a zygote, the nuclei of both cells fuse. Thus, after fusion twice as many chromosomes exist in the one new cell, the zygote, as existed in

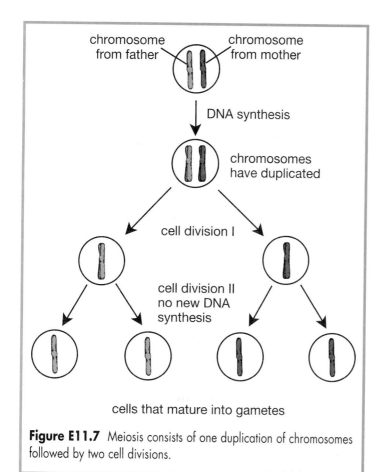

chromosome from father
chromosome from mother

DNA synthesis

chromosomes have duplicated

cell division I

cell division II
no new DNA synthesis

cells that mature into gametes

Figure E11.7 Meiosis consists of one duplication of chromosomes followed by two cell divisions.

Today we understand how chromosomes move during meiosis, and modern geneticists benefit because meiosis explains Mendel's second key discovery, the **principle of segregation**. Because meiosis results in the segregation of alleles, each gamete receives only one allele for a given gene in any chromosome pair. These gametes then can join with gametes from the opposite sex during sexual reproduction, and the resulting zygote will have a normal diploid set of chromosomes.

Meiosis provides a cellular explanation for the principles of independent assortment of genes and for segregation of alleles. It also provides a means of predicting the frequency of genotypes when organisms of known genotype mate and produce large numbers of offspring. In the example shown in Figure E11.9, parent guinea pigs that are heterozygous for a short hair gene (*Gg*) produce equal numbers of *G* and *g* gametes through meiosis. When matings occur, the gametes join in random combinations. Because ½ of the ova contain the *G* allele (and ½ of these will be fertilized by sperm that also carry the *G* allele), we would expect that ½ of the offspring (½ × ½ = ¼) will have the homozygous genotype *GG*.

By the same reasoning, fertilization of the other ½ of the ova—those that carry the *g* allele—by sperm carrying the *g* allele should result in about ¼ of the offspring having the homozygous genotype *gg* (long hair). The rest of the offspring should be heterozygotes, half of them coming from unions of *G* ova with *g* sperm and the other

either the ovum cell or the sperm cell alone. If chromosome numbers were not reduced from diploid to haploid during meiosis (gamete formation), then chromosomes would double in number with each generation, and these extra chromosomes would cause the organism to develop abnormally or die.

Rather than doubling, the number of chromosomes is in fact reduced because in meiotic cell division, the number of chromosomes is doubled and then divided in half *twice*. This process results in four cells with one set of chromosomes each. Figure E11.7 illustrates what happens to one chromosome pair in a gamete-producing diploid cell. Just before meiosis begins, DNA synthesis occurs, and each chromosome in the pair is doubled. During the first meiotic division, the two doubled

chromosomes separate, resulting in two offspring cells, each with one doubled chromosome. During the second meiotic division, the doubled chromosome in each cell separates, resulting in a total of four offspring cells, each with one chromosome. Of course, most organisms have more than one chromosome, so each cell formed in meiosis has one *set* of chromosomes. Figure E11.8 shows the events of meiosis in detail.

Meiosis and segregation. In his nineteenth-century monastery garden, Gregor Mendel worked out the principles of simple inheritance before the cellular mechanisms of meiosis were known, a truly astounding intellectual accomplishment. His discoveries were made solely on the basis of well-designed experiments, keen observations, and strict record-keeping.

diploid parent cell

a. This cell has two pairs of chromosomes.

beginning of meiosis prophase 1

b. Just before this diploid cell begins meiosis, DNA synthesis occurs and each chromosome is duplicated.

crossing over

c. The pairs of duplicated chromosomes become closely aligned and join in several places. At these junctions, equivalent pieces of the chromosome pair may exchange places. This exchange process, called crossing over, results in the switching of alleles. Because the chromosomes involved in the exchange originally came from different parents, a new combination of information now exists.

metaphase 1

d. The joined chromosomes line up along the middle of the cell. Cytoplasmic fibers attach to each duplicated chromosome.

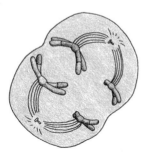

anaphase 1

e. The cytoplasmic fibers pull apart each pair of duplicated chromosomes during the first cell division.

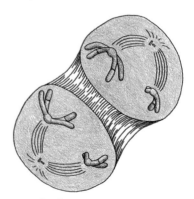

telophase 1

f. Each new cell resulting from this division contains two doubled chromosomes, one from each pair.

Figure E11.8 The stages of meiosis This figure illustrates the events of meiosis for a cell that has two pairs of chromosomes.

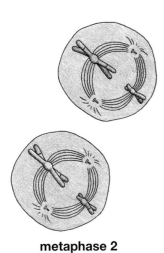

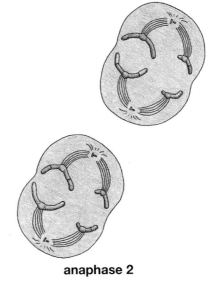

metaphase 2

anaphase 2

g. A second cell division now takes place with no further DNA synthesis. During this cell division, there are no matching chromosome pairs. Instead, the doubled chromosomes line up in single file.

h. Fibers pull apart each doubled chromosome.

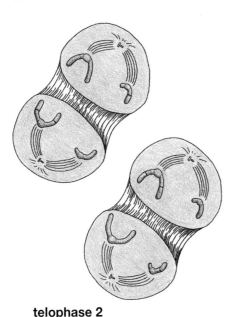

telophase 2

i. Each of the resulting four offspring cells has a single set of chromosomes and is haploid. Mature human gametes form from the products of meiosis.

half from *g* ova and G sperm. (The dominant trait is always listed first, regardless of which parent passed it on.) Of the resulting offspring at one mating, ¾ should have short hair (½ *Gg* plus ¼ *GG*) and ¼ should have long hair (*gg*), which produces a 3:1 phenotype ratio.

When many people first learn about allele segregation, they jump to the conclusion that this principle makes it possible to predict the outcome of crosses exactly. Segregation, however, only tells us the *likelihood* of a given result. Because each fertilization involves the union of a single ovum and a single sperm out of many possible gametes, each union is a separate event governed by the laws of probability. It is similar to flipping a coin. For example, imagine that the parent guinea pigs in Figure E11.9 have a litter of four. Both parents are heterozygous (*Gg*) for the short hair allele. If we examined many such litters, we probably would find *on the average* that ½ of the littermates had the *Gg* genotype, ¼ the *gg* genotype, and ¼ the *GG* genotype. It is possible, however, that three or even all four littermates *in any given litter* might turn out to be GG, just as it is possible to toss four heads

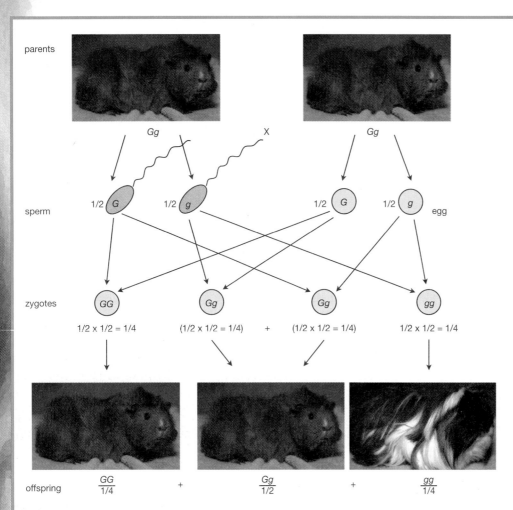

parents

Gg   X   Gg

sperm   1/2 (G)   1/2 (g)   1/2 (G)   1/2 (g)   egg

zygotes   (GG)   (Gg)   (Gg)   (gg)

1/2 x 1/2 = 1/4   (1/2 x 1/2 = 1/4)   +   (1/2 x 1/2 = 1/4)   1/2 x 1/2 = 1/4

offspring $\frac{GG}{1/4}$ + $\frac{Gg}{1/2}$ + $\frac{gg}{1/4}$

Figure E11.9 The regularity of allele segregation makes prediction of offspring possible. If the genotypes of the parents are known, then the gametes that result can be known as well. The possible combinations of gametes yield predictable genotypes and phenotypes of offspring.

The Role of Variation in Evolution

The continuity of genetic information as it passes, through inheritance, from one generation of a species to the next is impressive. Such continuity is possible because genes, for the most part, remain unchanged when they are passed from parents to offspring during reproduction. Nevertheless, examples of variation can and do arise. Indeed, for evolution to occur, sources of genetic variability on which natural selection can act must exist. There are two clear sources of such variability. One is the physical modification of genetic material, a process known as mutation. The other is recombination, in which new combinations of alleles are created when chromosomes cross over during meiosis (refer back to Figure E11.8).

in a row with a coin. When large numbers of matings are considered together, however, the frequencies of different genotypes in the offspring usually come fairly close to those predicted by allele segregation, just as the ratio of heads to tails will be fairly close to 50:50 if a coin is tossed 100 times.

Mutation. Mutations of alleles appear to be quite rare under natural conditions. In fact, the rate at which mutations occur is so low that if you examined 100,000 corn kernels for a particular phenotype such as altered color or shape, you would find only a few. Excessive mutations would have a negative impact on continuity because most mutations reduce or entirely eliminate the activity of the allele. Indeed, it is a mutation in the Factor VIII gene that causes hemophilia A and, in this case, the mutation is maintained through inheritance. Thus, the disease that this mutation causes can be passed down through inheritance as well.

Keep in mind, however, that only mutations in the genetic material of a gamete can be inherited by offspring. In other words, a mutation in one of your skin or stomach cells cannot be transmitted through inheritance. The hemophilia A allele that affected Czarevich Alexis apparently arose as a mutation in one of the two gametes that gave rise to his great grandmother, Queen Victoria.

Even though mutation might appear to be a negative thing, it is the ultimate source of all new genetic information in evolution. Thus mutations are critical for generating the diversity on which natural selection can act. Anything that raises the frequency of mutations increases genetic diversity. Unfortunately it also increases the proportion of individuals who may be born with birth defects or other unfavorable phenotypes (see Figure E11.10). Radiation and certain chemicals, which increase the rate at which mutations occur, frequently raise concerns about the environmental effects of pollutants.

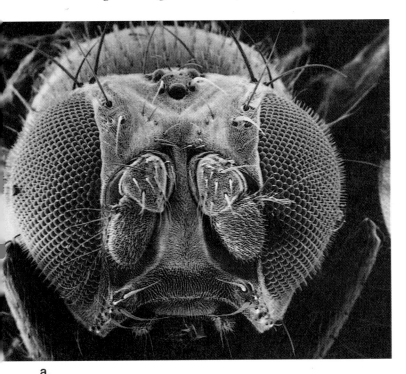
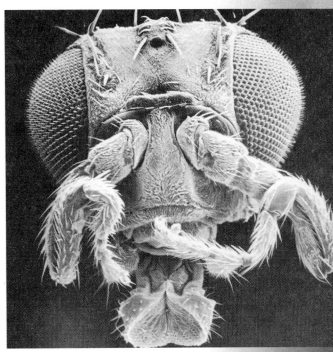

a.   b.

Figure E11.10 Most mutations are harmful. a. Normal fly. b. Mutant fly. In this mutant fly, the antennae structures have developed as legs.

Mutations generally are viewed in a negative light, but they have found a valued place in the research laboratories of geneticists. Scientists intentionally treat certain organisms—such as yeast, bacteria, and fruit flies—with toxic chemicals and radiation. They do so to mutate particular genes that they are interested in studying or to find new genes. One example of the research process is the search for a new yeast gene.

If a scientist has identified a process that he or she is interested in researching, such as discovering how chromosomes are pulled apart during meiosis, he or she can mutate thousands of

MUTATIONS CAN BE USED AS TOOLS FOR BIOLOGICAL RESEARCH

yeast cells and look for those cells that fail to separate their chromosomes during meiosis. Then, using genetic engineering techniques, the scientist places small pieces of DNA from a nonmutated organism into the mutant organism in the hope that the additional DNA will replace the mutated gene and serve as a functional copy of the gene (in this case, a gene that controls the separation of chromosomes). When the scientist finds a mutated organism that now functions properly (it once again can separate its chromosomes), he or she can assume that the change is due to the small piece of DNA that was added. Because it is possible to keep track of these small pieces of DNA, the scientist has, in effect, found the gene responsible for the previously defective process.

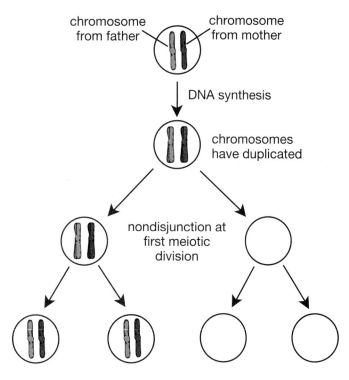

chromosome from father

chromosome from mother

DNA synthesis

chromosomes have duplicated

nondisjunction at first meiotic division

cells that mature into gametes

Figure E11.11 Nondisjunction Compare this sequence of events with normal meiosis in Figure E11.8.

Recombination. Because mutations are rare, they are responsible for introducing a relatively small amount of variation into most species. The major sources of genetic variation are independent assortment; segregation; and the crossing over, or recombination, that occurs during meiosis. Crossing over can result in new combinations of alleles on the two chromosomes of a matching pair, and the resulting unique chromosomes now can assort independently and carry their new allele combinations with them.

The frequency of recombination depends on how far apart adjacent genes lie on the chromosome. If two genes are located far apart, a crossover between the two genes is more likely. If the genes are close together, however, crossovers between them occur only rarely. Two genes so close together that they do not sort independently are said to be linked genes. Tightly linked genes—two genes that are *very* close together—usually appear in the same gamete because they almost never are separated by recombination. Linkage is an exception to the rule that genes assort independently.

The processes of meiosis and recombination are not perfect, and occasionally mistakes occur that can be fatal to the

offspring. If a chromosome is abnormal or if problems occur during crossover, the two members of that chromosome pair may not segregate from each other into the cells formed at the first division of meiosis. As Figure E11.11 indicates, the result then is that both members of the chromosome pair can end up in one offspring cell, and the other offspring cell gets neither. At the second division of meiosis, the doubled chromosomes separate in the usual way. This process, called nondisjunction, results in the production of four gametes. Two of the gametes have two copies of that particular chromosome (these gametes are diploid for that chromosome rather than haploid). The two other gametes have no copies. Fertilization of normal gametes by these gametes produces offspring that have either three copies or only one copy of that chromosome.

The imbalance caused by an extra set of all the genes on this third chromosome always produces multiple disruptions of the individual's phenotype. These disruptions usually have severe results. In humans individuals with three copies of one chromosome survive only if the extra copies are of the X or Y chromosomes or chromosome 21. Individuals who have three copies of any of the other chromosomes either die as embryos or very shortly after birth. Individuals who have three copies of chromosome 21, although they survive, exhibit a collection of phenotypes called Down syndrome. These individuals generally show some degree of mental and physical disability, ranging from slight to severe; are prone to leukemia and heart disease; and have a shorter than normal life expectancy.

Variation in the fossil record. The fossil record shows that continuity across many generations is an

Figure E11.12 Trilobite The trilobite's actual size is that of a paper clip. Comparison of fossils from sediments of different geological eras shows that there was little change in this species across thousands of years.

E166

ancient and fundamental property of living systems. For example, hard-shelled invertebrates such as trilobites (see Figure E11.12) and snails, which have been preserved in great numbers by the deposition of sediment, look recognizably the same across hundreds of thousands of years. Through sufficiently long periods of time, however, changes in organisms become clear. The pace of this change is often so slow that we can observe the changes only by following the organisms across very long periods of time, such as millions of years.

A good example of slow and gradual change across time is the evolutionary line that gave rise to modern horses. As Figure E.3.3 (page E33) shows, an early ancestral form of the horse was an animal about the size of a small dog, which ran on feet that had four toes each. If we saw this animal today, we might not immediately recognize it as an ancestor of the horse. As we study more recent fossil horses we can see structural adaptations in them that cause them to more closely resemble modern horses. Through millions of years, the feet of these ancestral horses changed until only a single large toe in the form of a hoof remains in the more recent horse ancestors (see Figure E11.13 and relate it to Figure E3.3). With each change in size and foot shape, we begin to see more and more resemblances to the modern horse. Thus, when one views all of the intermediate forms, the evolutionary connection between the modern horse and its original ancestor becomes obvious.

How did these types of evolutionary changes arise? The answer is that natural selection acted on the natural variation in a population. Variations that improved the chances of survival and reproduction in a species tended to be maintained. Such variations were adaptive in that they enhanced (or at least did not reduce) the ability of the organisms possessing them to deal with their environment.

Natural selection affects all traits, those that show discrete variation such as cystic fibrosis and gender as well as those that show continuous variation such as height and leg length. Figure E11.14 demonstrates that leg length shows a distribution among individuals that resembles a bell when graphed. Individuals with leg lengths clustered around the average value are the result of selective processes that favored survival and reproduction of their ancestors. That is why there are so many individuals near the middle range. Individuals that show *extreme* divergence from the average or norm (for example, very short or very long legs) are present in much smaller numbers because these extreme traits apparently were disadaptive; that is, these traits actually decreased the chances of survival and reproduction. For instance, horses with very short legs might have been slow runners and thus less likely to escape predators than horses with longer legs. Conversely, horses with very long legs might have been less able to tolerate the stresses of running, which may have led to more leg breaks, than horses with shorter legs. Regardless of the specific trait, all cases of

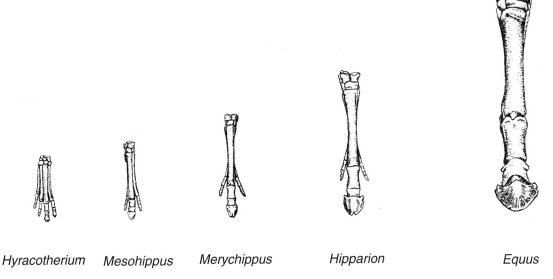

Hyracotherium   Mesohippus   Merychippus   Hipparion   Equus

Figure E11.13 The evolution of species across time During the evolution of the horse, the overall form changed through long time periods. Specialized structures such as the foot also changed. The modern horse's foot contains only a single toe, or hoof. What other features of the legs have changed?

evolutionary change result from the same two steps: the production of variation in each generation and the selection of traits, which tests each new variation in the struggle to survive and reproduce.

Even though changes in species, such as the changes in the horse, can occur gradually, not all evolutionary changes in organisms require millions of years. In 1972 the evolutionary biologists Stephen Jay Gould and Niles Eldredge proposed a theory known as **punctuated equilibrium** to explain the sudden appearance of new species in the fossil record. In punctuated equilibrium, one species changes to another species in a relatively brief geological time span, perhaps as short as several tens of thousands of years. One way that species might change so rapidly is related to geographical isolation. For instance, if one species of organism is confined to one specific, stable habitat, then the pressures of natural selection will favor relatively uniform adaptations within that species. If this species escapes its confined habitat, however, the offspring with different traits will be free to colonize habitats that best suit their unique differences.

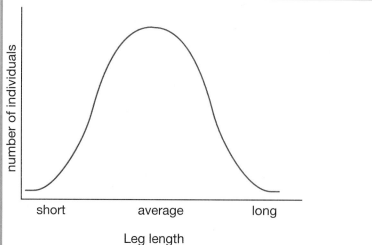

Figure E11.14 Continuous variation If a trait that varies continuously, such as leg length, is examined in a large population, it becomes obvious that most individuals cluster around the average value for that trait. Relatively few individuals exhibit the extreme ranges of the trait. Such traits are called multifactorial because they are caused by many genes interacting with many environmental factors. How many other traits that exhibit a continuous pattern of variation can you identify?

Punctuated equilibrium helps explain how new species evolve. Whether gradual or punctuated, all evolution is influenced by selective pressures acting on natural variation in populations.

Genetic Complexity

Can all phenotypes be explained simply by following one or two genes through meiosis? Is an understanding of segregation of alleles sufficient to explain continuously varying traits such as behavior? Yes and no. Huntington disease has many complex effects on human health and behavior, and its transmission is quite easy to follow genetically. But scientists cannot yet explain *how* a mutation in this one gene can cause so many effects. What about behavioral or physical phenotypes that we consider to be normal? Can genetics explain why you usually are happy while your best friend usually is sullen, or why

your brother is significantly thinner than you are?

It might seem strange to consider genetics when thinking about behavior, but if it is true that an organism's genotype and the environment interact to produce an organism's *complete* phenotype, then we would expect genetics to contribute to all sorts of complex behaviors. This expectation is true in principle but difficult to determine precisely in practice.

The influence of genotype on complex behavior. Direct evidence that genetics can influence behavior first was discovered in the 1930s by scientists who were studying the

maze-running ability of mice. Maze-running is a complex behavior that involves many components, such as speed, memory, and decision making. The scientists found that a given population of mice could run the mazes successfully when given enough time, but as expected some mice made many wrong turns whereas others made very few. The scientists mated "maze-dull" mice with other maze-dull mice from this population, and they mated "maze-bright" mice with other maze-bright mice to see if the ability to run the maze poorly or well, respectively, could be inherited. Figure E11.15 shows that after several generations, two

UNIT FOUR CONTINUITY: Reproduction and Inheritance in Living Systems

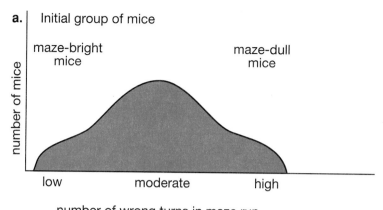

a. Initial group of mice

maze-bright mice   maze-dull mice

number of mice

low   moderate   high

number of wrong turns in maze run

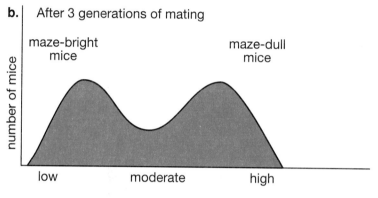

b. After 3 generations of mating

maze-bright mice   maze-dull mice

number of mice

low   moderate   high

number of wrong turns in maze run

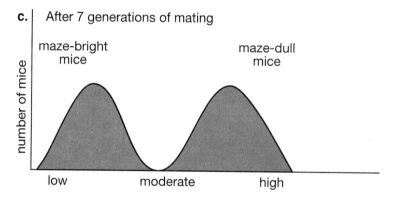

c. After 7 generations of mating

maze-bright mice   maze-dull mice

number of mice

low   moderate   high

number of wrong turns in maze run

Figure E11.15 A genetic component to behavior a. If scientists test a large group of mice for its ability to run a maze successfully, they find a wide range of behaviors. This graph shows the typical distribution for a trait exhibiting continuous variation. b. and c. If scientists choose maze-bright mice from this group and mate them with other maze-bright mice, they eventually end up with a population of consistently maze-bright mice with only a narrow range of behavior. Likewise, a population of uniformly maze-dull mice can be bred if maze-dull mice are mated with other maze-dull mice from the starting population. This experiment illustrates that complex behaviors have genetic components.

distinct populations of mice existed—maze-dull mice and maze-bright mice. Thus the scientists were able to demonstrate that genetic variation for a complex behavior existed and could be passed from one generation to the next. With so much genetic variation, it becomes clear that *normal* behavior from a population standpoint is really just an average of the tremendous diversity that exists among the individuals.

What distinguishes complex traits, such as behavior, intelligence, and personality, from simple traits, such as cleft chin or hemophilia? Most scientists think that two factors are involved: the environment and the number of genes. You already have seen that the environment can affect the phenotype of genetically identical plants grown at different altitudes. Skin color may offer one example of a trait that is affected by the number of genes—in this case, genes that control the production of skin pigments called melanins. How might these genes interact to produce the large variation in skin colors exhibited by populations around the world?

A MODEL FOR HOW MULTIPLE GENES MIGHT INFLUENCE THE COMPLEX TRAIT OF SKIN COLOR

Imagine that two genes, designated by alleles *C* or *c* and alleles *S* or *s*, are involved in skin color. Only capital-letter alleles result in the production of melanin. Thus the more melanin alleles of the capital-letter type that a person inherits (either *C* or *S*), the darker his or her skin will be.

There are nine possible genotypes when you consider all of the allele combinations generated by independent assortment and segregation. Figure E11.16 illustrates a genetic model in which the effect of having a capital-letter allele is more melanin. Notice that this is *not* simply a case of dominant and recessive alleles because the effects of *two distinct* genes must be taken into consideration. Models such as this one are additive; expression of the genes involved in that trait add together to produce a given phenotype (two capital-letter melanin alleles result in a darker color than only one capital-letter melanin allele). Keep in mind that environmental influences will have some impact on phenotype even when more than one gene is involved.

Genotype	Number of Melanin Alleles	Skin Color Phenotype
ccss	0	very light
Ccss or *ccSs*	1	light
CCss or *CcSs* or *ccSS*	2	medium
CCSs or *CcSS*	3	dark
CCSS	4	very dark

Figure E11.16 A model for the interaction of two genes that cause skin color

The tremendous diversity of phenotypes seen in any given species should be easier to understand now that you see how many complex genetic interactions are possible. Genes may have multiple alleles that contribute to genetic diversity at a single gene level as in the case of ABO blood types. Different genes can interact with each other to produce complex phenotypes, such as behavior and skin color. In addition to all of this variation at the genotypic level, the environment varies from place to place and exerts different pressures on the diverse individuals living there.

For example, some geneticists propose that environmental factors influenced the evolution of skin color genes to compensate for the varying amounts of sunshine in different parts of the world. In northern climates where there is less sunshine, too much skin pigment is a disadvantage because the pigments filter out ultraviolet light and humans require ultraviolet light to synthesize vitamin D. Thus some scientists propose that, through natural selection, the frequency of alleles that do not produce melanin increased among populations living in northern areas that tend to be cloudy and have short days in winter. This natural selection preserved the populations' ability to synthesize vitamin D. Similar interactions between genes and the environment ultimately account for all of the diversity of life on earth.

Incomplete Dominance

Figuring out an organism's genotype from its phenotype is not always straightforward. For example, a person who does not have cystic fibrosis may have either of two possible genotypes, *Cc* or *CC*, but the phenotype (no disease) is the same in either case. For some traits, however, it is possible to distinguish between individuals who have heterozygous genotypes and those who have homozygous genotypes on the basis of their phenotype.

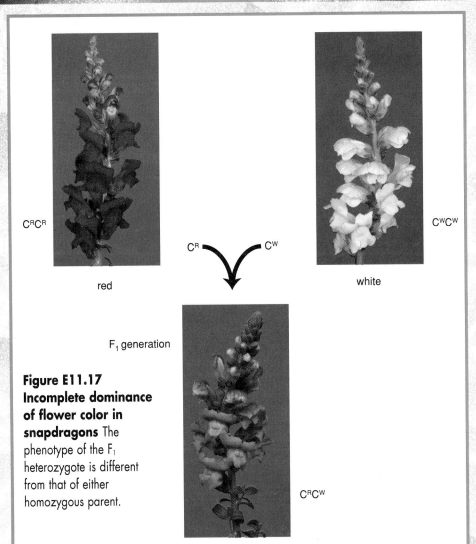

C^RC^R

red

C^R C^W

C^WC^W

white

F₁ generation

**Figure E11.17
Incomplete dominance
of flower color in
snapdragons** The
phenotype of the F₁
heterozygote is different
from that of either
homozygous parent.

C^RC^W

pink

represents a hemoglobin gene, and the ^s superscript designates which allele is represented.) The oxygen-carrying hemoglobin molecules in affected individuals' red blood cells clump when the oxygen supply is low, causing the cells to lose their flexibility and assume an abnormal, sickle shape. If a large number of these cells block small blood vessels and stop the flow of nutrients and oxygen, the individuals may develop bouts of illness that include fever and severe pain in their limbs and joints. Eventually vital organs may be damaged, and the individuals may die. Individuals who have the heterozygous genotype Hb^AHb^s normally are as healthy as Hb^AHb^A individuals (even though they produce some sickle hemoglobin). Under certain conditions (such as exposure to reduced oxygen pressure at high altitudes), they may develop sickling of their red blood cells and mild symptoms. They are said to have sickle cell trait.

Sometimes genotypes are written using superscripts above the gene symbol to designate different alleles of that gene. In snapdragons, for example, plants that have the genotype C^RC^R produce bright red flowers and plants that have the genotype C^WC^W produce white flowers. Plants that have the genotype C^RC^W bear flowers that are neither pure red nor pure white, but pink (see Figure E11.17). Traits in which the heterozygous genotype produces a phenotype that falls somewhere between the two homozygous phenotypes are described as showing **incomplete dominance**.

An example of a human trait that shows incomplete dominance is sickle cell disease, an inherited disorder that is most common in individuals whose ancestors lived in equatorial Africa, the Middle East, or the southern Mediterranean. Individuals who have the homozygous genotype Hb^sHb^s suffer from sickle cell disease. (In this genotype symbol, Hb

Scientists have discovered an interesting association between sickle cell disease and malaria, a common tropical disease that is caused by a mosquito-borne parasite. Individuals who have sickle cell trait (Hb^AHb^s *heterozygotes*) are less likely to be infected by the malarial parasite than are healthy, non-sickle cell individuals (Hb^AHb^A *homozygotes*). In other words, people who live in regions where malaria is prevalent have a better chance of avoiding malaria if they have sickle cell trait. It is thought that the shape of the sickled cells inhibits infection by the malarial parasite. Ultimately, this interaction between parasite and human means that people who carry one sickle cell allele and live in areas where malaria is common have a selective advantage over those who lack the allele. The trait increases their chances of living long enough to reproduce, and in this way, natural selection has preserved the allele for sickle cell disease.

Genetic Information Is Stored in Molecular Form

Serving as a juror at a murder trial in which lawyers present DNA evidence represents a rather special case in which some knowledge of the structure and function of DNA might prove useful. Very few of us, however, will face the particular dilemma that confronts the jury in such a courtroom situation. Nevertheless, whether we realize it or not, many of us already are affected by the same rapid advances in scientists' understanding of DNA that make DNA testimony in the courtroom possible. Moreover, advances in DNA technologies will affect most of us in the future.

To understand how your life may be affected by these advances, you need to understand that DNA contains information that is critical to the structure and function of each of the cells of your body. In fact, the instructions encoded in DNA play a major role in determining how your body operates.

DNA instructions are important for the continuity of humans and other living systems. Organisms depend on the accurate transmission of genetic information from one generation to the next. The basic structure of DNA and the processes by which DNA is copied help ensure the accurate transmission of genetic information from parents to offspring (and from a cell to its offspring cells). In studying DNA's role in genetic transmission, we actually are studying the molecular basis of reproduction.

Figure E12.1 expands on the genetic continuity ideas of Figure 11.8 by looking at the inheritance of genotype and the production of phenotype at a molecular level. The arrows that lead from the box labeled *genotype* to the word *replicate* and back to *genotype* represent the cyclical events of reproduction that maintain the genetic continuity of the species. The term *replicate* refers to making a copy of a DNA molecule. The replication of DNA allows organisms to keep a set of genetic instructions for themselves and also to pass on sets to their offspring.

DNA is important not only to offspring but also to each *individual* organism. To see why, we must consider the processes by which cells use the information they contain. The use of genetic information is known as gene expression, and the result of **gene expression** is an organism's phenotype. Figure E12.1 also illustrates these events. The arrows that lead from the box labeled *genotype* to the word *expression* and to the word *phenotype* illustrate the flow of information as genes are expressed. Ultimately the instructions stored in DNA are used to build, maintain, and regulate the cells of all living organisms. In this way the information stored in DNA is a little like the information stored in the operating system of a computer. A computer cannot operate without operating system instructions, and living systems cannot function without the instructions for life, which are encoded in DNA.

As you develop a deeper understanding of how living systems rely on DNA, you will better understand why it is critical that DNA is transmitted accurately from one generation to the next. You also will understand how

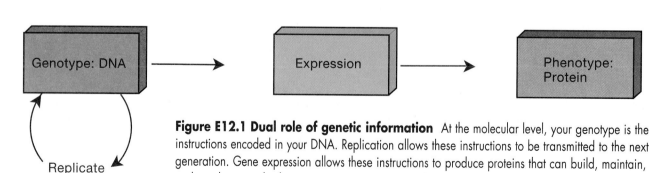

Figure E12.1 Dual role of genetic information At the molecular level, your genotype is the instructions encoded in your DNA. Replication allows these instructions to be transmitted to the next generation. Gene expression allows these instructions to produce proteins that can build, maintain, and regulate your body.

scientists' growing knowledge of DNA already is influencing many aspects of our lives. For instance, DNA now offers hope of early diagnosis and treatment of diseases such as cystic fibrosis, improvements in agriculture such as disease-resistant crops, and even the establishment of guilt or innocence in criminal cases.

DNA MAY BE THE STUFF OF GENES!

New York (1944) The effort to describe the molecular composition of those mysterious particles that scientists call genes moved a giant step forward this week. A new report indicates that a chemical called *deoxyribonucleic acid (DNA)* may be the so-called "transforming" substance discovered more than 10 years ago by Frederick Griffith. Griffith observed that mice died when exposed simultaneously to live, rough-surfaced, nonlethal bacteria and dead, smooth-surfaced, lethal bacteria. He then observed that bacteria in the tissues of the dead mice had been *transformed*; the bacteria now contained both the smooth-surface and lethal characteristics. This transformation—the change in the phenotype of the living bacteria from rough-surfaced to smooth-surfaced and from nonlethal to lethal—seems to occur when some information-rich substance moves from the dead bacteria to the live bacteria.

The identity of this transforming substance, however, remained unknown until this week. Oswald Avery, Colin MacLeod, and Maclyn McCarty—investigators at the Rockefeller Institute—reported a set of experiments that demonstrates that the substance causing the change actually is DNA. This discovery suggests that genes, the particles thought to carry the information that determines the physical and chemical characteristics of a cell, also are made of DNA. Scientists do not know much about this powerful substance, but this week's announcement likely will intensify efforts to describe its structure and function and to explain how it is able to bring about such important effects in cells.

DNA Structure and Replication

The fact that DNA is required for the building, maintenance, and regulation of the cells of all living organisms means each new cell must receive a copy of the genetic material from its parent. How is this copy made? The complete process of replicating the genetic material of a cell is quite complex. The key to understanding this complex process, however, is very straightforward. You simply have to understand a few things about how a cell's genetic material is structured and organized.

In humans (as in all eukaryotes), the genetic material consists of long linear DNA molecules that are packaged tightly in chromosomes. In each of our cells (except gametes and red blood cells), we have 23 pairs of chromosomes. Each chromosome, in turn, contains one long DNA molecule and many protein molecules, which are organized into beadlike clusters. As Figure E12.2 shows, DNA wraps tightly around these clusters to form nucleosomes, which pack together to form the condensed and compact structure we see when we look at a dividing cell under the microscope.

Because the information storage capacity of genes lies ultimately in the DNA and not in the protein components of chromosomes, the question of how a cell copies its genetic information during reproduction actually is a question of how a cell copies its DNA. Copying DNA accurately, much like communicating a message, is easier when some form of physical **template**, or pattern, is used. Games like "pass the message" demonstrate how easily inaccurate copying can occur when there is no physical record, such as a written note, of the transmitted message. Perhaps you can imagine that a physical template also would help carry out the huge task of accurately duplicating genetic information. In fact, this is exactly how biological systems accomplish this task: the DNA molecule itself serves as a template for information transfer. That is, the molecule that contains the genetic information acts as a pattern for its own replication.

DNA is able to serve as its own template because it is organized in a very regular manner. The drawing of DNA shown in Figure E12.4a reveals that a single DNA molecule is a double-stranded structure. The two strands are twisted together into a helical form. This aspect of DNA earned its nickname: the double helix.

Examining these two strands more closely reveals that each of these two strands (Figure E12.4b) is

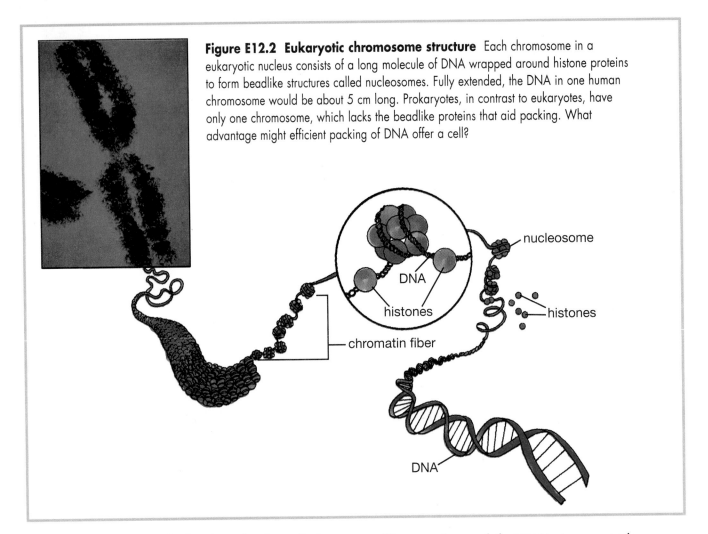

Figure E12.2 Eukaryotic chromosome structure Each chromosome in a eukaryotic nucleus consists of a long molecule of DNA wrapped around histone proteins to form beadlike structures called nucleosomes. Fully extended, the DNA in one human chromosome would be about 5 cm long. Prokaryotes, in contrast to eukaryotes, have only one chromosome, which lacks the beadlike proteins that aid packing. What advantage might efficient packing of DNA offer a cell?

composed of a series of subunit molecules called nucleotides. The energy-carrier molecule ATP is a nucleotide. A long strand of nucleotides bonded together is called a **nucleic acid**; thus DNA is an example of a nucleic acid.

A detailed look at the nucleotides that make up each strand of DNA (Figure E12.4c) reveals even more regularity. Each nucleotide is composed of the same three parts: a nitrogen base, a deoxyribose sugar, and a phosphate group. The sugar and phosphate portions do not vary from one nucleotide to the next. The first irregularity in DNA is found in the nitrogen bases. As Figure E12.4d shows, a DNA nucleotide may contain one of four different nitrogen bases. The four unique bases are like the letters in a word. Just as the order of letters in a word conveys information to someone who understands the code that we call a written language, so the order of nitrogen bases along one strand of DNA conveys information to any part of the cell capable of translating the code.

If you understand that DNA stores genetic information in the sequence of the nitrogen bases along one strand of DNA, then you already understand a great deal about the structure of genes at the molecular level. In fact, only one question remains to be answered to complete your basic knowledge of DNA structure: How is it that one DNA molecule can be replicated to make another identical molecule?

The answer to this question is that there are predictable interactions between the nitrogen bases. On close inspection of the DNA double helix, we find that the interaction between nitrogen bases on opposite strands is not random. Instead, a large base on one strand always bonds to the same small base on the opposite strand. Figure E12.5 shows this aspect of DNA structure. Note that the pairing that occurs between nitrogen bases is absolutely specific in DNA strands: adenine (A), a large base, always bonds with thymine (T), a small base; and guanine (G), a large base, always bonds to cytosine (C), a small base. Scientists refer to the recognition and specific pairing of these bases as **complementarity**.

Complementary base pairing explains how it is possible for DNA to act as a template for its own replication. Like other biosynthesis reactions, replication is carried out by specific enzymes. In essence, replication occurs as these enzymes separate the two DNA strands, read the sequence of nucleotides on one strand, and build a new, complementary strand by adding one nucleotide at a time to the new strand. Because A always bonds with T and G always bonds with C, the sequence of bases in the old strand *determines* the sequence of bases in the new strand. In other words, each base that is added to a growing chain during replication must *complement* the base in the old strand with which it will pair. In this way any DNA molecule can serve as a template for a new copy of the genetic information that it encodes.

Figure E12.6 illustrates the process of replication. As shown in the figure, the double strands of the DNA molecule first separate, which allows the replication enzymes to build a new strand of DNA to match each of the old strands. This separation of the double helix, which is critical to its ability to replicate, is possible because the attractive forces that hold the two strands together are relatively weak hydrogen bonds. Hydrogen bonds are much weaker than the strong covalent bonds that hold adjacent nucleotides together in each strand, but they are strong enough to allow one nucleotide to attract and hold its complementary base-pairing partner. This base-by-base addition process continues until there are two molecules of DNA where previously there had been one; each molecule contains one old strand and one new complementary strand.

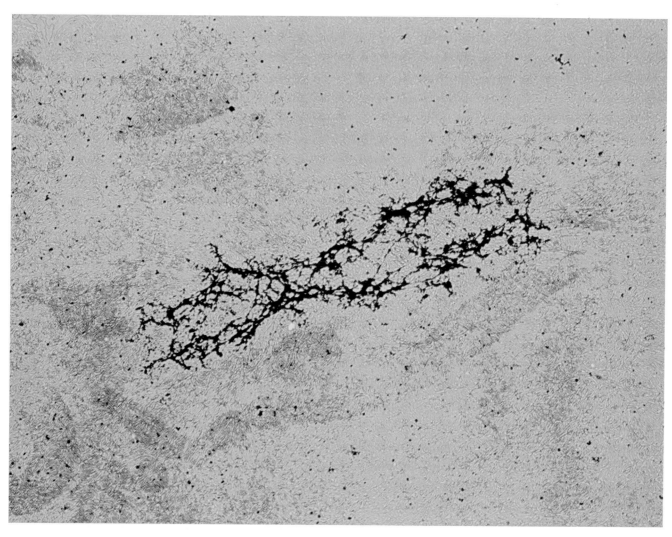

Figure E12.3 Human DNA This human chromosome, seen through an electron microscope, was treated with a substance that disrupted the chromosome's structure, releasing the DNA.

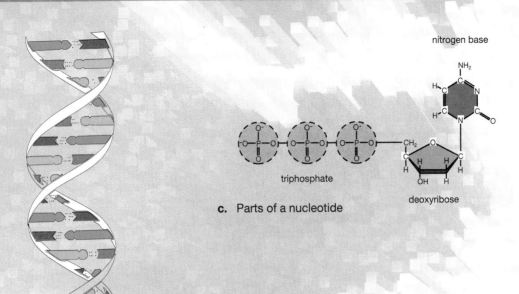

nitrogen base

c. Parts of a nucleotide

triphosphate

deoxyribose

a. The double helix

5'
—OCH₂
3'
adenine

CH₃
5'
CH₂
3'
thymine

5'
CH₂
3'
guanine

5'
CH₂
3'
cytosine

nucleic acid

symbol for nucleotide

b. Nucleotides are covalently bonded to form a strand known as a nucleic acid.

d. Nucleotides may contain one of four different nitrogen bases. The sequence of nucleotides along the strand encodes the cell's genetic information.

Figure E12.4 Structure of DNA a. The double helix. b. Nucleotides are covalently bonded to form a strand known as a nucleic acid. c. Parts of a nucleotide. d. Nucleotides contain one of four different nitrogen bases. The sequence of nucleotides along the strand encodes the cell's genetic information. How many different sequences could be made using just 10 nucleotides in a DNA strand?

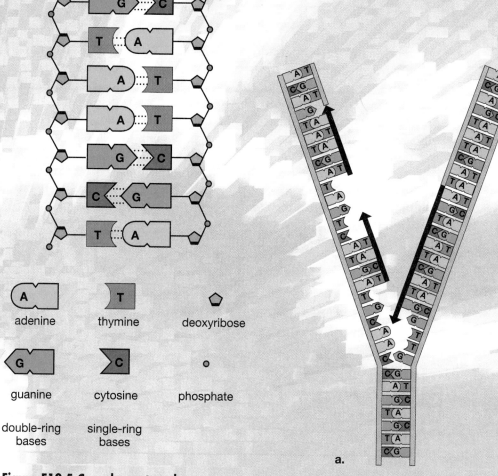

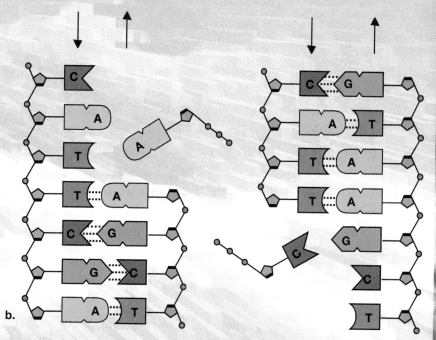

adenine

thymine

deoxyribose

guanine

cytosine

phosphate

double-ring bases

single-ring bases

Figure E12.5 Complementary base pairing in DNA Notice the pattern of how the bases pair with each other. The dots between base pairs represent hydrogen bonds, which hold the two strands in a double helix together. Are there any differences in hydrogen bonding between G and C and between A and T?

a.

b.

Figure E12.6 DNA replication a. A replication fork at which two new DNA strands are being synthesized. b. Details of nucleotide addition. DNA replication enzymes add nucleotides one at a time to each of the growing strands. In eukaryotic cells the process of DNA replication occurs in the nucleus, but in prokaryotes replication occurs in the cytoplasm.

Replication Errors and Mutation

Templates provide a very accurate way of transmitting information, whether it be written language or genetic information in DNA. Even with the use of a template for replication, however, errors do occur. In language these errors can be found as misspellings in books and articles. In DNA these errors are called mutations. Sometimes the replication enzymes mistakenly skip a base, and the new strand that is formed is missing a base. When this strand is replicated, the error is copied onto a new second strand, making a DNA molecule that is missing a nucleotide at that position. This is called a deletion mutation. Sometimes the replication enzymes mistakenly add the wrong base to a position, substituting one small base for the other or one large base for the other. This is called a substitution mutation. Although the complementary pairing normally seen in DNA is disrupted at the point of substitution, if the DNA later is replicated, the newly synthesized strand will contain an appropriate base pair for the substitution mutation (Figure E12.7). In this way the mutation is preserved and perpetuated. Despite the existence of specific

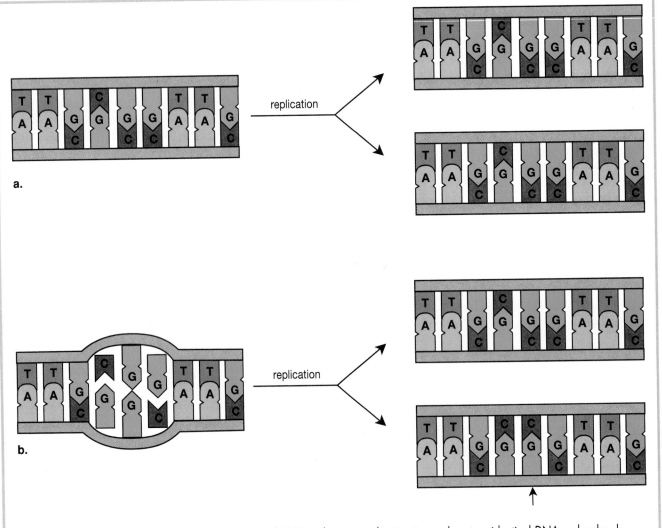

Figure E12.7 Preserving a mutation a. Normal DNA undergoes replication to produce two identical DNA molecules. b. Mutated DNA (note that a C from the normal DNA has mutated to G) undergoes replication to produce two new DNA molecules only one of which is identical to the normal DNA. In this way replication preserves mutation (arrow).

enzymes that repair mutations, some mutations become a permanent part of a cell's genetic material.

Mutations can occur in the DNA of any cell in the body of an organism. If mutations occur in cells that give rise to reproductive gametes, then the mutation will be passed along to offspring, perhaps becoming a lasting change in the genetic information of that species. In the case of single-celled organisms that make new individuals by simple cell division, any mutation may be passed along to new individuals. Ironically, the fact that the template method of copying tends to increase accuracy also means that, once a mutation occurs, it will be copied accurately as well.

If errors are copied accurately during replication, does this mean that every alteration in DNA structure gives rise to a lasting change in a species' genetic information? No, it does not because some mutations, such as the mutations that cause serious birth defects, are so harmful that they may kill the organism long before it reaches reproductive maturity. In these cases no more copies of the mutation will be made. Less serious but still harmful mutations may lead to phenotypes that are disadvantageous to survival and reproduction. For instance, a mutation that results in a frog with a short tongue may limit the frog's success in catching insects, which in turn may prevent it from growing large and strong and mating successfully. In these cases the forces of natural selection will tend to reduce the frequency of that genotype in the population.

Changes that *may* be perpetuated in the genetic material of a species are those that are advantageous to an organism possessing them, neutral in their effect or only mildly disadvantageous. Some neutral or mildly harmful mutations may be advantageous in a different environment. Others may become advantageous at some later point if changing external conditions cause the phenotype associated with the new genotype to become an advantage for survival of the individual. For example, individuals who carry one allele for the sickle cell trait have a selective advantage if they live in parts of the world where the malaria pathogen is present. One effect of this advantage is to increase the number of individuals in the population who carry an allele that under other circumstances would be considered rather harmful.

London, England (1953) The most recent issue of *Nature* hit scientists' desks this week. This issue contains a landmark paper proposing a physical structure for DNA, the substance that makes up the genetic material of living organisms. Coauthored by James Watson and Francis Crick, this paper already has the scientific world buzzing with enthusiasm and optimism.

WHAT'S NEW? WHY THE FUSS ABOUT WATSON AND CRICK?

To be honest, though, your humble science club's *What's New?* correspondent didn't see what the fuss was all about. I mean, I thought science was an *experimental* endeavor. How could a paper that only proposes a *model* be exciting? Where is the evidence that these two mavericks actually are right? Besides, I would think that coming up with a model would be easy—there must be hundreds of different ways to suggest how DNA is put together. So what's the big deal with *their* model?

To get an answer to my questions, I called an old friend of mine, Pete, at the university. Pete is a high-level science type, though you wouldn't know it from his no-pain/no-strain lifestyle. Anyway, did he set me straight! First, he pointed out that a good model *explains* all of the known characteristics of the system that you are studying. Take the Watson and Crick model as an example, he said. It explains how DNA can do all the things the genetic material appears to be able to do. Not only can DNA *encode* information, but it also can *store* that information in a stable manner through time. DNA also duplicates with very high fidelity and yet has sufficient structural flexibility to produce the full range of phenotypic variation that we see on the planet.

Then Pete pointed out that a good model also is *consistent* with all the available experimental evidence. In the case of DNA, this includes the evidence that DNA is shaped like a double helix, that the nitrogen bases are stacked inside the molecule in a regular pattern, and that the big nitrogen bases and the small nitrogen bases are present in the molecule in a 1:1 ratio. And finally, as if I weren't dead in the water already, Pete insisted that a good model is *testable*. He said, "Just look at all the predictions that the new DNA model makes that you could test: that a newly replicated DNA molecule would have one old strand and one new strand; that the two strands at any point on the molecule are complementary, but not identical; that . . . ."

"Okay, okay," I said. So I've learned something about modeling and science. And if Pete and his friends are right about where the study of DNA is going to take us, this is just the beginning of what I'm going to learn. In fact, all of us will be learning a lot in the next few decades. So, hats off to model-building. . . . Got any LEGOS™?

The Expression of Genetic Information

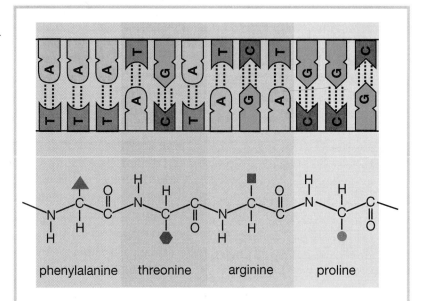

Figure E12.8 Relationship between DNA sequence and protein sequence Each strand of DNA consists of a series of nucleotides bonded together in a particular order. The protein that is specified by this DNA consists of a series of amino acids bonded together in the order encoded by the sequence of nucleotides in the DNA. A set of three nucleotides in the DNA codes for one amino acid in the corresponding protein. The genetic code is examined in more detail in the essay *Translating the Message in mRNA* (page E182).

When an error in replication occurs, the nucleotide sequence of a DNA molecule is altered, and the resulting mutations may be passed on during the next round of DNA replication. Knowing how a change in DNA structure (essentially a change in genotype) can be physically preserved, however, does not tell us what effect the altered DNA structure has on the organism. To see this effect, we must look at the organism's phenotype and discover whether it also is altered. In other words, we must look to see whether the mutation caused a change in a gene that is used to control some characteristic of the organism.

The process by which a cell uses genetic information is called **gene expression**. The first step in developing an understanding of how gene expression occurs is to realize that most genes contain information that is required to build proteins and that these proteins carry out critical biochemical and structural activities in the cell. These activities include a number of housekeeping functions that almost all cells continually must perform to remain alive. Just as you must replace burned-out light bulbs, take out trash, and grocery shop to maintain a normally functioning household, housekeeping chores in the cell include building and repairing various parts of the cell, eliminating wastes, and breaking down glucose for energy. In addition to these basic chores, however, most cells also conduct very specialized types of activities. For example, plant leaf cells harvest light

energy and use it to build sugars, and muscle cells contract to allow movement of your limbs, your heart, and your digestive system. Regardless of the activity involved, however, the information that a cell needs to build all of its proteins is contained in its DNA.

If most genes dictate the production of proteins, then the question of how a cell expresses genetic information as protein really is a question of how a nucleotide sequence (A, C, T, and G) in DNA can direct the formation of proteins, which are sequences of amino acids. The diagram in Figure E12.8 both illustrates and answers this question. Notice that there is a relationship between DNA structure and protein structure: both nucleic acids and proteins are long

UNIT FOUR CONTINUITY: Reproduction and Inheritance in Living Systems

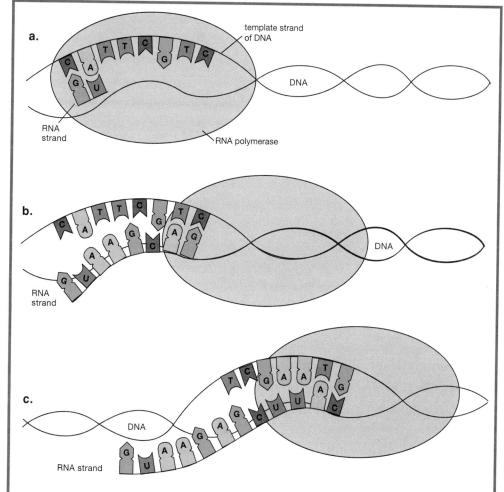

Figure E12.9 Transcription a. An enzyme called RNA polymerase sythesizes RNA from a template strand of DNA. The process of transcribing DNA into RNA is very similar to the process of replication except that only one RNA strand is made. b. Note that as the enzyme moves along the DNA, the double helix is unwound, and specific base pair interactions form between the DNA and RNA. c. The DNA behind the enzyme reforms a double helix. Eventually, when the enzyme that is building the RNA reaches the end of the gene, the DNA and the RNA are released. Transcription, like replication, occurs in the nucleus of eukaryotic cells.

expression. Regardless of the type of information a particular gene contains, in most cases the first step in making this information available for use in the cell is the synthesis of another nucleic acid, ribonucleic acid (**RNA**). Structurally, RNA is almost identical to DNA. The major differences are that RNA contains a ribose sugar instead of a deoxyribose sugar and that thymine (T), one of the bases in DNA, does not occur in RNA. Instead, a related base, uracil (U), is found at positions complementary to adenine (A).

Like DNA, RNA is built according to the information that is available in the DNA template. This first step in gene expression, which is the production of RNA from the DNA template, is called **transcription** (Figure E12.9). From the chemical similarity of RNA and DNA, you probably can imagine how DNA serves as a template to make a strand of complementary RNA. The new RNA nucleotides are arranged through base-pairing interactions in an order that corresponds to the DNA template, just as new DNA nucleotides are arranged during DNA replication. In fact, the process of transcription is, in many ways, very similar to the process of replication.

As Figure E12.10 shows, cells can make three types of RNA. The first type, called **mRNA** (messenger RNA), has the crucial role of carrying the very specific information that is necessary to make a particular protein from the DNA in the cell's nucleus to the site of protein synthesis in the cell's cytoplasm. This form of RNA logically is called messenger RNA

molecules made of repeating subunits arranged in a linear fashion. Because both nucleic acids and proteins are linear, a code in the sequence of nucleotides can specify the order of amino acids in the protein. That is, the sequence of nucleotides in a gene determines the sequence of amino acids in a protein. This is important because the sequence of amino acids in a protein specifies that protein's shape, and proteins will function properly only when they have folded into a very specific shape.

Genetic Information Is First Expressed as RNA. To see how this genetic code determines the sequence of amino acids in a protein *and* to see how the cell physically assembles these amino acids into a chain, we need to look at some of the specific steps of gene

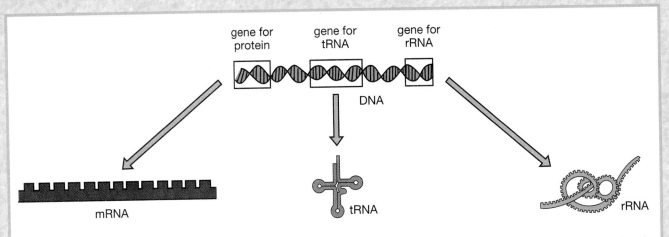

Figure E12.10 Types of RNA DNA encodes three types of RNA. The information encoded in mRNA will be translated directly into protein. The tRNA and the rRNA participate in the reactions that assemble proteins.

because its role is simply to carry a transcript (or message) of the DNA-encoded information to the place in the cell where that information can be translated (or read).

In contrast, the second and third types of RNA, called **tRNA** (transfer RNA) and **rRNA** (ribosomal RNA), function in the actual process of assembling amino acids to make proteins. These two forms of RNA are transcribed from genes that are not expressed as proteins. Instead, as Figure E12.10 shows, the final product of these genes is RNA. These RNAs play a key role in converting the information in mRNA into protein.

Translating the Message in mRNA

To express a gene for a protein, the cell first makes a molecule of mRNA. This mRNA then acts like a blueprint for building a house or a machine. Cells translate the mRNA blueprints, which are written in the language of nucleic acids, into specific sequences of amino acids that make up proteins, which are the cellular equivalent of structures and machines (refer to Figure E12.11). The name given to the process of converting the genetic code in an mRNA sequence into an amino acid sequence is **translation**. Just as you cannot use the information in a message written in an unfamiliar language until

it is translated into a familiar language, your cells cannot use the information encoded in mRNA until it has been translated into proteins.

Before we look at the actual steps of protein synthesis (translation), let's examine some of the basic characteristics of the code that links a nucleotide sequence to the amino acid sequence in a protein. This code is called the **genetic code**. Proteins are built from some 20 different amino acids, and each amino acid must be identified specifically by a coding system within the mRNA. The code cannot be a simple one-to-one correlation of nucleotides to

amino acids or the code could only specify (code for) four different amino acids (one for each of the four different nucleotides). Likewise, the code cannot involve a two-to-one correlation (two nucleotides encoding one amino acid), or the code could specify only 16 amino acids. Instead, the code involves a three-to-one correlation. In other words, the genetic code uses three sequential mRNA bases to identify each amino acid. Each triplet (three-nucleotide combination) is called a **codon**.

Unraveling the genetic code's details—determining exactly which amino acid is specified by

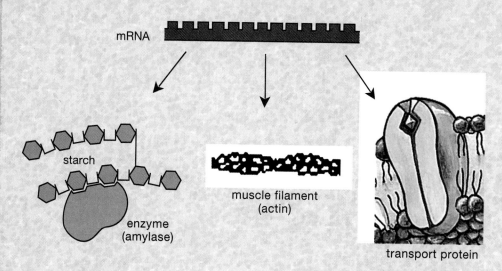

mRNA

starch

enzyme
(amylase)

muscle filament
(actin)

transport protein

Figure E12.11 Translating mRNA into protein The information encoded in mRNA is translated into proteins, which then can function in the cell as enzymes, structural components, or a variety of transport molecules. Translation takes place in the cytoplasm. In eukaryotes this means that the mRNA first must be transported out of the nucleus.

which codon—was one of the most exciting series of discoveries ever made in the field of biology. Now humans actually can read the information encoded in the genetic material of each species. If you study Figure E12.12 carefully, you will notice that using all four bases, three at a time, results in 64 possible codon combinations ($64 = 4^3$). Because there are only 20 amino acids, it became clear to the pioneering genetic code researchers that some amino acids have more than one codon. The triplet codes for each amino acid are listed in Figure E12.12.

If you examine the table closely, you will notice that some of the possible triplet base combinations do not correspond to any amino acid. Several triplet base combinations, each beginning with U, signal a *stop* in translation (that

First Letter	Second Letter				Third Letter
	U	C	A	G	
U	phenylalanine	serine	tyrosine	cysteine	U
	phenylalanine	serine	tyrosine	cysteine	C
	leucine	serine	stop	stop	A
	leucine	serine	stop	tryptophan	G
C	leucine	proline	histidine	arginine	U
	leucine	proline	histidine	arginine	C
	leucine	proline	glutamine	arginine	A
	leucine	proline	glutamine	arginine	G
A	isoleucine	threonine	asparagine	serine	U
	isoleucine	threonine	asparagine	serine	C
	isoleucine	threonine	lysine	arginine	A
	(start) methionine	threonine	lysine	arginine	G
G	valine	alanine	aspartate	glycine	U
	valine	alanine	aspartate	glycine	C
	valine	alanine	glutamate	glycine	A
	valine	alanine	glutamate	glycine	G

Figure E12.12 The genetic code The code letters represent bases in mRNA. The words in the boxes are the names of the 20 amino acids most commonly found in proteins. To use the code, follow a codon's three nucleotides by using the rows and columns labeled *First, Second, Third Letter* until you arrive at the corresponding amino acid. For example, GGA codes for glycine. How many of the amino acids have more than one codon?

is, the end of the protein). Another special codon is AUG, which not only specifies the amino acid methionine, but also is a signal to *start* translation (that is, to begin building the protein).

One of the most remarkable aspects of the genetic code is that it is nearly universal. For instance, although prokaryotes and eukaryotes have many basic structural differences, including

how the genetic material is organized, they still use the same genetic code. This basic similarity is an important piece of evidence supporting the theory of a common origin of all life forms.

Bethesda, Maryland (1961) It looks as though one of the greatest challenges in scientific cryptology soon will be conquered, thanks to the pioneering work of two modern-day detectives from the National Institutes of Health—M.W. Nirenberg and J.H. Matthaei. These scientists, intent on cracking the genetic code, announced last week the first decoded results from their landmark research: three mRNA nucleotides in the sequence UUU correspond to the amino acid phenylalanine, and the mRNA triplet CCC corresponds to the amino acid proline.

NEWS FLASH! WHITE-COATED SLEUTHS DECIPHER GENETIC CODE

This may not sound like landmark research to you, but because the genetic code lies at the heart of all living systems, this breakthrough truly is a remarkable accomplishment. The existence of this code—the sequence of bases in an organism's DNA—has been known since the mid-1950s, but until now scientists had no idea how to read it. The logic behind Nirenberg and Matthaei's work was simple: to decipher the genetic code, all you have to do is build an artificial in-the-test-tube protein-making system and then give it an artificial message that you *wrote*. If you know the message that you give the system and if you can retrieve and analyze the protein that the system makes from that message, then you can *decode* the code.

Although the process sounds simple, its execution represents an amazing feat. To read even these first two words (UUU and CCC), Nirenberg and Matthaei had to create a test-tube system that actually would *build proteins to order*. To do this, they extracted, from dead bacteria, enzymes and other biochemical machinery that participate in assembling proteins. For example, among many other components, their test-tube system contained molecules of rRNA and of tRNA, both known to be central to the assembly process. Then these intrepid investigators had to build a set of artificial mRNAs, messages that contained known nucleotide sequences. Only after both of these tasks had been accomplished were they able to start deciphering.

And what does all of this mean to you? Well, if you have any UUUs or CCCs in your genetic information (and you do), it means that we now know what those codons mean to your cells—we now can *read* them. As we break more and more of the code, who knows, one day we may be able to pull all of the mRNAs out of your cells and read your genes. What would our molecular sleuths say about that?

Cellular Components in Protein Synthesis

Understanding that mRNA codons correspond to amino acids is important to understanding the linearity of gene expression. The actual process of identifying the appropriate amino acids, aligning them properly, and joining them together is quite complex and involves a number of cellular components, as Figure E12.13 illustrates. The principal components responsible for translating the information encoded in mRNA into a protein are called **ribosomes**. Ribosomes are large structures, made of rRNA

and protein, that travel along a strand of mRNA and read the triplet codons (Figure E12.13a). Ribosomes can read any mRNA, and so they can produce an unlimited number of different proteins. For each mRNA strand, however, the specific order in which the amino acids are put together to form a protein is dictated by that mRNA's particular codon sequence.

A second important component of protein synthesis is the cell's population of tRNA molecules

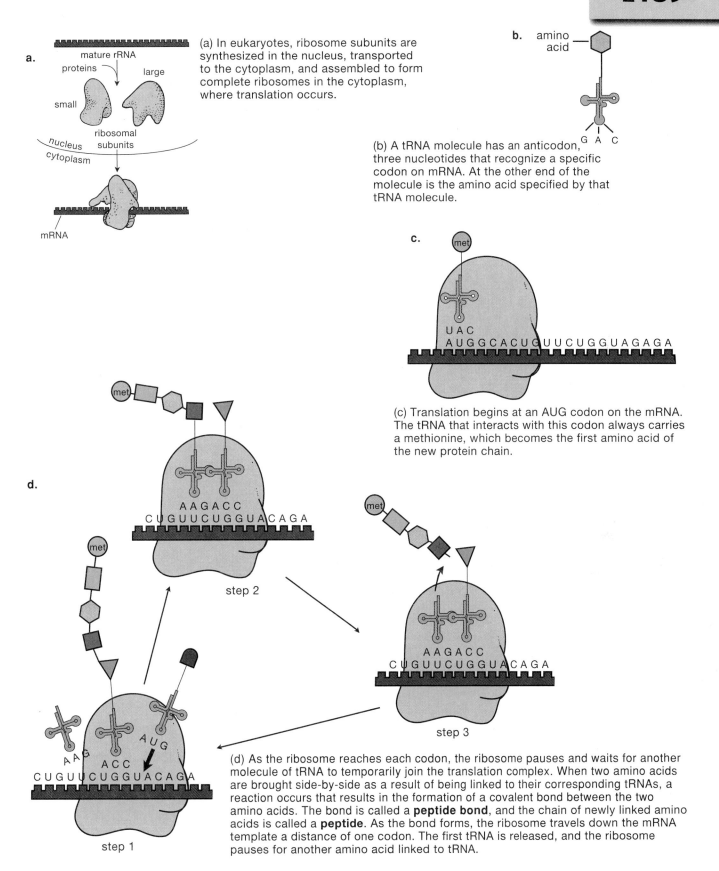

a.

mature rRNA
proteins
large
small
ribosomal
subunits
nucleus
cytoplasm
mRNA

(a) In eukaryotes, ribosome subunits are synthesized in the nucleus, transported to the cytoplasm, and assembled to form complete ribosomes in the cytoplasm, where translation occurs.

b. amino acid

G A C

(b) A tRNA molecule has an anticodon, three nucleotides that recognize a specific codon on mRNA. At the other end of the molecule is the amino acid specified by that tRNA molecule.

c.

met

U A C
A U G G C A C U G U U C U G G U A G A G A

(c) Translation begins at an AUG codon on the mRNA. The tRNA that interacts with this codon always carries a methionine, which becomes the first amino acid of the new protein chain.

d.

step 2

A A G A C C
C U G U U C U G G U A C A G A

step 3

A A G A C C
C U G U U C U G G U A C A G A

step 1

A A G
A C C
A U G
C U G U U C U G G U A C A G A

(d) As the ribosome reaches each codon, the ribosome pauses and waits for another molecule of tRNA to temporarily join the translation complex. When two amino acids are brought side-by-side as a result of being linked to their corresponding tRNAs, a reaction occurs that results in the formation of a covalent bond between the two amino acids. The bond is called a **peptide bond**, and the chain of newly linked amino acids is called a **peptide**. As the bond forms, the ribosome travels down the mRNA template a distance of one codon. The first tRNA is released, and the ribosome pauses for another amino acid linked to tRNA.

(see Figure E12.13b). These molecules are important because they recognize both specific mRNA codons and the amino acids that match those codons. One portion of a tRNA molecule interacts with the amino acid that corresponds to a particular codon. Another portion of the tRNA molecule then interacts with the codons in the mRNA. This interaction is a base-paring association between the nucleotides of the tRNA and the nucleotides of the mRNA. The base pairs form according to the same rules of complementary pairing that occur between nucleotides in DNA. As Figure E12.13c-d illustrates, it is actually the tRNA molecules that bring about the translation of the language of nucleic acids into the language of proteins.

Bond formation between the lengthening chain of amino acids and each new amino acid continues until the ribosome reaches a special codon that does not specify an amino acid. These codons are called **stop codons**. Stop codons cause the ribosome to pause indefinitely because there are no matching tRNAs. At this point a release factor—which is itself a protein—binds to the ribosome, and synthesis of the growing amino acid chain stops. When translation stops, the ribosome separates from the mRNA, and the completed protein folds into a shape determined by its amino acid sequence.

Each protein that is produced by translation plays some role in the function of a cell, tissue, organ, or organ system. For instance, if particular alleles encode the information for the protein hemoglobin, then the protein produced through the translation of the corresponding mRNA should function in oxygen transport in red blood cells. The ability of any protein to function properly, however, depends on whether the protein is folded properly.

It is at the level of a protein's final folded form that the effect of mutations becomes clear. Mutations that alter one or several amino acids can cause the protein to fold into a form that is nonfunctional (or inactive) or only partially functional. This is how a single amino acid mutation in hemoglobin can result in sickle cell disease. When you think of the change that this mutation causes at the molecular level, you can see why genotype is so important in determining phenotype.

As is the case with most other complex processes in living organisms, protein synthesis is a highly regulated process. The amount and type of protein produced in cells vary according to the life stage of an organism (even to the time of day) and according to the particular specialized cell types in a multicellular organism. This regulation happens at several levels. For example, the amount of protein in a cell can be regulated by how often a particular mRNA is translated. Likewise, the amount of mRNA in a cell can be regulated by how often a particular gene is transcribed. In this way transcriptional regulation also affects translation. Through precise regulation of gene expression, individual cells are able to maintain homeostasis in response to changing conditions.

San Francisco, California (1973) The study and practice of biology were radically and irreversibly changed today with the announcement by S. Cohen of Stanford University and H. Boyer of the University of California San Francisco of the first successful attempt to build and to clone a truly *recombinant* DNA molecule. What makes this accomplishment so remarkable is that this DNA molecule, made by joining together (or *recombining*) pieces of DNA from two completely different sources, never existed before. Yet today, it not only exists, but it exists in thousands and thousands of copies, and researchers can make thousands and thousands more at will.

NEWS FLASH! EXTRAORDINARY NEW TECHNIQUE CHANGES BIOLOGY FOREVER

When congratulated on their accomplishment, Cohen and Boyer readily acknowledged their debt to many other scientists whose hard work and important discoveries made creating and cloning a recombinant molecule possible. One key advance in just the last few years was the 1970 discovery by H. Smith of Johns Hopkins University of the first **restriction enzyme**. Restriction enzymes are able to recognize and to cut perfectly a specific sequence of nucleotides in DNA. Another key advance was the simultaneous isolation by several research groups of the enzyme *DNA ligase*. DNA ligase allows scientists to fasten back together fragments of DNA that have been cut by restriction enzymes, and the source of the DNA does not even matter. For instance, scientists can ligate bacterial DNA with yeast DNA to create recombinant molecules.

UNIT FOUR CONTINUITY: Reproduction and Inheritance in Living Systems

Conceptually, Cohen and Boyer's technique was really quite simple: they just cut DNA from two different sources with the same restriction enzyme, then mixed the DNA together, added the ligating enzyme to connect pieces to each other, and then inserted the recombinant molecules into living bacteria. As the bacteria reproduced, the tiny recombinant DNA molecules also multiplied, creating—literally within hours—a colony of reproducing bacteria from which scientists can extract as many copies of the recombinant molecule as they will ever want or need.

It is impossible to overestimate the effect that these techniques ultimately will have on all of us. Already scientists around the world are talking of isolating and inserting a wide range of human genes into bacteria, not only to study their expression and regulation, but also to harvest the products that might be formed. Imagine an unlimited supply of human insulin to use in the treatment of diabetes or unlimited quantities of human growth hormone to use in the treatment of dwarfism. Imagine, as well, inserting the genes for a wide range of desirable characteristics into crop plants (bigger cucumbers? frost-resistant strawberries?). And who knows? One day we may even be able to remove particular genes from peoples' bodies and replace them with other genes.

Of course, new technologies never come without some potential problems, and scientists already are pointing out the need to find beneficial and safe ways to use our new ability to move DNA from one organism to another. Recombinant techniques are very different from *selective breeding* techniques, in which we mate carefully selected organisms of the same species in a deliberate effort to heighten a set of desirable characteristics. All of the genes controlling those characteristics come from the same species. The deliberate combining of genes from very *different* species, however, is playing with genes and with evolution in a new way, and we may not be able to predict exactly where this will take us. Our world is a different place today from what it was yesterday, and it probably is important that all of us—scientists and nonscientists alike—keep that in mind.

Manipulating Genetic Material

In the last 40 years we have experienced an enormous increase in our understanding of the structure and function of genetic material and in our ability to work with it in the laboratory. In fact, our increasing ability to study and to manipulate molecular processes in a variety of species has led to a revolution in the ways that scientists conduct research, that pharmaceutical firms make medicines, and that many industries conduct their day-to-day business.

Although the technologies are new, our fundamental interest in studying and manipulating genetic information is not new. For centuries, humans have bred plants and animals selectively to produce organisms with desirable combinations of characteristics. Traditional methods of agricultural breeding were the earliest forms of **genetic engineering**, a process designed to artificially control the genetic makeup of an organism. In selective breeding humans cross plants or animals having desirable traits to produce new generations. In this way the expression of the desired traits becomes more and more prominent with successive generations, and less desirable traits sometimes can be eliminated.

In recent years scientists have developed more powerful techniques for examining genetic material at the molecular level and for selectively changing the genotypes of organisms. The

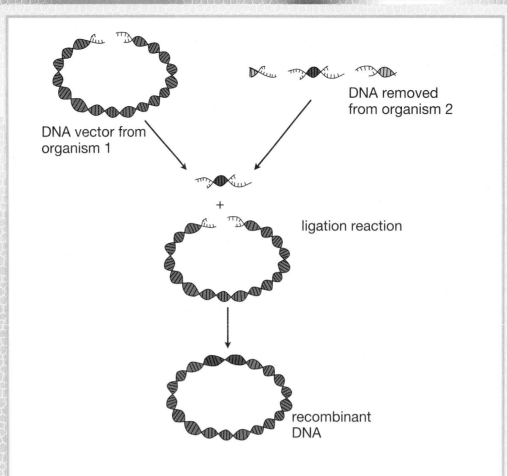

DNA vector from organism 1

DNA removed from organism 2

+

ligation reaction

recombinant DNA

Figure E12.14 Recombinant DNA Molecular biologists isolate DNA from separate sources and combine them in unique combinations. DNA from Organism 2 can be joined to a small circular molecule of DNA from Organism 1 in an enzyme reaction called ligation. The circular DNA molecules—called plasmids or vectors—are found in many bacteria. Because scientists can move vectors in and out of bacterial cells relatively easily, they serve as convenient carriers of DNA and allow the production of recombinant DNA molecules.

technique of DNA fingerprinting that you read about in the opening story is a relatively new technique that allows scientists to examine DNA in a very detailed manner. Many other new techniques allow scientists to alter the genetic information of a species in much more direct and extensive ways than the older methods of selective breeding. It is now possible to introduce into an organism genes that neither parent possessed. It even is possible for scientists to remove genes from one organism and introduce them into an unrelated organism—one that does not normally possess those genes. DNA that results from the combination of DNA fragments from different DNA strands is called **recombinant DNA** (Figure E12.14). Genetic engineering deals with the use of recombinant DNA to conduct research on how organisms function at the molecular and cellular levels and with the practical application of the understanding gained from that research. For instance, researchers use recombinant DNA technology to produce plants and animals with characteristics that could not be generated through selective breeding alone. Cotton plants that produce natural pesticides are an example. Another example of the use of recombinant DNA technology is the production of medicines that are valuable to human health, such as insulin and human growth hormone.

Recombinant DNA technology and the related techniques that scientists use today to study DNA and to move it from one organism to another already have affected the lives of humans and will influence us even more in the future. Such manipulations also influence other organisms in ways that may change how we view the earth's

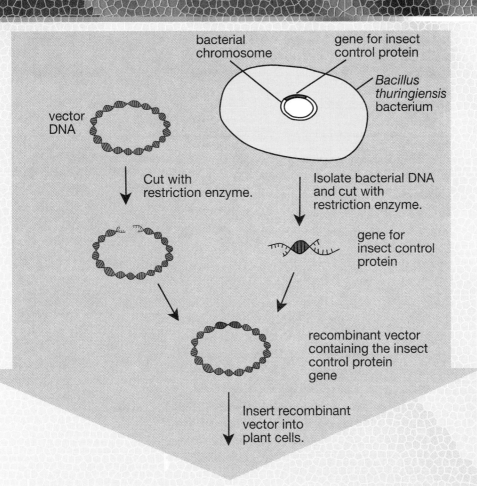

bacterial chromosome

gene for insect control protein

Bacillus thuringiensis bacterium

vector DNA

Cut with restriction enzyme.

Isolate bacterial DNA and cut with restriction enzyme.

gene for insect control protein

recombinant vector containing the insect control protein gene

Insert recombinant vector into plant cells.

nucleus

plant cell

Insect control protein gene is now part of plant cell chromosome.

plant chromosome

insect control gene

A whole cotton plant is grown from one altered cell. Every cell in the plant produces the insect control protein, and the plant can protect itself from bollworm attack.

Figure E12.15 Genetic engineering of cotton The toxic insect control protein gene of *Bacillus thuringiensis* is isolated using the same DNA-cutting restriction enzyme that is used to cut the DNA vector. After combining these two pieces of DNA, scientists insert the recombinant DNA into a plant cell and regenerate an entire plant from that one cell. The plant is now able to protect itself from the bollworm.

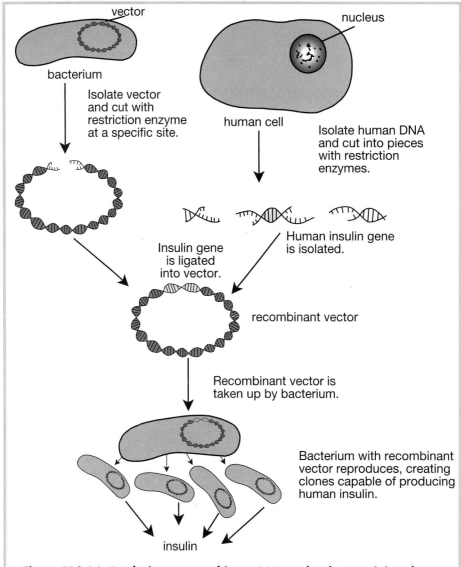

vector

bacterium

Isolate vector and cut with restriction enzyme at a specific site.

nucleus

human cell

Isolate human DNA and cut into pieces with restriction enzymes.

Insulin gene is ligated into vector.

Human insulin gene is isolated.

recombinant vector

Recombinant vector is taken up by bacterium.

Bacterium with recombinant vector reproduces, creating clones capable of producing human insulin.

insulin

Figure E12.16 Producing a recombinant DNA molecule containing the human insulin gene for expression in bacteria

diversity of life. For example, deliberate genetic engineering that is focused on creating new combinations of DNA sequences, by inserting bacterial genes into plants or by inserting human genes into animals, represents a profound change in the way the gene pool—the genetic composition of a population—evolves. Generally the gene pool of a particular species changes only very slowly, through rare, naturally occurring mutations and through natural selection. **Molecular biology**, the branch of biology that deals with the

methods and techniques used to construct recombinant DNA, has given scientists the ability to alter the gene pool of certain species overnight. In fact, it now is possible to produce gene combinations that nature may have *never* produced.

The Technology behind Genetic Engineering. One example of how scientists are using the technologies of DNA manipulation to address a specific practical problem involves the cotton plant. Cotton plants often are attacked by a pest called a bollworm, which

damages cotton crops and costs millions of dollars each year to control. Researchers, however, have known for a long time that if a bollworm eats the common bacterium called *Bacillus thuringiensis*, or *B.t.* for short, a protein produced in the bacteria is partially digested in the worm's gut and poisons the worm. Because this protein is so effective, for years farmers have sprayed *B.t.* bacteria on their cotton crops to discourage the bollworms from eating the crops. This protective measure has its drawbacks though: sunlight

E190

degrades *B.t.*, and rainfall easily washes it off the plants.

Through genetic engineering, however, researchers have overcome these drawbacks. Figure E12.15 shows how scientists isolated the gene that codes for the poisonous bacterial protein and transferred the gene into the cotton plant. The new cotton plants thus contain recombinant DNA, and they are able to produce a bacterial protein, *B.t.* toxin, in their leaves. These plants now are able to protect themselves from damage because, when a bollworm begins nibbling on the leaves, it ingests the protein and becomes sick. The same protein also can be produced by edible plants such as potatoes because this protein is not toxic to humans, animals, and most insects. A benefit of such technology is that farmers can use less insecticide, decreasing the amount of toxic chemicals that enter the water supply and food web.

Plants or animals that contain genes from unrelated species, such as cotton plants containing bacterial genes, are called *transgenic*. Transgenic organisms are widely used in research and industry. Other organisms altered by genetic engineering techniques can be just as important, however, even if they are not transgenic. Consider the genetically engineered tomato, which recently was approved for sale in the United States. Tomatoes normally ripen—become red and flavorful—and soften at the same time. This situation encourages farmers to pick the tomatoes while they are still green and ship them to stores before they soften;

the less desirable alternative is to ship ripe tomatoes that are easily damaged in transit. Unfortunately, green tomatoes do not ripen well after they are picked and often have little flavor even after they have turned red.

When scientists discovered that the ripening and softening processes were not the same, that is, that different genes and thus different biochemical pathways are used for the two processes,

---

Genetic engineering also has a tremendous effect on human health. Insulin, the peptide hormone that is required by diabetics, at one time could be obtained only from the pancreases of cattle and hogs.

---

they were able to target the softening process and modify it. The result was a genetically engineered separation of the ripening and softening pathways so that tomatoes now can be ripened on the vine, preserving their flavor, yet transported to the market without damage and rotting.

Genetic engineering also has a tremendous effect on human health. Insulin, the peptide hormone that is required by diabetics, at one time could be obtained only from the pancreases of cattle and hogs. Insulin produced from these

animals was available in limited supplies, was expensive, and was not very effective for some individuals because the insulin was not identical to human insulin. By recombining the human insulin gene with bacterial genes, however, researchers have been able to produce human insulin in large bacterial fermentation systems. This process is an economical method for producing authentic human insulin (Figure E12.16).

Researchers currently are working on ways to treat certain human genetic disorders using recombinant DNA that has been constructed from harmless forms of viruses and functional copies of human genes. Human patients who suffer from a disease that is caused by a missing or defective gene can have the replacement gene inserted into their cells. In such procedures the harmless virus is only a carrier that transports the new copy of the gene into the host's cells. This technique is called **gene therapy**, and it may provide medical science with a powerful method of treating the source of genetic defects, rather than only the symptoms.

Ethical and Public Policy Issues Related to Genetic Engineering

Rapid advances in genetic technology have raised important, interesting, and sometimes controversial new issues in the areas of ethics and public policy. For example, now that human gene therapy is possible, it is appropriate to consider whether the genetic engineering of humans is acceptable, who should make the related ethical and public policy decisions, and what the long-term effects of genetic engineering might be. A concern that has an impact on all of these areas, however, is whether the public will be knowledgeable enough to participate in making decisions about genetic engineering. If the people do not understand the science behind such technological innovations, they will be unable to evaluate the issues objectively and to arrive at well-reasoned conclusions. A scientifically literate population is crucial to the democratic processes of our society.

How one *applies* an understanding of science to decision making and ethical choices is just as crucial as knowing the science, and we can learn something about this process from philosophers and ethicists. Ethics is the study of what is right and what is wrong, what is good and what is bad, applied to the actions and character of individuals, institutions, and society. In short, ethical analysis is the conscious analysis of the justifications for our decisions.

One of the biggest ethical dilemmas currently confronting ethicists and researchers in the United States involves human gene therapy. Scientists and physicians have received approval to conduct gene therapy on somatic cells (body cells) but not germ-line cells (cells that give rise to gametes). The critical difference is that genetically engineered DNA introduced into somatic cells cannot be transmitted to future generations. The genetic material of germ cells, however, whether natural or genetically altered, is heritable. If we proceed with human germ-line therapy, will this lead to eugenics—the deliberate effort to *improve* the human condition by controlling gene distribution? Will gene therapy come to resemble cosmetic surgery, with individuals seeking to modify behavior and brain function as well as physical characteristics? How will our definition of *genetic disorder* change as a consequence of gene therapy? All of these questions demand careful consideration from people who understand this powerful new technology.

> How one *applies* an understanding of science to decision making and ethical choices is just as crucial as knowing the science, and we can learn something about this process from philosophers and ethicists.

Once ethicists, scientists, and the public have analyzed an ethical dilemma, they may choose to act on their decisions by trying to affect public policy. Public policy is a set of guidelines or rules developed by governmental agencies. These guidelines or rules can be in the form of laws, which result from new legislation, or in the form of judicial decisions, which result from the decisions made in court cases. It is important to note that public policy also can be set when government agencies do *not* act. By not setting rules or laws, the government permits individuals or institutions to act in any manner they choose with respect to the issue in question.

With so many rapid advances in genetic engineering technology, public policy debates on the subject are becoming common. Politicians, scientists, and the public must decide how they want to use this new technology. Should the government regulate recombinant DNA research? Does the public have a right to be informed and to have a voice in decisions about research that might involve ethical as well as technical issues? Who should make the decisions about what types of research are desirable and what risks are acceptable?

UNIT FOUR CONTINUITY: Reproduction and Inheritance in Living Systems

Safety is an issue that commands particular attention in genetic engineering and involves both ethical and public policy considerations. Safety is an ethical issue insofar as it involves the usual risk assessment (or cost/benefit analysis) that accompanies any emerging technology. Safety worries concerning the accidental release of engineered organisms have decreased as scientists have developed more effective laboratory containment procedures. For example, the microorganisms most often used in research are mutants that have such highly specific nutritional requirements that they cannot survive outside of a laboratory. Safety is a public policy issue insofar as it commands the attention of legislators and regulators who wish to respond to situations with potentially serious consequences. For example, regulators now are concerned about the deliberate release of genetically altered organisms, such as the recombinant cotton plants that express *B.t.* toxin. Can we predict how these organisms will interact in the complex ecological balance of nature? Could

dangerous new organisms or viruses inadvertently be produced?

Ethical and public policy issues are intimately linked to each other due to their shared concerns for achieving beneficial consequences and protecting people's rights. Of course, the interests of individuals and groups may differ, resulting in disagreement over ethical and policy decisions. Discussions about the deliberate release of genetically altered organisms illustrate this tension well. In response to safety concerns and advice to proceed with caution, regulators established guidelines that dictate the level of containment and the size of the area that may be treated with genetically altered organisms. There is political pressure, however, to relax some of the restrictions so that companies in the United States can compete more effectively with foreign companies. Any new public policy decisions to relax the restrictions should be based on a thorough examination of the anticipated scientific and ethical consequences.

The Long and Short of Development

That nasty pimple finally is disappearing. You're coming down with a cold. You've gained 2 inches in height in the last six months. You undoubtedly have noticed that your body is changing constantly. Many of these changes are short-term internal adjustments that ensure homeostatic balance. Homeostatic changes tend to be relatively rapid and reversible, occurring in seconds or minutes or hours. They allow conditions in our bodies to remain within rather narrowly defined limits.

Other changes occur at a slower pace, across a longer period of time. These changes are not directly involved in maintaining homeostasis, and they are not rapid or easily reversed. Instead, they are relatively slow and generally permanent. This change process is called **development**, and it begins at fertilization and ends at death.

Some developmental changes are dramatic, such as the formation of a beating heart in a 4-week-old human embryo; adjustments in body size, proportion, thinking ability, and coordination in a growing child; and the complex physical and emotional changes that occur at puberty. Other changes may seem less dramatic and even less welcome, such as the replacement of skin and blood cells every day of our lives or the slow changes in physical appearance as we approach old age. Even these changes, however, are important and natural aspects of the developmental process.

Development is not just a human experience. In fact, development occurs in all multicellular organisms, from tiny marine invertebrates to huge sequoias. Frog eggs turn into tadpoles that grow larger, lose their tails, and become frogs. Trees grow larger and develop branching limb and root systems (Figure E13.1). And as humans, rats, chimpanzees, and other mammals age, their hair (or fur) turns gray, and their skin tone is lost.

The two most important changes that we see when we examine the early stages of development are **growth** (the organism becomes larger through an increase in the number of cells) and **differentiation** (the cells of the organism become different from one another in both structure and function). In most organisms that reproduce sexually, the fertilized ovum (the zygote) grows into a multicellular **embryo** that grows larger and larger. As this embryo grows, specialized parts begin to appear. A human baby, for example, is not born as a large, shapeless mass of cells that are all identical to the zygote. Instead, a baby is a precisely formed organism with skin, nerve, liver, and blood cells already located in specific places and already performing distinct and important functions. Figure E13.2 illustrates some of the stages that occur during early mammalian development.

Growth and differentiation not only characterize early development, they also are

Figure E13.1 Development in Pinon pine tree (Pinus edulis) What evidence of development do you see in these photos?

UNIT FIVE DEVELOPMENT: Growth and Differentiation in Living Systems

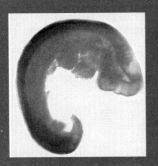
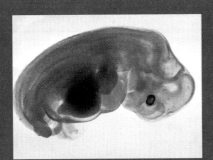

Figure E13.2 Pig embryos These images of pig embryos show many common features of mammalian development. The embryos show increasing development from left to right.

important events in the development that occurs after the so-called embryonic period (that is, after birth or hatching). For instance, these events are the processes by which a child's body grows and changes. Likewise, growth and differentiation are critical for the maintenance and repair of body systems that become damaged through accidents or illness.

In later life another developmental process becomes important as well. This is the process of **senescence**, or aging. Although the underlying biology of aging is not well understood, developmental biologists generally agree that some of the progressive and irreversible loss of function that occurs with aging is genetically controlled. For example, scientists have identified a gene in fruit flies that

appears to limit the flies's life spans. If this is true, then development truly can be defined as a set of changes that occur continually from fertilization to death. The central questions of development have to do with how this orderly pattern of developmental change takes place in organisms and how developmental changes are related to evolution and biological diversity.

The Cell Cycle and Growth Control

Have you ever wondered why beetles are small and whales are large? At hatching or birth, both are considerably smaller than at adulthood, yet they grow to a size that characterizes their particular species. How is their growth and final size determined?

Because nearly all organisms grow during their lifetimes, growth is regarded as an important developmental process; therefore, understanding development requires that we understand growth. In multicellular organisms growth occurs primarily through an increase in the *number* of cells, not through an increase in the *size* of cells. This is because the processes of cellular transport—diffusion and osmosis—place upper limits on how large cells can become. These physical limitations have influenced development in a way that favors more cells rather than large cells. As a result, an increase in the size of an embryo reflects the increase that has occurred in the number of cells that compose it.

Simple observation of your own body suggests that growth is a highly regulated process. Think, for example, of the changes in size and proportion that already have taken place in your own body. In most cases, growth takes place according to a predictable schedule and, with some minor exceptions, occurs evenly from one side of the body to the other. Not only does the production of new cells allow growth, it also allows for the replacement of skin and blood cells, which become damaged or die during the normal events of life. Such replacement activity, along with growth, is usually precisely controlled.

Before we consider how the production of cells is *controlled* during development, we must look at how cell division produces new cells. This process of cell division, called **mitosis**, produces offspring cells that are genetically identical to each other and to the parent cell. Figure E13.3 illustrates the distribution of genetic

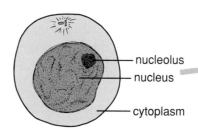

nucleolus
nucleus

cytoplasm

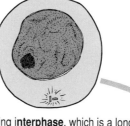

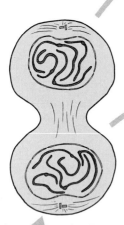

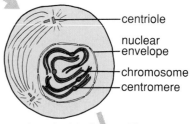

centriole
nuclear envelope
chromosome
centromere

a. During **interphase**, which is a long and active phase of the cell cycle, materials required for the next cell division are synthesized. For example, DNA and chromosomes are duplicated in the nucleus and cell structures such as mitochondria are made in the cytoplasm. The cell grows.

f. During **telophase** the chromosomes approach the opposite ends of the cell and group together. A new nuclear envelope is synthesized around the chromosomes. The cytoplasm begins to divide and a new cell membrane forms. (In plant cells, a new cell wall is laid down between the two new cells.) The new cells enter interphase.

b. As **prophase** begins, the long thin chromosomes coil and become shorter and thicker. Each chromosome now appears as a doubled structure joined at a centromere. The centrioles, which were duplicated during interphase, begin to move to opposite ends of the cell. (In plant cells there are no centrioles, but the events of mitosis otherwise occur as described here.)

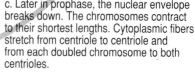

centriole-to-centriole fibers

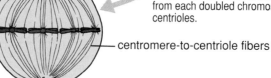

e. The doubled chromosomes separate during **anaphase**. The new chromosomes are pushed and pulled to opposite ends of the cell by the cytoplasmic fibers.

c. Later in prophase, the nuclear envelope breaks down. The chromosomes contract to their shortest lengths. Cytoplasmic fibers stretch from centriole to centriole and from each doubled chromosome to both centrioles.

centromere-to-centriole fibers

d. During **metaphase** the doubled chromosomes line up along the middle of the cell. Cytoplasmic fibers now are attached to each doubled chromosome at the centromere.

Figure E13.3 The phases of mitosis
How does mitosis compare with meiosis, the process of cell division that precedes the formation of gametes?

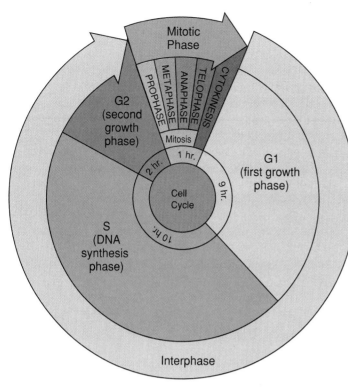

Figure E13.4 The cell cycle The times given for each phase represent the approximate times for a liver cell grown in the lab.

During most of this nondividing interphase, the cell is synthesizing RNA, protein, and other macromolecules in preparation for cell division. Another important function that takes place during this time is DNA replication, which doubles the cell's genetic material and cytoplasm, providing sufficient material to distribute to the two offspring cells produced by mitosis.

If a population of cells does not divide, growth can occur only by a limited increase in cell size. Controlling cell growth is important in some tissues such as adipose (fat-storing) tissues. Once your body produces a basic number of fat cells, these cells do not multiply in number, although they can become larger by incorporating more fatty substances.

In most tissues, controlling growth means controlling cell division, that is, controlling the occurrence and rate of mitosis. Cell division is affected by when, where, and how fast mitosis occurs, if it occurs at all. There are two major levels of control over cell division. The first level, called *internal control*, involves substances inside the cell that regulate the timing of the specific phases of the cell cycle. This internal regulation is critical to division. For example, if a cell were to enter mitosis before it replicated its DNA, there would not be a full set of chromosomes to give each offspring cell, and both of the cells produced would die.

A second level of cell division control involves environmental factors that affect a cell's reproductive behavior. These signals are *external*, but they trigger internal events in the cell that initiate or suppress the events of the cell cycle. During normal cell division,

material during mitosis. During the four phases of mitosis, the chromosomes duplicate, and then the duplicated chromosomes line up, separate, and are redistributed into two newly formed offspring cells. Other parts of the cell also are distributed between these offspring cells. The result of this type of cell division is two cells, each containing a copy of the same genetic information and the same types of subcellular compartments that were contained in the parent cell.

As shown in Figure E13.4, mitosis is only a small fraction of the **cell cycle**, the actual life cycle of a typical cell. The phase of the cell cycle where mitosis occurs is called the M phase. In a population of normally dividing cells, the bulk of a cell's life is spent in a phase between cell divisions, or mitoses.

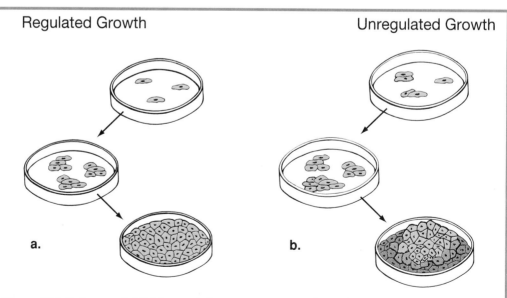

Figure E13.5 Contact inhibition a. Cells experiencing normal growth regulation stop dividing when they make contact with other cells. b. Cells that have lost this regulation will continue to divide and pile up on one another.

external signals affect internal control mechanisms, which in turn regulate division and growth.

One example of external control is *contact inhibition*. When normal cells divide, they eventually reach a population density or degree of crowding that causes the cells to stop dividing (Figure E13.5). This happens even though nutrients may be plentiful in the environment. Contact inhibition prevents overcrowding of cells within a particular organ or area of the body. Sometimes this inhibition must be relaxed to allow healing. When your skin is cut, the cells at the edge of the wound begin to divide and slowly cover the bare spot. Once the open wound is repaired, however, cell division and cell movement stops again. Contact inhibition is an important example of growth control, and problems arise when this control is lost. A cell that suffers mutations in the genes that regulate contact inhibition will divide without regard to external signals and eventually will form a tumor.

Hormones are another form of external growth control. Recall that hormones are produced by one set of cells but exert their effects on other cells, often in distant locations of the body. Growth hormone, for example, is critical in coordinating the proportional growth of the body. As its name suggests, growth hormone stimulates growth (that is, part of its function is to stimulate cells to divide). An excess of growth hormone before puberty results in gigantism. Likewise, lack of sufficient growth hormone produces dwarfism. Although growth hormone's most obvious effects are exerted on the long bones of the limbs, it also appears to be a major external signal by which growth all over the body is controlled.

COORDINATING GROWTH

Imagine engineers and mechanics replacing the engines, wings, and cockpit of an airliner *while the plane was airborne*. As difficult and absurd as that may seem, controlling the growth of a living organism is not so different. Internal and external controls on growth operate during every phase of an organism's development. During the embryonic period, rapid cell division produces millions of offspring cells, which slowly give shape and substance to an embryo's arms, legs, and internal organs. This growth must be coordinated constantly because an embryo is not like a machine, which does not work until all of its parts are in place. Instead, the embryo functions from the very moment that it exists, and it *keeps* functioning even as it is constantly changing. In some ways, controlling an embryo's growth is like remodeling an airplane as it flies.

Coordinating growth requires some complex events as an organism develops. For example, because an embryo's cells are alive, they need a regular supply of oxygen and nutrients, and many organisms have mechanisms that bring these substances inside the embryo and distribute them to the growing cells. Chicken eggs contain a special set of blood vessels that lie just under the shell and that also surround and penetrate the yolk. These blood vessels connect to the developing circulatory system of the growing embryo; they carry oxygen that is absorbed across the shell and

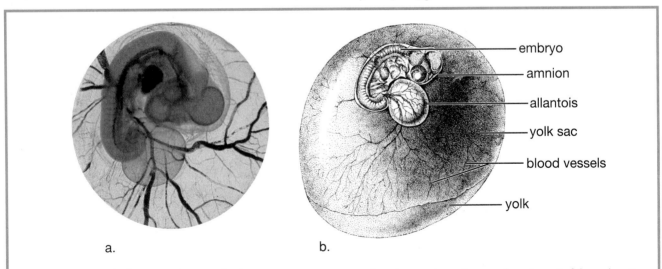

a.   b.

Figure E13.6 Which structures are involved in transporting oxygen and nutrients to the embryo and wastes out of the embryo?

UNIT FIVE DEVELOPMENT: Growth and Differentiation in Living Systems

nutrients that are absorbed from the yolk into the body of the embryo where they are needed (see Figure E13.6). The umbilical cord and the placenta serve similar functions in the human embryo.

As an embryo's tissues and organs grow and take shape, its circulatory system also grows and changes. This process is a good illustration of the interaction between internal and external growth controls. Through careful analysis of many experiments, scientists have found that different types of cells reproduce when different sets of proteins are present in the cellular environment. When the local concentration of these proteins, or growth factors, rises, cells in a formerly nondividing state begin to divide. For example, as the brain of a vertebrate embryo begins to develop, the new brain cells secrete a specific growth factor. This external growth signal promotes internal changes in cells at the tips of nearby blood vessels. These blood vessel cells undergo rapid cell division and grow toward the source of the growth factors, building a circulatory network that is important to the immediate survival of the cells secreting the growth factor. This network also will be important once development is complete. Scientists have discovered a whole set of growth factors that work during embryonic development and after birth. Typically, these substances have powerful effects even at low concentrations, and these effects are specific to the cells that release the factor and to the tissue that responds.

This constant remodeling, balancing, and regulation of growth continues throughout life. Look again at Figure 13.2. Notice how large the newborn's head is in proportion to its arms and legs. Then compare the proportions of the newborn with your own (or those of the adolescent in the same figure). Clearly, your arms and legs *must* have grown much faster than your head and torso. And even after we have achieved our full growth, our cells still divide in a controlled manner. This means that cells lost through normal wear and tear or through injury are replaced at a parallel rate. If too little division occurs, injuries go unrepaired. If too much division occurs, a tumor can develop. Although scientists still do not understand all of the mechanisms involved in the control of cell division, it is becoming increasingly clear that proper regulation occurs only when both internal and external mechanisms work together in a coordinated fashion.

DIFFERENTIATION AND THE EXPRESSION OF GENETIC INFORMATION

Early research into developmental events often involved painstaking observations of the structures of all types of developing organisms. These observations revealed that all embryos of a particular species go through apparently identical stages in the same sequence, and specific tissues and organs become visible and then functional at specific times. One of the first conclusions that scientists drew from this work was that the changes that developing organisms undergo must be under very precise control. We can make the same observation about development later in life: although people vary in exactly *when* they reach various stages (puberty, for example), most of us go through every stage in the same sequence.

These observations raise the question of how the processes of development are regulated and coordinated. In the case of humans, the union of the ovum and the sperm eventually gives rise to an adult composed of trillions of cells, and that adult's body continues to change and to develop until death. What affects how, where, and when each part of our body develops as it does?

Part of the answer to this question is that the original 46 chromosomes that were present in the zygote (23 from the ovum and 23 from the sperm) contained all of the basic instructions needed to produce a complete human. In other words, the basic information by which all cells grow and differentiate is located in the zygote's DNA. Through mitosis this information is copied for each offspring cell as it is produced.

The fact that all of the developing cells in the body contain all of the instructions for producing a complete human does not explain how cells become *different parts* of that human during development. Some cells, for example, become long and striated as they develop and build extended filaments of protein that allow them to contract (muscle cells). Other cells secrete digestive enzymes that will help reduce a lunch of a turkey sandwich and a piece of fruit to useful nutrients (pancreatic cells), and still other cells become flat and scaly and produce the pigment that is responsible for skin color (skin cells).

Knowing that virtually all of the cells in the body contain the same DNA also does not explain how a differentiated cell of a particular type tends to remain that same type. Most cells continue to produce proteins and display functions that are characteristic of that cell type. We don't have to worry that our heart cells suddenly will start growing hair or that our skin cells will start oozing digestive enzymes. Instead, these cells likely will spend their entire lives working as they always have worked.

Figure E13.7 shows several types of human cells. If all of these cells have the same genetic information, why do so many different and stable cell types form? Why are we not 6-foot-tall livers, and why do we never become 4-foot-wide brains? The general answer to these questions is that different cell types use different portions of their genetic material. Although most of the cells of the body receive exactly the same genes, *environmental influences* both inside and outside the cell cause certain genes to turn on (become active through transcription) and other genes to turn off

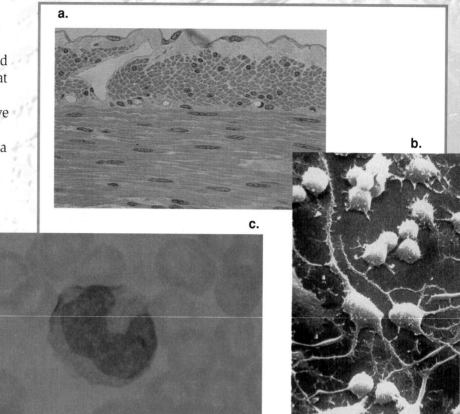

Figure E13.7 Different cells, same DNA a. Smooth muscle cells (X160). The nearly circular structures near the top of the image are smooth muscle cells in cross section; the long cells at the bottom are in longitudinal section. b. Nerve cells (X2500). The long cell structure extending toward the top of the photo from the cell body is an axon. The shorter structures extending downward are dendrites. c. Blood cells (X710). White blood cells such as this monocyte are part of the immune system. What differences do you see among the cells illustrated?

(remain or become inactive). The particular pattern of "on" genes and "off" genes determines cellular differentiation.

As differentiation proceeds, cells use more specialized genes and, therefore, are more restricted in their products and functions. Under normal circumstances a fully mature skin cell never will send an electrical signal to the brain the way a nerve cell would, even though it has the genetic information that theoretically would allow it to do so. Its inability to perform this function reflects the fact that the genes

required to accomplish this function are not active in the skin cell. Instead of generating electrical signals, the skin cell displays different functions (such as pigmentation) that are encoded by a different set of genes. These different functions are consistent with the environmental influences that have affected the cell's development and reflect the genes that are active within it.

A great deal of research has focused on understanding the mechanisms by which different patterns of genes are turned on and off in cells and how these patterns of gene expression

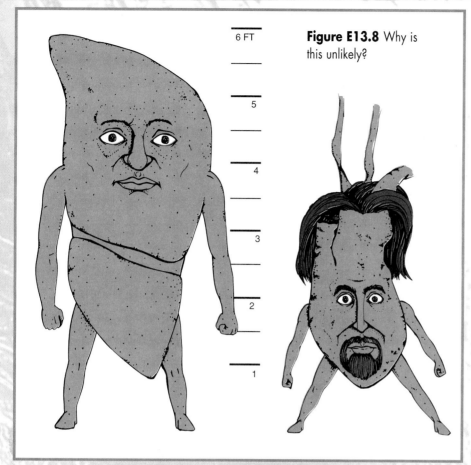

6 FT

Figure E13.8 Why is this unlikely?

have accumulated through more than 100 years of such studies is beginning to answer many important questions, including the central question of differentiation: Why don't all cells with the same DNA do the same thing?

One possibility is that the environments *inside* these cells differ, which can lead to different patterns of gene expression. For example, there might have been a substance in the cytoplasm of a cell that, as a result of cell division, ended up in one offspring cell and not the other. The substance might have changed the activity of the genes in the cell that received it. Another possibility is that something *outside* one of the cells—something secreted by a neighboring cell—affected that cell's pattern of gene expression. In either case, these influences from inside the cell or outside the cell might have affected the pattern of one cell's gene expression and led to its specific differentiation.

As it happens, experimental results from a wide variety of organisms and laboratories have provided important evidence for *both* explanations of

remain stable across time. Most of the earliest embryologists concentrated on *describing* development (describing, for example, the changes that occurred in cells during differentiation). Eventually, scientists went beyond describing *what* cells were doing and began investigating *how* cells went about doing it. One of the ways they tried to answer their questions was to *manipulate* embryos—to shake them, poke them, turn them upside down, and even to treat them with all types of chemicals to see how these actions would affect development. The experiments of Roux and Driesch are good examples of this type of *experimental* embryology. The knowledge that scientists

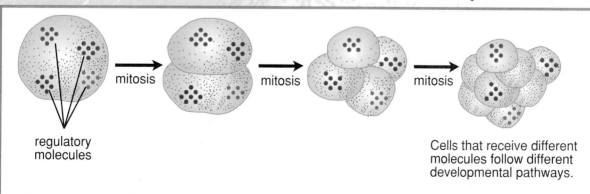

regulatory molecules

mitosis

mitosis

mitosis

Cells that receive different molecules follow different developmental pathways.

Figure E13.9 Cytoplasmic regulation Regulatory molecules are unequally divided during the development of some organisms. This can lead to different developmental fates for different cells.

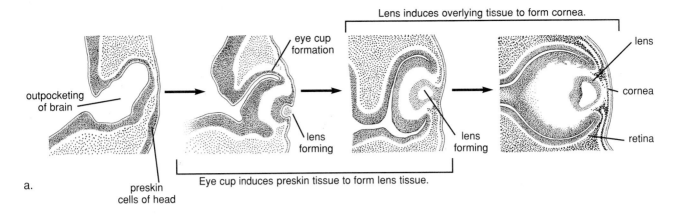

a.

outpocketing of brain

preskin cells of head

eye cup formation

lens forming

lens forming

Lens induces overlying tissue to form cornea.

lens

cornea

retina

Eye cup induces preskin tissue to form lens tissue.

Figure E13.10 Induction a. In the early development of a human's eye, an outpocketing of the brain called the eye cup induces early skin cells of the head to become the lens of the eye. The lens, in turn, induces skin cells to become a cornea. b. Regulatory molecules released by the inducing tissue can affect the responding tissue in different ways.

Later Induction

Early Induction

responding tissue

b.

inducing molecules

inducing tissue

sweat gland in skin

tooth in gum

limb in vertebrate

feather in bird skin

differentiation. In some species molecules that can regulate gene expression are located in different portions of the zygote's cytoplasm. The division of this cytoplasm into offspring cells distributes these molecules into some cells but not into others. As these molecules begin to influence gene activity, the cells that received them follow one path of development, and the cells that did not receive them follow another. Figure E13.9 illustrates how this mechanism of differentiation might occur.

In cases of outside influences on differentiation, substances released by one group of cells can cause a neighboring group of cells to differentiate in a particular way. This process is called **induction**, and an example of it is illustrated in Figure E13.10. Notice that the presence of the inducing tissue (the outpocketing of the brain) causes the cells that normally would have made skin to follow a different course of development. In this case induction leads to the formation of an eye. In the absence of substances released by the inducing tissue, these same cells would produce skin. Induction is important in the development of many human organs and systems, such as the nervous system, the eyes, the kidneys, and the pancreas.

Although the basic processes of development (growth and differentiation) are straightforward, development, nevertheless, involves a sophisticated system of genetic instructions and controls. What happens when these genetic instructions or control systems change? Usually, the result is a change in phenotype of the organism, and these changes, in turn, can lead to the evolution of new and different organisms.

Cloning People?

The idea that all of your cells contain all of the basic information required to produce a complete human (at least in physical form) raises the interesting possibility that one day you may be able to make an appointment with a local laboratory and ask them to *clone* you. Cloning uses the information present in one of your cells to build a whole new you. Such a procedure is not likely to be possible for humans for a long time if, in fact, it ever becomes possible.

In theory the idea sounds plausible: the lab would take a nucleus out of one of your cells and place it into a fertilized ovum from which the zygote nucleus had been removed. The fertilized egg would be ready to start development; the only thing your nucleus would provide is the instructions. Many complicating factors make such a cloning experiment more difficult than it first might appear. For example, it is not at all clear that the nucleus from a differentiated cell would be capable of providing the necessary instructions for development. In addition, the aged genetic material present in the donor nucleus might contain mutations that would cause development to proceed abnormally.

Cloning has been achieved in some plants (for example, carrots and tobacco plants) and in some animals (for example, frogs and toads). It has not been achieved in mammals, in part because mammalian systems have proven more difficult to work with in this regard and also because such experiments have been restricted for ethical reasons. Although this news may be disappointing—after all, it means that we are not likely to be able to make another you—it does bring with it the consolation that we also are not likely to be able to clone anyone else, at least not for a very long time.

Development and Birth Defects

The parents wait while the doctor examines their daughter immediately after she is born. Although the mother is physically exhausted from hours of labor and the father is mentally exhausted from the tension of waiting and from supporting her efforts, both are eager to interact with their new child. The doctor smiles and hands the child to the mother. Everything seems to be fine.

Not all delivery room scenarios go so smoothly. Despite the highly regulated and precise expression of the genetic instructions for development, sometimes things go wrong. Each year about 3 percent of all babies born in the United States have significant problems at birth, and many of these are problems in development. Often couples feel guilty about the challenges their newborns face, and they may ask their doctors if they did anything to cause the problem.

Although the doctor may not always be able to answer that question, in some cases the specific source of the developmental problem can be determined. Developmental problems can be categorized into two general groups: those that result from errors in the genetic plan and those that result from disruptions in the expression of that plan.

In the first category, certain mutations in the DNA inherited from either the mother or father (or, rarely, created in the embryo itself) can disrupt the sequence of events during the growth and differentiation of the embryonic cells. This disruption may result in permanent and even life-threatening changes in the offspring. Indeed, many changes in development cause a spontaneous abortion (miscarriage), sometimes so early that a woman never even realizes she was pregnant. As many as 20 percent of pregnancies may terminate in this manner. Other disruptions allow development of the embryo to full term in spite of the developmental errors, which results in a baby who is challenged physically or mentally or in both ways.

Changes in the genetic plan for development may occur if an

individual or a growing embryo is exposed to an agent that can cause mutations. One such agent is radiation. When your dental assistant makes an X-ray of your teeth, she or he usually covers much of the rest of your body with a leaded apron to limit your exposure to radiation. The radiation level of a dental X-ray is quite small, but it is a good idea to limit risk by limiting exposure. This precaution is particularly important for women because their gametes, the ova housed in their ovaries, are present from birth through their reproductive years. Thus, the effects of mutation-causing agents on a woman's ova are cumulative. (Males produce new sperm continuously, so genetic damage is less of a problem.) Likewise, if a pregnant woman is exposed to radiation at sufficient levels, alterations in the DNA in some early cells of a developing embryo might occur. This exposure could misdirect development (refer to Figure E13.11).

In the second category, developmental errors can occur if the *expression* of genetic material is disrupted. This can happen even if the genetic instructions for development are not mutated. Because development is highly regulated, there are many points at which an error could have lasting effects. For example, exposure of the embryo to chemicals can interrupt signals that are essential for the proper formation of specific tissues. Normally, the uterus provides protection for the growing embryo, and the placenta, which

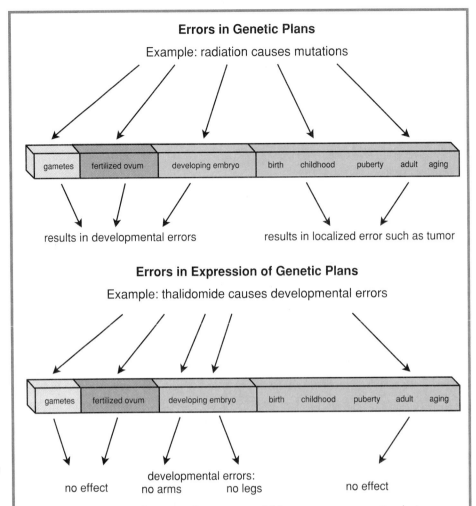

Figure E13.11 Developmental stages at which errors can occur Developing humans are vulnerable to developmental problems at different times and in different ways. Notice that changes may arise from a combination of genetic and environmental factors. Errors are either alterations in the genetic plan for development (in a gamete or in a developing embryo), or they are disruptions in the expression of that plan.

allows nutrient exchange from mother to embryo, helps provide special controlled conditions for the growing offspring (see Figure E13.12). However, toxic substances and even pathogens that enter the mother's blood supply may reach the embryo through the placental connection. The major organ systems become established during the first three months of development, and the embryo is particularly vulnerable to toxic substances during this early part of a pregnancy.

One of the most dramatic examples of this vulnerability occurred during the early 1960s. An unusually large number of extremely malformed babies were born in Western Europe, Japan, Canada, and Australia. Medical scientists began trying to find a cause for this developmental problem by collecting information from the mothers of the affected children. A pattern emerged: all the mothers had taken a mild sedative known as thalidomide to help control the nausea brought

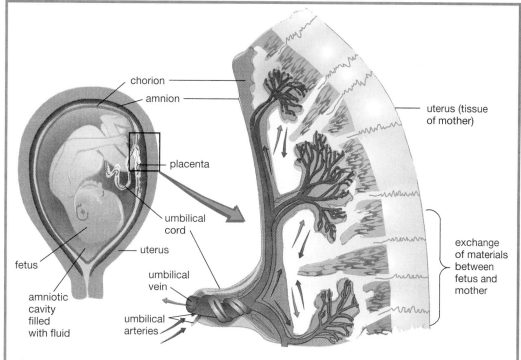

Figure E13.12 Developing embryo in uterus Part of the placenta is enlarged to show the circulation of the mother and the fetus. The two systems do not mix, but the exchange of nutrients and gases occurs between the maternal blood supply in the uterus and capillaries that deliver blood to the fetus.

on by pregnancy. Although the drug has almost no side effects for the mother, it has devastating effects on the development of the embryo if it is taken between the third and fifth weeks of pregnancy. As shown in Figure E13.13a, during the first part of this vulnerable period, even a single dose of thalidomide can result in the improper formation or the absence of arms in the offspring. An exposure a week later causes malformation or absence of legs, and other abnormalities can be mapped to other periods in development as well.

By the time doctors realized the effect of taking thalidomide during pregnancy, many children already had been born with serious problems. This striking example demonstrates that a developing embryo is *not* fully protected by the uterus and

placenta. More commonly used drugs or other factors also can do serious harm to the developing embryo. For example, oral progesterone taken at high levels can cause serious problems. (Recall from your study of human reproductive technologies the common form of progesterone that women might take.) Some nonprescription drugs such as aspirin may increase risk of miscarriage or otherwise threaten an already weakened embryo from properly developing to full term. Certain diet pills containing dextroamphetamine also have been implicated as a possible cause of birth defects if the mother takes them during certain times in pregnancy.

Damage to a developing embryo also can occur if a pregnant woman exposes herself to cigarette smoke or alcohol. In the case of smoking, the effect of

nicotine (from the cigarette smoke) on blood circulation is thought to result in low birth weight and a generally weakened infant. Alcohol produces a wide range of defects. (Damage to the embryo from alcohol consumption by the mother is called fetal alcohol syndrome and its effects are shown in Figure E13.13b.) When a pregnant woman drinks alcohol, it enters her blood stream and then that of the embryo. Amounts of alcohol that an adult can tolerate with limited damage can have an extreme effect on the delicately regulated developmental processes of the embryo. The developing brain is particularly sensitive to damage from alcohol. Scientists do not know for certain whether there is a minimum amount of alcohol that the mother can drink without harming the baby. Even limited drinking during pregnancy, however, appears to contribute to the danger of low birth weight and to an increased risk of serious birth defects and mental retardation. Many women choose to give up alcohol for nine months to avoid these problems or the possibility of caring for a severely handicapped person.

We have seen that environmental hazards can threaten the health of a newborn in several ways. They can mutate

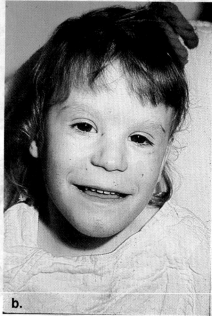

a.

b.

Figure E13.13 a. The effects of thalidomide on leg development b. Small child affected by fetal alcohol syndrome
Facial abnormalities and mental retardation are typical of children damaged in utero by alcohol.

genes in the gametes of the parents or can directly interfere with a healthy developmental scheme prior to birth. Some of these errors are beyond the control of the parents, whereas others directly reflect the lifestyle choices of people who choose to become parents.

Cancer: Unregulated Growth

The pediatrician shows the parents into her office to discuss the physical examination she has just made of their 7-month-old daughter. The child looks pale and is terribly weak. She has a fever, and the doctor can feel that her spleen is enlarged.

The doctor tells the parents that the baby will have to be hospitalized for tests to determine which of several possibilities explains their daughter's condition. As they sit down, the mother looks the doctor squarely in the eyes and asks, "Is leukemia the worst thing it could be?"

"No," the doctor replies. "Leukemia is the best thing it could be."

The parents are shocked because they are old enough to remember a time when leukemia, like almost any cancer, meant certain death. However, they are reassured to learn that many forms of childhood leukemia now can be treated and the

likelihood of success for these treatments is relatively high.

When the tests are done, the diagnosis comes back: their daughter Lauren does have leukemia. Fortunately, the pediatrician is also a specialist who treats cancer. Under the expert guidance of this doctor and a team of nurses, Lauren is treated chemically to control the growth of cancerous blood cells.

Now, at age 7, Lauren (Figure E13.14) is a healthy and energetic child, free of leukemia. All her parents have to worry about is what they would worry about for any child: whether or not she is careful while riding her bicycle or hiking in the mountains.

The period of time between fertilization and birth is not the only time that a human is susceptible to developmental errors. Indeed, such errors can occur

UNIT FIVE DEVELOPMENT: Growth and Differentiation in Living Systems

Figure E13.14 Lauren, a leukemia survivor, is now a healthy, energetic child.

throughout life. One condition that can occur even in a mature adult is the unregulated growth of cells that have lost some or all of their differentiated characteristics. Unregulated growth can occur in many different tissues and organs and is generally known as **cancer**. There are many different types of cancer, such as skin cancer, breast cancer, prostate cancer, lung cancer, and the type that Lauren had, a cancer of the blood cells known as leukemia.

Cancer gets its name from the Latin word for crab. It was so named because years ago physicians noticed that many cancerous growths were shaped much like crabs, with leg-like appendages extending from a central mass. To understand the nature of cancer, we first must examine what happens when cells no longer respond normally to controls on growth. These abnormal cells may divide often, and they are not inhibited from growth by contact with other cells (contact inhibition) as normal cells would be. The cancerous cells can pile up on one another, forming a tumor.

These abnormal cells are said to be *transformed*. If caught early, transformed cells often can be destroyed by scavenger cells of the immune system such as macrophages. If transformed cells escape the immune system's defenses, they can become a tumor. And if the tumor cells have the ability to release destructive enzymes that help them invade other tissues or migrate to other areas of the body and start new tumors, they form malignant tumors. Some tumors lack this ability and remain localized in one area, surrounded by connective tissue. They are said to be benign.

The ability of cancer cells to spread to new areas of the body is called **metastasis**. Figure E13.15 shows how metastasis can occur. Metastasis is the major source of

danger from a cancerous growth because a migration of cells can result in the production of new tumors in other parts of the body, which in turn can damage many organs.

For many years research scientists and physicians have focused attention on understanding the causes and progression of developmental errors that result in cancer and on finding ways to treat it. Basic scientific research looks for the cellular and biochemical mechanisms that bring about malignancy. As new evidence for the causes of cancer become known, this knowledge is used in applied research programs that search for new treatments or preventive measures. Through this combination of basic scientific inquiry and application to medicine, we know that cancer has a number of causes and there are a variety of treatments available.

In 1911 a young scientist named Peyton Rous performed an experiment that supplied evidence about the cause of cancerous transformations. He crushed cancerous cells from a chicken tumor and filtered out any remaining cells. He then was able to use the liquid from the crushed cancer cells to produce tumors in other chickens. Today we know that the liquid contained a virus, now called *Rous sarcoma virus* (refer to Figure E13.16). This virus can infect animal cells and transform them into cancer cells. Although it took more than 40 years for the scientific community to fully accept this idea, and more than 50 years for Rous to be awarded a Nobel prize for his work, many other cancer-causing viruses have since been identified.

In recent years scientists have learned about a special group of viruses, known as retroviruses, that have RNA genes rather than DNA genes. Many of these viruses can cause animal tumors because they carry a special type of gene, called an **oncogene**, that has the power to convert a normal animal cell into a transformed, cancerous cell. When a retrovirus infects a cell, a DNA copy of the viral genome is made, and this viral DNA is incorporated into the cell's DNA. When the viral oncogene is activated, unregulated cell growth occurs. The adaptive advantage to the virus is that its genome is replicated many times as the unregulated cells divide.

Additional genetic and biochemical experiments in the 1970s and 1980s led scientists to discover an unexpected connection between retroviruses and normal cells. Strangely enough, many viral oncogenes are almost perfect matches with genes that are normally found in animal DNA. Many of the animal counterparts of the viral genes code for regulatory proteins that affect cell growth and adhesion (see Figure E13.17). The alteration of these genes as they occur in the viral form causes them to be abnormally active. This can cause abnormal growth or alter adhesion of the host cell. (Most viruses do not carry oncogenes, and thus most viruses do not cause cancer in animals. For example, the influenza virus does not cause cancer.) Normal cellular genes for growth regulation can be altered in the absence of viruses as well. In these cases damage to normal regulatory genes causes a loss of growth control.

Metastasis

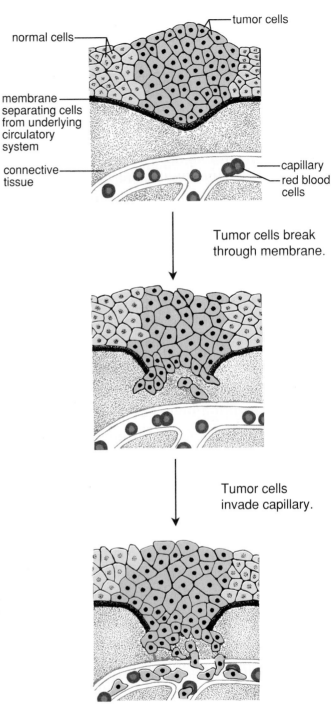

Tumor cells break through membrane.

Tumor cells invade capillary.

Figure E13.15 Metastasis involves the dangerous spread of cancer cells from a localized tumor to new areas in the body.

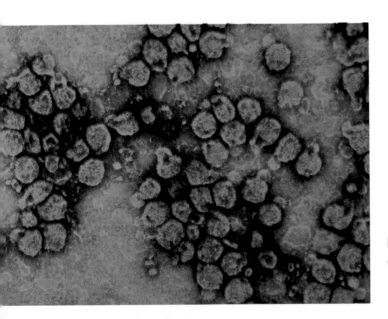

Figure E13.16 *Rous sarcoma viruses* (X85,000)

As more information about human chromosome structure and gene sequence comes to light, a number of mutant genes are being identified that appear to play a role in human cancer. For example, a mutation in a recently identified gene on chromosome 17, *BRCA1*, increases the risk of breast cancer as well as ovarian cancer in some women. The presence of the mutant gene causes about 5 percent of all breast cancer cases in women. (The genetic risk factors associated with other breast cancers are not fully understood.)

The breast cancer gene *BRCA1* occurs in about 1 in 800 women. One clue that helped researchers identify this mutant gene was the recognition that breast cancers occurred in several very close relatives in several families. This suggested a genetic basis for susceptibility to this cancer. As a blood screening test to identify the presence of this cancer gene becomes available, questions arise about who should be tested and how they may deal with the knowledge that they have inherited a high risk of early breast cancer.

Inheritance of a cancer gene, however, does not guarantee that the disease will occur. The gene may only increase susceptibility, and a second event, such as an additional mutation caused by an infection or by radiation, may be needed for cancer to occur. The risk of cancer also is offset by the action of a healthy immune system, which constantly searches for and destroys cells that it identifies as abnormal.

A similar combination of events may give rise to cancer in a person who has not inherited a particularly high risk of cancer. Cigarette smoke, for example, contains several compounds that greatly increase the risk of cancer as well as other ailments. Compounds such as nicotine can produce cancer-causing mutations in cellular DNA. In addition, other components of the smoke, such as tar, impair the action of the immune system in the lungs and esophagus. If the immune system is no longer efficient in destroying transformed cells, the risk of getting cancer increases.

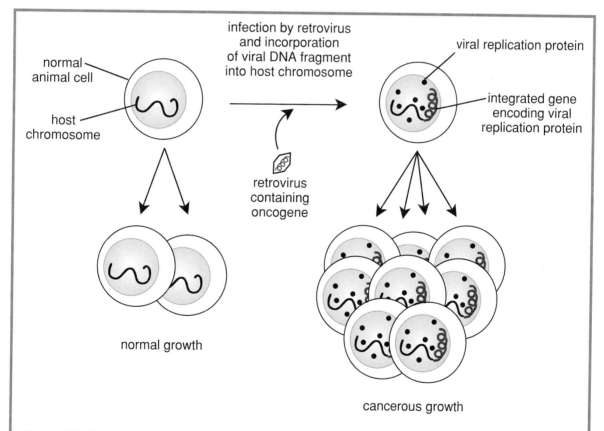

Figure E13.17 Cancer can result from alterations in normal genes whose products regulate cell growth or from infection by certain viruses. Cancer-causing genes, called oncogenes, can originate from changes in normal cellular genes or can enter the cell from an invading retrovirus. How might evolution explain the similarity between the DNA sequences of viral oncogenes and normal regulatory genes

As science builds an understanding of the mechanisms that underlie cancer, and as technology offers many new and effective treatments, our culture's view of cancer also changes. People begin to recognize that one of the best ways to avoid cancer is to avoid behaviors that either damage the immune system or increase the risk of cells becoming transformed. The choice to sunbathe only with sun-block protection is the choice to avoid damaging UV radiation and thus limit the risk of skin cancer. The choice not to smoke greatly reduces the risk of lung cancer (as well as heart disease and emphysema). Even if some transformed cells escape the body's immune system and cancer results, there are medical treatments that often can stop the dangerous growth and reduce the danger of metastasis.

One medical treatment is to surgically remove the cancerous tissues. The success of this treatment relies heavily on early detection, while a tumor is still localized. Modern technology is extremely helpful in early detection. For example, Magnetic Resonance Imaging, mammography (Figure E13.18), and ultrasound offer nonsurgical methods of searching for tumors while they are still small. Low-tech methods such as breast self-exams also are valuable in early detection. Finding a small breast tumor very early puts survival rates up near 90 percent, whereas late detection drops survival rates to about 10 percent. Other approaches for treating cancer rely on therapies that attack cells and kill them. The trick is to kill cancer cells without extensively damaging

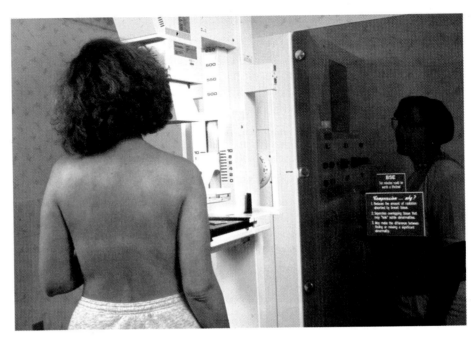

Figure E13.18 Mammography, a type of X-ray technology, is used to detect breast cancer. What other technologies are you aware of that help to detect or treat cancer?

noncancerous tissue. Radiation therapy relies on aiming the radiation at the affected tissue, thus minimizing extraneous damage. Chemotherapy involves taking toxic chemicals that can kill cells that are undergoing mitosis. Many normal cells will be affected, but the strategy relies on the fact that cancer cells have unregulated growth and therefore may be dividing more often than other cells. With this approach, it is not surprising that the patient feels many unpleasant and even dangerous side effects. The immune system also may be impaired temporarily by chemotherapy.

As the body of scientific and technical knowledge increases, our approach to treating cancer also changes. New therapies are being designed to *specifically* target cancer cells. The hope for these specific therapies is that they will be able to kill cancer cells without the risk of damage to healthy cells.

Patterns of Development

Development can be thought of as a series of important events that occur during the life of an organism. Some of these events occur in the lives of almost all multicellular organisms. For example, development in all sexually reproducing organisms begins with fertilization and ends with death. Development also typically involves both growth and differentiation, processes that produce a multicellular organism that looks quite different from the fertilized ovum from which it began.

Other developmental events occur only in *some* organisms. In fact, a remarkable number of different developmental plans exist among multicellular organisms. For

E210

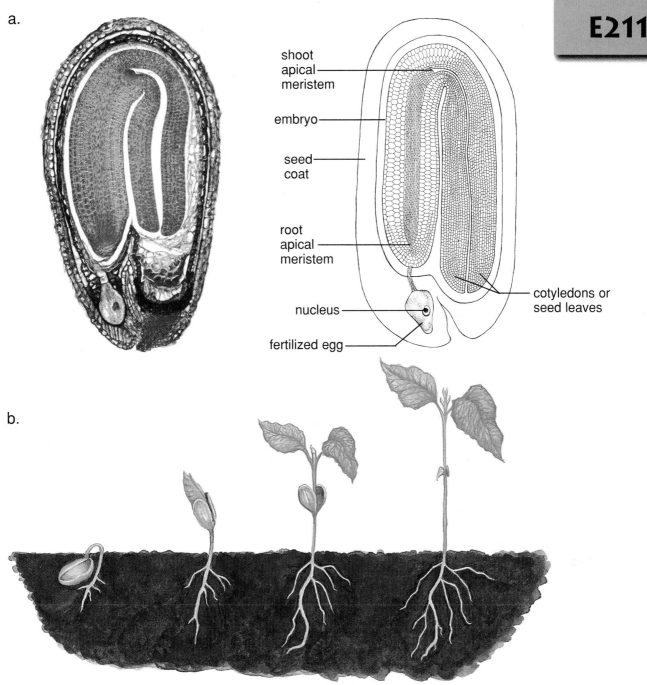

Figure E13.19 Development in a flowering plant a. When pollen grains, which contain sperm cells, come into contact with the female reproductive organs of a flower, fertilization of an egg can occur. The zygote then grows and develops into an embryo. In a seed, as shown here, the embryo has temporarily stopped developing. Seeds are well adapted for survival; many can survive being eaten and eliminated. b. Under favorable environmental conditions, a seed can germinate. When this happens, development resumes and a new plant develops from the embryo. A root develops from the root apical meristem in the embryo, and the cotyledons help provide food for the plant until the leaves develop.

example, growth is a basic feature of most developmental plans, but different types of organisms grow quite differently. Some organisms—for example trees—never completely stop growing; such organisms are said to show indeterminate growth. In contrast, some organisms such as humans stop getting larger (at least we stop getting *taller*) and are said to have determinate growth.

Indeterminate growth is particularly apparent in many perennial plants such as giant sequoia trees. These huge trees begin life as tiny seeds weighing less that 1/5000 of an ounce. Successful seedlings grow rapidly, and young trees begin producing a substantial number of seeds by the time they reach the size of a pine tree (at this point, the sequoia is likely to be about 100 years old).

Even after these trees reach sexual maturity, they do not stop growing. Healthy sequoias normally attain their full heights only at ages approaching 1000 years and might grow even taller except that these older, giant trees tend to lose their growing tips to lightning. Even so, truly ancient giants, such as the 2500-year-old General Sherman tree at Sequoia National Park, continue to produce each year abundant cones, seeds, new wood, leaves, and branches that amount to a mass equal to the wood of a tree 18 m (60 ft) tall and 0.5 m (1½ ft) in diameter. A similar if not so spectacular pattern of continued growth also is found in other types of trees and in nonwoody plants, such as tomatoes and cucumbers.

One of the reasons that plants can show indeterminate growth is that, unlike most animals, plants retain groups of essentially undifferentiated cells, called meristems, for as long as they live. Meristems act like permanently embryonic cells. Cell division in meristems near the tips of shoots and roots produces a constant supply of new cells that differentiate into all of the cell types needed for growth of the branches or the roots. It is the undifferentiated cells of meristems that scientists use to clone certain plants. Figure E13.19 illustrates development in a growing plant.

Even some animals (for example, many types of invertebrates, most types of fish, and some mammals) continue to increase in size for as long as they live. This growth usually slows markedly once they attain sexual maturity. More typical of most animal groups is the pattern of development in which the zygote develops into a single individual with a defined size and number of limbs and organs. In such organisms this size does not change much after some defined point in development. This form of development is seen in birds and in many mammals. Although some additional growth may occur after the individual reaches an age at which sexual reproduction is possible, this increase usually is minor compared with the individual's overall growth.

Another example of a developmental plan that occurs only in some organisms is metamorphosis. Organisms that undergo metamorphosis start life with one body plan and later develop a quite different plan. In most cases metamorphosis from the first plan (for example, a caterpillar) to the second plan (a butterfly) occurs before the organism has reached sexual maturity. In vertebrates metamorphosis occurs in several species of amphibians, such as frogs. Metamorphosis also occurs in a number of invertebrate organisms, such as insects and echinoderms (such as the sea urchins in Figure E13.20).

An especially important event in development is the attainment of sexual maturity. Under the pressures of natural selection, the final test of any organism's developmental scheme is how well it assures the organism's reproductive success or, more specifically, the perpetuation of its genes. From the time that an organism originates as a zygote until it reaches reproductive maturity, selection acts on the sequence of its developmental events. Developmental patterns that ensure that the individual survives and matures to reproduce successfully are maintained; developmental plans that do not assure reproduction are not maintained. In this way the developmental patterns of species respond to the forces of evolution just as any other characteristic might.

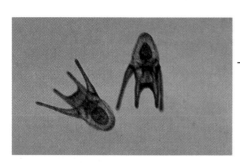

larvae   metamorphosis ⟶   adult

Figure E13.20 Sea urchin metamorphosis a. Sea urchin larvae (X150) before metamorphosis b. Adult sea urchin (8–20 cm across)

a. sea star

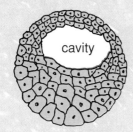

b. frog

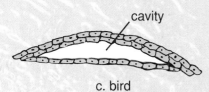

c. bird

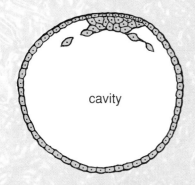

d. mammal

Figure E13.21 Developmental patterns The similarity of this early developmental phase among diverse organisms reflects an evolutionary connection. Which organism would you expect to most resemble the early developmental phases of a salamander, and why?

Not surprisingly, the diversity in developmental plans that we see among organisms of different types mirrors the diversity that we see among the organisms themselves.

This observation brings us to the second of the scientists' questions about development: What clues does development reveal about the process of evolution?

We already have started to answer this question by noting that all development involves the basic processes of growth and differentiation. This uniformity in basic developmental processes reflects the evolutionary relatedness of all types of life. There are, however, other evolutionary patterns that we can identify. One such pattern is that the earliest rounds of cell division in many animals give rise to embryos of the same basic form. This form consists of cells grouped around an internal cavity, although as Figure E13.21 shows, the exact shape and structure of this cavity varies with the animal involved. In more closely related animals, such as all vertebrates, subsequent developmental stages also show some similarities. In closely related vertebrates, such as mammals, the developmental similarities are even more obvious.

Such observations suggest that there is a general developmental plan for all vertebrate embryos and that features that distinguish the different classes of vertebrates result from modifications of this plan. Exactly what types of modifications are possible in a developmental scheme? How might these changes result in a changed phenotype or perhaps a new species? Can such changes ever lead to valuable differences that might be perpetuated through reproduction?

One way to answer these questions is to look for evidence that the same process, occurring in slightly different ways, can lead to changes in structures. A particularly important change that often occurs in development is a change in the *timing* of developmental events. A much-studied example of this is the tiger salamander, *Ambystoma tigrinum*. Throughout warmer regions of this tiger salamander's range, individuals pass their early lives as aquatic larvae that have external gills and flattened tails for swimming. When these organisms undergo metamorphosis, they become burrowing animals, with lungs and rounded tails, that live on land. The organisms become sexually mature only after these metamorphic events.

In colder regions these same tiger salamanders also have adapted to live permanently in streams and lakes. The major evolutionary change is that the tiger salamanders reach sexual maturity and adult size even though they never undergo metamorphosis (see Figure E13.22). Although this might seem to be a major developmental alteration requiring many genetic differences, the actual change turns out to be quite simple. In tiger salamanders that live in warmer regions, metamorphosis is triggered by a surge of thyroid hormone that occurs before the organisms reach sexual maturity. For tiger salamanders that live in colder regions, however, the cold environment prevents production of thyroid hormone in the amounts required to stimulate metamorphosis. This means that the animals simply continue to grow in size until they become sexually mature. As a result, these tiger salamanders retain gills and other adaptations of the larvae and remain aquatic organisms throughout their lives.

In terms of evolution, this is significant because a single environmental signal determines

E213

a.

b.

Figure E13.22 Changes in timing can lead to quite different results in different developing systems. a. In cold regions, tiger salamanders never receive the developmental signal that triggers metamorphosis. Thus, the salamander retains its aquatic larval form as it sexually matures. b. In warm regions, tiger salamanders produce a hormone that triggers metamorphosis *before* sexual maturity is reached. This causes the salamander to transform into a land-dwelling organism.

internal changes that affect the developmental path the organism will follow. In the case of *Ambystoma tigrinum*, flexibility permits adaptation to a specific environment. Developmental pathways can become fixed, however, as the life cycle of the Mexican salamander, *Ambystoma mexicanum*, illustrates. These salamanders never undergo metamorphosis in nature, and thus they live their entire lives in the larval form. The many similarities between these two species suggest that evolutionary change from one species to another may proceed through changes that alter the course of development.

Human Development 101

I am just about to begin observations for my human development class. I am both excited and apprehensive. I am already tired of reading the textbook and am really excited about getting to observe—I think it will be a great way to learn. So why am I apprehensive? I guess maybe it's because I won't be observing fast-growing plants or Daphnia as I'm used to doing from my biology classes—I'll be observing people!

My professor has been great about helping

us get ready. She has explained to us all of the important things to keep in mind. I think it's interesting that public ethical standards for working with children were established only after World War II when officials discovered the unethical and inhumane experiments that Nazi physicians had conducted on both children and adults. Because of this, officials developed the first written guidelines to protect people who participate in research projects. Then, in the late 1960s government officials in the United States established their own set of specific guidelines to protect human subjects. And in 1982 the

American Psychological Association published a set of ethical standards for conducting research with children.

I think the first important issue is respect for all participants. All participation must be completely voluntary. This means that if people do not want us to observe them or to include them in a study, then we must respect their wishes. Also, the participants can decide to drop out of the project at any time. And with children, it's important to remember that even if the parents have given permission, if the child does not want to participate, we must respect the wishes of the child—not just the parent.

When making observations, a complete, detailed written record is essential.

It's pretty obvious that you can't do anything that would harm a participant physically or emotionally, but some of the other reasons for the regulations aren't as obvious.

I remember that the ethical guidelines also say we must explain the purpose of our research or observations to the participants. Also, the information that we obtain is confidential. This means that we should never discuss personal information about a participant with anyone outside the project. When social scientists publish research papers, they never disclose the participants' identities.

The participants also have the right to ask questions about the observations or the research. It's important to remember that they may want to know the results. It's critical to remember that the rights of the participants come before the rights of the researcher.

O.K., I think I'm ready for my first day of observations.

Personal Journal Entry: 28 October

Well, I thought I was ready, but I wasn't. I was supposed to make observations of cognitive growth and development—I didn't even remember what that meant. So, I went back to my notebook from class and read through my notes. I found out that **cognitive growth** refers to the processes that humans use to learn about and adapt to their environment. These processes include perceiving, remembering, imagining, and reasoning. It makes sense that we get better at this as we grow older. My cognitive skills are better now than when I was 5, and they were better at 5 than when I was 5 months old.

So, what exactly do I look for to find examples of cognitive growth? Maybe I need to ask the children to do something, to solve some problem, to do a puzzle, or I could just ask questions to find out about their cognitive abilities.

Let's see, what are some ideas?

For the youngest children:

Ask them to point to things of specific colors.

Ask them to point to pictures of different animals.

For children a little older:

Ask them to write something or draw something.
Listen to how well they use language, how long their sentences are, and stuff like that.

For children in late childhood:

Ask them to solve simple math problems.

Listen to their sentences and their use of language.

Ask them to read.

For adolescents:

Ask them to discuss some book they just read or a movie that they just saw.

Ask them to solve a puzzle like Tangoes.®

Ask them to solve a more difficult math problem.

For adults:

Ask them to discuss a book or a movie.

Ask them to describe what they do in their jobs.

For the elderly:

Ask them what they remember about their families.

Ask them about the people in the photographs around the room.

Ask them to play a board game or card game.

A special note on children:

I remember a neat activity about the idea of conservation. You take a tall, narrow container of water, and then, with the children watching, you pour the water into a shallow cake pan or something like that. Then you ask the children which one had more water. Young children will say that the tall, narrow container had more water in it. But at some point in their cognitive growth (I get it now), their abstract thinking improves, and the children will realize that the amount of water is the same. I think that is so cool.

Goal: Maybe tomorrow I can try to find out at what age that happens. Complete, clear notes will be important. Maybe I will revise my forms so they are easier to use.

Personal Journal Entry: 29 October

Observing physical development is sort of straightforward. I mean, you actually can watch people do things. I asked the kids if I could measure how tall everyone is. They thought it was great and enjoyed helping me. Sometimes they were even a little bit competitive about who would get to help. (This is when I realized that, although I thought I was recording data about physical development, my partner was recording a lot of stuff about social and emotional behavior. That was great, and we didn't even have that planned! It helps to be ready for anything.)

Our professor explained that when we observe adolescents, we should measure only height not

weight because adolescence is a time when people generally are very sensitive about their weight.

Even though I thought observing physical development was more straightforward, I realized that I wasn't getting a very detailed picture. It would be great to be able to observe the kids on the playground. Maybe tomorrow I will ask them to skip and see who can do it and who can't; I should ask my mom how old I was when I started to skip.

Personal Journal Entry: 3 November

I just noticed that my notes on emotional growth and development aren't as complete as I would like them to be, but this type of development is a hard thing to observe. Is what I think someone is feeling really what the person is feeling, or am I misinterpreting him or her? This is hard. I need to try to separate what I see and hear from what I think that means—I need to separate the observation from the interpretation or inference. Sometimes when you record observations, it is a good idea to place brackets around any phrases that represent interpretations.

Should I ask or just observe? Maybe I could ask about what different emotions like anger, excitement, love, and loneliness mean and listen to what they think. (Remember—

this is not the same as observing the emotions.)

Social growth is pretty easy to observe with children in groups. As I was watching a group of children, it was easy to separate the sociable ones from the reserved ones. But again, I tried to write down my observation, not my interpretation. I could observe how many children were willing to share toys, how often they would make eye contact with others, and how much time they preferred to play alone.

I also began to notice how the size of the group that you are observing is important. If you don't have a large enough group (at least three), then you really limit the amount of social interaction that happens. But if the group is too large (more than six or seven), then it is too hard to keep track of everybody.

Social behavior will be harder to observe in adolescents and adults because their own awareness of themselves is higher and there is more concern about how others will perceive them. For these reasons they are more cautious in their interactions. How can I find out? What questions can I ask? Maybe I can ask them to estimate how much time they spend each week (outside of school and work) by themselves, with one other person, two other people, with more than two other people.

O.K. It looks like I have a place to start again. Observing is fun, but harder than I thought—it creates more questions.

Growing Up through Life's Phases

Learning to ride a bicycle, watching your young cousin take her first steps, helping a grandfather walk down the sidewalk with his cane—through simple experiences of your own and experiences of those around you, you have come to realize that human life proceeds through a series of phases that have a biological as well as an experiential basis. Virtually every human culture seems to

recognize benchmarks, such as learning to talk, learning to walk, going through puberty, selecting a mate, having children, and growing old. Let's look at these benchmarks more closely.

Infancy includes roughly the first year of life. Most human languages have one or more special words for individuals in this period of life. In English, we say *baby*; in French, *bébé*; and among the Trobriand

Average age in months	Physical activity
12	walks alone but may prefer crawling
11	stands alone, climbs stairs
10	walks with support
9	pulls self up, side steps along furniture
8	grasps in effort to pull self up
7	crawls, sits alone
6	sits unsupported up to half an hour
5	rolls over, moves by rocking and twisting
4	controls hands, rolls from side to side
3	controls head, bats at objects
2	vocalizes, grasps voluntarily
1	moves arms and legs reflexively
0	suckles, lifts head briefly

Figure E14.1 Physical activities during the first year of life This chart presents guidelines for the average ages at which infants are able to perform various activities during the first year of life. The age at which specific individuals will be able to accomplish each activity will vary.

Islanders near New Guinea, *wayway* until crawling and then *pwapwawa* until walking. During this period an enormous number of physical and mental changes take place in rapid succession. During the first year baby teeth begin to erupt, and physical growth is rapid—the child normally weighs three times its birth weight by the end of the first year. The brain grows to 80 percent of its adult size by the end of the second year and control of locomotion and manipulation improves markedly.

Figure E14.1 shows the average age at which a number of physical actions first appear. These are only *average* ages; many first-time parents become needlessly upset if their infant fails to keep "on schedule." As shown in Figure E14.2, infants in the United States often walk by the time they are 1 year old. In other societies the development of sitting and walking is influenced by various cultural practices. For example, among the Aka in Africa, infants tend to sit and walk several months earlier than infants in America. In the Aka culture, infants receive more physical stimulation and early, direct training to sit and walk. On the other hand, among the Ache in South America, family members hold the infants almost constantly, and they do not encourage the young to begin walking. Infants in this culture do not begin walking until they are 22 to 24 months old.

Figure E14.2 In the United States, an infant's first unassisted steps usually occur at about 1 year.

From birth to age 2, growth of the brain and the accompanying mental development are rapid. An important aspect of this development is growth in the infant's awareness of the people with whom he or she is in regular contact. This interaction results in **attachment**, the development of close emotional ties to the regular parent or caregiver. Again, the way in which attachment takes place is influenced by the cultural setting. In 1988 a group of researchers studied the parental patterns in five different cultures. Figure E14.3 demonstrates some of the differences that they found between a group of parents in Boston and a group among the Gusii in Kenya. Gusii mothers, like mothers in many developing societies, are concerned with the survival of their young. By holding them and keeping them close, the mothers also keep their children safe. In addition, the Gusii believe that children cannot understand language until they are close to 2 years old, and therefore they rarely talk to their babies. During conversations in general, the Gusii do not look at each other; it is a cultural norm to avert your gaze when you are talking with someone. Consequently, the Gusii mothers make very little eye contact with their babies. In Boston, on the other hand, infants spend more time in infant seats and playpens than in their mothers' arms. This practice may reflect the mothers' need for independence during the day and perhaps also the value of teaching independence to their young.

Childhood is the next phase of life. Again, most human societies associate individuals in this period with one or more specific terms: *child* in English, *gwadi* in the Trobriand Island language, *enfant* (not to be confused with the English "infant") in French, or *ki yakka pirika* ("it eats with adults now") in the language of the Ainu people of Northern Japan. During the decade or so of childhood, physical growth continues at a moderate pace, with a growth spurt just before adolescence. The first permanent teeth usually erupt during the seventh year. During childhood the skull grows mostly within bones of the facial region, as illustrated in Figure E14.4.

Other important changes also take place. Physical coordination and language skills improve both through deliberate teaching by adults and from less formal practices. For example, physical games and verbal activities, such as telling or listening to jokes and stories, contribute to this development. In the United

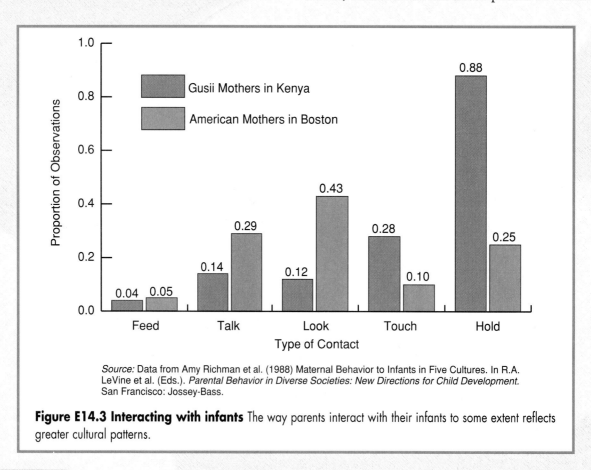

Source: Data from Amy Richman et al. (1988) Maternal Behavior to Infants in Five Cultures. In R.A. LeVine et al. (Eds.). *Parental Behavior in Diverse Societies: New Directions for Child Development.* San Francisco: Jossey-Bass.

Figure E14.3 Interacting with infants The way parents interact with their infants to some extent reflects greater cultural patterns.

States, parents engage in a lot of face-to-face interactions and conversations with their young children (the graph in Figure E14.3 represents this point). It is likely that young children in such settings will begin talking and interacting socially earlier than in cultures where talking to young children and interacting face-to-face with them are less common practices.

The growing child also continues to develop its own unique identity and personality. The child discovers his or her unique personal set of likes and dislikes in everything from foods to hobbies and formal studies. An important part of a healthy childhood is steady progress in the development of personal judgment and an internal sense of what *is* or *is not* appropriate behavior toward others.

Studies by child development specialists have shown that, across cultures, the type of economic activity that predominates in a society influences its child-rearing values and practices. For example, in hunting and gathering societies where adults work independently much of the time, self-reliance and autonomy are highly valued, and the child-rearing style is permissive with little use of punishment. On the other hand, in agriculturally based societies where the adults work under close supervision, conformity and obedience are highly valued, and the child-rearing style is more restrictive and controlling.

With puberty the individual enters the period called *adolescence,* which is recognized by most human societies as a distinct milestone. Secondary sexual characteristics develop, such as the appearance of hair in the pubic region and under the arms, and the voice deepens. In females the pelvic area broadens, and in males muscular development becomes pronounced. In early adolescence girls generally are temporarily taller than boys due to an early growth spurt, although males grow to be, on the average, five inches taller than females by the end of adolescence.

The onset and duration of puberty vary widely among individuals. Puberty is primarily under hormonal control, but it can be influenced by socioeconomic conditions and nutrition. As adolescents are experiencing the obvious changes associated with the onset of sexual development, a number of other physical alterations occur, as summarized in Figure E14.5.

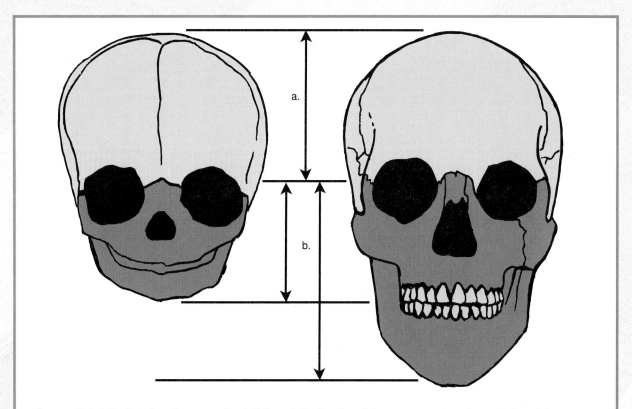

Figure E14.4 Facial development in childhood The heights of the cranium in the infant and the adult are indicated by distance (a). Distance (b) indicates the length of the bones of the facial region. It is clear that after infancy, most of the growth of the skull occurs in the facial bones, especially the jaw bones.

Physical Development in Adolescents

Boys	Age Span	Girls	Age Span
Beginning of growth spurt; growth of scrotum, testes, and penis; sparse, light-colored, slightly curled pubic hair	11-13	Beginning of growth spurt; breasts and nipples elevated (breast buds); sparse, light-colored, slightly curled pubic hair	10-11
Maximum physical growth; deepening of voice; maturation of scrotum, testes, and penis; ejaculation of semen; sparse facial hair; pubic hair darker, denser, and more curled	13-16	Maximum physical growth; deepening of voice; further breast enlargement; rapid growth of uterus, vagina, and ovaries; vaginal secretion acidic; menstruation begins; pubic hair darker, denser, and more curled	11-14
Slowdown or cessation of physical growth; further deepening of voice; penis adult size; darkening of scrotum; increased body and facial hair	16-18	Slowdown or cessation of physical growth; breasts adult size; clitoris mature; adult-type pubic hair	14-16

Figure E14.5 Physical Development in Adolescents

Mental capacity changes, too. Thinking skills improve, including the ability to deal with abstract concepts and to engage in critical thinking. In the United States, this development often is accompanied by an interest in examining and challenging the statements and dictates of authority from the adult world, which includes parents. Many individuals develop a strong drive toward gaining social acceptance from their peer group by presenting the "right" appearance and behavior. Striking a balance between questioning and conformity is difficult. This balance is perhaps the most difficult part of adolescence for every generation, and it marks a key step in the development of each person's individual identity.

Adulthood begins when individuals reach maximum physical stature, usually in the later teens. Adulthood does not mark the end of development, however. In most people social, emotional, and cognitive development can and does continue long after physical growth has ceased. Key events that occur during the adult lives of many individuals include developing a long-term pair bond with another individual, learning and practicing some skill to support themselves, creating a new household, producing children, and parenting those children through all the stages of their own development. Each of these activities clearly calls for the development of many new social and cognitive skills that individuals can acquire only gradually as they practice them. In traditional societies people often establish families

close to their parents and other relatives and follow the same occupations throughout their adult life. In contemporary Western societies, however, people can and do continue to change their homes, occupations, and even family affiliations well into their adult years.

Biologically, the adult years represent the period of aging. This process of aging begins when adults are in their twenties and involves a slow change and deterioration of all the body's systems. Muscles slowly lose strength and mass, skin becomes less resilient and cuts heal more slowly, bones and tendons become gradually more brittle, and the peak performance of major organ systems declines. For people physically gifted enough to become professional athletes in vigorous sports, these changes usually compel retirement from top-level competition by age 40. For most people, however, the changes of aging are slow and gradual enough that they can maintain much the same level of activity through three or four decades of adult life.

Physical Growth Influences Mental Growth

Physically, humans are animals that possess the same basic biological functions and organs as their relatives among the primates. They are also highly social creatures and are able to learn a more complex range of social interactions than almost any other creature on earth. Because of this ability, growth and development in humans is not just a physical process but also a cognitive, social, and emotional one. Nevertheless, the capacity of humans for cognitive growth has a clear biological basis. Because the human birth canal would not be big enough to accommodate a baby's head if its brain were full-sized, humans show a very large amount of brain growth after they are born. The growth of the brain allows humans to acquire the capacity for the most complex cognition and learned behavior of all living creatures.

The frontal lobes of the cerebrum grow rapidly around age 2, with another increase between ages 5 and 7. This is accompanied by a growth spurt in the circumference of the head. The development of an insulating myelin sheath around neurons in the cerebrum is nearly complete by age 7. This sheath speeds neural transmission between different parts of the brain. By this age children's brains generally have developed a level of complexity that is comparable to that of adults' brains.

When the growth or structure of the human brain is disrupted by genetic or developmental disorders or physical accidents, the capacity of affected individuals for growth in mental abilities is delayed or permanently limited. One serious example of an injury that results in this loss of ability is the shaken-baby syndrome. This brain injury, which can occur when a frustrated parent or caregiver severely shakes an infant, often results in permanent retardation or even death. At the other end of life, conditions such as Alzheimer's disease, stroke, or injuries that produce degeneration of the brain frequently are accompanied by deterioration of mental abilities.

Cognitive growth refers to the increase in an individual's ability to construct a mental picture of the external world and understand the relationships of objects in it. Cognitive growth also appears to parallel the physical growth of the brain. Most developmental psychologists agree that the brain develops in stages, but they hold different opinions about the precise stages in the process. For example, Jean Piaget is well-known for his theoretical contributions to developmental psychology.

Cognitive Stages Developed by Piaget	
Cognitive Stage	**Description**
Sensorimotor	**0-2 years**: infants learn through direct experience; do not understand things that exist outside their own actions
Preoperational	**2-7 years:** children develop the ability to use symbols, but have difficulty understanding multiple classification systems; sometime between 5-7 years, a qualitative improvement occurs in the child's ability to organize information logically and coordinate information from several sources
Concrete Operational	**7-12 years:** by 7-8, most children can reason about objects and events that they can perceive; this stage is characterized by the ability to solve problems involving cause and effect, ordering, multiple classification, and numbers
Formal Operational	**12+ years:** as adolescents, they begin to think logically about abstract or imagined concepts

Source: *Child Development: Individual, Family, and Society* by L. Fogel and G. Melson © 1988. St. Paul, MN: West Publishing Company.

Figure E14.6 Cognitive Stages Developed by Piaget

a. When children of two years are given a drawing tool, they generally use simple repetitive movements to draw scribbles. Children of three begin to control their scribbling by observing the lines and concentrating on where they put them. Soon after this, most children begin to realize that their scribble can represent things: "that's my cat" or "this is the sun." They often begin drawing first and only name it as they work. Often they will say the same drawing "represents" something different if asked about it later.

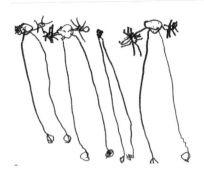

b. Children of three and one-half to four begin to use particular symbols to represent the same thing every time they draw. A common example is when a child draws a person with only a head and limbs and no body.

c. Children of five begin to add other features to their drawings of humans and even then might not always include a body. As children begin to depict real activities in their drawings, they often exaggerate particular features that are important for the action they are representing.

Figure E14.7 Drawings by children of different ages

d. Around six or seven children begin to represent actual scenes and add more detail such as stripes on clothing and shoelaces. Even at this point their drawings tend to be built of simple geometric shapes. In one experiment children were asked to draw a person using different media such as clay, pre-cut paper shapes, and various drawing instruments. The children's representations were similar regardless of the medium they used.

Trained as a biologist in Switzerland, Piaget later became interested in human development. He suggested that all complex forms of knowing develop through the interaction between the individual and the environment. He divided childhood into a series of developmental stages that he defined according to the type of knowing the child uses, as summarized in Figure E14.6.

Cognitive growth in children is well-illustrated by language development. Language development in children is not merely a matter of acquiring a larger vocabulary along with a grasp of the finer points of grammar. Evidence suggests that language development proceeds by a similar series of steps in children of many different cultures. At around 1 year, infants begin to utter single words. Most of these are nouns that simply identify things the infant sees. These nouns can represent names, questions, or demands, depending on the intonation, as in "toy," "toy?" or (demand) "TOY!" Between 18 and 24

months, children proceed to the stage of two-word utterances, which allow them to make a surprising variety of statements, such as "book here," "sit chair," "allgone milk," and (a favorite remark concerning many different things in the infant's world) "MY ball." Three-word sentences follow shortly and allow the child to express more complex ideas or demands such as "daddy read book." As language acquisition continues, growing children begin to learn the particular grammatical rules of their own language, and this process generally follows a sequence that is similar among children in all cultures. In English, for example, children learn at a young age that they can form a question by inverting the word order, as in "Can he walk on the road?" instead of "He can walk on the road." It takes longer, however, to be able to form a question that begins with a question word, so children will continue to say, "Why he can walk on the road?" for some time after they learn the basic rule.

Interestingly, the developing human brain appears to be able to acquire language most easily very early in life. Young children who live in situations in which more than one language is constantly being spoken appear to acquire multiple languages with relative ease. It is much more difficult for people to learn new languages later in life, such as in the teens. Even highly intelligent people who learn a second language in adulthood almost never gain the same ease and fluency with a second language as people who learn one as children. Moreover, studies on rare cases of children who were separated from other people throughout childhood and consequently were not exposed to language have suggested that they never can develop complex language skills.

Cognitive growth also can be expressed as visual development. We can observe this through the artwork that children produce at different ages, as illustrated in Figure E14.7.

Physical Growth Influences Social and Emotional Growth

The social growth of humans is the best known type of mental development. The types of social behavior that individuals show become more complex as they grow from infancy to maturity. Psychologists often describe social growth as a process in which individuals gradually develop from seeing other people only in terms of their own immediate wants and needs to seeing themselves as defined by give-and-take relationships with others. Social growth has a physical basis, however. Humans who suffer from an impairment in physical development of the brain frequently are unable to develop mature social skills. For

example, in fetal alcohol syndrome, the growth of the brain is disrupted in the developing fetus because the mother drank alcoholic beverages during pregnancy. The normal convolutions of the brain may fail to develop, and affected children frequently are unable to develop mature social skills or other types of advanced mental functions.

The first social interactions in a person's life are the bonding with parents and other members of the immediate family. As the child grows, his or her social contacts reach beyond the immediate family to include playmates and other adults (refer to Figure E14.8). As

Figure E14.8 Play is so central to the life of a young child that the years from 2 through 5 are often referred to as the play years.

they reach maturity, most individuals develop an extensive circle of friends and acquaintances, which include teachers and other authority figures. Through their interactions with these people, they acquire the appropriate social skills that include not only a desire to please and help close relatives and friends but also a general sense of social responsibility toward all other people.

Interestingly, evidence indicates that such responsible behavior is governed by a specific portion of the human brain. Individuals who experience accidents that damage a region located in the frontal lobes but leave the rest of the brain intact sometimes retain normal reasoning powers but lose all sense of social responsibility and appropriate behavior toward others. A remarkable case that illustrates this point vividly is the case of Phineas Gage, an engineer who lived in the 1800s. In an industrial accident, a metal rod was propelled through the frontal lobes of his brain (refer back to Figure E1.3). Although he survived the accident, the once congenial, dependable worker was now irresponsible and foul-mouthed.

Like cognitive and social growth, emotional growth in humans has a biological basis and occurs progressively as a child's brain matures. In many types of mental retardation, emotional growth is delayed or limited. These conditions often result in individuals who become physically mature but remain somewhat childlike in their emotional development. Moreover, older individuals who are affected by disorders such as Alzheimer's disease frequently undergo emotional regression and may become childlike.

Emotional growth normally begins in a child's relationship with its parents or caregivers and is gradually extended to a wider circle of individuals. With successful emotional growth, a child learns to express negative feelings in a nondestructive manner and comes to accept the fact that some of his or her wishes and desires cannot be satisfied immediately—or indeed sometimes at all. A key point in the process of emotional growth occurs at puberty, when each individual begins to experience the very powerful drives and emotions connected with sex. For some young people, this stormy period is complicated further by the discovery that their sexual orientation is not the same as most of those around them. Because this often happens at the same age when the desire to fit in and be like one's peers is at its peak, such young people may experience special hardships.

All Phases of Life Require Self-Maintenance

Because both our physical and mental capacities have a biological basis, our ability to develop and sustain them requires self-maintenance throughout life. The most obvious and familiar example of self-maintenance is physical fitness. Anyone who has participated in either competitive or recreational sports has discovered that the body's ability to perform increases with training, practice, and a good diet. Moreover, anyone who has interrupted a regular routine of training and a healthy diet for any reason has found, when attempting to resume a sport, that performance level definitely drops off when fitness declines. In recent years we also have realized that the body's ability to carry on routine activities is improved by regular exercise and a healthful diet throughout life.

An individual's environment influences whether that individual reaches his or her genetic potential. The environment is especially influential during the periods when the body and brain are growing and developing. We already have noted that mothers who drink even small amounts of alcohol while they are pregnant may produce a child who is affected by fetal alcohol syndrome. Many other agents—including tobacco and certain prescription and illicit drugs—also can affect a developing fetus unfavorably.

Once born, growing children also are at risk. Some children who go through periods of severe starvation during early childhood, particularly those whose diets lack protein, never reach full physical or mental growth, even if food is plentiful thereafter. In fact, even far less severe situations can affect the attainment of biological potential. During middle childhood, average heights of Nigerian boys from wealthy families were significantly greater than those from poor families. Likewise, the average height of the population of Japan increased throughout the twentieth century as the people have adopted a diet richer in protein along with other environmental changes.

Evidence indicates that the social environment also may affect the development of mental capacities such as emotional and social development. Some research has been done on babies who have been kept in impersonal institutional

environments where their physical needs are met, but where there is little time for the attendants to cuddle them or talk to them. These studies have indicated that such infants do not thrive or develop as well as babies given comparable physical care, but more attention. In fact, children who spend their entire childhood in such situations often appear to be stunted in their emotional and social growth, and studies suggest that many of them never are able to form normal relationships in later life.

It is clear that the environment influences an individual's potential. It is also true that an individual's genetic makeup influences his or her potential. Researcher's have studied identical twins (that is, individuals who have the identical genetic makeup), both those who have been raised apart and those who have been raised together. To the extent that genes influence the development of an individual, you would expect that identical twins, even those raised

apart, would demonstrate a high degree of similarity in various traits. Based on research done with twins and other research as well, scientists now think that physical characteristics are more strongly influenced by an individual's genetic plan than intellectual abilities and personality traits are. Scientists also have found that some traits such as creativity and social attitudes seem to be more influenced by environmental factors than genetic ones. Indeed, both genes and the environment are influential during the entire life span. It seems, however, that environmental influences may become somewhat less important with age.

Once the body and brain mature, self-maintenance involves keeping physically and mentally active, eating balanced meals, and avoiding activities or behaviors that accelerate degenerative changes as the body begins to age. When they are still in their teens or twenties, people

Figure E14.9 Seniors exercising Exercising throughout life provides both physical and psychological benefits.

sometimes make damaging lifestyle choices and acquire harmful habits, such as mental and physical inactivity, heavy drinking, smoking, use of drugs, excessive exposure to the sun, or eating large quantities of fatty foods. At these ages, their bodies are able to tolerate some of these activities without showing many obvious ill effects. Through time, however, any of these behaviors may cause cumulative damage to one or more major organ systems.

Many of the changes once thought to be unavoidable aspects of old age, such as the weakening of joints, bones, and muscles and the loss of mobility, can be slowed or even reversed to some extent by programs of exercise and nutrition. Apparently, regular exercise and good eating habits throughout life also may help maintain the immune system's activity into advanced years, which lowers susceptibility to disease. Moderate exercise for fitness and attention to good nutrition should be lifelong priorities for everyone, as illustrated by the seniors exercising in Figure E14.9.

Interestingly, there is a growing body of evidence that mental capacities also may benefit from continued mental exercise throughout life. A number of psychological studies have suggested that people who continue to challenge their minds retain a much higher ability for dealing with new information and adjusting to new situations and challenges as they age. Two activities that are known to improve general mental abilities are playing bridge and learning a foreign language.

There is a common theme to most of these aspects of self-maintenance; at some level, they are at least partially under the conscious control of the individuals themselves. In the case of a fetus or a young child, the individual is dependent on the care and good judgment of the adults who are responsible for him or her. As individuals approach maturity and demand more control over their own actions, however, they also are expected to take responsibility for making intelligent decisions about maintaining their mental and physical health.

Culture: The Great Shaper of Life

Personal Journal Entry: 5 April

Well, I'm getting the hang of observing humans. I think I probably make the best observations with infants and young children and then with older adults. I think it is probably hardest to observe people closest to my own age—I find myself beginning to compare my own thinking and behaviors with theirs and forget to focus on them.

I just got a letter from Malcolm, a friend of mine who is in graduate school in California. He has an even bigger challenge. He is studying anthropology and is doing some field work in New Guinea and the surrounding area. He was describing to me how hard it is to study people from another culture. He says that when you are just beginning to learn about another culture, it is natural to fill in the gaps in your understanding with what is familiar from your own culture. He points out, however, that by doing so, you often develop an inaccurate picture of the culture you are trying to learn about.

I remember from my anthropology course last semester that it is important to try to put your own cultural values aside and learn about another culture from the perspective of how certain behaviors and attitudes work within that other culture—to whatever extent that is possible. Of course, this is tricky, but Malcolm explained in his letter how cool it is to learn not just about the differences, but that, in spite of some distinct differences, humans living all over the globe have many things in common.

I also remember learning about the work of Urie Bronfenbrenner. In 1979 he proposed an ecological approach to the study of development. As the diagram in Figure E14.10 indicates, Bronfenbrenner points out that an individual is embedded in a set of nested environments. The environments influence each other and together influence the development of an individual.

UNIT FIVE DEVELOPMENT: Growth and Differentiation in Living Systems

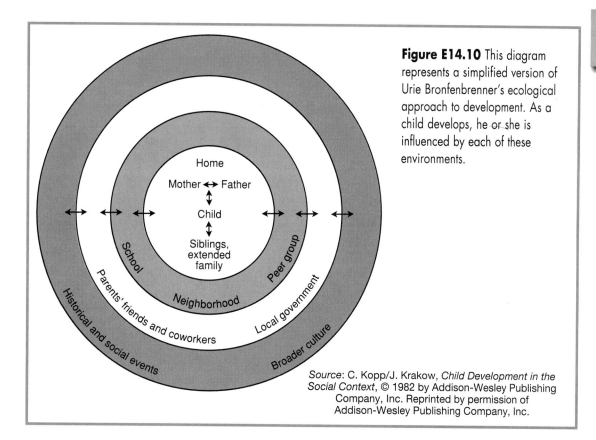

Figure E14.10 This diagram represents a simplified version of Urie Bronfenbrenner's ecological approach to development. As a child develops, he or she is influenced by each of these environments.

Home

Mother ↔ Father

Child

Siblings, extended family

School

Parents' friends and coworkers

Neighborhood

Peer group

Local government

Historical and social events

Broader culture

Source: C. Kopp/J. Krakow, *Child Development in the Social Context*, © 1982 by Addison-Wesley Publishing Company, Inc. Reprinted by permission of Addison-Wesley Publishing Company, Inc.

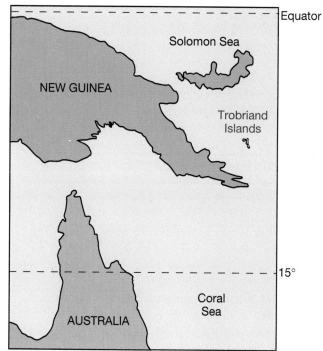

Equator

Solomon Sea

NEW GUINEA

Trobriand Islands

15°

Coral Sea

AUSTRALIA

Figure E14.11 The Trobriand Islands The Trobriands are a group of small islands off the east coast of New Guinea.

Personal Journal Entry: 15 May

I just got another letter from Malcolm. He now is spending time with the modern-day Trobriand Islanders (see Figure E14.11). That reminded me of the traditional Trobriand society that we had studied in our class, so I sent him the following copy of my notes from class last semester.

Anthropology 220   11 November

Trobriand Islanders

In general, traditional societies are societies that have not been influenced by the modern or Western cultures around them. Because we live in an extremely mobile world, most traditional cultures have had contact with modern cultures, but the timing and the extent of this contact have varied.

In traditional Trobriand society, children were usually weaned from breast-feeding by the time they were 2. They usually were weaned by being sent to sleep with their father or their maternal grandmother. When the children turned 4 or 5, they began to spend time with a children's group that had quite a bit of independence, and they stayed with this group until puberty. On any given day, the children might remain with their parents or go with their group as they chose.

When the children stayed at home, the parents gave the children miniature tools and showed them how to plant crops and do other adult work. When the boys reached 6, their maternal uncles (their mothers' brothers) took more and more charge over their training. After the boys reached puberty, they still ate at home, but began sleeping in special bachelors' huts with several other boys their own age. This was also the time when the boys began to participate in the regular occupations of adult men.

After girls reached puberty, they often joined their current boyfriends in the bachelors' huts at night, but could still sleep at home whenever they wanted. At this time, they began to do more adult women's work in the home.

After several years of informal relationships, a pair of adolescents would form a long-term relationship and begin to appear together in public. This act was a signal that they were ready to marry. If the girl consented, the family would build them a hut of their own near a maternal uncle of the boy, and the couple would begin adult life together.

In cases of divorce, which were usually at the woman's request, all of the children would stay with their mother. In this society the grandparents had little to do with the training of children, and respect for elders was not highly valued.

Personal Journal Entry: 16 May

As I was looking through the notes from my class to send to Malcolm, I came across notes about other cultures as well. I read through them and again was struck by the different ways in which people approach life. The following is a copy of my notes.

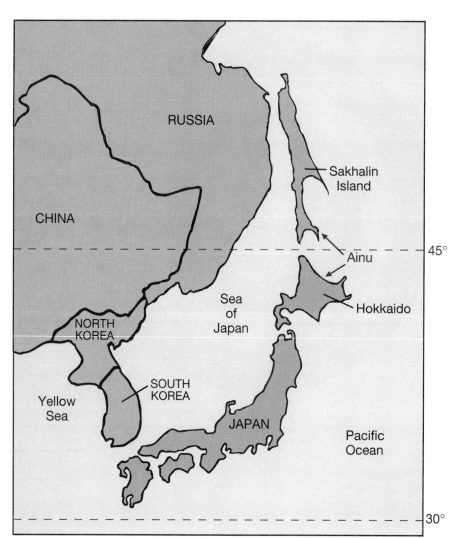

Figure E14.12 The Ainu lived in small coastal settlements in far northern Japan and on the southern tip of Sakhalin Island.

Anthropology 220 18 November
Ainu—People of Northern Japan

The Ainu were a traditional hunting and fishing people who lived in small coastal settlements in far northern Japan (see Figure E14.12). Mothers often breast-fed their children until the age of 4 or 5, and families traced their descent through the female side.

Small children played with carved fish, toy boats, and other toys that resembled objects that they would use later when they became adults. Older boys played with miniature hunting weapons, and girls played with dolls. Both boys and girls lived at home during childhood. Usually, the grandparents would instruct the children in the proper behavior and the duties expected of people in Ainu society. One of the most important behaviors was respect for elders.

Figure E14.13 When Ainu girls began to mature, they started receiving tattoos. The Ainu believed that the tattoos would keep evil spirits away.

(The Ainu believed that tattoos in conspicuous places would keep evil spirits at bay.)

Young couples usually courted with little involvement of their parents and usually married by the time they were 16 or 17. When a child was born, both the new father and new mother would spend a period of time at home and would not engage in their usual activities. Either partner could end the marriage if he or she wanted; if this occurred, the daughters would live with their mother, and the sons would live with their father.

When boys were about 5, they were allowed to watch the men prepare for fishing expeditions. At about 12 or 13, they began going out in the ocean with the adult men. When boys turned 15, they started wearing their hair as the adult men did: they began to let it grow long and to grow a mustache and beard as well. Girls began learning household chores when they turned 5 or 6. At 13, the girls began to receive facial tattoos. A few years later, when the tattoos were complete (see Figure E14.13), they put on adult women's clothing.

Gusii—People of Kenya

Among the traditional Gusii of Kenya (see Figure E14.14 for location), people lived in farming homesteads as extended families. Usually one man was the head of these extended families, and the family might include his wives and their children as well as his married sons and their wives and children. People traced their descent entirely through the male line.

Generally, children were weaned when their mother became pregnant again, which might mean anywhere from 1 to 3 years. Children stayed close to home and had duties to perform. As soon as they turned 5, the girls began to take care of the infants, and the young boys helped with the cattle. Both boys and girls worked in the fields when they were about 6 or 7 years old.

The grandparents generally were friendly and good-humored, but the children's own father was a strong authority figure whom the children both respected and feared. When the boys

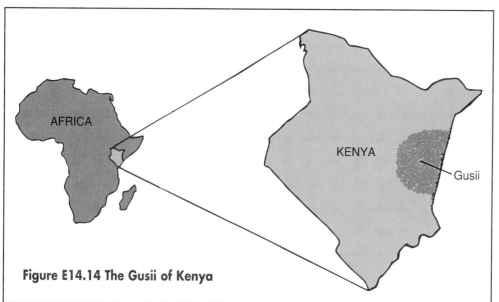

Figure E14.14 The Gusii of Kenya

AFRICA

KENYA

Gusii

turned 7 or 8, they began living in a separate children's house on the homestead.

Girls formally were initiated into womanhood when they were only 8 or 9, as soon as they began to show a strong interest in women's duties. Boys formally were initiated into manhood when they were between 10 and 12. After special initiation ceremonies, there would be a period of seclusion and then a formal public appearance. At this point the boys and girls would be welcomed as adult members of society.

After a few years of brief relationships, a young man would choose a girl as a prospective wife and would send someone to inquire about her background and to find out the formal price (in cattle) that her family would expect in return for her hand in marriage. If everyone agreed and the young man's family had paid the price, the groom and his family would come and take the bride to his home. It was a tradition that the bride would put up a fight rather than go willingly.

After a girl was married, she would remain with her new husband for a month, but then she would return to her father's home for about two months. During this time, she could request that the marriage be dissolved if she were truly unhappy. She also might leave her husband later, especially if she did not become pregnant within a year or so. In all cases, the children from a marriage legally belonged to the father.

The Amish in America

The Amish are a present-day traditional culture that exists in North America side by side with modern American culture (refer to Figure E14.15). These people are of Swiss and German descent and speak a particular dialect of German. They refer to all members of the modern American culture as "the English."

They reject many aspects of the dominant culture and prefer simplicity. In general, Amish people avoid any use of electric power other than some batteries they use in the home. They travel in horse-drawn buggies and do not own or operate automobiles (although they will hire or accept rides in cars and buses when necessary). They farm using horse-powered cultivating methods or else are carpenters or craftspeople.

Figure E14.15 Amish communities like this one in Lancaster County, Pennsylvania, exist throughout North America.

Amish children dress almost as miniature adults as soon as they are old enough to be out of diapers (see Figure E14.16). They are regarded with great affection by the adults. Children are encouraged to play but also to be responsible for many chores around the farmstead.

At around 5, the children begin formal schooling. In the United States, the Supreme Court has recognized the right of the Amish to run and administer their own system of schools, which are staffed by Amish teachers. These teachers are usually young women who are not yet married.

The children study English, practical arithmetic, High German (which is necessary to read their religious books), geography, basic science, and several other subjects. The goal of all of their studies is to provide the children with the knowledge that they will need to function successfully in Amish society and to deal with the surrounding culture. The Amish consider their society a community of equals and they frown on pride. In school, then, they expect the bright students who finish their lessons quickly to help the slower schoolmates with their work so that the whole group can advance together. The Amish believe that formal education should end with the eighth grade to avoid the possibility of the children becoming too worldly.

When Amish children are not in school, they spend time playing and learning about the skills of adult life. It is not unusual to see a young Amish boy of 10 driving a five-horse hitch of big draft animals and plowing a field by himself. And Amish girls of

UNIT FIVE DEVELOPMENT: Growth and Differentiation in Living Systems

Figure E14.16 At a very young age, Amish children are dressed in styles similar to the adults.

the same age are expected to handle a big garden or care for a group of younger children.

When their formal education is over, Amish youngsters work full time at home where they learn agricultural skills. The Amish believe that a person must be a mature adult, usually in his or her early 20s, in order to freely choose to accept formal baptism into the Amish community. (Formal baptism means a commitment to completely reject the forbidden aspects of the outside culture.) Therefore, young people between 16 and 22 have more freedom to experiment with the outside culture than either young children or baptized adults have. This period is described by a Germanic term that roughly translates as "running-around-time." Young Amish men often spruce up their buggies in ways that never would be tolerated later and may dress in clothes like those of the "English" teenagers. These young men often work at town jobs in construction. Several friends even may own a car, which they generally keep somewhere other than at home.

Girls often make and wear brightly colored dresses that would be completely unacceptable at other ages and may hold jobs as waitresses or work in stores. Groups of young people or courting couples may travel into cities to attend films, sporting events, or even dances. Their elders are aware of what is going on but, in general, regard it as a necessary period in the young people's development. Those young people who finally decide not to receive baptism, but instead join the mainstream culture, usually remain on good terms with their families and often return to visit. Those who accept baptism but later decide to leave, however, are shunned.

When young Amish people marry, they take up a lifestyle like their parents and raise a family of their own. Amish elders are highly respected and generally turn over the responsibility of running the farm when the youngest child is old enough to take over. At this time the elders move from the main house into a small one that is located nearby. They take an active role in raising and teaching their grandchildren and are always on hand to lend advice and help around the farm.

Personal Journal Entry: 19 May

Because I have been making observations of humans and learning about many related aspects of human growth and development, it has been interesting to read over my notes from anthropology. It has been clear to me that fundamental aspects of human biology are responsible for both similarities and differences in the way humans develop. It now is becoming clear to me that all human cultures have developed ways of responding to the milestones and the processes involved as humans develop. But, because of the specifics of culture and environment, these ways of responding show diversity in expression.

Interdependence Involves Limiting Factors and Carrying Capacity

Newspaper articles worldwide report that famines continue to occur in Africa and that new animals are being added to the endangered species list. What are the causes of these events? Are they caused by human mismanagement, or do some occur naturally?

Scientists agree that we can find some of the answers by studying the concepts of limiting factors and carrying capacity. A **limiting factor** is anything that can slow, or limit, the growth of a population. The combination of limiting factors in a given habitat influence the **carrying capacity** of that habitat—that is, the maximum population of a particular species that the habitat can support.

The carrying capacity of an environment is not necessarily fixed. It changes as environmental conditions change. The limiting factors that influence the carrying capacity may be biotic (living) or abiotic (nonliving) or both. Biotic factors include food supply and other organisms; abiotic factors include space, raw materials, and climate.

Climate is the prevailing weather conditions in a given area through long periods of time. Weather conditions result from abiotic limiting factors that affect all organisms. These weather factors include temperature, sunlight intensity, precipitation (rainfall, snowfall, and fog), humidity (amount of moisture in the air), and wind. Although you can measure

each factor alone, each affects the others, and together they affect the population size of organisms living in the area. Climate and particular weather conditions can affect the population size directly by presenting optimal conditions for the growth and reproduction of various plants or by presenting conditions that are adverse to growth and reproduction. The abundance of or the lack of certain plants then translates into an abundant or a limited food supply for a chain of other organisms. In this way climate and vegetation help determine the carrying capacity of a particular environment for various organisms (see Figure E15.1 for an example).

You might be asking yourself, How can temperature be a factor that limits the size of a population? In the northern part of the United States, the first heavy frost of the fall kills almost all adult mosquitoes. When the population of insects drops, the population of organisms that feeds on mosquitoes is affected. In this example the limiting factor of temperature affects the mosquito

Figure E15.1 The snowy owl (*Nyctea scandiaca*) inhabits arctic tundra regions and in winter is always searching for food in this sparse environment. When the winter food supply is especially limited, the snowy owl can be seen as far south as Colorado.

population directly by killing most mosquitoes. By limiting other organisms' food supply, however, temperature also affects the carrying capacity of the winter environment for those organisms.

The same type of interactions occur between various wildflower populations and the insects that pollinate them. As temperatures drop in the autumn, the wildflowers die and the insects either die, become dormant, or migrate. In both cases temperature

UNIT SIX ECOLOGY: Interaction and Interdependence in Living Systems

Figure E15.2 Water is a limiting factor in the savanna in East Africa, where only grasses and scattered trees can grow. These plants, in turn, support grazing animals. The grazing animals move on when the rainfall in particular locations decreases.

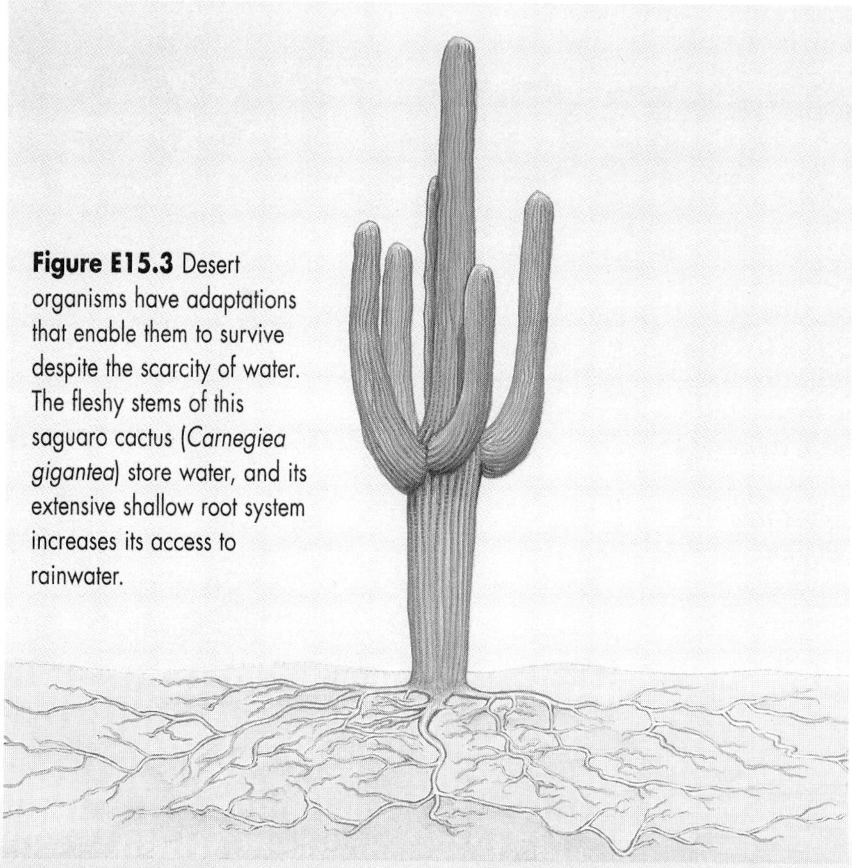

Figure E15.3 Desert organisms have adaptations that enable them to survive despite the scarcity of water. The fleshy stems of this saguaro cactus (*Carnegiea gigantea*) store water, and its extensive shallow root system increases its access to rainwater.

has caused limited population growth for a period of time.

Water is a requirement of all living organisms, and thus it is a limiting factor in every ecosystem in the biosphere. Almost all chemical reactions that organisms carry out to stay alive take place in water, and water molecules themselves are a part of many chemical reactions. Although a few organisms can survive by becoming inactive when there is no water, most organisms die. Because water is essential for all living organisms, it is a limiting factor that influences the carrying capacity of virtually every ecosystem. As illustrated in Figure E15.2, water limits the carrying capacity of a savanna in East Africa and consequently limits the distribution of the organisms.

Because of other weather conditions, rain and snow alone may not ensure that an environment has enough water. Wind increases the rate of water evaporation. Low humidity in the desert also results in an increased evaporation rate. Thus, high rates of evaporation can affect the survival of certain organisms, even when the precipitation is adequate. In some cases a high rate of evaporation together with low rainfall may permit only plants with special features to survive. This situation severely limits the carrying capacity of the desert. For example, desert plants often have thick, succulent stems that can store water and extensive root systems that can absorb large quantities of water during infrequent rainfalls (see Figure E15.3). Some animals such as the kangaroo rat can survive without drinking water because they use the water that is released during cellular respiration and they produce a nearly solid urine.

Light also can be a limiting factor. The penetration of light into the ocean determines the depth to which photosynthetic organisms can grow, generally not beyond the depth of 180 m

a.

b.

Figure E15.4 a. In one population experiment, researchers provided mice with more than enough food. b. As a result, the population grew dramatically. Space then became a limiting factor that resulted in a high death rate among young mice.

(585 ft). If silt or algae growth reduces the clarity of the water in a pond or lake, plant growth is limited. In dense rain forests, the tallest trees spread their leaves and take most of the light. The ground below them is shaded, preventing other plants from reaching great heights. Light affects animals indirectly because the amount of light influences the number of plants that can grow, which then influences the carrying capacity of the environment.

Space is another abiotic limiting factor for populations. Although every individual needs living space, some organisms need more space than others. For example, individual corn plants

GROWING, GROWING,

By just looking around you, you can observe plenty of evidence that populations grow—the guppies in your fish tank have more guppies, the song birds in your yard have young each spring, the stray cats in the neighborhood seem to have a litter of kittens every couple of months. But have you ever thought about how populations grow?

In 1798, after years of thinking about just this question, Thomas Malthus (pictured in Figure E15.5) reported that in ideal conditions, populations tend to grow **exponentially**. In exponential growth both the total population size and the rate of increase rise steadily across time. Charles Darwin later wrote in his book, *On the Origin of Species* (1859), that "There is no exception to the rule that every organic being naturally increases at so high a rate, that, if not destroyed, the earth would soon be covered by the progeny of a single pair." Biologists

Figure E15.5 Thomas Malthus, an English economist who lived from 1766 until 1834, gave serious thought to how populations grow.

grow well when they are planted close together. A mountain lion, on the other hand, usually requires many square kilometers for its range. The mountain lion needs this much space to find enough food to sustain itself. You may remember that the higher a given organism is on the energy pyramid, the fewer of that organism the environment can support. Therefore, the amount of space needed by all organisms at trophic levels above the producers is linked primarily to a biotic factor—the availability of food energy.

The available space is affected by **population density**, the number of individuals in relation to the space the population

occupies. For example, in an experiment at the University of Wisconsin that is illustrated in Figure E15.4, researchers gave mice in cages more food than they needed each day. As the mice reproduced, the density of the population increased, and the cages became very crowded. Some female mice stopped taking care of their nests and young. Mice continued to be born, but many newborn mice died from neglect. Eventually, the death rate of the young mice reached 100 percent, which kept the population density from increasing further. In a similar experiment conducted in England, the death rate of young mice was not affected. Instead, the birthrate

declined almost to zero. In this experiment the extremely low birthrate kept the population density from increasing. Space, as a limiting factor, affects all living populations and the carrying capacity of particular ecosystems for those populations.

GROWN

discovered that *any* population growing exponentially can *potentially* approach an *infinite* population size in a relatively short amount of time if the population is provided with necessary resources such as food, water, space, and protection from other organisms.

When populations grow exponentially, the population growth curve takes on a J-shaped appearance due to the continuous increase of a larger and larger population size, as illustrated in Figure E15.6. You can observe this principle of population growth when you examine a fast-breeding species such as the common housefly or a slow-breeding species such as the elephant. Biologist L.O. Howard discovered that, if the environmental conditions were ideal, a housefly population beginning with just one adult female could increase to more than 5.5 *trillion* flies in seven generations or just *one year* (refer to Figure E15.7). Even the rather slow-breeding elephant possesses the capacity for explosive population growth. Charles Darwin calculated that an elephant population consisting of

just two elephants (a single breeding pair) could grow to 19 *million* elephants in only 750 years.

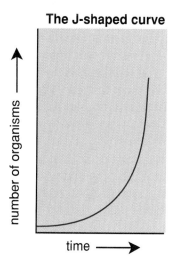

The J-shaped curve

number of organisms ⟶

time ⟶

Figure E15.6 The J-shaped curve is characteristic of populations that are growing exponentially.

Predicted Population Growth of the Common Housefly in One Year	
Generation	Population size
1	120
2	7200
3	432,000
4	25,920,000
5	1,555,200,000
6	93,312,000,000
7	5,598,720,000,000

Figure E15.7 Predicted population growth of the common housefly in one year This prediction is based on the following observations and assumptions: an average female fly lays 120 eggs at a time, about half of the eggs develop into females, there are seven generations in one year, and individual flies live for one generation.

Source: Data from E. J. Kormondy (1984) *Concepts of Ecology, third edition.* Englewood Cliffs, NJ: Prentice Hall Publishers.

In addition to realizing that, under ideal conditions, populations tend to increase in size exponentially (that is 1, 2, 4, 8, 16, 32, 64, 128), Thomas Malthus realized that the food supply tends to increase only arithmetically (for example, 1, 2, 3, 4, 5, 6, 7). Most of the world's organisms serve as the food supply for other organisms that are higher on the food chain. Because of this and other limiting factors such as space, populations in general do not continue to grow exponentially. When a food supply is limited, the carrying capacity of the environment for the populations that feed on that food supply is limited.

Although technological advances in agriculture have helped food production generally keep pace with human population growth, many scientists are concerned that a time will come when food production will not meet the requirements of the world's rapidly increasing human population—a time when the human population will exceed the carrying capacity of the earth.

In theory, populations could continue to grow exponentially if they are provided with ideal conditions and unlimited resources. In the real world, as we have just seen, conditions are not ideal and resources are in limited supply, so we do not observe ongoing exponential growth. **Logistic growth** represents a pattern of population growth that scientists have observed in real populations in natural conditions. The logistic growth curve has the shape of a flattened S (see Figure E15.8), rather than

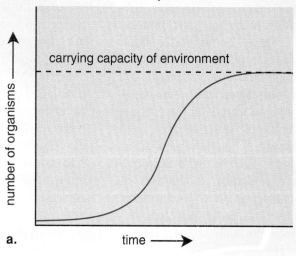

The S-shaped curve

a.

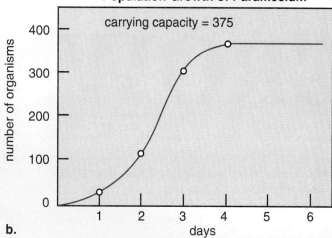

Population Growth of Paramecium

b.

Source: From G.F. Gause (1934) *The Struggle for Existence.* Baltimore: Williams and Wilkins.

Figure E15.8 a. An idealized S-shaped growth curve b. Population growth curve for the single-celled organism *Paramecium* Limiting factors force this population of organisms to level off.

the J-shape that is characteristic of exponential growth. In exponential growth the population continues to increase across time, but in logistic growth the population levels off at the carrying capacity of the environment.

UNIT SIX ECOLOGY: Interaction and Interdependence in Living Systems

Legend has it that the game of chess was invented by a mathematician who worked for an ancient king. The king was so delighted with the game that he asked the mathematician to name his own reward. Telling the king that he was a humble man who wished a humble reward, he asked that a single grain of wheat be placed on the first square of the chessboard, two grains of wheat on the second square, four grains on the third square and so on, doubling the number of grains on each square until all 64 squares on the chessboard were filled. The king granted his request and ordered the Master of the Royal Granary to begin counting out the grains. But before the task was anywhere close to being finished, the king handed his kingdom over to the mathematician.

Inquiring Minds

Drs. Rita and Ricardo, colleagues at the university, answer readers' questions about population dynamics.

Dear Dr. Rita,

I am a biology student who has just completed an investigation about the interactions that occur within a small aquatic environment. I began to wonder what would happen if I had only guppies in my container. My container is really quite small. What if the guppies had hundreds of babies? How long do you think the guppies could live in the container?

—Michael Johnson

Dear Michael,

Your investigation sounds interesting, and your questions are good ones.

All populations, including the guppies that you describe, have a **growth rate** that is characteristic of the population. This growth rate is defined as the rate at which the group would grow if food and space were unlimited and individuals bred freely. In your case the size of the container is truly a limiting factor. If you had only guppies in the container and they bred freely while you gave them ample food, the jar probably would fill with guppies.

As the population became denser, the amount of waste produced by all the guppies would pollute the water to the point that they likely would die off due to an accumulation of ammonia, which is a toxin. They also might die as a result of some disease spreading among them. Remember, the population's natural growth rate, along with changes due to various environmental conditions, determines the size of the surviving population.

Dear Dr. Ricardo,

At school we are learning about how populations grow. A friend of mine said, "Imagine that you have two fish in a small pond and that they double their number every day. You also know that by day 30 the pond will be completely full. So on what day will the pond be half full?" I thought and thought about this question. My first guess was after 15 days. Then my friend had me work through the problem. She asked, "How many fish will you have on day 2?" I said, "Let's see, doubling two fish equals four fish." She then asked, "OK, then how many fish will you have on day 3?" I thought about it and realized that I would have eight fish. Then I realized that if I was doubling the number of

fish each day, the pond would be half full on day 29. What else can you tell me about this kind of population growth?
—Erica Sarkis

Dear Erica,

I like the puzzle your friend posed to you. Let's look at exponential growth in other populations. One yeast cell can give rise to thousands of cells in just one day. One *Escherichia coli* bacterium dividing every 20 minutes in an environment with unlimited food and space could give rise to 40 septillion (4×10^{24}) bacteria in about 24 hours. A population with a growth rate of 10 percent will double in much less time than a population with a growth rate of 5 percent. This is true, however, only if the generation time is the same in each case. For example, if you have two different bacteria and one reproduces every 5 minutes and the other reproduces every 10 minutes, then the second type of bacteria will double more slowly than the bacteria that reproduces every 5 minutes. The amount of time it takes for a population that is growing steadily to double in size is called its **doubling time**.

Since the 1970s the rate of human population growth may have decreased from about 2 percent to about 1.63 percent. So, in recent years the doubling time for human population may have decreased slightly. Even so, we probably will grow to 12 billion people by the year 2020. The doubling time for the human population is currently about every 43 years.

Dear Dr. Rita,

I am still confused about what factors influence population size. Please help!
—David Cho

Dear David,

The size of a population is influenced by a number of things. Suppose a biologist counted 700 ponderosa pines on a hill in Colorado in 1980. In 1990, when the biologist counted the trees again, there were only 500. In other words, there were 200 fewer trees in 1990 than in 1980; this represents a decrease in the population of ponderosa pines. This change in population may be expressed as: −200 trees per 10 years = −20 trees per year. To the biologist,

this means that each year there were 20 fewer trees in the population. Keep in mind, however, that this rate is an average. It is unlikely that the trees disappeared on such a regular schedule. All of the trees may have been lost in one year due to fire, or the decrease may have been caused by selective cutting during several years.

What does the decrease of 200 pine trees in 10 years represent? Because pine trees cannot wander away, they must have died or have been cut down. In this situation the decrease represents the death rate, or **mortality** rate, of the pine population. Mortality is not the only change that can affect a population. Although some of the pines may have died, some young pine trees may have started to grow from seed. Death decreases a population; reproduction increases it. The rate at which reproduction increases the population is called the birthrate, or **natality**.

Organisms that can move have two other ways to bring about a change in population size. If you studied a pigeon population in a particular city park, you might discover that a certain number of pigeons moved into the park in one year and a certain number left the park. **Immigration** occurs when one

Endless Interactions

You have just finished a delicious bowl of chicken soup for lunch. As you clean up, you wash out the bowl and throw away the can. A garbage truck will haul the can to the dump where it will lie among other trash. A female fly might lay her eggs in the can, and her offspring might fly from the dump, feed on some decayed food, and then rest on top of someone's peanut butter sandwich. Each of these interactions is dependent on the one before it. Most of the time these interactions go unnoticed.

All populations, including human populations, interact with one another in a complex web of relationships. The set of interacting populations present at one time in one place is called a community. In your community there may be dogs, cats, trees, weeds, and humans that interact. When you mow your lawn or your dog bites the mail carrier, for example, the interaction

is very direct. Much of the time, however, interactions are indirect.

One type of direct interaction is a **predator/prey** interaction. In this direct interaction, one type of organism (the predator) eats the other (the prey). The predator benefits from the relationship, but the individual prey does not. An animal can spend its life as a predator and then abruptly become the prey. A snake, for example, may prey on ground squirrels and then become prey for a hawk. The snake depends on a large population of ground squirrels. The hawk, then, depends directly on the snake and, in this case, indirectly on the quantity of ground squirrels. The hawk, however, also may prey directly on the ground squirrels. The limiting factor in this case is the food supply.

Although the individual ground squirrel does not benefit, the population of ground squirrels may benefit

or more organisms move into an area where others of their type can be found. Immigration increases the population. **Emigration** occurs when organisms leave the area. Emigration decreases the population. In any population in which individuals can move, natality and immigration increase the population, and mortality and emigration decrease the population. Thus, the size of a population is the result of the relationships among these four rates.

Figure E15.9 Four factors that affect population size The size of this population of red fox is affected by the number of fox pups that are born, the number of fox that die, the number that leave the area, and the number that move into the area.

by being held in check by the snake population. There also might be a selective advantage to the ground squirrel population because the snakes would tend to prey on the slower, weaker squirrels. Consequently, the best-adapted ground squirrels would be more likely to escape becoming prey and thus would survive to reproduce.

Competition is another type of direct interaction. This type of interaction between organisms benefits neither one. Organisms compete for limiting factors, such as space, food, sunlight, nutrients, and water (see Figure E15.11). Competition among organisms may increase as a particular resource becomes scarce.

Figure E15.10 For food, this red-tailed hawk (*Buteo jamaicensis*) depends directly and indirectly on a number of organisms in its habitat. Note the rodent in its talons.

Figure E15.11 The melaleuca tree (*Melaleuca quinqueneruia*) was introduced to Florida from Australia and has been overwhelming the cypress in the everglades.

When tadpoles live in densely packed areas, the competition for available space and food resources increases. As a result, these tadpoles remain tadpoles longer, they suffer higher death rates, and they develop into smaller frogs. In communities of wolves, competition leads to the establishment of social hierarchies. A high position in the pack gives the individual an advantage in terms of obtaining food, mates, shelter, and other resources. Individuals of any species that compete successfully can survive to reproduce and pass on their genetic material to future generations. In this way, populations adapt through time to changes in the environment.

Competition also can occur between different species. If the species' needs are similar and resources are scarce, then more competition will exist. In New Guinea four species of pigeons rely on a particular fruit tree for their primary food source. The four types of pigeons differ in size, and the smaller pigeons have adapted to feed on the fruit of the lower branches.

An experiment involving paramecium demonstrates another example of this form of adaptation. Researchers placed two species of paramecium in the same test tube, which was filled with liquid food. The researchers formulated a hypothesis that one of the species would die out as a result of competition. In fact, both species survived and thrived. One species fed on the food that settled at the bottom of the tube, and the other fed on the food suspended in the liquid. The two species of paramecium, as well as the four species of pigeons, occupied separate **niches**. The niche of an organism refers to its role in the community: what it eats, what organisms eat it, where it lives, and what indirect

Figure E15.12 The mistletoe growing in this pine tree is a parasitic plant.

UNIT SIX ECOLOGY: Interaction and Interdependence in Living Systems

relationships it has with other organisms. As a result of occupying different niches, the two species were able to feed and reproduce without interfering with each other even though they lived in the same place, or **habitat**.

Competition is not the only relationship that exists among populations of different species that live in the same habitat. When different species of organisms live in direct, physical contact with one another, the relationship is called **symbiosis**. For example, certain species of fungi live on the roots of many plants. The fungi absorb nutrients from the soil and secrete an acid that makes the nutrients available to the plant. At the same time, the fungi are nourished by photosynthetic products from the plant. The fungi also absorb water and protect the plant against various pathogens in the soil. This type of symbiosis that benefits both organisms is called **mutualism**. Another example of mutualism is the interaction between a certain type of bacteria and the cow in whose intestines the bacteria live. The cow benefits because the bacteria digest the cellulose in the plants that the cow eats, and the bacteria benefit because the cow provides a steady supply of food.

A common form of symbiosis is **parasitism**. In parasitism one organism (the parasite) lives on or in another organism (the host) and uses it as a food source. The host usually remains alive during the interaction, although parasites may weaken a host to the extent that it becomes susceptible to disease or becomes an easier prey for predators. A leech, for example, clings to a turtle's skin and sucks its blood. Other parasitic microorganisms in the turtle absorb food directly from its blood. Interactions such as these eventually may weaken the turtle. In humans, tapeworms absorb food directly from the intestines, where they live.

Plants also may have parasites and large microorganisms may have smaller parasitic microorganisms in them. Parasites may be molds, microorganisms, and even other plants such as the mistletoe shown in Figure E15.12. All viruses are parasitic because they require a host to reproduce and acquire energy.

These examples illustrate a few of the many different types of interactions that occur among organisms. Some of these interactions, such as mutualistic relationships, are beneficial for both organisms. Other interactions are not as positive and result in some degree of harm to one member of the interacting pair, as in parasitic relationships. Each of these interactions is influenced by limiting factors such as light, temperature, food supply, and space. Because of these interactions, there is a web of interdependence among living things.

The Ecology of Kirtland's Warbler

The Kirtland's warbler (*Dendroica kirtlandii*, pictured in Figure E15.13) is a long-distance traveler that winters in the Bahamas and migrates to the jack pine forests of Michigan in the summer to breed. The bird has very specific nesting requirements: it builds its nest only on the dead, ground-level branches of younger jack pine trees that grow in a specific type of soil found only in central Michigan. Trees younger than 6 years or older than 21 years of age do not provide the proper branches, and the warblers do not use them for nesting. Male birds are territorial and defend up to 80 acres of forest against the intrusion of other male Kirtland's warblers. In spite of these nesting limitations,

Figure E15.13 Kirtland's warbler is about 15 cm (6 in) long and is known for its extremely limited breeding range.

Figure E15.14 a. Jack Pine (*Pinus banksiana*) b. Red Pine (*Pinus resinosa*) c. White Pine (*Pinus strobilus*) When Europeans settled in Michigan, they began logging red and white pine but found the Jack pine unsuitable for lumber.

when European settlers arrived in Michigan in the mid-1800s, jack pines grew over a large portion of the state, and Kirtland's warblers were abundant. The population of the warbler continued to increase, and biologists estimate that the size of the Kirtland's warbler population reached its peak during the late 1800s.

The jack pine (*Pinus banksiana*) is a tree with craggy, flaky bark and short needlelike leaves. It is described as a fire-adapted species because it persists only in areas where forest fires occur periodically. Jack pine cones open and release seeds only when they are heated to very high temperatures as would happen in a fire. Scientists think that the presence of dead branches at ground level in younger trees promotes the spread of fire. In addition, the jack pine requires full sunlight for growth, and fires open

up areas to sunlight by removing plant cover.

When the Europeans began settling in Michigan, they began cutting down trees for lumber. They cut primarily red pine and white pine but not jack pine because it did not provide high-quality lumber. Naturally occurring fires roared through the logged areas because of the buildup of cuttings from the logging operation. Loggers sometimes

Yellowstone Burning: Is Fire Natural to Yellowstone National Park?

Fire is an important and powerful natural agent of change in many ecosystems, and it influences the makeup and structure of the biotic community. For example, the grasslands of North America and Africa, the conifer forests of northern and western North America, and the chaparral of the southwestern

United States each have been shaped by fire, and all contain plants that exhibit adaptations to fire. Fire can become an important ecological factor in ecosystems whenever sufficient organic material is available to burn, a source of ignition such as lightning is present, the terrain permits the spread of fire, and dry weather conditions make the organic material combustible.

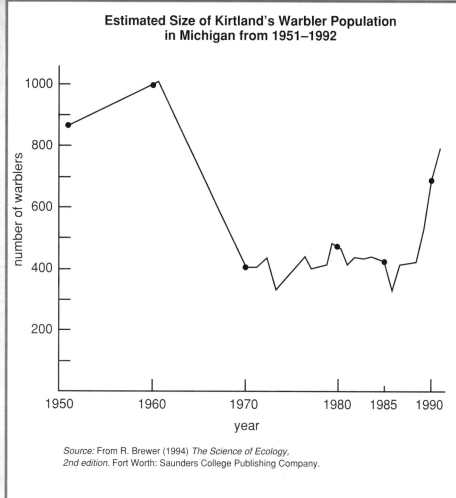

Estimated Size of Kirtland's Warbler Population in Michigan from 1951–1992

Source: From R. Brewer (1994) *The Science of Ecology, 2nd edition.* Fort Worth: Saunders College Publishing Company.

Figure E15.15 This graph depicts the estimated population size of Kirtland's warbler in Michigan from 1951–1992.

cleared jack pine areas by setting fire to them to provide access to other areas of red and white pine.

The country's attitude toward fire changed during the 1920s, and officials created forest fire suppression programs. These programs were in full swing by the 1930s and 1940s. They resulted in a decrease both in the amount of forest that was affected by fire and in the size of individual fires.

During the early 1960s, there was a dramatic reduction in the population of Kirtland's warblers. Concerned that the bird might become extinct, biologists initiated a species recovery program that involved deliberate controlled burning of forest areas that contained jack pine. When scientists again studied the population size of the warblers in the early 1990s, they found that the population was increasing.

One of the current management goals for Yellowstone National Park is to maintain the natural character of this ecosystem. But this was not always the case. The park (refer to Figure E15.16) was founded in 1872, and beginning in 1886, park administrators adopted a general policy of fire suppression. At this time officials assumed that all fires were destructive. In the early 1970s, ecologists began to recognize that fires can be part of a natural, dynamic process in some ecosystems. Soon, the federal government changed to a "prescribed natural burn" policy for many national forests and parks, including Yellowstone. Officials decided to let fires that began naturally from lightning strikes continue to burn as long as the fires did not jeopardize property, people, or important cultural and natural sites.

This natural burn policy came to public attention when large-scale fires started in Yellowstone during the summer drought of 1988. The fires attracted much public attention and spectacular images of burning forests were on television and in newspapers across the country. Responding to public pressure and their own concerns about the scale of the fires, park officials abandoned their natural burn policy soon after the fires started and began trying to fight them. In the end more than one-fourth of the 2.2 million acre park had burned. Biologists, government officials, park visitors, and people across the nation wondered whether the fires of 1988 were a natural part of the Yellowstone ecosystem, possibly the abnormal result of a fire suppression policy that had been in effect for a long time, or just an unusual occurrence.

Biologists William Romme and Don Despain tried to answer this question by reconstructing the fire history of Yellowstone. They began by constructing a map of the fires that burned in 1988 (see Figure E15.17). The two researchers were able to determine the fire history of an area by studying fire-scarred trees in a study plot that covered 15 percent of the park. From their work, they were able to estimate the years of occurrence and the range of fires in their study area during the past 350 years. Figure E15.18 shows the extent of fires in their study area in 50-year intervals.

Extent of 1988 Yellowstone Fires

Source: From P. Schullery (1989) The fires and fire policy. *BioScience* **39**: 686-694.

Figure E15.17 In 1988 much of Yellowstone was devastated by a series of fires. The orange areas on this map show the extent of those fires.

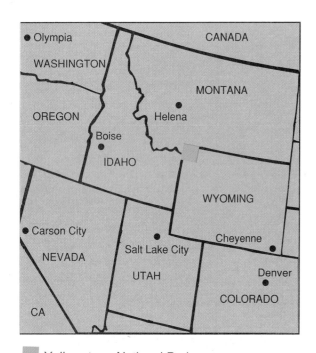

Yellowstone National Park

Figure E15.16 Yellowstone National Park Yellowstone National Park was founded in 1872 and consists of 2,219,823 acres in northwestern Wyoming. Its boundaries extend somewhat into Montana and Idaho.

Forests of the Past

It's easy to think about how ecologists make observations of an area from year to year and study changes in an ecosystem that occur in a matter of days, as in a fire, or in a matter of decades. But how can ecologists learn about ecosystems that existed thousands of years ago?

In certain locations ecologists can determine the history of an area by examining the types of fossil pollen found in the different layers of sediment that have been accumulating across time on the bottom of a lake or bog. The pollen from plants growing in an area is blown or washed into a body of water, and it gradually sinks to the bottom along with other sediment. Ecologists can

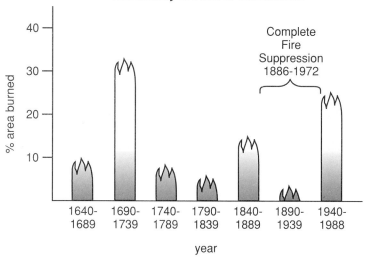

The History of Fires in Yellowstone

Source: From W.H. Romme and D.G. Despain (1989) Historical perspective on the Yellowstone fires of 1988. BioScience **39**: 695-699.

Figure E15.18 The history of fires in Yellowstone This graph shows the percentage of area that burned during each half-century interval from 1640 to 1988.

extract a core sample from a lake by using a coring device (shown in Figure E15.19). They extend the coring device many meters into the lake bottom and extract a core sample that includes sediment that was deposited there across many thousands of years. The ecologists then begin the tedious work of identifying the fossil pollen grains that are present in each different layer of the core. Following this, they must determine the age of the different layers through radiometric dating methods.

Finally, the ecologists are able to create a fossil pollen profile and then use this profile to reconstruct the history of the vegetation that inhabited a particular site. Figure

E15.20 shows a fossil pollen profile for the different types of trees that grew in the northeastern part of the United States during the past 15,000 years, and Figure E15.21 shows a profile of the trees that grew in the northern Midwest of the United States during the past 20,000 years.

Figure E15.19 Biologist obtaining a sediment sample from a lake Biologists use a special coring device to obtain a sample of sediment from the bottom of the lake. This sample will contain pollen from the species of plants that used to grow there.

Fossil Pollen Profile for
Northeastern United States

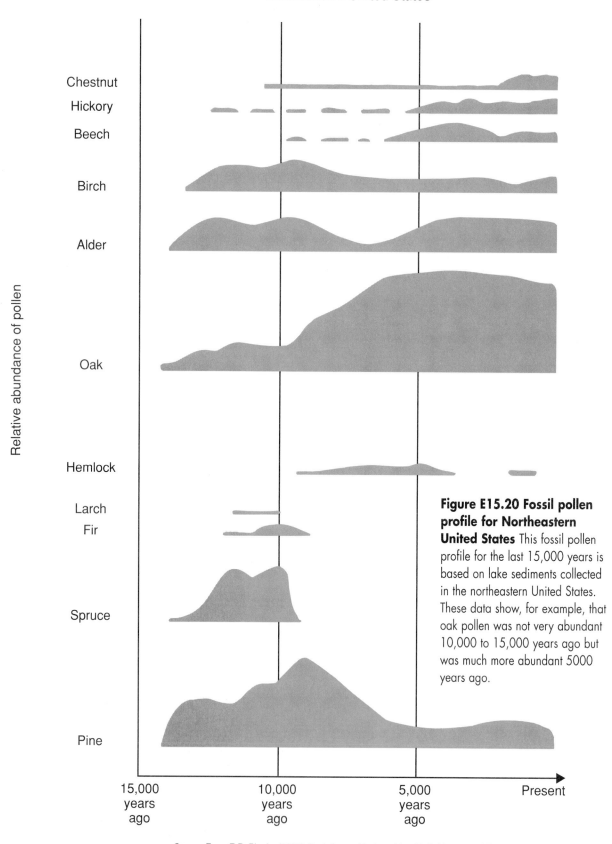

Relative abundance of pollen

Chestnut
Hickory
Beech
Birch
Alder
Oak
Hemlock
Larch
Fir
Spruce
Pine

Figure E15.20 Fossil pollen profile for Northeastern United States This fossil pollen profile for the last 15,000 years is based on lake sediments collected in the northeastern United States. These data show, for example, that oak pollen was not very abundant 10,000 to 15,000 years ago but was much more abundant 5000 years ago.

15,000 years ago   10,000 years ago   5,000 years ago   Present

Source: From E.R. Pianka (1985) *Evolutionary Ecology*. New York: Harper and Row.

Changes in Climate and Vegetation in Northern Midwest United States

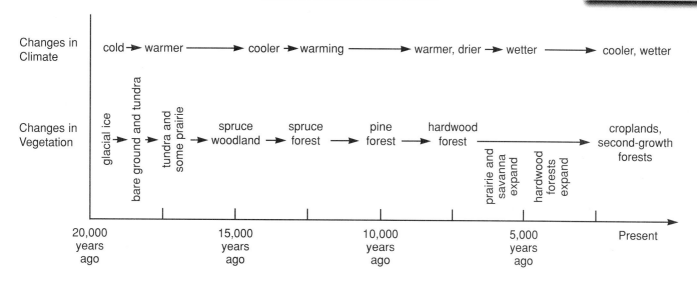

Source: From R. Brewer (1994) *The Science of Ecology, 2nd edition.* Fort Worth: Saunders College Publishing Co.

Figure E15.21 Changes in climate and vegetation in Northern Midwestern United States This chart represents a generalized description of climatic conditions and plant communities present in the northern Midwest of the United States during the last 20,000 years. Scientists obtained the information on plant communities from fossil pollen analysis.

From Puddle to Forest

Imagine that on a hike in the mountains, you come across a lovely beaver pond. You notice several beavers swimming about, hard at work repairing their dam. You also notice cattails and other rushes along the edges of the pond. Week after week you visit the pond, and week after week it looks pretty much the same. Is it possible that small changes that are hardly noticeable are taking place? If you returned several years later, you might notice that the beavers have cut down more trees, or a different type of grass is growing alongside the pond or different wildflowers are growing. Changes such as these would be more obvious and probably lead to changes in other aspects of the ecosystem. In fact, change in ecosystems is an ongoing process, but how long does change take?

Ecological succession is a dynamic process of change in the characteristics of any ecosystem across time. These changes result from complex interactions between biotic and abiotic factors, and the types of change vary depending on the type of ecosystem.

Succession that occurs in a body of water such as the beaver pond described before is called aquatic succession. Ecologists can demonstrate aquatic succession by recording the changes in the biotic and abiotic characteristics of the pond ecosystem during the course of many years.

Biologist Edward Kormondy once studied the changes that occurred in beach ponds located along the southern shore of Lake Erie. These ponds form when storms create sandbars that isolate small bodies of water from the rest of the lake. Ponds such as these can persist for days, months, or even years until subsequent storms wash the sandbar away or fill the pond with sand. The ponds are typically 100 to 250 meters (325 to 812 ft) long, 20 to 40 meters (65 to 130 ft) wide, and about one meter (3.25 ft) deep. Kormondy located four ponds of different ages (ranging from 1 to 75 years) and studied the differences that existed in the types of plants and animals living in and around the ponds. The ponds

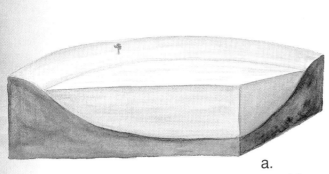

Chara
Chara

pondweed
Potamogeton

a.
1 year old

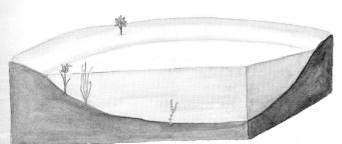

b.
5 years old

bayberry
Myrica pensylvanica

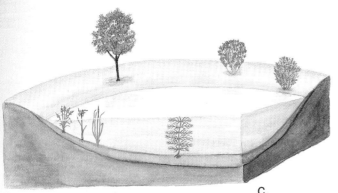

c.
50 years old

willow
Salix

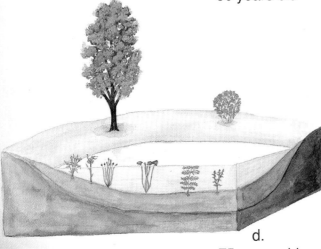

d.
75 years old

cattail
Typha angustifolia

bulrush
Scirpus

spikerush
Eleocharis

water lily
Nuphar advena

bluejoint
Calamagrostis canadensis

Figure E15.22 Beach ponds that are a. 1 year, b. 5 years, c. 50 years, and d. 75 years old Notice the change in the types of plants inhabiting the ponds. The abundance of each type of plant also changes, but this feature is not shown. Use the drawings to identify the types of plants present in each pond.

water milfoil
Myriophyllum

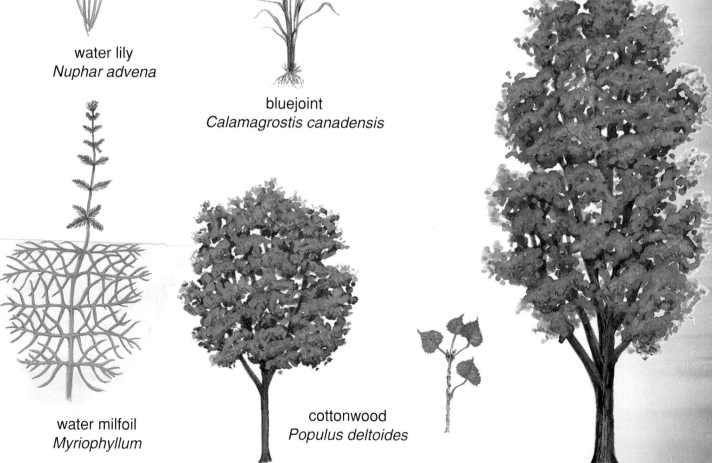

cottonwood
Populus deltoides

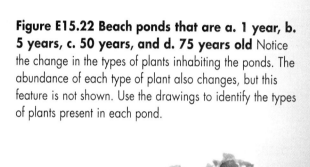

Mollusk Species	Ponds			
	A	B	C	D
Gastropoda				
 Viviparidae				
Viviparus contectoides	-	-	-	-
V. malleatus	-	-	-	-
 Physidae				
Physa gyrina	-	+	+	+
P. integra	-	+	-	+
 Lymnaeidae				
Lymnaea columella	-	-	-	+
L. humilis	-	+	+	+
L. palustris	-	+	+	+
Succinea ovalis	-	-	+	-
 Planorbidae				
Gyraulus deflectus	-	+	+	-
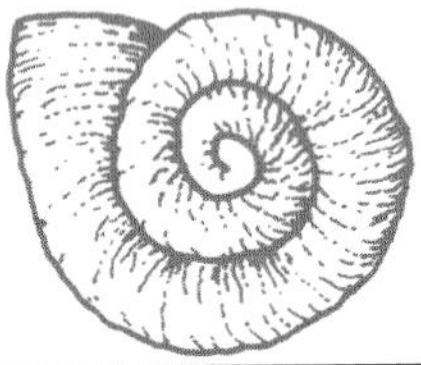 *G. parvus*	-	+	+	+
Helisoma trivolvis	-	-	-	-
Menetus exacuous	-	-	-	-
Pelecypoda				
Unionidae				
Ligumia nasuta	-	-	-	-
Sphaeriidai				
 *Pisidium casertanum*	-	-	-	+
P. variabile	-	-	-	-
 *Sphaerium lacustre*	-	-	-	-
S. partumeium	-	-	-	+
S. securis	-	-	-	+

Source: From E.J. Kormondy (1969) Comparative ecology of sandpit ponds. *American Midland Naturalist* **82**: 28–61.

Figure E15.23 Mollusks collected from beach ponds of different ages In his study Kormondy found different types and numbers of mollusks in each pond. Pond A is 1 year old, Pond B is 5 years old, Pond C is 50 years old, and Pond D is 75 years old. A (+) means the species is present, a (-) means the species is absent.

progressed from the newly formed open water stage to a semiterrestrial stage (see Figure E15.22). Because all of the ponds were located on the same small island, Kormondy felt that they represented the different stages of succession that an individual pond would undergo during the course of 75 years.

As part of his studies, Kormondy documented differences in the animals that lived in and around each pond. Figure E15.23 indicates the species of mollusk that inhabited each pond. Some species of plants and animals were present during most successional stages, whereas others were present only during certain stages.

One way to measure the changes that have occurred in an ecosystem is to compare the species that inhabit a community at different stages of succession, that is, at different points in time. Ecologists use simple quantitative measures to determine how similar two communities are: they look at both the presence and absence of plant or animal species in each community. Such a measure is called a community similarity index. One such index

uses the following formula: CSI = $2c \div (s_1 + s_2)$ where CSI = the measure of community similarity, s_1 = total number of species present in community number 1, s_2 = total number of species present in community number 2, and c = number of species present in both communities. The value of CSI can range from 0 when no species are present in both communities to 1.0 when every species is present in both communities.

Figure E15.24 illustrates this formula. If there are 20 plant species present in a 50-year-old pond (s_1 = 20), 18 plant species present in a 75-year-old pond (s_2 = 18), and 16 plant species present in both ponds (c = 16), then the measure of community similarity = (2)(16) ÷ (20 + 18), or 0.84. This number indicates that there is an 84 percent similarity in the types of plants that inhabit the ponds. This also means that there is a 16 percent difference in the plants that inhabit them. By comparing the makeup of ponds of different ages, ecologists can determine how much change occurs as ponds age.

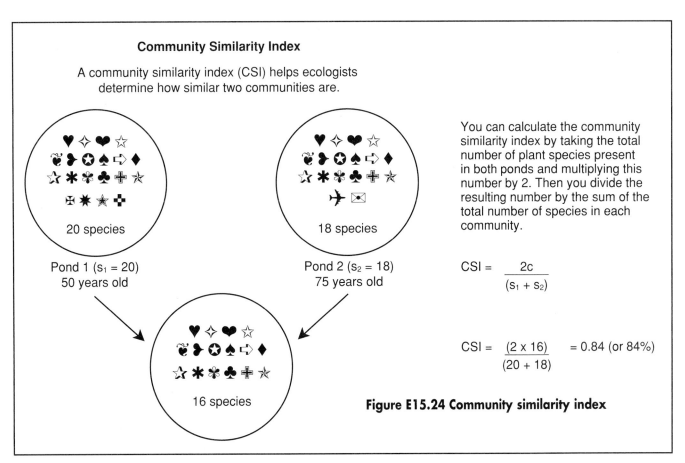

Figure E15.24 Community similarity index

Environmental Ethics and You

Environmental ethics is a branch of ethical study that considers how people *ought* to relate to the natural world. The different ethical perspectives people have about humanity's relationship to nature are based on different values, beliefs, and attitudes. People's values, beliefs, and attitudes about nature are formed by a variety of influences including the overarching values and attitudes of their culture. Peoples' attitudes about nature also are influenced by their personal experiences, the people they talk to, and the books they read. People's values taken together influence how they behave toward the environment, and this has a major impact on the state of the natural world we see around us.

Below are brief descriptions of two rather distinctive ethics. How similar or different is your personal environmental ethic from the two views described here? What factors have influenced your environmental ethic? What other environmental ethical models are you aware of?

The Human-Centered Environmental Ethic. People who support a human-centered environmental ethic take the position that humans should dominate the natural world because they are fundamentally different from all other life forms and are somehow superior to or separate from nature. From this perspective, nonhuman life forms are commodities that have no value other than their usefulness to humankind. People holding this ethic view the earth primarily as a collection of natural resources that humans can use to promote their own economic growth and prosperity. People who support this position also think that the earth is vast and has many resources in unlimited or abundant supply. They point out that human history has been characterized by continual progress and that, through technology, people find solutions to all problems (including those of resource depletion and pollution), and progress continues.

The Deep Ecology Environmental Ethic. The deep ecology ethic is a life-centered ethic. People who support this position think that all life forms on earth have intrinsic value regardless of their usefulness to humankind. From this perspective, humans have no right to reduce the earth's biological richness and diversity except to satisfy fundamental needs. People who support this position think that the earth's resources are in limited supply and that these resources are for all life forms, not just for humans. The deep ecology ethic takes the

Ultraviolet Light and the Ozone Layer

Reclining on the beach, soaking up the rays, the warmth of the sun feels good on your back. You undoubtedly appreciate the fact that the sun is a welcome feature of the environment in which we live. Sunlight is essential for life, but excessive exposure to sunlight can be harmful. The component of sunlight that is most harmful to living tissues is ultraviolet light. For example, in humans low doses of ultraviolet light are important in the synthesis of vitamin D in our skin (a deficiency of vitamin D can cause rickets in growing children), but high doses can cause DNA damage, skin cancer, premature aging, cataracts, and even damage the immune system.

The effects of ultraviolet light on DNA are relatively complex. The most important damage that it causes, however, occurs when the ultraviolet light triggers the formation of new chemical bonds between certain bases on the same strand of the DNA—for example, between thymine (T) and thymine (T) or between thymine (T) and cytosine (C). These new bonds distort the shape of the DNA molecule (see Figure E16.1) and can block both DNA and RNA synthesis. This result is so dangerous

E252

Human-
centered
ethic

Deep
ecology
ethic

position that both human and nonhuman life will continue to flourish only if humans greatly reduce their rate of population growth. People who hold this position think that human

interference in the nonhuman world has been excessive and human impact must be minimized. From this perspective, people must make economic, technological, and ideological

changes that promote a sustainable lifestyle rather than one that seeks the highest standard of living.

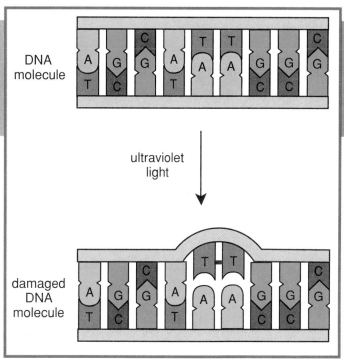

Figure E16.1 Ultraviolet light can cause new bonds to form between adjacent bases in a DNA molecule.

that as few as one such distorted section can be lethal to the cell involved.

A cell that has been damaged by ultraviolet light will die unless the damage is repaired. If the damage is repaired accurately, the cell will live and will be unchanged. Figure E16.2 illustrates one process by which DNA repair can take place. On the other hand, if the damage is repaired, but the repair process alters the DNA sequence at the point of repair, the cell will live but will be changed (that is, it will carry a mutation). If this mutation occurs in certain genes, the cell may become cancerous, even though the specific damage that resulted from ultraviolet light was repaired. As you might guess, the genes that control the cell's ability to repair DNA damage accurately are very important genes.

A delicate balance exists, then, between the need for sunlight (some sunlight must reach the earth's surface if life is to continue) and the need to be protected from sunlight (excessive ultraviolet light can damage life seriously). As you may know, far more ultraviolet light reaches the earth's atmosphere than ever penetrates to its surface. **Ozone**, a chemical in the upper atmosphere, filters out most of this destructive component of sunlight and thus protects living things from experiencing too much DNA damage. Ozone (O_3) is a three-atom type of oxygen that forms in the upper atmosphere when sunlight causes oxygen molecules (O_2) to split into single oxygen atoms; these atoms join with other oxygen molecules to form ozone.

The total amount of ozone in the upper atmosphere depends on the balance between ozone-producing processes and a variety of ozone-destroying processes. Some of these ozone-destroying processes are triggered by phenomena such as volcanic activity; others involve chemical reactions between ozone and various polluting gases that are produced by human activity at the earth's surface. These polluting gases rise through the atmosphere and increase the rate of ozone destruction. Because of the importance of ozone's filtering action on ultraviolet light, continued production of gases may have serious consequences for life on earth.

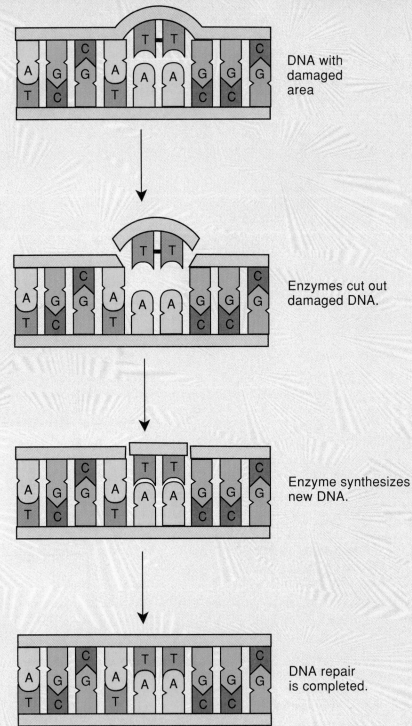

DNA with damaged area

Enzymes cut out damaged DNA.

Enzyme synthesizes new DNA.

DNA repair is completed.

Figure E16.2 Living systems have mechanisms that allow the repair of damage caused by ultraviolet light.

The Farmers Are Saved!

Throughout the late 1800s and early 1900s, insects were wreaking havoc on the crops of the western and southern United States. The fluted scale insect was eating up the fruit trees in California; the boll weevil was decimating the cotton crop in the South; and the Colorado potato beetle (pictured in Figure E16.3) was destroying potato plants from Colorado all the way to Illinois. Obviously the farmers were distressed; their livelihood was being ruined by these pests. Consumers also were feeling the effect, because the shortage of fruits and vegetables was driving up the price of food.

To combat the problem, the U.S. Bureau of Entomology (entomology is the study of insects) set out to find a way to reduce the population of these destructive insects. The government hired a hard-working entomologist named C.H. Tyler Townsend to study the problem of the boll weevil (see Figure E16.4) and to suggest a method to eradicate it. Townsend studied the life cycle of this beetle and determined that its reproductive, eating, and hibernation patterns made it susceptible to a low technology method of insect control. The bureau recommended that farmers plant their cotton in widely spaced rows and remove fallen plant material from between the plants. This method would reduce the weevil's summer reproduction by creating harsh conditions for newly hatched larvae, which develop between the rows under fallen cotton leaves. In addition, the bureau advised the farmers to plow the plants under the soil after the first cotton harvest in early fall. This strategy would deprive the

Figure E16.3 Colorado potato beetle, *Leptinotarsa decemlineata* (about 13 mm or 0.5 in long)

Figure E16.4 Boll weevil, *Anthonomus grandis* (about 6 mm or .25 in long)

weevil of its food source for hibernation; whatever small reserves of food the weevils stored before plowing would be insufficient to survive the winter.

This plan had several drawbacks, however, and the farmers objected. They argued that if they plowed under the plants after the main harvest, they lost their chance at a second late fall harvest. The extra cotton from this small secondary harvest meant a little extra money for the farmers, and they were unwilling to sacrifice it. In addition, these control measures required extra work on the part of the farmers. Despite assurances from the bureau that the short-term loss of a secondary harvest would be compensated by a larger harvest the next year (because the weevils would not destroy as much cotton), the farmers argued that the risks were too great. What if a natural disaster, such as a flood, destroyed their entire crop the next year? In this case the first-year losses would not be offset by second-year gains. The farmers refused to adopt the recommendations.

Instead farmers tried various control methods that ranged from the ingenious (a vacuum to suck the bugs from the plants) to the harebrained (passing an electric current through an entire cotton field). Nothing seemed to work until the bureau discovered that a chemical, calcium arsenate, poisoned the insects quite effectively. A simple solution to the boll weevil problem finally had been found! Calcium arsenate production in the United States increased from 23,000 kg (50,000 pounds) in 1918 to 4,550,000 kg (10,000,000 pounds) in 1920.

Figure E16.5 Cotton farming in the early 1900s

Breakthrough Insecticide Benefits the Boys Overseas and Farmers Alike

1946. Nearly three decades after the introduction of crop-saving chemical pesticides to farming, a truly remarkable compound has been approved for use in the United States. DDT, the insecticide that some scientists are calling miraculous, is potent, inexpensive, long-lasting, and very toxic to many different insects, yet it has low toxicity in mammals. This is good news for farmers because pesticides, which include herbicides (for killing weeds), fungicides (for killing fungi), and insecticides (for killing insects), have become common weapons in the arsenal that farmers use to combat the organisms that threaten their livelihood.

Since the first American tests of DDT in 1942, the use of this chemical in farming has grown steadily, and there is now talk of expanding the use of DDT to benefit farmers and nonfarmers alike. Because DDT kills mosquitoes and houseflies, as well as farm pests such as the potato beetle, DDT eventually may be used to rid neighborhoods of these bothersome pests.

DDT's effectiveness in killing so many different types of insects led many scientists during the war to wonder about potential benefits that DDT might offer, and the dedicated troops of our World War II military were the beneficiaries of these experiments. After exposure studies demonstrated that DDT had very low toxicity for humans, scientists decided to use this chemical to protect troops serving in areas where they were exposed to insect-borne diseases such as typhus and malaria. By killing the insects that might have infected these troops with numerous pathogens, DDT saved many American military personnel from illness and death (refer to Figure E16.6). Today, some entomologists are confident that, with DDT, they finally will be able to eliminate all insect pests and, therefore, all diseases transmitted by insects. No health problems have been associated with DDT, although scientists have noticed that this chemical accumulates in the body and can be passed to infants through mother's milk.

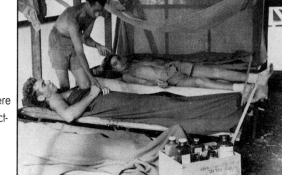

Figure E16.6 Troops during World War II were plagued by insect-borne diseases such as typhus and malaria.

Conservationists in Hysterics over DDT

23 June 1958, *The Daily Telegraph*

A handful of people from several towns in our area have begun a campaign to rid our communities of DDT, the insecticide that has saved so many of our beautiful trees from Dutch elm disease. These radicals claim that there are fewer songbirds in the neighborhood. A Girl Scout troop on a hike in the woods claimed that robins and several other birds were found dying of seizures. A number of people have said that since DDT tree spraying began, cats appear to be dying in greater numbers than usual.

These reports appear to be unfounded; government studies of DDT have repeatedly shown that exposure to DDT is not toxic. The weakness of the claims against DDT has been demonstrated further in several recent legal decisions. In Long Island, New York, a number of local residents sued the Secretary of Agriculture and the Agriculture Commissioner to prevent the spraying of DDT in their community, on the grounds that the chemical might have long-term health risks. Witnesses for these residents cited a handful of scientific studies that suggest DDT has harmful effects on wildlife. They also made outrageous claims that DDT might cause cancer and several other diseases.

Fortunately, however, the side of reason won the day.

Witnesses for the defendants testified that DDT was, in fact, a valuable public health tool. DDT has been credited with saving as many as 7 million people from insect-transmitted diseases, including American troops who served in the tropics. In addition, the defense maintained that any health problems that might arise as a consequence of long-term, low-level exposure to DDT would have been detected by government studies in which subjects were exposed to high levels of the chemical for short periods of time.

Cases like this one prove once again that DDT is safe. We must not let antichemical hysterics in our area ban the use of this beneficial pesticide, or we risk losing even more of our trees to gypsy moths and other hungry insects.

DDT Banned!

1972. The chemical pesticide DDT (dichlorodiphenyltrichloroethane) has been banned by the Environmental Protection Agency (EPA). Since the early 1940s, DDT and many other pesticides have been used widely for the control of pests. The apparent benefits of DDT, however, now must be viewed in the light of recent evidence that indicates that DDT may have some disturbing side effects on the ecology of wildlife exposed to it. It appears that DDT is long-lasting in the environment and can enter the food chain and accumulate in the fatty tissues of fish and birds. Many scientists have documented dramatic declines in bird populations as a result of DDT, and several species are approaching extinction.

These findings have raised concerns about the potential long-term health risks to humans exposed to DDT and other pesticides.

Figure E16.7 The reproductive success of bald eagles and other species of birds was constrained by the widespread use of DDT during the mid-1900s.

Systems Analysis

When a situation arises that is so complex that predictions seem at best unreliable and at worst impossible, scientists turn to a method of analysis called systems analysis. In systems analysis, scientists try to understand enough about the interactions of complex situations that they can reliably predict the effect that certain changes will have on the system.

Weather provides an example of a complex system that is relatively easy to understand in the short term; for example, dark black clouds usually lead to rain. The interactions that produce weather are collectively so complex, however, that it is nearly impossible to understand or predict weather across large distances or times. Ecosystems, the collections of living and nonliving elements that produce interdependent networks of living systems, also can be viewed as examples of complex systems.

The first step in the analysis of any complex phenomenon is to identify the components of the system under study. The **components**, or collection of *things*, that make up an ecological system include living (biotic) as well as nonliving (abiotic) things. For example, an ecologist might describe a pond in terms of

its biotic components—its microscopic organisms, plant life, and fish, but she also might describe the pond in terms of its abiotic components—the pH of the pond water, the nutrient and oxygen levels, and the depth of sunlight penetration. These components also include elements outside of the pond; thus, the pond system is not isolated from the world around it. In this case, the pond's pH is influenced by the pH of rainwater and the pH of anything that might wash into the pond; the nutrients are determined by the amount and type of food that washes or falls into the pond and the types of organisms that decay there; and the penetration of sunlight is affected by the amount of shading the pond receives (for instance, is it surrounded by large trees?) and the stillness of the water. All of these components influence and define the system.

Once the components of a system are identified, most scientists try to make the system more manageable by *limiting* the number of components that are included in the analysis. For instance, our pond ecologist might organize the analysis of the pond to include only the effects of several different nutrients on the number and size of perch in the pond. The scientist who establishes

Figure E16.8 Weather is a complex system, as this computer model of a thundercloud demonstrates.

such a structure recognizes that outside factors, such as pH and sunlight, also influence the fish, but she chooses not to take these into consideration because they further complicate an already complex study.

Limiting the number of components is necessary because it focuses the scientist's thinking on a particular area of interest. Developing and testing hypotheses is much easier when the focus of the investigation is narrow. Unfortunately, imposing a limited structure somewhat reduces the accuracy of the analysis. Most systems analysts try to organize their system in a way that is simple enough to be manageable and complex enough to be accurate.

The last essential step before beginning any analysis of a complex system is that the scientist understands as much as possible about the initial behavior of the system. If the initial behavior of the system is not understood, then it is difficult to interpret how the system will react when one of its components is altered. In the pond example, if the ecologist does not already know that perch populations decline each fall (because a particular nutrient that they require becomes scarce), then she might mistakenly attribute the decline to a

recently observed, but very slight, change in pH. (In fact, the pH change might influence the perch population, but this effect may be small compared with the annual fluctuations caused by varying nutrient levels.)

Once a limited analysis is complete, scientists may try to combine the results of several analyses to construct a more complex, and more accurate, understanding of the whole system.

From this brief description of systems analysis, you can see that it is important to know as much as possible about the behavior of a system before attempting to analyze the effect that *changes* in particular components might have on the system. For this reason, ecologists collect large amounts of data in an attempt to understand how one structured set of interactions affects another. Despite this effort, ecologists realize that they never will have all of the information that would be necessary to make an error-free prediction. To compensate for this limitation, many ecologists choose to be conservative in their predictions—that is, they generally assume that intervening in ecosystems will result in some unpredictable consequences.

Figure E16.9 A pond is a good example of a complex system that interacts with the world outside.

Page numbers in bold print indicate item was in bold in the text; italics indicate a chart, table, graph, or illustration.

A

Abiotic matter, **E132**

Abortion, E155

Abstinence, E154

Acid rain, 402

Active transport, E55, E56, *E56*

Adaptations, 64, **E27**, *E27*; desert, E46, *E46*; to environment, E45–46

Acquired immune deficiency syndrome. *See* AIDS.

Adenine, 264

Adenosine diphosphate, *See* ADP.

ADP, *E114*; cell energy, E120; in photosynthesis, *E125*

Adenosine triphosphate, *See* ATP.

Adrenal glands, *E62, E64*

Aerobic energy release, **E104**, *E104*

Aerobic respiration, **E118**, *E121*

AIDS, 134, E81, E149–50, *E150*

Ainu people of Northern Japan, E228–29, *E228, E229*

Air, particles in, 123

Air pollution, 403

Alcohol, 90, E106

Alga, first plant, E43

Algae, E43

Allantois, *E198*

Alleles, **E156**, E156–57, *E157*; complex traits and, *E160*; genetic variation, 252; heterozygous, E157, *E157*; homozygous, E157, *E157*; mutation in, E164; segregation during meiosis, E161–164, *E164*

Allergies, E81

Alveoli, E68, *E69*

Alzheimer's disease, E221, E224

Amino acids, E91, *E97*, E128; in sickle cell hemoglobin and mRNA codons in gene expression, E184, *E185*; repair of muscle tissue, 153; sequence in proteins related to DNA structure, E180–81, *E180*

Amish people of America, E230–31, *E230, E231*

Ammonia, E58

Amnion, *E198*

Amoeba, E143

Amylase, **152**, *E96, E98*; pancreatic, *E96*

Anabolic steroids, E105

Anacharis plant, *E126*

Anaerobic energy release, **E104**, E121

Analysis of an ecosystem, E258–259, *E258, E259*

Anaphase, mitosis, *E196*

Anasazi Indians, E129–130

Animals, E38–39

Anole lizard, *E33*

Anorexia nervosa, E98–99

Antibiotics, **54**; bacterial resistance to, 54, E77, *E77*

Antibodies, **127, E79**, E79, *E80*

Antibody-mediated response, **E79**, E79, E80

Antigens, **127, E79**, E79, E80

Antigen response, *E82*

Antihistamines, E81

Anus, *E96*

Aortic arch, *E57*

Appendix, *E96*

Apple tree, *E40*

Aquarium, *334*

Aquatic succession, E247, E251, *E248–249, E250, E251*

Archaebacteria, E40, *E41*

Arteries, E57, *E57*; renal, E59, *E59*

Arterioles, E57

Artifacts, **50**; from Pueblo Bonito, *E131*

Asexual reproduction, 217, **E143**, *E143*

Aspen (tree), 215, *215*

Astronauts, dietary and metabolic needs of, *394*

Atomic theory, E29

Atoms, **166**, E108, *E108*

ATP, **E113–114**, *E114*, E116–117, *E116, E118,* E119–120; energy and, E120; in carbon fixation, E126, *E126,* E127; in cellular respiration, *E128*

Attachment, **E218**

Australopithecines, E22

Autoimmune diseases, **E83**

Autosomes (nonsex chromosomes), 246

Autotrophs, **176**

Avery, Oswald, E173

Axon, E11, *E11*

B

Bacteria, *E38*; on human skin, E78, *E78*; resistance to antibiotics, E77, *E77*

Bacteriophage, 63, *63*

Bald eagle, *E257*

Baltica, *40, 41*

Basal metabolic rate (BMR), **E105**, E105

Bats, E129; little brown, *E130*

Bayberry, *E248*

B-cells, **127, E79**, *E80*; plasma B-cells, **127**; memory B-cells, **127**

Beaver, *E132*, E132–133

Beef production, *190*, E140

Behavior, E106; influence of genotype, E158–69, *169*

Bering Sea, *325*

Bighorn sheep, E151, *E151*

Bile, E98; duct, *E96*; salts, E98

Binary fission, **217**

Biochemical homologies, E47–48

Biological classification schemes, 66, *68–69*, E48–49

Biological structures, E99
Biology, unifying principles of, 11, 61
Biology in medicine, scoring rubric for recognizing, *385*
Biomass, **E138**
Biosphere, **E135**
Biosynthesis, **180**
Biotic matter, **E132**
Biotin, *E94*
Bipedal movement, **E7**, *E7*
Birth control methods, *219*; side effects, E149
Birth defects, E203–06; alcohol and, E205–06; drugs and, E205; genetic errors, E203–04, *E204*; genetic expression errors, E204–06, *E204*; smoking and, E205; thalidomide, E204–05, *E206*
Bladder, *E58, E144*
Blood, E57; loss, homeostasis and, *E73*; pressure, **115**, *116*; transfusion, E81–82; types, human, E82, *E82*
Blood vessels, E57, *E57*; embryo, *E198*
Bluejoint, *E249*
Body temperature, **115**; regulation, E71
Boll weevil, *E255*; low-technology pest control, E255; DDT as insecticide, E256
Brainstem, **E4**, *E4*, E63
Brain: amphibian, *24*; bird, *24*; fish, *24*; human, 24–25, *24*, E9–13, *E9, E10*; impact of growth on human development, E221; mammal, *24*; plasticity of, E10; regions of, *E4*
Breakdown reactions, E127
Breathing, mechanism of, E68–70, *E69*; rate, 116, E70
Bronchus (bronchial tubes), E68, *E69*
Bronfenbrenner, Urie, E227, *E228*
Brongniart, Alexander, E18
Bryophyllum leaf, E143
Budding, **217**
Buffer, **107**
Bullfrogs, *198*
Bulrush, *E249*

C

Calcium arsenate, E256
Calcium, *E95*
Calorie, 173, E91
Calorimeter, 173
Cancer, E81; abnormal cell growth, E206–07; detection of, E210; and DNA genes, E208; mutant genes, E209, *E209*; pesticides and, *360*; prevention, E210; and RNA genes, *E208*; treatment, E210; viruses and, E207
Capillaries, E57
Capillary bed, *E57*
Carbohydrate loading, E106
Carbohydrates, E91
Carbon atom, *E108*, E109
Carbon cycle, E134, *E134*
Carbon dioxide, E68, E118, *E119*; in carbon fixation, E126–127, *E126*

Carbon fixation, **E126**, *E126*
Carbon-14 dating, E18–19, *E19*
Cardiac outputs, marathoner and non-athlete comparison, 162
Cardiovascular disease, E92
Carotid artery, *E57*
Carrier, **E113**
Carrying capacity, **E232**
Cattail, *E249*
Cell, *E34*, E51; diploid, **E160**, E160–61; haploid, **E160**; model of, 291
Cell cycle, **E196**, E196, *E196*
Cell division, control over, E197–198, *E197–98*
Cell-mediated response, **E79**, E79, *E80*
Cell membrane, E52–53, *E53*
Cell structure, E51–52, *E51*
Cell theory, E29
Cell types, human trachea, 294
Cellular differentiation, E199–202, *E200*; mechanisms for, E201–02, *E201, E202*
Cellular respiration, 176, **E116**, *E116*, E120–22
Cerebrum, growth of, E221
Centriole, *E196*
Centromere, *E196*
Cerebellum, 24, *24*, **E4**, *E4*; sheep, *23*
Cerebrum, 24, *24*, **E4**, E4, E9, *E9*; sheep, *23*
Cervix, **E145**, *E144*
CFCs. See chlorofluorocarbons.
Chain, Ernest, 55
Chara alga, *E44, E248*
Chemical bonds, **E108**, E108–109, *E109*
Chemical reactions, E109
Chewing, E97–98
Child development, ecological approach, E226, *E227*
Chimpanzees, 19, *19*, E2–3, *E3*
Chitin, *E102*, E104
Chlamydia, *E150*
Chloride, as vital sign, 116; and CFTR protein, 389–90, *389*
Chlorine, *E95*
Chlorofluorocarbons, 350, *351*, 351; ozone destruction, *351*
Chlorophylls, **E124**, E124–25
Chloroplasts, **E124**, *E125*
Cholera pathogen, cystic fibrosis and, 390–391, *391*
Cholesterol, 116
Chromatin fiber, *E174*
Chromatium okenii, *E42*
Chromosomes, **E156**, E156, *E156, E196*; autosomes, 246; diploid, **E160**, haploid, **E160**; human, *E175*; human diploid set, 246; meiosis, E160–64, *E161*; mitosis, *E196*; models of in diploid animal cell, *242*; results of extra copy, E166; sex chromosomes, 246; structure, *E174*
Cigarette smoking, 135, E84
Circulatory system, gas exchange system and, E68, *E69*; human, E57, *E57*
Class, **E49**, *E49*
Classification, **66**
Classification of organisms, 68–69; early models, *E38*
Click beetle, *E39*

Cliff dwellings, *E130*

Climate, E232; changes in, *E247*

Cloning, E203

Coastal Zone Color Scanner, 401

Codon, **E182**

Cognitive growth, E215, E215–16, E221–22

Cognitive stages, *E221*

Collared lemming, *E71*

Collie dogs, *238*

Colorado potato beetle, *E255*

Common ancestor of apes and humans, *E22*

Community, **185**, *E35*, E238; competition in, E240, *E240*

Community similarity index, E251, *E251*

Competition, **E28, E239**

Complementarity, as pairing of nitrogen bases in DNA, **E174**, E174–75

Complement proteins, **127**

Complex carbohydrates, E92

Components, **E258**

Compost, 194

Conditioning, E105–06

Conservation of energy, **E112**

Consumable biomass, *E138*

Consumers, **E136**

Contact inhibition, *40, 41, 44, E197*

Continental drift, **E17**

Continents, changes in, during earth's history, E16–17; *40, 41*

Controlled experiments, designing them, **82, 83**

Converted, **E110**

Coppens, Yves, 34

Corpus callosum, E9–10, *E63*

Cottonwood, *E249*

Covalent bonds, **E113**, *E113*

Cretaceous Earth, 41

Crick, Francis, 261, E179

Cricket, *222, 397, 398*

Criteria, **67**

Criteria for projects, 27

Crossing over, 245, *E162–63*

Crown gall, *399*

Cultural celebrations: childbirth, E144; marriage, E153–54, *E153*; sexual maturity, rites of initiation, E153–54

Cultural evolution, **50**

Culture, **E13**

Cultures, E226–231; Trobriand Islanders, E227–28, *E227*; Ainu of Northern Japan, E228–29, *E228, E229*; Gusii of Kenya, E229–30, *E229*; Amish in America, E230–31, *E230, E231*

Cuvier, Georges, E18

Cycling of matter, E132–135

Cystic fibrosis, allele, 390; CFTR protein and, 389, *389*, effect of cholera on, *391, 391*; inheritance pattern, E158–59, *E160*; pancreas and, *390*; pedigree of, 389, *389*

Cytoplasm, E117, in mitosis, *E196*; role in DNA replication, *E177*, E181, *E183*

Cytoplasmic regulation, *E201*

Cytosine, 264

D

Dalton, John, E29

Daphnia, 8

Darwin, Charles, 47, E23–28, *E23*; on population growth, E234, E23

Data, **6**

DDT, *355, 357*, E256–257; as insecticide, E256, 257; chemical structure, *357*; decomposition, *357*, concentrations in food chain, *359*; effect on environment, E257; guidelines for a systems analysis of, *357*; impact on bird eggshells, *360*; production after Word War II, *257*; to combat insect-borne diseases, E256

Decomposers, **E136**

Deep ecology environmental ethic, **E252–253**

Deep time, **36**

Dendrites, E11, *E11*

Deoxyribonucleic acid (DNA), 256, **E21**, *E21*, E34, *E34*, E52–53; complementary base pairing, E175, *E177*; discovery as transforming substance, E173; in eukaryotic chromosome, *E174*; fingerprinting, 258; gene expression and, *E207*; genetic plan, E156; genetic engineering and, E187–91, *E188, E189, E190*; human, *E175*; ligase enzyme, E186–87; physical structure, E179–80; relation to protein structure, E180–81, *E180*; replication, E173–75, *E176, E177*; replication errors, E178–79, *E178*; role in genetic information, E172–73, *E172*; role in RNA production, E181, *E181*; structure, models of, *262–263*, E173–77, *E176, E179–80*; synthesis during meiosis, E161, *E161, E162*; viral antigens and, E83

Descending aorta, *E57*

Descent with modification, E24

Determinate growth, E211

Development, **E194**; flowering plant, *E211*; phases of humans', E216–E220, *E217, E218, E219, E220*; sexual maturity, E212

Developmental biology experiments, 288, *288*

Developmental errors, 297; and birth defects, E205–06; cancer and, E207

Developmental patterns, E213

Developmental plants, E211–12

Devonian Earth, 41

Diabetes mellitus, *E73*

Diagnosis, **E76**

Dialysis tubing, 88

Diet, **E91**; comparison of types, *E106*; effect on muscle glycogen and muscle endurance, *162*; lifestyle choice, E226

Dietary fiber, E92

Differentiation, **E194**, E194–195

Differentiated plant tissue, 293, *293*

Diffusion, **E54**, *E54*

Digestive enzymes, *E96*, E97

Digestive system, E91, *E96*

Dihybrid cross, **244**

Diuretic, **91**

Divisions, **E49**, E49

DNA. *See* Deoxyribonucleic acid.
Dodo bird, *E16*
Dominant trait, **E158**
Doubling time, **E238**
Driesch, Hans, 287; experiment of, *288*
Dyslexia, 203

E

Earth, geological phases, *40*, 41
Earth's history, time line of, *37*
Earthworm, *188*, E102, *E102*, E135
Ecological succession, **E247**
Ecosystem, **185**, *337*, E258; measuring changes in E247–251, *E248–249*, *E251*; river as ecosystem, 324; variety of, *337*
Ectothermic regulation of body temperature, E71
Egg, chicken, 82, E98–99, *E198*
Eldredge, Niles, E168
Electrocardiogram, E76–77, *E76*
Electrocardiograph, E76
Electron transport system, E116, **E117**, E117, E119–120, *E119*
Electrons, *E108*
Element, *E108*
Elephant, African, *E36*, *E142*
Embryo, **E194**, *E195*; growth of, E198–99, *E198*; mouse, *E67*; pig, *E195*
Embryologist, **42**
Emigration, **E239**, *E239*
Emotional growth, E224; biological basis, E224; effect of mental retardation or Alzheimer's disease, E224
Endocrine system, **E62**, E62–64
Endoskeleton, **E103**
Endothermic chemical reaction, *E110*
Endothermic processes, **E109**, *E110*
Endothermic regulation of body temperature, E71
Energy, and matter expended in training and racing, *160*, 166; expenditure and exercise, *161*; nutrients and, *161*; use and locomotion, *395*; conversion, E110–114, *E111*; production in cells, E120–22, *E121*; pyramid, *E141*; regulation in cells, E120–121, *E121;* resources, world production of, *378*; storage, *E112*
Energy transfer, E112; trophic levels, E141, *E141*
Entropy, **E54**, *E54*
Environmental ethics, E252–253
Environmental indicators, *371–381*
Enzymes, **152**, E94, E96, **E97**; in chemical reactions, E113, *E114*
Epididymis, **E145**, *E145*
Epilepsy, E10
Escherichia coli, *E36*, E238
Esophagus, *E96*, E98
Essential amino acids, E91
Essential elements, E91, E94, *E95*
Estrogen, **E147**, *E148*
Ethical analysis, process of, E87
Ethiopia, 34, *34*
Ethologists, 221

Eukaryote, **E36**; chromosome structure, *E174*; DNA replication, *E177*; translation of mRNA in, *E183*
Eukaryotic chromosome structure, *E174*
Evidence, **35**
Evolutionary biologists, 41, 38
Evolution, behavior and, E25, E28; of brain, 20; cultural, 50; evidence for, *33*, E3, *E20*, E23–E26, E28, E30, E32, *E33*, E35, E117, E122, E164, E195, E202, E209, E212–E214; fossil evidence, 35; genetic evidence, E25, E30, E122, E164, E165, E187, E195, E202, *E209*, E212; geologic evidence, 33, E24, E26, E30; molecular evidence, 248, 249, E25, E35, E117, E212; organisms' developmental pattern, E212–E214, *E213*; of plants, 65, 73, E25, E28, E60; of primates, 20, *E3*, *E21*, *E22*, E221; role of variation in, *E20*, E25, E26, E28, E30, E32, *E33*, E164, *E213*, E214; selective pressures, E28–E29; time line of major events, *E42*; theory of, 47
Evolutionary tree, horse, *E33*
Exercise, effect on muscle structure, *162*; for lifelong fitness, E226
Exoskeleton, *E102*, **E103**, E104
Exothermic reactions, **E109**, *E109*
Experiments, controlled, background on, 83
Exponential population growth, **E234**, example of, E237, E238
Extensor muscle, *E100*
Extinct animals, E16–17
Extinction, **E17**

F

Factors affecting immune system function, E84–85, *E85*
Family, *68*, **E49**, E49
Fats, E91, E92
Fatty acid, *E97*
Feedback, **E62**, E62–64
Feedback control, **E66**, E66
Fermentation, **E121**, *E121*, E122
Fertilization, **217**, of ovum, *212*, E147
Fetal alcohol syndrome, E205, *E205*; effect on social skills, E221
Fever, infections and, 388
Filaments, E101, *E101*
Filtration, **E59**
Fire, E115; as ecological factor, E242
Fish, homeostasis and, *E74*
Fitness, *144*, E88–90; resources required for, E89; levels, *E90*
Five kingdom scheme for organisms, E37, *E37*
Fleming, Sir Alexander, 55
Flicker, *E27*
Florey, Sir Howard, 55
Folic acid, *E94*
Follicle, *E144*
Follicle stimulating hormone (FSH), **E146**, E147, *E147*, *E148*
Food, *165*; E91–95, **E91**
Food pyramid, U.S.D.A., *E92*

Food supply, world, *373–376*; and population growth, E236; as limiting factor, E236

Food web, **E136**, *E137*

Foot, gorilla and human, *E8*

Forest fires, *338*, E242–244, *E244, E245*; suppression programs, E243

Fossils, E15–17; dating methods, E17–19, *E18*; formation, E15–16, *E16*

Fossil pollen, E244–246, *E245*; results of studying, *E247*

Fragmentation, 217

Franklin, Rosalind, 261

Frontal lobes, *E3*; effect of development on humans' social skills, E224

Fruit flies, 245

Fuligo septica, E41

Fungi, E39–40

G

Gage, Phineas, E2–3, *E4*, E224

Galapagos Archipelago, E20

Gall bladder, E98

Gametes, 217; from meiosis, 243; **E142**, *E143*, E142–143, E160

Gammarus, 8

Gas exchange, experiment, *E36*; human system, E57, *E69*, E68–70

Gene, **E156**; for blood type, *E157*, E156–157

Gene expression, **E172**, *E172*, **E180**, E200–02

Gene therapy, **E191**

Genetic code, **E182**, E182–84, *E182, E183*; deciphering of, E184; universality of, E184

Genetic disorders, cystic fybrosis, E158–159; human, 253; Huntington disease, E158, *E159*

Genetic engineering, **E187**; ethical and public policy issues, E192–93; technology of, E190–91, *E188, E189, E190*

Genetics, and behavior, E168–169, *E169*; environmental impact, E155

Genetic variability, E164–168, *E165, E168*

Genetic variation, complex behavior, E169; evolution and, E167; sources, E166, *E166*

Genetic information, role of, *E172*

Genital herpes, *E150*

Genotype, *241*, **E155**, E155–156; behavior, E168–169, *E169*; DNA and, E172, *E172*; for Huntington disease, *E159*; relation to phenotype, *241*

Genus, 68, **E48**, E49

Geologist, **39**

German measles, 386

Gestation, 217

Giant panda, E47–48, *E47*

Giant water bugs, *301*

Gila monster, 107, *107*

Glands, *E64*

Glomerulus, **E58**, E58–59, *E59*

Glucose, **88, 150**; in cellular respiration, 176; E91, E92, *E116*, E117, *E118*; as vital sign, 116

Glucose test strips, **88**

Glycogen, E104; levels in muscles related to effort, *160*

Glycolysis, **E116**, *E116*, E117, *E118*, E128

Gombe Stream Chimpanzee Reserve, 19, E2–3, *E3*

Gomphosphaeria, E43

Gondwana, *40*, 41

Gonorrhea, *E150*

Goodall, Jane, 18, *18*, 19, E2–3

Good story, criteria for, *341*

Gould, Stephen Jay, E168

Gopher, *E48*

Grasshopper, *E102*

Griffith, Frederick, E173

Grip, E5, *E5*

Growth, **E194**, E194–195; cognitive, E221–23, *E221, E222*

Growth control, of cells, E195, E197–198, *E197*; contact inhibition, *E197*, E198; external control, E197–98; internal control, E197

Growth rate, **E237**

Guanine, 264

Guard cells, *E51*; plant, E61, *E61*

Guinea pigs, *E164*

Gusii people of Kenya, E229–30, *E229*

H

Habitat, **E241**

Hamstring muscles, *154*

Hands, grip, E5, *E5*; primate, *E6*

Health hazards, sexual activity and, E149–150

Heart, E57, *E57, E62, E64*

Heat energy, E109

Helper T-cells, **128**, E79, *E80*

Hemispheres of the brain, E9, *E9*

Hemoglobin, normal and sickle, 271, *271*; sickle cell disease and, 267–268

Hemophilia, 230; in Czarevitch Alexis' family, 230, *231*

Heritable, **E47**

Heterotrophs, **176**

Histamines, E81

Histones, *E174*

Histogram, 249

HIV infection, E149–150

HMS Beagle, 47, E23, *E23*

Homeostasis, **E60**, 60–61, *E61*; disruption of, E73, *E73*; maintenance of, E72–73; feedback, E65

Hominids, 35, E22

Homologies, **E20**, *E20*, E47

Homo sapiens, E48

Hooded oriole, *E39*

Hormones, **E62**, *E63–64*; origins and effects of, *E63–64*; sexual reproduction, E146–148, *E147*

Horned lizard, E46

Horse, proposed evolutionary tree, *E33*

Hospital equipment and procedures, E34

Hot springs, *E41*

Human body, as a compartment, E50; energy use in locomotion, *395*; temperature ranges, *387*

Human-centered environmental ethic, **E252**

Human development, cognitive growth, E221–23, *E221, E222*; effect of individual's environment, E225; effect of social environment, E225; influence of genetic makeup, E225; language skillls, E223; responsibility for self-maintence, E224–226, *E225*; social growth, E223–24; studies on twins', E225

Human genetic disorders, *253*

Human physical performance, E88–90, E107, *E88*

Hunger, E96

Hydra, E143

Hydrogen atom, *E108, E109*

Hydrogen carrier molecules, *E116*, E117

Hydrostatic skeleton, E101–102, *E102*

Hyperthermia, **117**

Hypertonic, **83**

Hyperventilation, E71

Hypothalamus, E62, *E62*, E63, *E63*; sexual reproduction and, E146, *E147*

Hypothermia, **117**

Hypothesis, **E30**

Hypotonic solution, **83**, *83–84*

I

Iceman mummy, 50–52

Imaging technology, *E76*

Immigration, **E238**, *E239*

Immune system, 120; components of, 127–128, E78–81, **E81**; factors affecting, *E85*

Indeterminate growth, E211

Incomplete dominance, **E171**, *E171*; sickle cell disease, E171

Independent assortment, **E158**, *E157*

Induction, **E202**, *E202*

Inferences, **35**

Infertility, E149

Ingen-Housz, E122

Inheritance, *239*

Insecticides, E255–257. *See also* DDT.

Insects, 398–99

Insulin, E191

Interphase, mitosis, *E196*

Intestine, *E57*

In vitro fertilization, **E149**

Ionic bonds, **E113**

Ions, **E53**

Isotonic solution, **83**

J

Jack pine, E242–243, *E242*

Johanson, Donald, 34

Jugular vein, *E57*

K

Kale, *E25*

Karyotypes, 246, *246*

Kidney, *E57*, E58–59, *E58*, *E59*, *E144*, *E145*

Killer T-cells, 128, E79, *E80*

Kinetic energy, **E111**, *E111*

Kingdom, *68*, **E49**, E49; Animalia, E38–39, *E39*; Fungi, E39–40, *E41*; Plantae, E39, *E40*; Prokaryotae (Monera), E38, *E39*; Protists, E40, *E41*

Kirtland's warbler, E241–243, *E241*; population size, E243

Kneecap, *154*

Kormondy, Edward, E247, E251

Krebs cycle, *E116*, **E117**, E117, E118–119, *E119*, *E128*

L

Lactic acid, *E104*, E122

Lake Erie, 9

Land resources, world usage, *377*

Language, E13–14

Language development, E223

Large intestine, *E96*, E98

Larynx, *E69*

Lateral prehension, *E5*

Laurentia, *40*, 41

Lavosier, Antoine, E115

Leaf structure, *E51*

LH, *E148*

Lichens, *E32*

Lifestyle choices, E227

Light, as limiting factor, E233–234

Limiting factor, **E232**, E232–234

Linkage, *245*

Linnaeus, Carolus, E49

Lions, *58*

Lipases, *E98*; pancreatic, *E96*

Lipid molecules, in cell membrane, E52–53, *E53*

Lipids, *148*

Liver, *E57*, E98, *E96*

Liver cancer, *298*

Living systems, characteristics of, E32–34

Lizard, *E72*

Logistic growth, **E236**; growth curve, *E236*

Lower epidermis, *E51*

Low-fat diets, E92

Lucy, the hominid fossil, *32*, 34, 35–36, E22

Lugol's iodine solution, **88**, 152

Lung, *E57*, E68, *E68*, E69, *E69*; healthy and damaged, *E84*

Lung cancer, E84

Luteinizing hormone (LH), E146, *E147*

Lymphocytes, **127**

Lymphokines, **127**, *E80*

Lysosome, *E51*

Lysozyme, E78

M

MacLeod, Colin, E173

McCarty, Maclyn, E173

Macaque monkey, E14, *E15*
Macromolecules, E97, *E97*
Macrophages, **128**, *E79*, *E79*, *E80*
Magnesium, *E95*
Malaria, and sickle cell disease, E171
Malthus, Thomas, E26, E234, *E234*, E236
Maltose, **152**, E98
Mammography, *E210*
Marriage, E154. *See also* cultural celebrations *and* other cultures.
Mars, *E31*
Mating behavior, human, 224; nonhuman, 221, E151–152, *E151–152*
Matthaei, J. H., E184
Mayr, Ernst, E26
Measuring pH, protocol for, 104
Medical technology, E75–77
Meiosis, 217, *241*, **E160**, *E161*, E160–165
Melaleuca tree, *E240*
Melanin, *E170*
Membrane, *E51*
Membrane protein, *E56*
Memory B-Cells, **127**
Memory of infection (immune cells), **E82**, E82
Mendel, Gregor, *243*, E158, E161
Menopause, **E149**
Menstrual cycle, **E147**, *E148*
Menstruation, **E147**, E147–148
Mental capacity, effect of mental exercises, E226
Meristems, E212
Mesophyll, *E51*
Metabolic pathways, *E128*
Metabolic waste, E58
Metabolism, **E127**, E127–128
Metamorphosis, E212, *E212*; sea urchin, *E212*; tiger salamander, E213, *E214*
Metaphase, mitosis, *E196*
Metastasis, **E207**, *E208*
Methane, *E109*; molecular models, *170*
Micrasteria, E41
Microstoma floccosa, E41
Mistletoe, *E240*
Mitochondria, *E51*, **E104**, E118, *E118*
Mitosis, **217**, **E195**, E195–197, *E196*
Model, **37**
Molecular biology, **E190**
Molecules, *E109*; behavior of, 88; size of, 88
Mollusks, *E250*
Monohybrid cross, **244**
Morgan, Thomas Hunt, 246
Mortality, **E238**, *E239*
Moss, *E40*
Motor cortex, E9, *E9*; map of, *E10*
Mount St. Helens, *197*
Mouse embryo, *E67*
Movement, methods of, *E8*
mRNA, **E181**, role in protein synthesis, *E182*; translated into proteins, *E183*
Mule, *E48*
Multiple sclerosis, E82
Muscles: fatigue, E104; human leg, 154, *154*; muscle fiber, *E101*; muscle structure, E101, *E101*;

human arm, E99, *E100*
Mutant genes, E209
Mutation, in DNA code for hemoglobin, 277, **E164**, E164–165, *E165*; in DNA replication, E178–79, *E178*
Mutation, by deletion of nucleotide, E178; effect on phenotype and genotype, E179; from exposure to ultraviolet light, E253–254, *E253*; in gametes, E179; preservation of, E178–79, *E178*; by substitution, E178, *E178*; tools in biologic research, E165
Mutualism, **E241**
Mycema lejiana, E41
Myelin sheath, E221
Myofibril, *E101*
Myosin, *E101*

N

NADH, E117, *E118*, *E119*
NADPH, E125; in carbon fixation, E126, *E126*, E127
Nasal cavity, E68, *E69*
Natality, **E238**, E239
Natural selection, **E26**, *E26*; genetic variation and, E167–168, *E167–168*
Negative feedback, **E64**, *E64*, *E64*, E65, **E66**, E66
Nephrons, **E58**, *E59*
Nerve cell, E11, *E11*
Nervous system, *E11*, E62
Neurons, **E11**, *E11*
Neurotransmitters, **E11**
Neutrons, *E108*
Niche, **E240**
Nicotine, E84
Nimbus-7 (satellite), 401
Nirenberg, M. W., E184
Nitrates, 402; pollution, *403*
Nondisjunction, *E166*, mistakes during meiosis, E166
Nonspecific barrier, E78
Nostoc, E39
Nuclear envelope, *E196*
Nucleic acid, **E174**; DNA as example of, E174
Nucleolus, *E197*
Nucleosome, *E174*
Nucleotides, E174, *E176*, *E177*; genetic expression and, *E180*, E181; mutations and, E178
Nucleus, *E197*
Nucleus, cell, E51, *E51*; atom, *E108*
Nutrients, E91
Nutrient tests, protocol for, 148–149

O

Octane, *E109*
Olfactory bulbs, 24, *24*
Oncogene, **E208**, *E209*
On the Origin of Species, E24

Optic lobes, 24, *24*
Order, *68*, **E49**, E49
Organisms, classification of, *68–69*
Organisms, hierarchy of, *68–69*
Ornitholestes, E16
Osmosis, **E54**, E54–55, *E55*
Ova (Ovum), 217, E142, E144, *E143, E144*
Ovarian cycle, *E148*
Ovary, *E62, E64*, **E144**, *E144*
Oviduct, **E145**, *E144*
Ovulation, **E147**
Oxygen, E68; appearance in atmosphere, E43; role in fire, E115; from photosynthesis, E125
Ozone, 346, *346*, **E254**; CFCs and, 351, *351*; production of, 348, *348*; ultraviolet light and, E254, *E254*
Ozone layer, 345, *345, 346*

P

Pacemaker, E75
Paleogeologist, **39**
Paleontologist, **39**
Palmar prehension, *E5*
Pancreas, *E62, E64*, E98; cystic fibrosis effect on, 390, *390*
Pangea, 41, E17
Pantothenic acid, *E93*
Parasitism, **E241**, *E240*
Parathyroid glands, *E62, E64*
Passive transport, **E55**, E55–56, *E56*
Pathogens, **E78**
Peas, *244*; inherited traits, 244
Penicillin, 55, *55*
Penicillium, **55**
Penis, **E145**, *E145*
Pepsin, **E96**, *E97*
Peptide, **E185**, E185
Peptide bond, **E185**, *E185*
Peranema, E36
Performance factors, E105–107
Peristalsis, E98
Pesticides, increase in crop losses since use, *359*; production in the United States, *358*; resistant pests, *359*. See also DDT.
Pesticides and cancer, 360
pH, 8, **102**; balance, 107; protocol for monitoring change in, 190; scales, *8, 103*
pH buffer, **107**
Phenotype, *241*, **E155**, E155; DNA and, E172, *E172*
Phloem, 293, *293*
Phosphorus, *E95*
Photosynthesis, **176**, E43, E122–123, E123–124, *E124*, 126; ATP production, *E125*; by products of, E125–126; protocol for, 178–179
Photosynthetic bacteria, *E43*
Phototrophic bacteria, *E42*
Phyla, **E49**, E49
Phylum, *68*
Physical anthropologist, **42**

Physical structure, E99
Physiologic data related to physical performance, *160–162*
Piaget, Jean, E221
Pig embryos, *E195*
Pineal gland, *E62, E63*
Pinon pine tree, *E194*
Pituitary gland, *E62, E63*
Placenta, E147
Plant evolution, E42–45
Plants, E39
Plasma B-Cells, **127**
Plasma membrane, *E56*
Pleistocene Earth, 41
Pollution, nitrate, *403*
Pond, as system, *E259*
Pondweed, *E248*
Population, **E45**, E45; control, 368; schematic map of world's, *331*; world, *371*; world growth rate, *371*
Population density, effect on birthrate, E235
Population growth, E234–236; exponential growth curve, E235, *E235, E236*; selected nations, *372*; world, *371, 397*
Population size, factors that affect it, E238, E238–239, *E239*; limiting factors, E232–238
Population trends, *371–373*
Potassium, *E95*; as vital sign, 116
Potential energy, **E111**, *E111*
Predation, **E28**, *E28*
Predator/prey relationship, 47, *E28*, **E238**
Prenatal care, 136
Priestley, Joseph, E115, E123
Primates, hands, *E6*; common ancestor, E22, *E22*; evolution of, 20, *E3*, E21, *E22*, E221
Principle of segregation, (genetics), **E161**
Processes of science, 11, 204
Producers, **E136**
Progesterone, **E147**, *E148*
Project criteria, 27, *27*
Prokaryote, **E37**; DNA replication, *E177*
Prophase, mitosis, *E196*
Prostate, *E145*
Proteases, *E98*; pancreatic, *E96*
Protein, 148, E91–92, *E97*
Protein synthesis, 272, *272*, E182–84, E184–86, *E183, E185*; mRNA and, E184, *E185*, E186; role of ribosomes, E184, *E185*, E186; role of tRNA, E184–86, *E185*
Protein translation, mRNA and, *E183*
Protists, E40
Proton diffusion, E120
Protons, *E108*
Puberty, E105
Puberty, female, E147; male, E146
Public policy, sexual practices, E154–155
Pulse, **115**; resting, *116*
Punctuated equilibrium, **E168**, E168
Pyruvate, E117, 118, *E118, E119*; anaerobic respiration, *E121*; fermentation, *E121*

Q

Quadriceps muscles, *E154*

R

Radioactive isotopes, **E18**
Radioactive-isotope fossil dating techniques, E18–19
Rate of breathing, **115**
Reabsorption, **E59**, E59–60
Recessive trait, **E158**
Recombinant DNA, **E188**, *E188*; development of the technique, E186–87
Recombination, **E166**, E166, *E166*
Rectum, *E96*
Recycling of matter, E135–136
Red blood cells, abnormalities in, 267–268, *268*
Red-tailed hawk, *E239*
Red Irish lord fish, *E27*
Reflex, **E12**, *E12*
Reflex arc, *E12*
Reindeer, 325, 325–326; population changes on St. Paul Island, *326*
Reproduction, **214**, *E49*, E142; a glossary, *217–218*; human, *E144, E145*, E144–146
Reproductive behavior, nonhuman, 221
Reproductive strategies, 225, E143
Reproductive system, human female, *E144*, E144–145; human male, *E145*, E145–146
Restriction enzyme, **E186**
Restriction fragment length polymorphism, 277
Retroviruses, E208
RFLP, *See* Restriction Fragment Length Polymorphism.
Retinol (vitamin A), *E93*
Rheumatoid arthritis, E82
Ribonucleic acid, **E181**; production of, E181–82, *E181*; role of DNA in production of, E181, *E181*; rRNA, **E182**, tRNA, **E182**, types of, E181–82, *E182*
Ribosomes, **E184**, E184–86, *E185*
Risk factors, 128; controllable and uncontrollable, E86
River bed ecosystem, *E138*
RNA. *See* Ribonucleic acid.
Rock hyrax, *E36*
Roses, *58*
Rous, Peyton, E207
Rous sarcoma viruses, *E208*
Roux, Wilhelm, 287, *287*; experiment, *288*
Rubella virus, occurence in the United States, 386, *386, E83*

S

Saguaro cactus, *E233*
St. Paul Island, 325–326, *325*; reindeer population on, *326*

Saliva, E98
Salivary amylase, E96
Salivary glands, *E96*
Salt concentration, regulation of, E72, *E72*
Sample size, importance of, *232*
Savanna, East Africa, *E233*
Scientific knowledge, 71
Scientific names, reason for, E48
Sea urchin metamorphosis, *E212*
Secretion, **E60**, E60
Seeds, E44–45, *E45*
Selective breeding, E27
Selective pressures, E28
Selectively permeable, **E53**
Self-maintenance, E224–26; physical fitness and, E224, E225–26, *E225*; mental exercise, E226
Semen, **E146**
Senebier, E122
Senescence, **E195**
Sensory cortex, E9, *E9*; map, *E10*
Septicemia, E74
Sequoia trees, E211–12
Sexual activity, and health hazards, E149–150
Sexual drive, E151, E152
Sexual intercourse, E146
Sexual maturity, hormones and, E146–148, *E147*
Sexual reproduction, **217, E142**
Sexual selection, E151
Sexually transmitted diseases, E149, *E150*
Shaken-baby syndrome, E221
Sheep brain, *23*
Shivering, E71, *E71*
Shock, **117**
Sickle cell disease, 267–268, *267*
Simple fat, *E97*
Simple sugar, *E97*
Skeletal muscle, E99
Skeleton, gorilla, *E7*; human, *E7*
Skin, E78
Skin color, model for, *E170*
Skull, *33*; early hominid, *43*; gorilla, *43*
Sleep, E107
Small intestine, *E96*, E98
Smith, William, E18
Snake caterpillar, *E27*
Sneezing, 120, *120*
Snowy owl, *E232*
Social growth, importance of play, *E224*
Social skills, effect of physical impairment of brain, E221
Sodium, *E95*; as vital sign, 116
Solar energy, E123, *E123*; in trophic levels, *E141*
Solute, **83**
Solvent, **83**
South Pole, ozone layer over, *346*
Space, as limiting factor, E234–235, *E234*
Species, *68*, E26, **E48**, E48–49
Spectrum of light energy, *E123*
Sperm, **217, E142**, *E143*, E145
Spiders, *325*
Spikerush, *E249*
Spirella voluntans, E39

Spores, **217**

Sporulation, **218**

Staphylococcus aureus, 55, *E49*

Starch, **88**, 148, *150*, **152**, E92, *E97*

Steller's sea cow, *E17*

Step test, protocol for conducting, *100–101*

Stereum complicatum, *E41*

Stickleback fish, *E152*

Stimuli, **E60**, E60–61

Stimulus, **E11**

Stomach, *E96*, E98

Stomates, *E51*, **E61**, E61

Stop codons, **E186**

Stratigraphy, **E18**, E18

Strawberries, *E49*

Streptococcus bacteria, *E39*

Stress, E107

Stressors, **E73**, E73–75, *E73*

Stroma, E124, *E125*

Substrates, **152**, **E97**

Sugar, 149

Sugar beet, *191*

Sugar cane, *191*

Sulfur, *E95*

Sun, 345, *345*; ozone production, *348*. *See also* solar energy *and* other related topics.

Sword fern, *E40*

Symbiosis, **E241**

Synapse, **E11**

Synthesis reactions, E127–128

Syphilis, *E150*

Systems analysis, E258–259

T

Taboos, cultural, E154

Tasmanian wolf, *E17*

Taieb, Maurice, 34

T-Cells, **128**; helper, **128**; killer, **128**

Technology, E107. *See also* medical technology *and* fitness.

Teeth, *E96*

Telophase, mitosis, *E196*

Temperature, 116; as limiting factor, E232–233, *E232*; protocol for measuring, 98; of human body, 99, *387*

Template, **E173**; in DNA replication, E178

Testable questions, definition of, 21

Testes, *E62*, *E64*, **E145**, *E145*

Testosterone, E105, **E146**

Thecamoeba cell, *E33*

Theory, **E30**, *E30*

Thermophile bacteria, *58*

Thylakoid, E124–125, *E125*

Thymine, 264

Thymus gland, *E62*, *E64*

Thyroid gland, *E62*, *E64*

Tiger salamander, E213, *E214*

Tip prehension, *E5*

Tobacco, E106

Tongue, E96

Topsoil, E135

Toxins, **E106**

Trachea, E68, *E69*; cell types, 294

Transcription, **E181**, E181–82, *E182*

Transgenic organisms, E191

Translation, **E182**; converting mRNA sequence into amino acid sequence, E182–84, *E183*. *See also* protein synthesis.

Triage, **112**; general guidelines, *116–17*

Triassic Earth, 41

Tricarboxylic acid (TCA) cycle. *See* Krebs cycle.

Trichonympha, *E41*

Trilobite, *E166*

Trobriand Islanders, E227–28, *E227*

Trophic levels, **E139**, *E139*; pyramid, *E140*, *E141*

Tropical rain forest, E135, *E136*

Tube sponges, *E39*

U

Ultraviolet light, and ozone, E254, *E254*; effect on DNA, E252–254, *E253*, *E254*

Unifying principles of biology, 12

Upper epidermis, *E51*

Urea, E58

Ureter, *E59*, *E58*, *E144*

Urethra, **E145**, *E144*, *E145*

Urinary system, E57, E58–60, *E58*

Urine, E58, E59

Uterus, **E145**, *E144*

V

Vaccines, 135–36, E83, *E83*

Vacuole, *E51*, **E55**

Vagina, **E145**, *E144*

Variables, **83**

Variation, genetic, 24, 247, *E25*, E99, E164, E202; in characteristics, *E25*, E28, E99; in organisms, 56, 247; inherited, E26; in population, E27, E28; role in evolution, E28, E164, E202

Vas deferens, E145, *E145*

Vascular tissue, E44, *E44*

Vegetative reproduction, **218**

Vein, *E51*, E57, *E57*; renal, E59, *E59*

Vena cava, *E57*

Ventricles, *E57*

Venules, E57

Vestigial structures, E20, *E21*

Viruses, cancer-causing, E207

Visible light, E123, *E123*

Visual development, E222, E223

Vital signs, glossary of, *115*; range of, *116*

Vitamins, E91, *E93–94*, E94; list of, *E93–E94*

W

Wallace, Alfred Russel, E24, *E24*

Waste, *E129*

Water, *E110*; as limiting factor, E233, *E233*; as nutrient, E91

Water cycle, E133, *E133*

Water lily, *E249*

Water milfoil, *E249*

Watson, James, 261, E179

Waxy cuticle, *E51*

Weather, as system, *E258*

Western gull, *E72*

Whales, evolution of, *393*; fossil evidence of ancestry, *393*

Wilkins, Maurice, 261

Willow, *E248*

Wolves, *E14*

World, schematic map of, *331*; population, *see* population growth; resources, *see* related topics.

X

Xylem, 293, *293*

Y

Yeast, *E49*; reproduction, 233–234

Yeast genetics, protocol for, 233–237

Yellowstone National Park, *338*, E243–244, *E244, E245*

Yolk, *E198*

Yolk sac, *E198*

Yucca, E130

Z

Zebra, *71*; African, *72*

Zygotes, **218, E143**, *E143*

ACKNOWLEDGMENTS

ENGAGE SECTION: **Opener** BSCS by Carlye Calvin; **Silhouette** BSCS by Carlye Calvin; *Daphnia* Bruce J. Russell/BioMEDIA Associates; *Gammarus* John D. Cunningham/Visuals Unlimited; **Scientist** BSCS by Carlye Calvin; **En.5** The Bettmann Archive.

UNIT 1

Unit Opener (fungus) R. Calentine/Visuals Unlimited; (bacterium) David M. Phillips/Visuals Unlimited; (bird) BSCS by Carlye Calvin; (slime mold) Richard Walters/Visuals Unlimited; (moss) BSCS by Carlye Calvin; **Unit Silhouette** Science VU/RUCUCSFGL.

CHAPTER 1: **Marketplace** BSCS by Carlye Calvin; **1.1** Hugo Van Lawick/Goodall Institute; **Chimp** Larry Lipsky/Tom Stack & Associates; **Rollerblader** BSCS by Carlye Calvin; **1.4** BSCS by Carlye Calvin; Geese Carlye Calvin; **E1.2** Jane Goodall/Goodall Institute; **E1.3** Science, Vol. 264, 20 May 94, Digital image: Dr. Hanna Damasio, Dept. of Neurology & Image Analysis, University of Iowa College of Medicine; **E1.5**(a-g) BSCS by Carlye Calvin; Guitar player Carlye Calvin; **E1.8** (top) Bert & Jeanne Kempers, (bottom) BSCS by Carlye Calvin; **E1.9** (left & right) BSCS by Carlye Calvin; **E1.15** Carlye Calvin; **E1.16** Carlye Calvin.

CHAPTER 2: **Opener** Institute of Human Origins; **Silhouette** Cabisco/Visuals Unlimited; Johanson The Bettmann Archive; **E2.3** BSCS by Carlye Calvin; **E2.9** The Bettmann Archive; **E2.11** The Bettmann Archive; **E2.12** BSCS by Carlye Calvin; **E2.14**(a) George D. Dodge/Tom Stack & Associates, (b) Carlye Calvin, (c) Randy Morse/Tom Stack & Associates; **E2.16** Warren Garst/Tom Stack & Associates.

CHAPTER 3: **Opener** Tammy Peluso/Tom Stack & Associates; **Silhouette** Dave B. Fleetham/Visuals Unlimited; **3.1**(a) Spencer Swanger/Tom Stack & Associates, (b) Corale Brierley/Visuals Unlimited; **Girl** BSCS by Carlye Calvin; **3.3** BSCS; **3.8** Chip Isenhart/Tom Stack & Associates; **E3.1** NASA; **E3.2** BSCS by Carlye Calvin; **E3.4** Carlye Calvin; **E3.5** Bruce J. Russell/BioMEDIA Associates; **E3.9**(a) Richard Holmes/Carlye Calvin, (b) William Grenfell/Visuals Unlimited; **E3.10**(a) Bruce J. Russell/BioMEDIA Associates, (b) David M. Phillips/Visuals Unlimited; **E3.13**(a, b) David M. Phillips/Visuals Unlimited, (c) John D. Cunningham/Visuals Unlimited; **E3.14**(a) Mike Bacon/Tom Stack & Associates, (b) Ken W. Davis/Tom Stack & Associates, (c) BSCS by Carlye Calvin; **E3.15**(a) Carlye Calvin, (b) Terry Donnelly/Tom Stack & Associates, (c) BSCS; **E3.16**(a) Ken Wagner/Visuals Unlimited, (b) R. Calentine/Visuals Unlimited, (c) Stanley Flegler/Visuals Unlimited; **E3.17**(a) M. Abbey/Visuals Unlimited, (b) A.M. Siegelman/Visuals Unlimited, (c) Richard Walters/Visuals Unlimited; **E3.18** Courtesy of Caitlin Burnett; **E3.20** R. Guerrero/Ward's Natural Science Establishment, Inc./Five Kingdom Slide Set; **E3.21** Lynn M. Margulis/Dept. of Biology, University of Massachusetts, Amherst; **E3.22** John D. Cunningham/Visuals Unlimited; **E3.23** Ripon; **E3.24**(a) BSCS by Doug Sokell, (b) BSCS by Carlye Calvin; **E3.25** Joe McDonald/Visuals Unlimited; **E3.26** Barbara von Hoffmann/Tom Stack & Associates; **E3.29**(a) Carlye Calvin, (b) E. Musil/Visuals Unlimited, (c) BSCS by Rich Tolman.

UNIT 2

Unit Opener BSCS by Carlye Calvin; **Unit Silhouette** BSCS by Carlye Calvin.

CHAPTER 4: **Opener** NASA/Airworks/Tom Stack & Associates; **Young man** BSCS by Carlye Calvin; **4.1**(left & right) BSCS by Carlye Calvin/Jerry Grant; **4.2** BSCS by Jerry Grant; **E4.1** BSCS by Janet C. Girard; **E4.2**(a) Richard H. Thom/Tom Stack & Associates; **E4.5** BSCS by Carlye Calvin; **E4.9**(b) P.M. Motta, A. Caggiati, G. Macchiarelli/Science Photo Library/Photo Researchers.

CHAPTER 5: **Opener** Colorado Historical Society, Fl5025; **Silhouette** BSCS by Carlye Calvin; **5.1** BSCS by Carlye Calvin; **5.2** BSCS by Carlye Calvin; **5.5** BSCS by Carlye Calvin; **5.9** Mike Bacon/Tom Stack & Associates; **E5.2** Cabisco/Visuals Unlimited; **E5.7** David M. Phillips/Visuals Unlimited; **E5.8** David Spears LTD/Science Photo Library/Photo Researchers; **E5.9** Fred Hossler/Visuals Unlimited; **E5.11** BSCS by Carlye Calvin; **E5.12** Steve McCutcheon/Visuals Unlimited; **E5.14** Joe McDonald/Visuals Unlimited.

CHAPTER 6: **Opener** Hank Morgan/Photo Researchers; **Silhouette** Courtesy of John A. Moore/Harvard Press; **Hikers** Spencer Swanger/Tom Stack & Associates; **Technology images**(a-c) SIU/Visuals Unlimited; **6.6** David M. Dennis/Tom Stack & Associates; **6.8** Gary Milburn/Tom Stack & Associates; **6.11**(left) Courtesy of Bill Berner, (center) BSCS by Carlye Calvin, (right) Thomas Kitchin/Tom Stack & Associates; **HIV** David M. Phillips/Visuals Unlimited; **Entrepreneur** BSCS by Carlye Calvin; **E6.3**(a) John Cunningham/Visuals Unlimited, (b) Tom Stack/Tom Stack & Associates; **E6.4** James Gremillion, DPM, MS; **E6.5**(a, b) Courtesy of Penrose Hospital, Co. Spgs., CO; **E6.6** Raymond B. Otero/Visuals Unlimited; **E6.7** David M. Phillips/Visuals Unlimited; **E6.8** Lennart Nilsson/Boehringer Ingelheim International GmbH; **E6.13**(a, b) O. Auerbach/Visuals Unlimited; **E6.16** Gerald Bellow/Colorado Springs Police Department; **E6.17** BSCS by Carlye Calvin.

UNIT 3

Unit Opener Bruce Berg/Visuals Unlimited; **Unit Silhouette** Don W. Fawcett/Visuals Unlimited.

CHAPTER 7: **Opener** (bowler, walkers) BSCS by Carlye Calvin; (windsurfer) Brian Parker/Tom Stack & Associates; **Silhouette** BSCS by Carlye Calvin; **Jet** BSCS by William J. Cairney; **7.2** (left) Courtesy of Dave Black, Colordo College Hockey, (center) BSCS by Carlye Calvin, (right) Bruce Berg/Visuals Unlimited; **E7.1**(golfer) BSCS by Carlye Calvin, (gymnast) Tom Stack/Tom Stack & Associates, (student) BSCS by Carlye Calvin; **E7.2** Allen B. Smith/Tom Stack & Associates; **E7.11**(a) S. Maslowski/Visuals Unlimited, (b) David Vessey; **E7.14**(a) Milton Rand/Tom Stack & Associates; **E7.16** BSCS by Carlye Calvin.

CHAPTER 8: **Opener** C. Allan Morgan; **Silhouette** George B. Chapman/Visuals Unlimited; **8.1** Jeff Isaac Greenberg/Visuals Unlimited; **8.2** BSCS by Carlye Calvin; **Teens** BSCS by Carlye Calvin; **8.9** SIU/Visuals Unlimited; **E8.4** BSCS by Carlye Calvin; **E8.10** The Bettmann Archive; **E8.13** Don W. Fawcett/Visuals Unlimited; **E8.16** BSCS by Carlye Calvin; **Library** BSCS by Carlye Calvin; **E8.21** George B. Chapman/Visuals Unlimited; **E8.23** Bernd Wittich/Visuals Unlimited; **E8.25** BSCS by Carlye Calvin.

CHAPTER 9: **Opener** John Shaw/Tom Stack & Associates; **Silhouette** John Shaw/Tom Stack & Associates; **9.1** BSCS by Carlye Calvin; **9.2** Carlye Calvin; **9.4**(a) Brian Parker/Tom Stack & Associates, (b) Eric A. Soder/Tom Stack & Associates; **9.5** BSCS by Carlye Calvin; **9.6** Thomas Rothe/Tom Stack & Associates; **9.7** Pat Armstrong/Visuals Unlimited; **E9.1** Bernd Wittich/Visuals Unlimited; **E9.2** Kerry T. Givens/Tom Stack & Associates; **E9.3** Rich Buzzelli/Tom Stack & Associates; **E9.5** Tom J. Ulrich/Visuals Unlimited.

EXPLAIN SECTION: **Opener** Ann Duncan/Tom Stack & Associates; **Silhouette** David M. Dennis/Tom Stack & Associates; **Children w/teacher** BSCS by Carlye Calvin; **The Thinker** The Bettmann Archive; **Scientist** Courtesy of Michael Dougherty.

UNIT 4

Unit Opener (technology) BSCS by Carlye Calvin, (parent and child) Carlye Calvin; **Silhouette** Irving Geis/Photo Researchers.

CHAPTER 10: **Opener** David M. Phillips/Visuals Unlimited; **Silhouette** David M. Phillips/Visuals Unlimited; **Forest** David Vessey; **10.3** BSCS by Jerry Grant; **10.4** Milton Rand/Tom Stack & Associates; **E10.1** Barbara von Hoffmann/Tom Stack & Associates; **E10.3**(a) A.M. Siegelman/Visuals Unlimited, (b) Cabisco/Visuals Unlimited, (c) James W. Richardson/Visuals Unlimited; **E10.9** W. Perry Conway/Tom Stack & Associates; **E10.11** J. Terrence McCabe.

CHAPTER 11: **Opener** Bill Bachmann/Photo Researchers, Jon Feingerish/Tom Stack & Associates; **Silhouette** Science VU/Visuals Unlimited; **Romanov family** The Bettmann Archive; **11.6** BSCS by Jerry Grant/Collies, Courtesy of Shelly Bergstraster, Wild Wind Collies; **11.9** BSCS by Jerry Grant; **11.12** L. Lisco, D.W. Fawcett/Visuals Unlimited; **E11.9** BSCS by Carlye Calvin; **E11.10**(a) Courtesy of Michael Dougherty, (b) F.R. Turner/Visuals Unlimited; **E11.12** Carlye Calvin; **E11.17** BSCS by Carlye Calvin.

CHAPTER 12: **Opener** Science/Visuals Unlimited; **Witness and lawyer** BSCS by Jerry Grant; **12.1** Larry Mulvehill/Photo Researchers; **12.2**(b) Science Source/Photo Researchers; **Jury** BSCS by Jerry Grant; **E12.2** Science/Visuals Unlimited; **E12.3** J.R. Paulson, U. Laemmli, D.W. Fawcett/Visuals Unlimited; **Strawberries** E. Webber/Visuals Unlimited; **E12.15** John Cunningham/Visuals Unlimited.

UNIT 5

Unit opener (photos 1-5) Media Design Associates, (photo 6) John D. Cunningham/Visuals Unlimited, (photo 7) Dave Fleetham/Tom Stack & Associates; **Unit Silhouette** Courtesy of the Carnegie Institution of Washington.

CHAPTER 13: **Opener** John D. Cunningham/Visuals Unlimied; **Silhouette** BSCS by Carlye Calvin; **13.2** Courtesy of Jean Chatlain; **13.3** BSCS by Carlye Calvin; **13.4**(a, b) Will Allgood, Mark Viner/Media Design Associates; **Hitler** The Bettmann Archive; **13.6** Lochlear Macleay, M.D.; **Water bug** Ken Lucas/Visuals Unlimited; **E13.1** BSCS; **E13.2**(a) Ward's Natural Science Establishment, Inc./Media Design Associates, (b) Cabisco/Visuals Unlimited, (c) John D. Cunningham/Visuals Unlimited; **E13.6**(a) Media Design Associates; **E13.7**(a) Fred Hossler/Visuals Unlimited, (b) David M. Phillips/Visuals Unlimited, (c) John D. Cunningham/Visuals Unlimited; **E13.13**(a) John Moss/Photo Researchers, (b) BSCS; **E13.14** Courtesy of Phil and Ellen Goulding; **E13.16** K.G. Murti/Visuals Unlimited; **E13.18** Science/Visuals Unlimited; **E13.19** BSCS; **E13.20**(a) Cabisco/Visuals Unlimited, (b) Randy Morse/Tom Stack & Associates; **E13.22**(a) Rudolf Arndt/Visuals Unlimited, (b) Victor Hutchinson/Visuals Unlimited; **Student** BSCS by Carlye Calvin.

CHAPTER 14: **Opener** Courtesy of Jean Milani; **Silhouette** Courtesy of Jean Milani; **14.1** Jon Feingerish/Tom Stack & Associates; **14.2** BSCS by Carlye Calvin; **Senior w/kids** BSCS by Carlye Calvin; **Students debating** BSCS by Carlye Calvin; **E14.2** Courtesy of Jean Milani; **E14.8** BSCS by Carlye Calvin; **E14.9** BSCS by Carlye Calvin; **E14.13** Bayard H. Brattstrom/Visuals Unlimited; **E14.15** D. Long/Visuals Unlimited; **E14.16**(top) D. Long/Visuals Unlimited, (bottom) Link/Visuals Unlimited.

UNIT 6

Unit Opener NASA; **Unit Silhouette** Tammy Peluso/Tom Stack & Associates.

CHAPTER 15: **Opener** Terry Donnelly/Tom Stack & Associates; **Silhouette** BSCS by Carlye Calvin; **Teen filling feeder** BSCS by Pam Van Scotter; **Ranch** Carlye Calvin; **15.2** Dominque Braud/Tom Stack & Associates; **Field of sunflowers** David Newman/Visuals Unlimited; **15.7** BSCS by Jerry Grant; **15.8**(upper left) J. Lotter/Tom Stack & Associates, (upper right) Brian Parker/Tom Stack & Associates, (lower left) Doug Sokell/Visuals Unlimited, (lower right) Doug Sokell/Tom Stack & Associates; **15.9**(a) Courtesy of Ward's Natural Science Establishment, Inc., (b) BSCS by Pam Van Scotter; **E15.1** Joe McDonald/Visuals Unlimited; **E15.2** David Vessey; **E15.4**(a, b) BSCS by Carlye Calvin; **E15.5** The Bettmann Archive; **E15.10** C. Allan Morgan; **E15.11** David S. Addison/Visuals Unlimited; **E15.12** BSCS by Carlye Calvin; **E15.13** Richard P. Smith/Tom Stack & Associates; **E15.14**(a) Walt Anderson/Visuals Unlimited, (b) John Shaw/Tom Stack & Associates, (c) Rod Planck/Tom Stack & Associates; **E15.19** David M. Dennis/Tom Stack & Associates.

CHAPTER 16: **Opener** Glenn M. Oliver/Visuals Unlimited; **Silhouette** Chip and Jill Isenhart/Tom Stack & Associates; **16.2** Carlye Calvin; **16.3**(a) Carlye Calvin, (b) Courtesy of U.S. Space Foundation; **UN delegates** The Bettmann Archive; **16.3** John Serrao/Visuals Unlimited; **E16.4** Rod Planck/Tom Stack & Associates; **E16.5** The Bettmann Archive; **E16.6** The Bettmann Archive; **E16.7** T. Kitchin/Tom Stack & Associates; **E16.8** UCAR/National Center for Atmospheric Research/NSF; **E16.9** BSCS by Carlye Calvin.

EVALUATE SECTION: **Opener** BSCS by Jerry Grant; **Silhouette** BSCS by Janet C. Girard; **Ev.10**(a) BSCS by Carlye Calvin, (b) Carlye Calvin; **Ev.15** Glenn M. Oliver/Visuals Unlimited; **Ev.16**(a) Jack Bostrack/Visuals Unlimited, (b) Gary Robinson/Visuals Unlimited; **Big band** The Bettmann Archive; **Rock concert** The Bettmann Archive; **Wetlands** Carlye Calvin; **Student** BSCS by Jerry Grant.

APPENDIX B: **B.8** John Bostrack/Visuals Unlimited; **B.10** BSCS by Jerry Grant.